PARTICLE PHYSICS
at the Silver Jubilee of
Lomonosov Conferences

Faculty of Physics of Moscow State University

INTERREGIONAL CENTRE
FOR ADVANCED STUDIES

Proceedings of the Eighteenth Lomonosov
Conference on Elementary Particle Physics

PARTICLE PHYSICS
at the Silver Jubilee of Lomonosov Conferences

Moscow, Russia   24-30, August 2017

Editor

Alexander I. Studenikin
Department of Theoretical Physics
Moscow State University, Russia
and
Joint institute for Nuclear Research (Dubna), Russia

World Scientific

NEW JERSEY · LONDON · SINGAPORE · BEIJING · SHANGHAI · HONG KONG · TAIPEI · CHENNAI

Published by

World Scientific Publishing Co. Pte. Ltd.
5 Toh Tuck Link, Singapore 596224
USA office: 27 Warren Street, Suite 401-402, Hackensack, NJ 07601
UK office: 57 Shelton Street, Covent Garden, London WC2H 9HE

British Library Cataloguing-in-Publication Data
A catalogue record for this book is available from the British Library.

PARTICLE PHYSICS AT THE SILVER JUBILEE OF LOMONOSOV CONFERENCES
Proceedings of the Eighteenth Lomonosov Conference on Elementary Particle Physics

Copyright © 2019 by World Scientific Publishing Co. Pte. Ltd.

All rights reserved. This book, or parts thereof, may not be reproduced in any form or by any means, electronic or mechanical, including photocopying, recording or any information storage and retrieval system now known or to be invented, without written permission from the publisher.

For photocopying of material in this volume, please pay a copying fee through the Copyright Clearance Center, Inc., 222 Rosewood Drive, Danvers, MA 01923, USA. In this case permission to photocopy is not required from the publisher.

ISBN 978-981-120-232-2

For any available supplementary material, please visit
https://www.worldscientific.com/worldscibooks/10.1142/11333#t=suppl

Desk Editor: Nur Syarfeena Binte Mohd Fauzi

Typeset by Stallion Press
Email: enquiries@stallionpress.com

**Moscow State University
Faculty of Physics
Interregional Centre for Advanced Studies**

EIGHTEENTH LOMONOSOV CONFERENCE ON ELEMENTARY PARTICLE PHYSICS

Moscow, August 24-30, 2017

Mikhail Lomonosov
1711-1765

Sponsors

Russian Ministry of Education and Science
Russian Foundation for Basic Research
Interregional Centre for Advanced Studies

Supporting Institutions
Faculty of Physics of Moscow State University
Joint Institute for Nuclear Research (Dubna)
Institite for Nuclear Research of Russian Academy of Sciences
Ministry of Education and Science of Russia
Bruno Pontecorvo Neutrino and Astrophysics Laboratory (MSU)

International Advisory Committee

E.Akhmedov (Max Planck Inst., Heidelberg),
V.Belokurov (MSU),
V.Berezinsky (GSSI/LNGS, Gran Sasso),
S.Bilenky (JINR, Dubna),
J.Bleimaier (Princeton),
M.Danilov (Lebedev Phys.Inst., Moscow),
A.Dolgov (Novosibirsk State Univ.,INFN-Ferrara & ITEP, Moscow),
N.Fornengo (Univ. of Torino & INFN),
C.Giunti (Univ. of Torino & INFN),
M.Itkis (JINR, Dubna),
L.Kravchuk (INR, Moscow),
A.Masiero (INFN, Padua),
V.Matveev (JINR, Dubna),
M.Panasyuk (MSU),
K.Phua (World Scientific, Singapore),
V.Rubakov (MSU & INR, Moscow),
J.Silk (Univ. of Oxford),
A.Skrinsky (INP, Novosibirsk),
A.Slavnov (MSU & Steklov Math.Inst, Moscow)
A.Smirnov (Max Planck Inst., Heidelberg)
P.Spillantini (INFN, Florence),
A.Starobinsky (Landau Inst., Moscow),
N.Sysoev (MSU),
Z.-Z.Xing (IHEP, Beijing)

Organizing Committee

V.Bagrov (Tomsk State Univ.), I.Balantsev (MSU),
V.Bednyakov (JINR, Dubna), L.Dubasova (ICAS, Moscow),
A.Egorov (ICAS, Moscow), D.Galtsov (MSU),
A.Grigoriev (MSU), A.Kataev (INR, Moscow),
K.Kouzakov (MSU), Yu.Kudenko (INR, Moscow),
A.Lokhov (INR, Moscow) - Scientific Secretary,
N.Nikiforova (MSU), A.Nikishov (Lebedev Physical Inst., Moscow),
S.Ovchinnikov (MSU), A.Popov (MSU), Yu.Popov (MSU),
P.Pustoshny (MSU), V.Ritus (Lebedev Phys. Inst., Moscow),
V.Savrin (MSU), K.Stankevich (MSU),
M.Sumin (MSU), V.Tsaturyan (MSU), V.Yakimenko (MSU),
A.Studenikin (MSU, JINR & ICAS, Moscow) - Chairman

**Moscow State University
Interregional Centre for Advanced Studies**

TWELFTH INTERNATIONAL MEETING ON PROBLEMS OF INTELLIGENTSIA
"The Future of the Intelligentsia"

Moscow, August 30, 2017

Presidium of the Meeting

J.Bleimaier (Princeton)
A.A.Fedyanin (MSU)
A.I.Studenikin (MSU, JINR & ICAS)
N.N.Sysoev (MSU)

FOREWORD

The 18th Lomonosov Conference on Elementary Particle Physics was held at the Moscow State University (Moscow, Russia) on August 24–30, 2017 under the patronage of Victor Sadovnichy, the Rector of the university.

The conference was organized by the Faculty of Physics of the Moscow State University in cooperation with the Joint Institute for Nuclear Research (Dubna) and the Institute for Nuclear Research of the Russian Academy of Sciences.

The conference was supported by the Ministry of Education and Science of Russia and the Russian Foundation for Basic Research. The Vice Minister of Education and Science academician Grigory Trubnikov opened the first session of the conference with the welcome talk.

A reasonable support in organizing the event was provided by the Interregional Centre for Advanced Studies and the Bruno Pontecorvo Neutrino and Astrophysics Laboratory (MSU).

One of the goals of the conference was to bring together scientists, both theoreticians and experimentalists, working in different fields. The conference has produced a high interest in the world scientific community and there were much more applications to present a talk at the conference from scientist all over the world. The amount of applications reasonably surpassed the capacity to allocate oral talks within the six working days of the conference, therefor 15 parallel sessions were organized at the conference. The physics programme of the 18th Lomonosov Conference has included review and original talks on a wide range of items such as neutrino and astroparticle physics, electroweak theory, fundamental symmetries, tests of Standard Model and beyond, heavy quark physics, non-perturbative QCD, quantum gravity effects, physics at the future accelerator. In total there were more than 400 participants with 176 talks (including 99 plenary 25–20 min talks, 71 session 20-10 min talks and 6 posters) presented by scientists from 36 countries.

A special Round Table discussion on "Frontiers in Particle Physic" was held during the closing day of the conference.

The 18th Lomonosov Conference celebrates the twenty-fifth anniversary since foundation of the Lomonosov conferences in 1992. The first talk of the conference was given by Marco Giammarchi (INFN Milan) who was among the participants of the first edition of the Lomonosov Conference in the year 1992. The title of the talk was "The Lomonosov Conference and physics of the last 25 years".

As early as in 1983 the first of the series of conferences (from 1992 called the "Lomonosov Conferences") was held at the Department of Theoretical Physics of the Moscow State University (June 1983, Moscow). The second conference was held in Kishinev, Republic of Moldavia, USSR (May 1985).

After the four years break this series was resumed on a new conceptual basis for the conference programme focus. During the preparation of the third conference (that was held in Maykop, Russia, 1989) a desire to broaden the programme to include more general issues in particle physics became apparent. During the conference of the year 1992 held in Yaroslavl (Russia) it was proposed by myself and approved by many participants that those irregularly held meetings should be transformed into regular events under the title "Lomonosov Conferences on Elementary Particle Physics".

Since the conference of the year 1993 (the 6th Lomonosov Conference on Elementary Particle Physics) all of the subsequent conferences of this series has been organized at the Moscow State University on a regular basis each odd year under the title the Lomonosov Conference on Elementary Particle Physics. A wide variety of interesting things, both in theory and experiment of particle physics, astrophysics, gravitation and cosmology, were included into the programmes. It was also decided to enlarge the number of institutions that would take part in preparation of the events.

Mikhail Lomonosov (1711-1765), a brilliant Russian encyclopaedias of the era of the Russian Empress Catherine the 2nd, was world renowned for his distinguished contributions in the fields of science and art. He also helped establish the high school educational system in Russia. The Moscow State University was founded in 1755 based on his plan and initiative, and the University now bears the name of Lomonosov.

There is a tradition to publish the conferences proceedings as a collection of review and research papers prepared by the conference speakers on the presented talks basis. It started in 1994 when publication of the volume "Particle Physics, Gauge Fields and Astrophysics" containing articles written by speakers of the 5th and 6th Lomonosov Conferences was supported by the Accademia Nazionale dei Lincei (Rome, 1994). Proceedings of the 7th and 8th Lomonosov Conference (entitled "Problems of Fundamental Physics" and "Elementary Particle Physics") were published by the Interregional Centre for Advanced Studies (Moscow, 1997 and 1999). Proceedings of the 9th, 10th, 11th, 12th, 13th, 14th, 15th, 16th and 17th Lomonosov Conferences (entitled "Particle Physics at the Start of the New Millennium", "Frontiers of Particle Physics", "Particle Physics in Laboratory, Space and Universe", "Particle Physics at the Year of 250th Anniversary of Moscow University", "Particle Physics on the Eve of LHC", "Particle Physics at the Year of Astronomy", "Particle Physics in the Year of Tercentenary of Mikhail Lomonosov", "Particle Physics in the Year of Centenary of Bruno Pontecorvo", "Particle Physics in the Year of Light") were published by World Scientific Publishing Co. (Singapore) in 2001, 2003, 2005, 2006, 2009, 2011, 2013, 2015 and 2017, respectively.

The next 19th Lomonosov Conference will be held at the Moscow State University on August 22–28, 2019.

Following the tradition that has started in 1995, each of the Lomonosov Conferences on particle physics has been accompanied by a conference on problems of intellectuals. The 12th International Meeting on Problems of Intelligentsia held during the 18th Lomonosov Conference (August 30, 2017) was dedicated to discussions on the issue "The Future of the Intelligentsia".

The success of the 18th Lomonosov Conference was due in a large part to contributions of the International Advisory Committee and Organizing Committee. On behalf of these Committees I would like to warmly thank the session chairpersons, the speakers and all of the participants of the 18th Lomonosov Conference and the 12th International Meeting on Problems of Intelligentsia.

We are grateful to Victor Sadovnichy, the Rector of the Moscow State University, Victor Matveev, the Directors of the Joint Institute for Nuclear Research, Leonid Kravchuk, the Director of the Institute for Nuclear Research of Russian Academy of Sciences, Andrey Fedyanin, the Vice Rector of the Moscow State University, Mikhail Itkis, the Vice Director of JINR for the support in organizing these two conferences.

We extremely appreciate continuous support of the Lomonosov Conferences from Nikolay Sysoev, the Dean of Faculty of Physics of the Moscow State University.

I would like to thank HansPeter Beck, Dmitry Denisov, Guido Drexlin, Nicolao Fornengo, Carlo Giunti, Takashi Kobayashi, Gennedy Kozlov, Yury Kudenko, Cristina Lazzeroni, Mikhail Libanov, Ida Peruzzi, Sergey Petrushanko, Piergiorgio Picozza, Leo Piilonen, Gioacchino Ranucci, Grigory Rubtsov, Stefan Schmitt, Uve Schneekloth, Kate Scholberg, Stefan Soldner-Rembold and Christian Spiering for their help in planning of the scientific programme of the meeting and inviting speakers for the topical sessions of the conference.

I would like to thank Alexey Lokhov, Konstantin Stankevich, Yana Zhezher, Sergey Ovchinnikov, Mikhail Sumin and Andrey Egorov, the members of the conference Scientific Secretariat. Furthermore, I am very pleased to mention Darya Grechukhina, Mikhail Markin, Fedor Lazarev, Anna Petrova, Artem Popov, and Pavel Pustoshny for their very efficient work in preparing and running the meeting. I am also very thankful to Lyulya Dubasova, Manya and Ivan Studenikin for their contribution.

I would also acknowledge Vladimir Arkhipov who made video records of the scientific sessions of the conference and interviews with the plenary speakers that are on view at the conference web site.

These Proceedings were prepared for publication at the Interregional Centre for Advanced Studies with support by the Russian Foundation for Basic Research. We appreciate fruitful cooperation with World Scientific Publishing Co. (Singapore) and we are thankful to Professor Kok Khoo Phua for the support of the conference.

Alexander Studenikin

CONTENTS

Eighteenth Lomonosov Conference on Elementary Particle
Physics — Sponsors and Committees   v

Twelfth International Meeting on Problems
of Intelligentsia — The Future of the Intelligentsia   vii

Foreword   ix

Physics and the Lomonosov Conference Series   1
M. Giammarchi

Neutrino Physics   11

Neutrino phenomenology   13
S. Goswami

Low-energy neutrino physics with liquid scintillator detectors   20
L. Ludhova

Review on solar neutrino studies Borexino   27
I. Drachnev

Recent results from Daya Bay neutrino experiment   32
J. Xu

Results from the OPERA experiment at LNGS   37
T. Matsuo

Cross-section measurements at the T2K experiment   42
L. Maret

Recent results from the NOvA experiment   46
J. Zálešák

The KM3NET neutrino telescope and the potential of a neutrino
beam from Russia to the Mediterranean Sea 53
D. Zaborov

The T2HKK Experiment and non-standard interaction 61
M. Ghosh, O. Yasuda

Status and physics potential of the JUNO experiment 65
G. Salamanna

The neutrino mass hierarchy from oscillation 71
L. Stanco

Neutrino mass ordering and neutrinoless double-beta decays 79
S. Zhou

Status and prospects of the search for neutrinoless double beta
decay of ^{76}Ge 83
K. T. Knöpfle

The CUORE experiment at LNGS 90
D. Chiesa

Probing the Majorana neutrino nature at the current and future
colliders 94
R. Soualah

Study of the ν_μ charged current quasielastic-like interactions in
the NOvA near detector 98
L. Stanislav

Search for supernova neutrino bursts with the LVD experiment 101
N.Yu. Agafonova

Cosmological relic neutrino detection with the PTOLEMY experiment 105
A. G. Cocco

Effects of BSMS on θ_{23} determination 110
C. R. Das et al.

The aging behaves and the small batch test of the 20″ MCP-PMTs 114
Y. Zhu et al.

ICARUS: From CNGS to booster beam 118
A. Guglielmi

Pursuit for optimal baseline for matter nonstandard interactions
in long baseline neutrino oscillation experiments 125
T. Kärkkäinen

Neutrino physics and leptonic weak basis invariants 128
M. N. Rebelo

Nonstandard neutrino interactions in a dense magnetized medium 135
A. V. Borisov

Non-conservation of the lepton current and asymmerty of relic
neutrinos 138
V. B. Semikoz, M. Dvornikov

Breakings of the neutrino μ-τ reflection symmetry 142
Z. Zhao

Neutrino clustering in the milky way 146
S. Gariazzo

High energy neutrinos and dark matter 150
A. Esmaili

A scale-invariant radiative neutrino mass generation and dark matter 154
S. Nasri

The coherent weak charge of matter *A. Segarra*	158
Signatures of neutrino magnetic moment in collective oscillations of Supernova neutrinos *O. Kharlanov, P. Shustov*	161
A decay of an ultra-high-energy neutrino $\nu_e \to e^- W^+$ in an extremely high magnetic field *A. Kuznetsov et al.*	165

Physics at Accelerators and Studies in SM and Beyond   169

Electroweak precision measurements in ATLAS *E. Soldatov*	171
Searches for supersymmerty with the ATLAS detector *M. Ronzani*	175
Hadronic resonance production with ALICE at the LHC *S. Kiselev*	180
Search for exotic charmonium-like states at COMPASS *A. Gridin*	185
Overview of the CEPC project *X. Shi*	189
Recent results on diffraction at HERA *S. Levonian*	193
The olympus experiment — Two-photon exchange in electron proton scattering *U. Schneekloth*	199

A few-body approach to low-energy atomic collisions involving
muons and antiprotons   206
R. A. Sultanov

Fractality of strange particle production in PP collisions at RHIC   211
M. V. Tokarev, I. Zborovský

Analysis of anisotropic transverse flow in Pb-Pb collisions at 40A
GEV in the NA49 experiment   215
O. Golosov et al.

Elliptic flow OF π^- measured with the event plane method in
Pb-Pb collisions at $40A$ GeV   218
J. Gornaya et al.

Production of nuclear fragments in the $^{12}C + ^{7}Be$ collisions at
intermediate energies   221
B. M. Abramov et al.

Muon particle physics program at J-PARC   225
S. Mihara

THE muon g-2 experiment at Fermilab   232
W. Gohn

Intense gamma radiation by accelerated quantum ions   237
N. Sasao

The international linear collider project   241
D. Jeans

Astroparticle Physics and Cosmology   249

The detection of gravitational waves: 50 years of experimental efforts 251
E. Coccia

Primordial black holes and cosmological problems   262
A. Dolgov

Solar modulation of galactic cosmic rays: Physics challenges for AMS-02   272
N. Tomassetti

Critical indices and limits on space-time dimensions from cosmic rays   277
Y. Srivastava et al.

Direct measurements of cosmic rays   282
P. Spillantini

Fast radio bursts: A new major puzzle in astrophysics   293
M. Pshirkov et al.

A candidate for an UV completion: Quadratic gravity in first order formalism   297
E. Alvarez et al.

Calocube: A novel approach for a homogeneous calorimeter for high-energy cosmic rays detection in space   300
G. Bigongiari

Conformal invariance and phenomenology of cosmological particle production   305
V. Berezin et al.

Problems of spontaneous and gravitational baryogenesis   309
E. Arbuzova, A. Dolgov

Directional detection of dark matter with a nuclear emulsion based detector   314
A. Alexandrov

DAMA/LIBRA Results and Perspectives 318
R. Bernabei et al.

Global geometry of the Vaidya space-time 328
V. Berezin et al.

Entering the cosmic ray precision era 332
P. D. Serpico

Current status of warm inflation 339
R. Rangarajan

Kinematics of spirals as a portal to the nature of dark matter 346
P. Salucci

Search for dark sector physics in missing energy events in the NA64 experiment 353
M. Kirsanov

Recent results from the DAMPE experiment 370
G. Marsella

Schwarzschild geodesics and the strong force 374
D. Grigoriou, C. G. Vayenas

Current state of dynamical dark energy versus observations 378
J. Solà

Magnetic field generation in dense gas of massive electrons with anomalous magnetic moments electroweakly interacting with background matter 385
M. Dvornikov

Vacuum polarization in cosmic-string and global-monopole background 388
Yu. V. Grats, P. Spirin

Use of pseudo-Hermitian models in promising researches the possible structure of dark matter *V. N. Rodionov et al.*	392
Composition studies with the Telescope Array surface detector *M. Kuznetsov et al.*	396
Performance of the MPD experiment for the anisotropic flow measurement *P. Parfenov et al.*	402

CP Violation and Rare Decays — 405

Recent results of rare B/D decays from Belle *M.-Z. Wang*	407
Search of cLFV processes with muons and others *K. Ieki*	412
Study of $K^+ \to \pi^0 e^+ \nu_e \gamma$ decay with OKA setup *A.Yu. Polyarush*	419

Hadron Physics — 423

Selected results from the ATLAS experiment on its 25th anniversary *F. Djama*	425
Review of the recent Tevatron results *S. Denisov*	432
Top pair and single top production in ATLAS *F. Fabbri*	443
Physics at the Compact Linear Collider *I. Božović-Jelisavčić*	448

The TOTEM experiment: Results and perspectives   456
E. Bossini

The ADAMO project and developments   464
V. Caracciolo

New Physics patterns in $b \to s\ell^+\ell^-$ transitions   468
B. Capdevila

Review of NA62 and NA48 physics results   475
R. Fantechi

Status of the KLOE-2 experiment at DAΦNE   483
P. Fermani

Latest tests of hard QCD at HERA   490
O. Zenaiev

xFitter   497
O. Zenaiev

Low energy nuclear reactions in normal and exploding batteries   500
Y. Srivastava et al.

Calculation of the R–ratio OF $e^+e^- \to$ hadrons at the higher–loop levels   504
A. V. Nesterenko

Local groups of internal transformations isomorphic to local groups of spacetime tetrad transformations   510
A. Garat

New Developments in Quantum Field Theory   515

Hadronization via gravitational confinement   517
C. G. Vayenas, D. Grigoriou

Switching-on and -off effects of external fields on the vacuum instability   525
T. C. Adorno et al.

The three-loop contribution to β-function quartic in the Yukawa couplings for the $\mathcal{N} = 1$ supersymmetric Yang-Mills theory with the higher covariant derivative regularization   534
V. Shakhmanov

Quantum field theory effects on the horizons of static metrics   537
M. Fil'chenkov, Y. Laptev

Problems of Intelligentsia   541

A future for the Intelligentsia   543
J. K. Bleimaier

Conference Programme   565

List of Participants   585

PHYSICS AND THE LOMONOSOV CONFERENCE SERIES

Marco Giammarchi [a]
Istituto Nazionale di Fisica Nucleare, Via Celoria 16, 20133 Milano, Italy

Abstract. The first conference in the Lomonosov Series took place in Yaroslav in 1992. Twenty-five years have passed since those days and Physics has greatly progressed in many ways. I will summarize here only the part of this story that is mostly relevant to this Conference, namely Particle Physics, Cosmology, Astrophysics and Theoretical Physics. I am of course aware that this is only a small part of Physics and human knowledge at large. Unluckily, even this tiny section will be covered in a very cursory way.

1 Introduction

The universe was very different when the Lomonosov Conference Series started, back in 1992. First of all, we had no dark energy (while we already had dark matter) and the universe at photon decoupling was still compatible with homogeneity. In Particle Physics, the top quark and the tau neutrino had not yet been discovered. Moreover, the Higgs (or BEH, Brout-Englert-Higgs) boson was only a prediction and neutrinos did not oscillate (and did not have mass).

This whole picture was going to change in the years of the Lomonosov Conference, in ways that I am now going to summarize.

2 Cosmology

The operation of the COBE satellite [1] (from 1989 to 1993) has brought us two pieces of fundamental knowledge. First of all, the Cosmic Microwave Background (CMB) radiation has a nearly ideal blackbody spectrum, freely streaming in the universe from the age of about 370,000 y, now red-shifted to 2.7 K by a $\simeq$1,100 redshift factor. Secondly, variations from temperature uniformity at the 10^{-5} level were seen (Fig. 1), mapping the seeds of structure formation in the Universe.

These stunning discoveries were closely followed by the experimental evidence of dark energy. The decisive step was obtained by studying the propagation of light coming from the Ia-class Supernovae; these are the "Chandrashekhar-limit" white dwarfs explosively burning their fuel and featuring a relatively well-defined series of physical processes. They constitute the best "standard candles" for cosmological distances. The results obtained by the Supernova Cosmology Project [2] and by the High-z Supernova Search Team conclusively demonstrated that dark energy constitutes $\sim$70% of the overall Universe energy budget.

---

[a] E-mail: marco.giammarchi@mi.infn.it

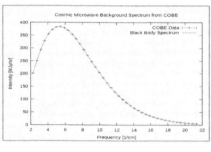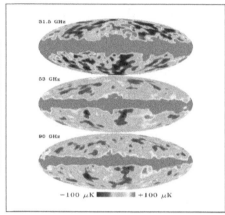

Figure 1: Left hand side: Observed CMB spectrum compared to ideal Blackbody spectrum. Right hand-side: temperature fields obtained at different frequencies by the instrumentation on board the COBE satellite, showing the variations from homogeneity.

These breakthroughs have been complemented by many other research programs, culminating in the present day precision cosmology with the results of the Planck satellite [3] and the so called Standard Cosmological (Concordance) Model [4], where both dark energy and dark matter play an essential role.

3 Gravitational Waves and Astrophysics

The Lomonosov Conference Series began when the Hulse-Taylor binary-pulsar studies were ending, with its astonishing (albeit indirect) confirmation of the existence of gravitational waves. Their direct detection constitutes a subject of investigation spanning from the 60's (with the *Weber style* cryogenic detectors) [5] to the present times of interferometric researches at LIGO and Virgo.

The milestone discovery, crowning this research, took place in 2015 and was announced in 2016. Gravitational waves were detected for the first time in the two interferometers of the LIGO setup, providing one of the most spectacular confirmations of General Relativity as a classical theory of gravitation; the event consisted in the coalescence of two black holes (called a BH-BH event) 1 billion of light years away [6], generating $\sim 10^{-22}$-amplitude ripples in the metric of spacetime and converting about three solar masses into gravitational energy (Fig. 2, left). This constituted one of the highest energy event ever observed by any human-made instrument and was followed in the next months by a few more detection of BH-BH collapse induced gravitational waves, including one that was also seen by the Virgo detector.

The capabilty of three detectors (LIGO and Virgo combined) to provide 3-d information about a gravitational wave source was immediately exploited

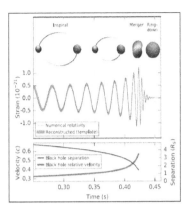 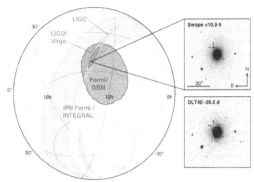

Figure 2: Left hand side: spacetime ripples are generated by BH-BH coalescence event. The figure shows at the bottom also the velocity and separation of the two black holes as a function of time. Right hand-side: the angular area identified by the gravitational interferometers (and by Fermi) and the following first optical evidence of Neutron Stars collision.

in August 2017 when, a few days before this edition of the Lomonosov Conference, the gravitational interferometers detected an event at the same time when the Fermi satellite issued a gamma ray burst automated alarm. The 3-d localization achieved by LIGO-Virgo[b] made it possible to conduct an optical search identifying within a few hours the location of the event by means of the telescopes located in Chile (Fig. 2, right).

The location of the event, in galaxy NGC4993 100 million light years away was confirmed by several optical counterparts and then followed by X-ray and radio evidence in the following days. This confirmed the view that the collapse involved at least a Neutron Star (and very likely two Neutron Stars of solar mass size). This historical event [7] is the first observation of Neutron Star collisions and the related kilonova phenomenon, a key place where to study the formation of heavy elements in the Universe. It marks the beginning of a new era: the MultiMessenger Astronomy, where gravitational waves, gamma and X-rays, optical and radio radiations are used together with neutrinos to get coordinated information about astronomical events.

4 Neutrinos

The last 25 years have been of amazing importance in the neutrino sector of Particle Physics. The long-standing hypotesis by Pontecorvo (of neutrino mass and oscillations [8]) has been put to the test in an unprecedented way.

---

[b]In addition to the angular information, the observed frequency of the gravitational wave makes it possible to estimate the intrinsic masses, therefore providing a distance estimator (in the present case somewhere between 30 and 60 Mpc).

The construction by R. Davis and J. Bahcall of the first solar neutrino detector opened a new era with the emergence of the so-called solar neutrino problem [9]. The deficit in electron-neutrinos detected in the experiment could be due either to an astrophysical problem (poor understanding of the physics of the Sun) or to new properties of neutrinos, namely neutrino mass and oscillation.

The deficit originally detected by Davis was confirmed by other experiments, both in radiochemical mode [10] and in real time [11]. These experiments also confirmed the energy-dependence of the deficit. The detection of the oscillation of neutrinos of atmospheric origin in 1998 by the Superkamiokande experiment [12] started to indicate the route to the solution of the problem: clearly, neutrinos have mass and oscillate. The SNO experiment on solar neutrinos was the first to be sensitive to both charged and neutral current events, thereby confirming without any doubt that electron neutrinos were converting into other neutrinos (muon or tau) during their path from the Sun to the Earth [13].

The solution of the (solar) neutrino problem then conducted to the full demonstration that neutrinos do indeed have a mass and do mix. Additional experimental confirmation came from the KamLand, OPERA and Borexino experiments [14] to complete the picture and prepare for the next news in what is the first clear glimpse of Physics beyond the original Standard Model.

The next step in neutrino oscillations was the demonstration that the parameter θ_{13} is different than zero. The value of θ_{13} is important because it couples atmospheric to the solar oscillations and (if different than zero) makes CP violation possible in the lepton sector. This parameter was measured to be different than zero in 2012 by the T2K, Day Bay and RENO experiments [15].

Neutrino Physics at the time of the Lomonosov Conference has been living a series of revolution [16] that has lead to the determination of the oscillation parameters listed in Fig. 3 and is now leading into the new era of determination of mass hierarchy and search for CP violation.

5 The Standard Model and the Higgs Boson

During the time of the Lomonosov Conference Series, Particle Physics also progressed in the frame of the Standard Model. First of all, we should in fact remember that back in 1992 we had no direct evidence of both the top quark and the tau neutrino.

The top quark was discovered at Fermilab by the CDF experiment in 1995 [17], while the tau neutrino was first observed in 2001 by DoNuT [18] and then confirmed by OPERA. The new millenium has then opened with the observation of all fundamental fermions in the Standard Model. Still something of decisive importance was missing: the Higgs (or Brout-Englert-Higgs, BEH) Boson [19]. This particle was predicted in the 60's to play the essential role of generating the masses for the Standard Model vector bosons in a way

$$U^D = \begin{pmatrix} 1 & 0 & 0 \\ 0 & c_{23} & s_{23} \\ 0 & -s_{23} & c_{23} \end{pmatrix} \begin{pmatrix} c_{13} & 0 & s_{13}e^{i\delta_{13}} \\ 0 & 1 & 0 \\ -s_{13}e^{i\delta_{13}} & 0 & c_{13} \end{pmatrix} \begin{pmatrix} c_{12} & s_{12} & 0 \\ -s_{12} & c_{12} & 0 \\ 0 & 0 & 1 \end{pmatrix}$$
$$\vartheta_{23} \sim \vartheta_{ATM} \qquad\qquad\qquad\qquad\qquad\qquad\qquad \vartheta_{12} \sim \vartheta_{SOL}$$

Parameter	best-fit	3σ
Δm^2_{21} [10^{-5} eV^2]	7.37	6.93 – 7.97
$|\Delta m^2|$ [10^{-3} eV^2]	2.50 (2.46)	2.37 – 2.63 (2.33 – 2.60)
$\sin^2 \theta_{12}$	0.297	0.250 – 0.354
$\sin^2 \theta_{23}$, $\Delta m^2 > 0$	0.437	0.379 – 0.616
$\sin^2 \theta_{23}$, $\Delta m^2 < 0$	0.569	0.383 – 0.637
$\sin^2 \theta_{13}$, $\Delta m^2 > 0$	0.0214	0.0185 – 0.0246
$\sin^2 \theta_{13}$, $\Delta m^2 < 0$	0.0218	0.0186 – 0.0248
δ/π	1.35 (1.32)	(0.92 – 1.99)
		((0.83 – 1.99))

Figure 3: Neutrino parameters, as defined in the Oscillation Matrix for Dirac Neutrinos (above). The table reports the 2017 Particle Data Group best values. The sign of Δm^2 refers to the case of *normal* (positive) or *inverted* mass hierarchy.

that is consistent with the electroweak gauge invariance. In addition, the same mechanism provides masses to the fundamental fermions by means of a suitable Yukawa coupling to the Higgs.

The Higgs Boson can be produced at the CERN LHC proton-proton collider, featuring a proton energy that has increased over time from 3.5 to 4 and then to 6.5 TeV. The production mechanisms of Higgs Bosons at LHC are g-g fusion, WW and ZZ fusion, t-$\bar{t}$ fusion, W and Z bremsstrahlung [20].

The announcement of the Higgs discovery was made on the 4th of July 2012, when the ATLAS and the CMS experiments showed evidence of the $H \to \gamma\gamma$ invariant mass peak above the background [21] [22] [23]. This discovery was soon followed by the identification of the cleanest decay: the *golden mode* four lepton channel (H to ZZ to four leptons). Figure 4 shows the invariant masses for both channels, at the time of this Conference, four years after discovery.

After 2012, the hunt was on to demonstrate that the observed boson (the only fundamental scalar of the Standard Model) is indeed the long sought Higgs Boson of the Electroweak Simmetry and the Standard Model [24]. This involves demonstrating that it is produced at the predicted level and (most of all) showing that its couplings to particles are the ones predicted by the model. This is shown in Fig. 5 for the couplings observed up to now, with a good agreement with the Standard Model predictions.

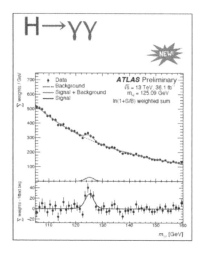
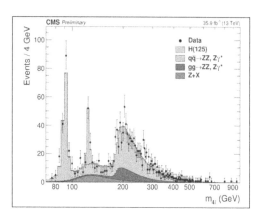

Figure 4: Left: ATLAS plot of the Higgs Boson in the two-photon final state. Right: evidence of the Higgs Boson decay to four leptons from CMS. Both plots refer to data up to 2017.

6 Theoretical Physics and Further Challenges

Theoretical Physics addresses today (as it has always had) enormous challenges. In spite of all the success of the Standard Models of Particle Physics and of Cosmology, several problems remains unsolved and many questions unanswered.

First of all, the fine-tuning and Higgs radiative correction needs to be addressed at high-energy, where new symmetries (like supersymmetry) can possibly play a role. This leads directly to the problem (or the possibility) of the Unification of the Quantum Forces. In addition, the issues of Dark Matter and Dark Energy need to be understood and clarified. At even higher energy scales the problem of harmonizing Gravitation with Quantum Theories (Quantum Gravity at the Planck energy) will have to be faced.

Problems like the actual value of the Cosmological Constant (seemingly unnatural by up to 120 orders of magnitude [25]) or of the mechanism triggering Cosmic Inflation Model [26] deeply question our overall understanding of nature.

It is of course impossible to discuss these problems with any completeness here. In addition, we should also somehow try to resist the temptation to use just Particle Physics and Cosmology as the most fundamental sciences.[c] In addition of being more in the spirit a polymath genius like Lomonosov himself, there are specific reasons for a more general approach to knowledge.

---

[c]Their epistemological relevance is, however, based on the fact that they study space and time at their most extreme scales.

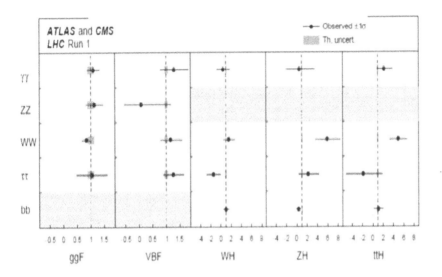

Figure 5: The table taken from the Particle Data Group 2017 shows the agreement between the discovered boson and the Standard Model Higgs/BEH Boson for the various cases of different production modes (horizontal axis) and decay mode (vertical axis).

First of all, we do not know whether the reductionist *ansatz* is valid, which means that we are unsure about the laws governing objects at scales far away from being fundamental in the quantum sense (or cosmological in the astronomical sense). Secondly, even assuming the reductionist viewpoint, fundamental questions arise that indicate boundaries or instrinsic limits of the Universe. Black Holes and Cosmological Expansion both indicate that the realm of Being is more vast than the observable Universe. Matter that is effectively outside space-time (like in a Black Hole) still exerts observable effects in our Universe. In the case of the expansion, we could literally loose contact with pieces of the Universe which have (in our Universe have had) some level of existence [27]. Last but not least, Quantum Mechanics questions the existence of physical quantities independently from the observer [28] and requires the use of concepts like the multiverse or megaverse [29] [30] to provide a more consistent understanding of reality (and ourselves). It is therefore interesting (and unavoidable) that Physics at its (likely) most fundamental level is dealing with philosophical (ontological) concepts. This would really have been in the spirit of Mikhail Lomonosov.

7 Conclusion

Particle Physics, Astrophysics, Cosmology, Theoretical Physics have immensely progressed from the time the Lomonosov Conference Series has started. In this very short review I sadly had abstained from mentioning very important developments, such as CP-violation and antimatter,[d] Cosmic Neutrinos and many others. Fortunately, they are covered in many ways in the Proceedings of the Lomonov Conference of the last 25 years.

Acknowledgments

I would like to thank Luciano Mandelli for many useful discussions.

References

[1] G.F. Smoot, 2006 Nobel Lecture.
[2] S. Perlmutter, Rev. Mod. Phys., **84** (2012) 1127.
[3] R. Adam et al., Astronomy & Astrophysics **594** (2016) A1. Talks by V. Lukash at the Lomonosov Conference 2001, K. Abazajian at the Lomonosov Conference 2003, A. Dolgov at the Lomonosov Conference 2005, N. Mandolesi, P. Natoli and A. Starobinsky at the Lomonosov Conference 2015.
[4] K.A. Olive and J.A. Peacock, *"Big Bang Cosmology"* (Particle Data Group 2017). Talks by A. Starobinsky at the Lomonosov Conference 2007 and at the Lomonosov Conference 2009.
[5] B. Schutz in *"A First Course in General Relativity"*, Cambridge University Press, 2009. Talks by V.M. Lipunov at the Lomonosov Conference 1995, V. Braginski at the Lomonosov Conference 1997 and F. Ricci at the Lomonosov Conference 2015.
[6] B.P. Abbott et al., Phys. Rev. Lett. **116** (2016) 061102. Talk by E. Coccia at this Lomonosov Conference.
[7] B.P. Abbott et al., Astroph. J. Lett. **848** (2017) L12.
[8] B. Pontecorvo, Soviet Physics JETP **6** (1958) 429. B. Pontecorvo, Soviet Physics JETP **7** (1958) 172.
[9] R. Davis, 2002 Nobel lecture.
[10] W. Hampel et al., Phys. Lett. B **447** (1999) 127. J.N. Abdurashitov et al., Phys. Rev. Lett. **83** (1999) 4686. M. Altmann et al., Phys. Lett. B **616** (2005) 174.
[11] S. Fukuda et al., Phys. Rev. Lett. **86** (2001) 5651.
[12] M. Koshiba, Rev. Mod. Phys. **75** (2003) 1011. Talks by Y. Fukuda at the Lomonosov Conference 2001 and by K. Ganezer at the Lomonosov Conference 2003.
[13] Q.R. Ahmad et al., Phys. Rev. Lett. **87** (2001) 071301.

---

[d]My very own field of research!

[14] K. Eguchi et al., Phys. Rev. Lett. **90** (2003) 021802. N. Agafonova et al., Phys. Rev. D **89** (2014) 051102(R). G. Bellini et al., Phys. Rev. D **89** (2014) 112007.
[15] K. Abe et al., Phys. Rev. Lett. **107** (2011) 041801. F.P. An et al., Phys. Rev. Lett. **108** (2012) 171803. J.K. Ahn et al., Phys. Rev. Lett. **108** (2012) 191802.
[16] Talks by V. Gavrin, S. Micheev, A. Smirnov and C. Giunti at the Lomonosov Conference 2003, J. Pulido, T. Kobayashi, N. Savvinov and J. Shirai at the Lomonosov Conference 2005, V. Gavrin, K. Sakashita and S. Micheev at the Lomonosov Conference 2007, V. Gavrin, J. Hartnell and M. Shibata at the Lomonosov Conference 2009. Talks by N. Barros, Y. Obayashi, A. Izmaylov, C. White, M. Mezzetto and E. Litvinovich at the Lomonosov Conference 2011, Y. Suzuki, S. B. Kim, M. Malek, S. Wojciki, N. Kitagaza and C. Giunti at the Lomonosov Conference 2013, C. Giunti and M. Batkiewicz at the Lomonosov Conference 2015.
[17] F. Abe et al., Phys. Rev. Lett. **74** (1995) 2626. Talks by P. Bhat at the Lomonosov Conference 1997 and E. Boos at the Lomonosov Conference 2005.
[18] K. Kodama et al., Phys. Lett. B **504** (2001) 218. Talk by A. Ereditato at the Lomonosov Conference 2011.
[19] P. Higgs, 2013 Nobel Lecture.
[20] Talks by S. Armstrong at the Lomonosov Conference 1999, P. Colas at the Lomonosov Conference 2001, V. Matveev and S. Gentile at the Lomonosov Conference 2003, N. Krasnikov at the Lomonosov Conference 2007, P. Jenni, D. Krofcheck, E. Ros, S. Shmatov, M. Savina, D. Galtsov, A. Cheplakov and V. Gavrilov at the Lomonosov Coference 2009, A. Gritsan at the Lomonosov Conference 2011.
[21] G. Aad et al., Phys. Lett. B **716** (2012) 1.
[22] S. Chatrchyan et al., Phys. Lett. B **716** (2012) 30.
[23] Talks by R. St. Denis and S. Choudhury at the Lomonosov Conference 2013.
[24] Talks by F. Mazzucato at the Lomonosov Conference 2001, G. Unel at the Lomonosov Conference 2007, A. Cakir at the Lomonosov Conference 2011, E. Masso, F. Scutti, Yu. Smirnov and E. Romero Adam at the Lomonosov Conference 2015.
[25] S. Weinberg, Rev. Mod. Phys. **61** (1989) 1.
[26] P. Steinhardt, Sc. Am., April 2011.
[27] R. Caldwell et al., Phys. Rev. Lett. **91** (2003) 071301.
[28] M. Giammarchi, Eur. J. Sc. and Tehol. **11** (2015) 155.
[29] M. Tegmark, Sc. Am., May 2003.
[30] L. Susskind, in *"Cosmic Landscape"*, Ed. Little, Brown and Company, 2005.

Neutrino Physics

NEUTRINO PHENOMENOLOGY

Srubabati Goswami [a]
Physical Research Laboratory, Ahmedabad-380058, India

Abstract. In this talk I will summarize the recent results from neutrino oscillation experiments and the current status of the oscillation parameters. A comparative account of the salient features of the global fits performed by the different groups will be presented. I will also address the future prospects for determination of the unknown parameters. The status of sterile neutrino scenarios in the light of recent results will be described briefly. How non-standard interactions can impact the hierachy sensitvity will also be discussed.

1 Introduction

The subject of neutrino phenomenology occupies a center stage in research in the arena of High Energy Physics. There are several interrelated subtopics that can come under the purview of this subject. One of the most popular among these is neutrino oscillation which implies neutrinos are massive and there is mixing between different flavour states. This points towards physics beyond the Standard Model of particle physics underscoring the importance of the topic. Neutrino oscillation was first proposed by Bruno Pontecorvo in 1857. Thus, the year 2017 marks the 60th year of neutrino oscillation. Since then series of experiments involving solar, atmospheric, accelerator and reactor neutrinos, spanning over four decades, have established neutrino oscillation on a sound footing. With more and more high statistics high precision experiments, the knowledge on the oscillation parameters have increased many-fold over the past years. In this talk I will mention the recent results that are published in this field and will summarize the current status of the oscillation parameters. An upto date comparison of the global fit results by various groups will be presented. I will further discuss to what extent the unknown oscillation parameters can be determined in future experiments. In addition, I will briefly describe the status of sterile neutrinos in the light of the recent results. The proposed future high statistics experiments also open up the possibility of exploring sub-leading effects due to new physics scenarios. I will discuss some results from recent studies on non-standard interactions (NSI).

2 Neutrino Oscillation

Neutrino Oscillation is a phenomenon where one flavour of neutrino can get converted to another flavour after travelling through some distance. This happens because the neutrino flavour states which are the eigenstates of weak interaction are not the same as the mass states which are the eigenstates of propagation.

---

[a] E-mail: sruba@prl.res.in

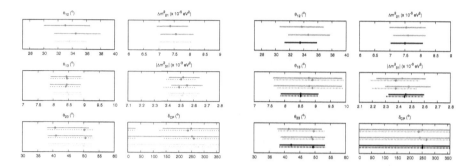

Figure 1: The best-fit values and 3σ ranges of oscillation parameters from different global fits. The left-hand plot depicts the status in 2017. The blue line is from [10], the yellow line is from Nu-fit [11] while the red line is from [12]. The solid (dashed) line is for NH (IH). The right-hand plot presents the status in 2014 from [13]. In this, the Nu-fit results are shown by black line.

This is possible if neutrinos posses non-degenerate mass. This creates a relative phase shift between the propagation states leading to an oscillatory pattern. The flavour mixing can be expressed mathematically as, $|\nu_\alpha\rangle = U_{\alpha i}|\nu_i\rangle$ where ν_α denote the flavour states with $\alpha = e, \mu, \tau$ and ν_i are the mass states with $i = 1, 2, 3$. U is the unitary mixing matrix known as the Pontecorvo-Maki-Nakagawa-Sakata (PMNS) matrix. The oscillation probability in vacuum from one flavour to another can be computed as

$$P_{\alpha\beta} = \delta_{\alpha\beta} - 4\sum_{i>j} \text{Re}(U_{\alpha i}U^\star_{\beta i}U^\star_{\alpha j}U_{\beta j})\sin^2(\Delta_{ij})$$
$$+ 2\sum_{i>j} \text{Im}(U_{\alpha i}U^\star_{\beta i}U^\star_{\alpha j}U_{\beta j})\sin(2\Delta_{ij}), \qquad (1)$$

where, $\Delta_{ij} = \Delta m^2_{ij} L/4E$; $\Delta m^2_{ij} = m^2_i - m^2_j$; L is the distance travelled and E is the energy of the neutrino. For neutrinos traveling in a medium, the neutrino electron coherent scattering modifies the masses, mixing angles and probabilities. For three neutrino flavours U is a 3 × 3 matrix parametrized in terms of three mixing angles and three phases. Neutrino oscillation can probe only one of these phases. This is the Dirac CP phase. This has a different sign for neutrinos and antineutrinos. Oscillation among two neutrino flavours do not depend on this phase. Thus to have a signature of CP violation in the lepton sector we need all three neutrinos to participate in oscillation and all the three mixing angles characterizing the mixing matrix to be non-zero.

3 Three Neutrino Oscillation Parameters and Their Current Status

The three flavour oscillation parameters are the three mixing angles θ_{12}, θ_{23} and θ_{13}, the two mass squared differences Δm^2_{21} and Δm^2_{31} and the Dirac CP phase δ_{CP}. From solar neutrino data it has already been established that $\Delta m^2_{21} > 0$.

However the sign of Δm_{31}^2 has not yet been determined and there can be two possibilities (i) normal hierarchy (NH) with $\Delta m_{31}^2 > 0$ and (ii) inverted hierarchy (IH) with $\Delta m_{31}^2 < 0$. The two other major unknown parameters are the octant of θ_{23} and the leptonic CP phase δ_{CP}. $\theta_{23} < 45^o$ corresponds to lower or first octant (LO) and $\theta_{23} > 45^o$ corresponds to higher or second octant (HO). Maximal CP violation occurs for $\delta_{CP} = \pm \pi/2$ while $\delta_{CP} = 0, 180°$ correspond to CP conservation.

The new data that have come in the last one year are (i) The 1230 days electron antineutrino spectrum from Daya-Bay [1]; (ii) First combined analysis of neutrino and antineutrinos in T2K with 7.4×10^{20} protons on target (pot). [2]; (iii) T2K muon neutrino and antineutrino disappearance results [3]; (iv) Disappearance results from NOνA [4]; (v) Appearance and disappearance analysis of NOνA [5]; (vi) Preliminary T2K neutrino results with 14.7×10^{20} protons on target [6].

The crucial parameter in three generation analysis is the third mixing angle θ_{13}. The precision of this parameter is controlled by the Daya-Bay reactor neutrino experiment. The current best-fit value of this angle from Daya-Bay measurement is given as $\sin^2 \theta_{13} = 0.0841 \pm 0.0033$ [1]. Interestingly the accelerator+ solar + KamLAND data also give the same best-fit value but the precision is poorer [7]. A non-zero value of θ_{13} allows one to measure δ_{CP}. Results on measurement of δ_{CP} has come from both T2K and NOνA experiment. NOνA collaboration has declared the results for only neutrino run with a p.o.t of 6.05×10^{20} [5]. They have reported two best-fit solutions (i) $\sin^2 \theta_{23} = 0.404, \delta_{CP} = 1.48\pi = 266.4°$ and (ii) $\sin^2 \theta_{23} = 0.623, \delta_{CP} = 0.74\pi = 133.2°$. These denote two degenerate solutions in opposite octants with different values of δ_{CP}. One of them is the true solution whereas the other is a degenerate solution with wrong-octant and wrong-δ_{CP}. Inverted hierarchy with θ_{23} in lower octant is disfavoured at greater than 90% C.L. for all values of δ_{CP}. The most recent T2K results [6] with statistics of neutrino mode almost twice that of antineutrino mode continue to give best-fit value close to 270°. $\delta_{CP} = 90°$ gets excluded with a higher sensitivity. It is intriguing to see the implications of $\delta_{CP} = 270°$ as measured in T2K from the point of view of parameter degeneracies [8]. For this value of δ_{CP}, there is no degeneracy for neutrinos(antineutrinos) for NH-HO (NH-LO). Thus if δ_{CP} is 270° is corroborated by both neutrino and antineutrino data – it would imply θ_{23} to lie in different octants. This problem can be overcome if if θ_{23} is maximal – for which both neutrino and antineutrino probability do not have any degeneracy. Thus an unambiguous hint of $\delta_{CP} = 270°$ coming from both neutrino and antineutrino channel will imply hierarchy to be normal and θ_{23} to be maximal. The value of θ_{23} from analysis of T2K data indeed comes out to be 45° MINOS data, on the other hand, gives best-fit θ_{23} in the lower octant [9].

An unified treatment of data is performed in global analysis [10–12] . The analysis by the Nu-fit group [11] does not include the atmospheric neutrino

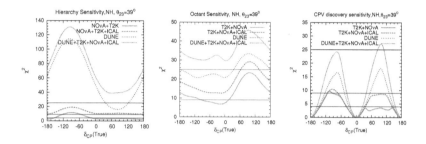

Figure 2: Future potential of hierarchy, octant and CP discovery.

data. The other two groups include atmospheric neutrino data partially. Figure 1 compares the best-fit values and 3σ ranges of these parameters from the analyses performed by the different groups. It can be seen from the figure that there is good agreement in the values of the parameters among the different groups. For θ_{23} the best-fit comes in lower octant for NH and in higher octant for IH. The global best-fit comes for NH although the statistical significance is still not very high. The $\Delta\chi^2$(IH-NH) is 0.8 [11], 2.7 [12] 3.6 [10] respectively. Maximal θ_{23} and $\delta_{CP} = 90°$ are disfavoured in all three analyses although to a different degree. The NOνA results play an important role in disfavouring maximal mixing. The analysis in [12] shows a greater exclusion of maximal mixing specially for IH. $\delta_{CP} = 90°$ is disfavoured to a greater degree for IH in [12]. For NH, on the other hand $\delta_{CP} = 90°$ is disfavoured to a greater extent in the analysis presented in [10]. Some of these differences can perhaps be attributed to the non-inclusion and/or different treatment of the atmospheric neutrino data. In Fig. 1 we also present the comparison of the global fit results of different groups as presented in ICHEP 2014 [13]. The main changes in three years are the following: (i) The best-fit θ_{23} for NH obtained in [12] is now in the first octant in agreement with the analysis by the other groups. (ii) The δ_{CP} has become more restricted with $\delta_{CP} = 90°$ getting disfavoured at 2σ in all the three global fits. (iii) There is a visible improvement in the precision of θ_{13}.

The future prospects include more data from the ongoing experiments T2K and NOνA . With their full planned run, NOνA +T2K can reach around 2σ hierarchy sensitivity which can enhance to 3σ near maximal CP violation. If one adds atmospheric neutrino data to these then the sensitivity can increase. For instance adding 500 kt-yr data from ICAL detector proposed by the INO collaboration the sensitvity can reach 3σ over the full range of δ_{CP} and 4σ for favourable values of δ_{CP} [14]. Among the future experiments DUNE can provide a very good hierarchy sensitvity due to larger matter effect. For instance even 10kt DUNE detector can reach more than 10σ hierarchy sensitvity in 10 years for NH, $\theta_{23} = 39°$ and $\delta_{CP} = -90°$. The octant sensitvity of DUNE for these values can be more than 4σ. ICAL detector can help to increase the CP sensitvity of T2K and NOνA in the wrong-hierarchy region [15]. DUNE can discover maximal violation of δ_{CP} at close to 4σ level. These features can be seen from Fig. 2.

4 Sterile Neutrinos: Current Status

Light sterile neutrinos became popular since the declaration of the LSND results [16] This indicated oscillations with mass squared difference $\sim$ eV^2. MiniBooNE experiment was planned to confirm the LSND results. The recent results from this experiment also confirmed oscillation with a scale $\sim$ eV^2 and existence of at least one light sterile neutrino [17] Other evidences in favour of sterile neutrinos come from the Gallium anomaly [18] and the deficit in the observed event rate of the electron antineutrino flux at reactor experiments with respect to the recalculated fluxes [19]. Light sterile neutrino scenarios are constrained from cosmology. However recent analysis of the Planck data has shown that eV scale light sterile neutrino scenario is admissible by deviating slightly from the ΛCDM model [20]. The minimal model consists of adding one sterile neutrino to SM. The most favoured from solar and atmospheric neutrino data as well as cosmology is the 3+1 scheme which contains three active neutrinos of mass in the sub-eV range and one sterile neutrino with mass in the eV scale. However, there is a severe tension between appearance and disappearance data in global sterile neutrino fits [21, 23]. In addition, recent combined analysis of MINOS, Daya-Bay and Bugey-3 have severely disfavoured the allowed area from global analyses [21]. The allowed area from analysis in [22] is disfavoured at 90% C.L. while only a very tiny region from analysis done in [23] remains allowed.

5 Non-Standard Interactions

Off late non-standard interactions have received a lot of attention specially in the context of the proposed next generation experiments. Effect of NSI's have been considered in the context of production, detection and propagation. The NC current Lagrangian relevant for the propagation NSI is given as,

$$\mathcal{L}_{\mathrm{NSI}}^{NC} = (\bar{\nu}_\alpha \gamma^\rho P_L \nu_\beta)(\bar{f}\gamma_\rho P_C f) 2\sqrt{2} G_F \epsilon_{\alpha\beta}^{fC} + \mathrm{h.c.} \qquad (2)$$

In presence of NSI, the matter potential can be expressed as,

$$V = A \begin{pmatrix} 1+\epsilon_{ee} & \epsilon_{e\mu}e^{i\phi_{e\mu}} & \epsilon_{e\tau}e^{i\phi_{e\tau}} \\ \epsilon_{e\mu}e^{-i\phi_{e\mu}} & \epsilon_{\mu\mu} & \epsilon_{\mu\tau}e^{i\phi_{\mu\tau}} \\ \epsilon_{e\tau}e^{-i\phi_{e\tau}} & \epsilon_{\mu\tau}e^{-i\phi_{\mu\tau}} & \epsilon_{\tau\tau} \end{pmatrix}, \qquad (3)$$

where, $A \equiv 2\sqrt{2}G_F N_e E$ and $\epsilon_{\alpha\beta}e^{i\phi_{\alpha\beta}} \equiv \sum_{f,C} \epsilon_{\alpha\beta}^{fC} \frac{N_f}{N_e}$, with N_f the number density of fermion f. The unit contribution to the (1,1) element of the V matrix is the usual matter term arising due to the standard charged-current interactions. Here, diagonal elements of the V matrix are real due to the Hermiticity of the Hamiltonian. Additional degeneracies can arise in presence of NSI. A generalized hierarchy degeneracy $\epsilon_{ee} \to -\epsilon_{ee} - 2$, $\delta_{CP} \to \pi - \delta_{CP}$, was studied in [24]. It can be seen from Eq. (3) that for the special case of $\epsilon_{ee} = -1$, the NSI effect cancels the standard matter effect. If in addition $\delta_{CP} = \pm\pi/2$, there exists an exact intrinsic hierarchy degeneracy in the appearance channel, which is independent of baseline and the neutrino beam energy [25]. Note that this degeneracy cannot be lifted even if ϵ_{ee} and δ_{CP} are precisely measured around these values. This result assumes more importance in the light of current data from the T2K experiment hinting at $\delta_{CP} \sim -\pi/2$. NSI induced matter effects has also been invoked to explain the discrepancy between best-fit values of θ_{23} in T2K and NOνA [26, 27].

6 Concluding Remarks

The oscillation parameters $\Delta m_{21}^2, \Delta m_{31}^2, \theta_{12}, \theta_{13}, \theta_{23}$ are stable in different global fits and have been determined with a precision $\lesssim 10\%$. The unknown parameters are hierarchy, octant and δ_{CP}. Best-fit value of θ_{23} are in lower octant for NH and in higher octant for IH with a slight overall preference for NH. Maximal mixing is disfavored in global fit governed by the NOνA data. Future high statistics experiments can not only determine these unknown parameters with a high confidence, they also open up the possibility of exploring new physics. The precision measurements of oscillation parameters can also test models of neutrino masses and mixings.

Acknowledgments

I would like to thank my collaborators K. Chakraborty, K.N. Deepthi, M. Ghosh, N. Nath, L.S. Mohan, S. Raut, K.N. Vishnudath for discussions and help. I would also like to thank E. Lisi and M. Tortola for discussions on global fits.

References

[1] F. P. An et al. [Daya Bay Collaboration], Phys. Rev. D **95** (2017) no.7, 072006.

[2] K. Abe et al. [T2K Collaboration], Phys. Rev. Lett. **118** (2017) no.15, 151801.

[3] K. Abe et al. [T2K Collaboration], Phys. Rev. D **96** (2017) no.1, 011102.

[4] P. Adamson *et al.* [NOvA Collaboration], Phys. Rev. Lett. **118** (2017) no.15, 151802.
[5] P. Adamson *et al.* [NOvA Collaboration], Phys. Rev. Lett. **118** (2017) no.23, 231801.
[6] Talk by M. Hartz, Kavli, IPMU, (2017) https://www.t2k.org/docs/talk/282/kekseminar20170804.
[7] E. Lisi, talk at Invisibles17 workshop, https://indico.cern.ch/event/591895/, (2017).
[8] M. Ghosh, S. Goswami and S. K. Raut, Mod. Phys. Lett. A **32** (2017) no.06, 1750034.
[9] P. Adamson *et al.* [MINOS Collaboration], Phys. Rev. Lett. **112** (2014) 191801.
[10] F. Capozzi, E. Di Valentino, E. Lisi, A. Marrone, A. Melchiorri and A. Palazzo, Phys. Rev. D **95** (2017) no.9, 096014.
[11] I. Esteban, M. C. Gonzalez-Garcia, M. Maltoni, I. Martinez-Soler and T. Schwetz, JHEP **1701** (2017) 087.
[12] P. F. de Salas, D. V. Forero, C. A. Ternes, M. Tortola and J. W. F. Valle, arXiv:1708.01186 [hep-ph].
[13] S. Goswami, Nucl. Part. Phys. Proc. **273-275** (2016) 100.
[14] A. Ghosh, T. Thakore and S. Choubey, JHEP **1304** (2013) 009.
[15] M. Ghosh, P. Ghoshal, S. Goswami and S. K. Raut, Phys. Rev. D **89** (2014) no.1, 011301.
[16] C. Athanassopoulos *et al.* [LSND Collaboration], Phys. Rev. Lett. **77** (1996) 3082.
[17] A. A. Aguilar-Arevalo *et al.* [MiniBooNE Collaboration], Phys. Rev. Lett. **110** (2013) 161801.
[18] C. Giunti and M. Laveder, Phys. Rev. C **83** (2011) 065504.
[19] G. Mention, M. Fechner, T. Lasserre, T. A. Mueller, D. Lhuillier, M. Cribier and A. Letourneau, Phys. Rev. D **83** (2011) 073006.
[20] P. A. R. Ade *et al.* [Planck Collaboration], Astron. Astrophys. **594** (2016) A13.
[21] T. J. Carroll, arXiv:1705.05064 [hep-ex].
[22] J. Kopp, P. A. N. Machado, M. Maltoni and T. Schwetz, JHEP **1305** (2013) 050.
[23] S. Gariazzo, C. Giunti, M. Laveder and Y. F. Li, JHEP **1706** (2017) 135.
[24] P. Coloma and T. Schwetz, Phys. Rev. D **94** (2016) no.5, 055005; Erratum: [Phys. Rev. D **95** (2017) no.7, 079903].
[25] K. N. Deepthi, S. Goswami and N. Nath, Phys. Rev. D **96** (2017) no.7, 075023.
[26] J. Liao, D. Marfatia and K. Whisnant, Phys. Lett. B **767** (2017) 350.
[27] S. Fukasawa, M. Ghosh and O. Yasuda, arXiv:1609.04204 [hep-ph].

LOW-ENERGY NEUTRINO PHYSICS WITH LIQUID SCINTILLATOR DETECTORS

Livia Ludhova [a]

Institut für Kernphysik, Forschungszentrum Jülich, 52425 Jülich, Germany

and

RWTH Aachen University, 52062 Aachen, Germany

Abstract. Liquid-scintillator based neutrino detection is a well established experimental technique of neutrino spectroscopy. Thanks to a high light yield of liquid scintillators, detection of even sub-MeV neutrinos is possible with a good energy and position reconstruction resolution. Here, we present a review of the detection principles, summary of the world liquid-scintillator based detectors, their scientific goals and latest results.

1 Introduction

Neutrino physics is a very active field, in which major discoveries were performed in a relatively recent period, as for example the discovery of neutrino oscillations (Nobel Prize 2015) or the discovery of non-zero θ_{13} mixing angle (Breakthrough Prize in Fundamental Physics 2016). Still, many questions concerning the basic neutrino properties, as well as neutrinos as messengers to study astrophysical objects, remain open, being the central topics of several running and planned experiments.

The energies of neutrino sources, both natural and human-made, span through several orders of magnitude. Liquid scintillators are a key medium in the detection of MeV neutrinos. Sources of MeV neutrinos have different origins and come to us with different baselines L. Solar neutrinos ($L \sim 1.5 \times 10^{11}$ m) are the only direct probe about the ongoing fusion reactions powering the Sun. They bring to us information about the interior of our closest star and give us a possibility to study the phenomenon of neutrino oscillations. The long lived radioactive elements in the deep Earth's emit electron flavour anti-neutrinos (geoneutrinos) and represent a unique tool to study the Earth's radiogenic heat, as well as abundances and distributions of the heat producing elements (HPE), as ^{232}Th, ^{238}U, and ^{40}K. Strong β radioactive sources placed next to large volume detectors provide a unique way to search for a light sterile neutrino through an observation of short distance neutrino oscillations, non compatible with the 3-flavour model. Diluting specific $\nu\nu\beta\beta$ decaying isotopes in liquid scintillators is a competitive way to search for $00\beta\beta$ decay and to test the Majorana vs Dirac character of neutrinos. Nuclear reactors represent the strongest human-made sources of neutrinos, providing a unique tool in the determination of oscillation parameters, neutrino mass hierarchy, and searches for sterile neutrinos.

---

[a] E-mail: l.ludhova@fz-juelich.de

2 Neutrino and Antineutrino Detection

An important advantage of the detection of MeV neutrinos with LS is a high light yield (LY). The highest LY so far achieved is 500 detected photoelectrons (p.e.) / MeV of deposited energy as in the Borexino detector, corresponding to 5% @ 1 MeV energy resolution. The scintillation light is isotropic and thus the intrinsic directionality, possible in the water Cherenkov detectors, is lost. Since it is not possible to distinguish the neutrino-related signals from the signals from natural radioactivity, the radiopurity of the LS and other construction materials is of primary concern. Since the pulse shape of the scintillation signal in LS depends on the particle type, particle identification in LS is possible, especially α/β discrimination, e^+/e^- being more challenging.

The detection of neutrinos ν of any flavor relies on their weak interaction elastic scattering

$$\nu + e^- \to \nu + e^- \tag{1}$$

with the electrons of the molecule of an organic liquid scintillator. Only a fraction of the neutrino energy E_ν is transferred to an electron and the interaction of the latter with the medium originates the fluorescence light that is detected, usually by surrounding photomultipliers (PMTs). The recoil spectrum is thus continuous even in the case of monochromatic neutrinos and it extends up to a maximum energy T_e^{max} given by

$$T_e^{max} = \frac{E_\nu}{1 + \frac{m_e c^2}{2 E_\nu}}. \tag{2}$$

A typical cross section for 1-2 MeV electron flavour neutrinos is $\sim 10^{-44}\,\mathrm{cm}^2$, while it is about 6 times smaller for μ and τ flavours. The reason for this is the fact, that in addition to the neutral current interaction (Z exchange) having the same cross section for all flavours, the ν_e scattering can proceed also through the W exchange charge interaction. The lowest energy threshold was achieved in Borexino detector by measuring solar neutrinos down to about 200 keV.

Electron antineutrinos are detected by the inverse-beta decay (IBD) reaction

$$\bar{\nu}_e + p \to e^+ + n, \tag{3}$$

having a typical cross section at few MeV of $\sim 10^{-42}\,\mathrm{cm}^2$. Liquid scintillators are used as proton-rich targets. Only antineutrinos with energies above 1.8 MeV, the kinematic threshold of this interaction, can be detected. The IBD interaction provides a powerful tool to suppress backgrounds, thanks to a possibility to require a space and time coincidence between the prompt signal and the delayed one. The positron comes to rest very fast and then annihilates emitting two 511 keV γ-rays, yielding a *prompt event*. The visible energy E_{prompt} is directly correlated with the incident antineutrino energy $E_{\bar{\nu}_e}$:

$$E_{prompt} = E_{\bar{\nu}_e} - 0.784\,\mathrm{MeV}. \tag{4}$$

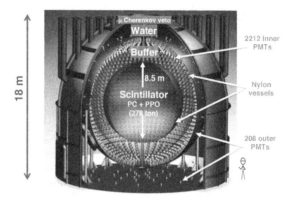

Figure 1: Scheme of the Borexino detector.

The neutron, produced in IBD reaction together with positron, keeps initially the information about the incident $\bar{\nu}_e$ direction. Thermalized neutron is typically captured on a proton after a relatively long time with $\tau = 200 - 250\,\mu$s (depending on scintillator), during which the directional information is mostly lost. When the thermalized neutron is captured on a proton, a 2.22 MeV de-excitation γ-ray is emitted, providing a coincident *delayed event*. In detectors of limited sizes, LS is doped with Gd having a large cross section for neutron capture. In that case, the τ is reduced to $\sim 30\,\mu$s and the total energy of several emitted gammas is 8 MeV.

3 World Liquid Scintillator Detectors

Neutrinos interact with matter only through the weak interactions and thus, their probability to interact is small. As a consequence, neutrino detectors have to be shielded against cosmic radiations and are often placed in underground laboratories under the shield of overburden rocks.

The Borexino detector (280 ton of LS) is placed in the world's largest underground laboratory, Laboratori Nazionali del Gran Sasso (LNGS) in Italy, under 1400 m of rock shielding. The detector has a characteristic onion-like structure, as shown in Fig. 1: the most radio-pure scintillator is placed in the center of the concentric shells, shielded by a layer of buffer and viewed by an array of PMTs. The most external part of the detector is the water tank, serving both as a passive shield against the external radiation, as well as an active muon Cherenkov veto. Borexino was designed to measure ^{7}Be solar neutrinos (0.862 keV), requiring unprecedented radiopurity. This detector contains the world's radio-purest LS and is the only experiment that provided the spectroscopic measurement of all pp-cycle solar neutrinos [1, 2]. It was also the first experiment to observe geoneutrinos with more than 3σ CL [3]. In a near future, ^{144}Ce/^{144}Pr antineutrino source will be placed below the detector in order to search for a light sterile neutrino [4].

The world's largest liquid scintillator detector in operation is KamLAND, having the target mass of 1200 ton and being placed in Kamioka mine in Japan in the depth of 1000 m. The detector was constructed to measure reactor antineutrinos with 260 km baseline and was the first to observe the neutrino oscillation pattern [5], having the best sensitivity to Δm_{12}^2. KamLAND was also the first to investigate geoneutrinos [6]. Currently, the KamLAND-ZEN experiment is dedicated to $0\nu\beta\beta$ search using a ^{136}Xe loaded scintillator.

The three experiments Double Chooz [7] in France, Daya Bay [8] in China, and RENO [9] in Korea are dedicated to measure the θ_{13} mixing angle detecting reactor antineutrinos at ~1-2 km baseline. Due to their limited size (max 20 ton), they all have in the core Gd-loaded scintillators. In order to suppress the systematic errors, each experiment has a near detector monitoring the flux of reactor antineutrinos constructed at a distance of few hundreds of meters from the reactor.

Several smaller scale detectors placed at a few meters distance from nuclear reactors are dedicated to the search for a light sterile neutrino: NEOS [10] in Korea, Setero [11] at ILL in France, and Prospect [12] under construction in Oak Ridge in the USA.

SNO+ [13] is a successor of the Water Cherenkov detector SNO, in which the water will be replaced with 1 kton liquid scintillator doped with Nd in order to search for $0\nu\beta\beta$ decay. It is currently the deepest detector of this kind, placed at the Sudbury mine in Canada 2070 m underground. The detector is currently filled with water that will be shortly replaced by the LS. Solar ^{8}B neutrinos and geoneutrinos are among further scientific targets of this project.

Jiangmen Underground Neutrino Observatory (JUNO) [14, 15] is a 20 kton LS detector under construction in China. This first multi-kton LS detector ever expects to start data taking in 2020, with the aim to determine the neutrino mass hierarchy (normal vs inverted). Thanks to its large mass, it has also a strong potential in measuring solar neutrinos, geoneutrinos, SN neutrinos, as well as in the discovery of the Diffuse SN background neutrinos.

A 4 kton LS detector dedicated to the measurement of solar (CNO), geo, SN, and DSNB neutrinos is planned to be built in the world's deepest Jinping laboratory in China [16].

4 Solar Neutrinos

The Sun is powered by nuclear fusion reactions in which protons form ^{4}He nuclei. The principal chain of reactions is called pp-cycle, in which are emitted pp neutrinos, representing the dominant component with continuous energy

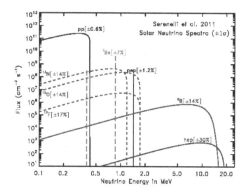

Figure 2: Energy spectrum of solar neutrinos.

spectrum up to 420 keV, mono-energetic ^{7}Be (862 and 384 keV) and *pep* (1440 keV) neutrinos, and ^{8}B neutrinos, extending up to ~15 MeV and having a very low flux. A small fraction of solar energy is expected to be produced in the fusion process catalyzed by the presence of heavier elements, in the so called CNO cycle, a dominant fusion process in heavy stars. So far, the CNO-neutrinos have not been observed. Figure 2 shows the energy spectrum of solar neutrinos.

Solar neutrinos help us to understand our star. The Standard Solar Models (SSM) have among other inputs (luminosity, mass, radius, opacity etc.) also the solar *metallicity*, the ratio of abundances of elements heavier than He to that of Hydrogen. Among SSM outputs are the ν-fluxes and the sound-waves speed profiles. A hot topic in solar physics is the so-called metallicity problem: the new 3D SSM [17] using new spectroscopic *high metallicity* data do spoil the agreement between the helioseismological data and the predictions of the older 1D SSM [18], using in input the older *low metallicity* spectroscopic data. Since the low- and high-metallicity SSMs predict different fluxes of neutrinos, their precise measurement can significantly help in solving this puzzle.

Solar neutrino studies can also contribute to our understanding of the nature of neutrino interactions, for example by studying the electron-neutrino survival probability P_{ee} as a function of energy as predicted by the LMA-MSW solution of neutrino oscillations [19].

Only LS-based experiments are able to perform the spectroscopy of all species of solar neutrinos. Recently, Borexino performed the first simultaneous precision spectroscopy of *pp*, ^{7}Be, and *pep* neutrinos [1, 20] as well as of ^{8}B neutrinos [2].

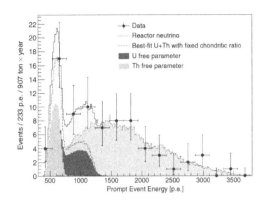

Figure 3: Prompt light yield spectrum in photoelectrons (p.e.) of 77 antineutrino candidates measured by Borexino and the best fit [21]. 1 MeV corresponds to ∼500 p.e.

5 Geoneutrinos

The main aim of geoneutrino studies is to determine the Earth's radiogenic heat, especially the unknown contribution from the mantle. Through the IBD interaction only ^{232}Th and ^{238}U geoneutrinos can be detected. Today, only two experiments succeeded to measure geoneutrinos: KamLAND and Borexino. Borexino provided a new update in 2015 [21], as demonstrated in Fig. 3. Within the exposure of $(5.5 \pm 0.3) \times 10^{31}$ target-proton × year, $23.7^{+6.5}_{-5.7}(\text{stat})^{+0.9}_{-0.6}(\text{sys})$ geoneutrino events have been detected. The null observation of geoneutrinos has a probability of 3.6×10^{-9} (5.9σ). A geoneutrino signal from the mantle is obtained at 98% confidence level. The radiogenic heat production for U and Th from the present best-fit result is restricted to the range 23 to 36 TW, taking into account the uncertainty on the distribution of HPE inside the Earth. The latest published KamLAND result, 116^{+28}_{-27} geoneutrinos detected with 4.9×10^{32} target-proton × year exposure, is from 2013 [22], including the period of low reactor anti-neutrino background after the Fukushima disaster in April 2011.

6 Medium Baseline Reactor Neutrino Experiments

The three experiments Double Chooz [7,24], RENO [9,23], and Daya Bay [8,25] have all measured the θ_{13} mixing angle. The latter provides the most precise value of $\sin^2 2\theta_{13} = 0.0841 \pm 0.0027(\text{stat}) \pm 0.0019(\text{sys})$ as well as that of $|\Delta m^2_{ee}| = [2.50 \pm 0.06(\text{stat}) \pm 0.06(\text{sys})] \times 10^{-3}\,\text{eV}^2$. Daya Bay provided also the highest statistics measurement of the reactor antineutrino spectrum through the detection of 1.2 million IBD interactions [26], confirming the integral deficit of about 6% of events as well as the excess of events in region

4-6 MeV with respect to the Huber + Mueller prediction. A possible explanation to the observed deficit and the so called reactor anomaly could be a possible overestimation of the IBD yield of ^{235}U by 7.8%, based on the Daya Bay observation of the evolution of the shape of the reactor spectrum as a function of the composition of the reactor [27].

References

[1] M. Agostini et al. (Borexino Coll.) arXiv:1707.09279 (2017).
[2] M. Agostini et al. (Borexino Coll.) arXiv:1709.00756 (2017).
[3] G. Bellini et al. (Borexino Coll.), Phys. Lett. B **687** (2010) 299.
[4] G. Bellini et al. (Borexino/SOX Coll.), JHEP **08** (2013) 038.
[5] S. Abe et al. (KamLAND Coll.), Phys. Rev. Lett. **100** (2008) 221803.
[6] T. Araki et al. (KamLAND Coll.), Nature **436** (2005) 499.
[7] Y. Abe et al. (Double Chooz Coll.), JHEP **01** (2016) 163.
[8] F. P. An et al. (Day Bay Coll.), Phys. Rev. D **95** (2017) 072006.
[9] J.H. Choi et al. (RENO Coll.),, Phys. Rev. Lett. **116** (2016) 211801.
[10] Y.J. Ko et al. (NEOS Coll.), Phys. Rev. Lett. **118** (2017) 121802.
[11] J. Haser for Stereo Coll., arXiv:1710.06310 (2017).
[12] J. Ashenfelter et al. (Prospect Coll.), J. of Phys. G: Nucl. and Part. Phys. **43** (2016) 11.
[13] M.C. Chen et al. (SNO+ Coll.), arXiv:0810.3694 (2008).
[14] F. An et al. (JUNO Coll.), J. Phys. G: Nucl. Part. Phys. **43** (2016) 030401.
[15] G. Salamanna for JUNO Coll.: Status and physics potential of JUNO experiment, in this volume.
[16] J.F. Beacom et al., Chin. Phys. C **41**, No. 2 (2017) 023002.
[17] A.M. Serenelli, W.C. Haxton and C. Peña-Garay Ap. J. **743** (2011) 24.
[18] M. Asplund, S. Basu, J.W. Ferguson, and M. Asplund Ap. J. Lett. **705** (2009) L123.
[19] S.P. Mikheyev and A.Y. Smirnov, Sov. J. Nucl. Phys. **42** (1985) 913; L. Wolfenstein Phys. Rev. D **17** (1978) 2369.
[20] I. Drachnev for Borexino Col.: Review of solar neutrino studies with Borexino, in this volume.
[21] M. Agostini et al. (Borexino Coll.) Phys. Rev. D **92** (2015) 031101 (R).
[22] A. Gando et al. (KamLAND Coll.) 2013 Phys. Rev. D **88** (2013) 033001.
[23] S.H. Seo for RENO Coll.: Review on RENO, in this volume.
[24] T. Matsubara for Double Chooz Coll.: Recent results from Double Chooz reactor neutrino experiment, in this volume.
[25] J. Xu for Daya Bay Coll.: Recent results from Daya Bay, in this volume.
[26] F. P. An et al. (Day Bay Coll.) Chin. Phys. C. **41** No.1 (2017) 013002.
[27] F. P. An et al. (Day Bay Coll.) Phys. Rev. Lett. **118** (2017) 251801.
[28] M. Gromov for Borexino/SOX Coll.: The SOX project, in this volume.

REVIEW ON SOLAR NEUTRINO STUDIES BOREXINO

I. Drachnev, M. Agostini, K. Altenmuller, S. Appel, V. Atroshchenko, Z. Bagdasarian, D. Basilico, G. Bellini, J. Benziger, D. Bick, G. Bonfini, D. Bravo, B. Caccianiga, F. Calaprice, A. Caminata, S. Caprioli, M. Carlini, P. Cavalcante, A. Chepurnov, K. Choi, D. DAngelo, S. Davini, A. Derbin, X.F. Ding, A. Di Ludovico, L. Di Noto, K. Fomenko, A. Formozov, D. Franco, F. Froborg,F. Gabriele, C. Galbiati, C. Ghiano, M. Giammarchi, A. Goretti, M. Gromov, C. Hagner, T. Houdy, E. Hungerford, Aldo Ianni, Andrea Ianni, A. Jany, D. Jeschke, V. Kobychev, D. Korablev, G. Korga, D. Kryn, M. Laubenstein, E. Litvinovich, F. Lombardi, P. Lombardi, L. Ludhova, G. Lukyanchenko, L. Lukyanchenko, I. Machulin, G. Manuzio, S. Marcocci, J. Martyn, E. Meroni, M. Meyer, L. Miramonti, M. Misiaszek, V. Muratova, B. Neumair, L. Oberauer, B. Opitz, F. Ortica, M. Pallavicini, L. Papp, N. Pilipenko, A. Pocar, A. Porcelli, G. Ranucci, A. Razeto, A. Re, A. Romani, R. Roncin, N. Rossi, S. Schonert, D. Semenov, M. Skorokhvatov, O. Smirnov, A. Sotnikov, L.F.F. Stokes, Y. Suvorov, R. Tartaglia, G. Testera, J. Thurn, M. Toropova,E. Unzhakov, A. Vishneva, R.B. Vogelaar, F. von Feilitzsch H. Wang, S. Weinz,24 M. Wojcik, M. Wurm, Z. Yokley, O. Zaimidoroga, S. Zavatarelli, K. Zuber, and G. Zuzel

Abstract. The Borexino detector is a large-volume scintillator that has entered the phase of active data taking on 15 May 2007. It has immediately demonstrated very high sensitivity to solar neutrino fluxes. It had the radiopurity level that was not previously believed to be achievable by a liquid scintillator detector. It has opened a new era in solar neutrino studies — spectral studies of solar neutrino fluxes. Among the achievements of Borexino experiment solar program are a precision measurement of ^{7}Be neutrino flux with uncertainty of 5 %, limit on its day/night asymmetry, first spectral measurement of pp-neutrinos, first evidence of monoenegretic pep neutrino line, ^{8}B neutrinos detection with the lowest visible energy threshold of 3 MeV and the best current limit on CNO neutrino flux.

The water extraction campaign that ended in December 2011 has decreased radioactive contamination of the detector even more. Data processing techniques were improved. The detector was thermally insulated in order to improve the fluid stability. As an outcome, quality of the data has significantly increased and that leads to new levels of sensitivity to all solar neutrino fluxes.

1 Solar Neutrino

The neutrino-producing reactions of pp-chain and CNO-cycle go through Coulomb barrier, so the probabilities depend upon internal solar construction and thus bring valuable information about it. The main components of the solar neutrino spectrum are pp-neutrino, generated in pp-chain in reactions of pp and pep merging (continuous spectrum with endpoint of 420 keV and monoenergetic line of 1.44 keV respectively) as well as in decays of ^{7}Be and ^{8}B (a couple of monoenergetic lines of 384 keV and 862 keV and a continuous spectrum with endpoint around 15 MeV) and reactions of CNO cycle, creating continuous spectra.

Studies of these components would bring light on both matter propagation (Micheev-Smirnov-Wolfenstein effect, decrease of electron neutrino survival probability due to solar matter propagation) and internal solar structure, i.e. the solar metalicity, bringing us to a better understanding of internal stellar structure. So such studies make significant impact on both astrophysics and neutrino physics.

2 The Borexino Detector

The Borexino detector is a liquid scintillator that detects neutrino through the reactions of elastic scattering of neutrino and inverse beta-decay induced by antineutrino. The detector is designed to be ultrapure. The construction is based on the concept of graded shielding such that the radio purity level increases towards the detector center. The main housing of the detector is a cylinder with a hemispheric top with a diameter of 18 m and height of 15.7 m and is made of highly radiopure stainless steel. Contained inside is a stainless steel sphere with a diameter of 6.75 m and thickness of 8 mm fixed in place by a stainless steel support structure. The space between the outer tank and stainless steel sphere is filled with ultrapure water and is equipped with 208 8-inch PMTs. It serves as a Čerenkov muon veto and is called the outer detector. The inner side of the stainless steel sphere is equipped with 2212 8-inch PMTs of the inner detector and the inner volume is filled with pseudocumene ($C_9 H_{12}$). The inner detector contains two transparent spherical nylon vessels with radii of 5.5 m (radon barrier) and 4.25 m (inner vessel, IV) located concentrically within the stainless steel sphere (see Fig. 1).

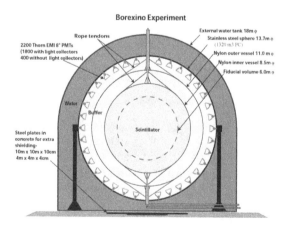

Figure 1: Principal scheme of the Borexino neutrino detector. Doping of PPO and DMP is shown by yellow and cyan colors respectively. Fiducial volume is shown in an arbitrary way and does not reflect the one used in the current analysis.

The scintillator volume inside the inner vessel has an admixture of PPO used for creation of Stokes shift. The scintillator outside the inner vessel is doped with DMP that quenches light production and decreases scintillation signals whose origin is not in the inner vessel.
The detector was carefully purified with various liquid handling procedures including water extraction campaign and shows exceptionally low level of radioactive impurities contained in the bulk of the inner vessel fluid.
A detailed description of the detector could be found elsewhere [1].

3 Measurement of Low-energy Neutrinos from pp-cycle

The spectral study of solar neutrinos with the Borexino detector is performed through the reaction of elastic scattering on electrons and is performed under conditions of nonzero background (see Fig. 2). The approach used for this study is a maximum likelihood fit of the spectrum with all possible neutrino and background components that enter or could possibly enter it. The background is strongly suppressed by the generic data selection algorithm including fiducial volume cut, muon and cosmogenic veto of 0.3 s after each muon crossing the stainless sphere. This procedure leaves us with relatively pure data with background contributions from ^{14}C, ^{210}Po, ^{210}Bi, external gammas from ^{208}Tl and ^{214}Bi, cosmogenic ^{10}C and ^{11}C and, possibly, from technogenic ^{85}Kr.

It is possible to improve sensitivity to neutrino components through discrimination of some of these components and/or setting bounds to their fluxes. In case of ^{11}C, which is a long-lived cosmogenic nuclide, produced by ^{12}C spallation with cosmic muons is possible to create a dynamic veto, based on the muon track and neutron detection position. Such veto, performed in terms of likelihood of ^{11}C content was applied together with a veto after showering process, dividing the dataset into ^{11}C-enhanced and ^{11}C-subtracted parts that

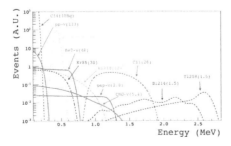

Figure 2: The Borexino detector spectrum components with their typical count rates. Values are expressed in terms of counts per day per 100 tons of liquid scintillator.

are fitted simultaneously. Since ^{11}C is a positron emitter and the annihilation gammas scintillate in a larger volume, likelihood of time-of-flight position reconstruction allows one to treat ^{11}C through pulse-shape discrimination, so a corresponding histogram is fitted simultaneously as well. Same operation is performed for radial distributions in order to increase sensitivity for external gamma background.

The spectra of all background components are well known, while all neutrino components could be derived through convolution with elastic scattering cross section. The detector response was modeled with high-precision Monte-Carlo simulation [3] or with an analytical model [2]. Two different approaches were used to cross-check the measured neutrino fluxes that show satisfactory agreement within their statistical uncertainties. As a result, we obtain an increase of neutrino flux measurement precision with respect to previous Borexino results.

4 The CNO Challenge

Generally speaking, extraction of CNO neutrino flux has no special issues with respect to the fluxes of pp-chain. The problem with its extraction arises due to the very strong correlation of CNO-neutrino signal with ^{210}Bi flux since they have similar spectral shapes. This fact limits available spectral information on the CNO neutrino flux. The only way to overcome such correlation is via a spectrum-independent establishment of ^{210}Bi content in the detector. It could be done with ^{210}Pb - ^{210}Bi - ^{210}Po chain since ^{210}Po is an alpha-emitter and it could be discriminated through pulse-shape analysis. The problem here is that ^{210}Po is out of dynamic equilibrium with ^{210}Bi and has an additional support term due to its convective transport to the detector fiducial volume. Such transport was strongly suppressed after the detector was equipped with thermal insulation yet this effect remains non-negligible. A way to overcome this problem lies in ^{210}Bi uniformity inside the fiducial volume (at least up to some precision) and determination of some volume part where the support term could be estimated. The effort in this direction is applied and could possibly bring an outcome. At the current state Borexino is entering the SOX phase, the search for sterile neutrino with ^{144}Ce - ^{144}Pr source, so the possibility to extend the solar dataset is limited, but there is still a strong hope to achieve the CNO neutrino flux estimation.

Acknowledgments

The Borexino program is made possible by funding from INFN (Italy); the NSF (U.S.); BMBF, DFG, (HGF, and MPI (Germany); RFBR (Grants No. 15-02-02117, No. 16-29-13014, No. 16-02-01026 and No. 17-02-00305), RSF (Grant No. 17-12-01009) (Russia); NCN Poland (Grant No. UMO-2013/10/E/ST2/00180);

FNP Poland (Grant No. TEAM/2016-2/17). We acknowledge the generous support and hospitality of the Laboratori Nazionali del Gran Sasso (LNGS).

References

[1] Alimonti et al. The Borexino detector at the Laboratori Nazionali del Gran Sasso. In: Nuclear Instruments and Methods in Physics Research Section A: Accelerators, Spectrometers, Detectors and Associated Equip- ment 600.3 (2009), pp. 568–593.
[2] M. Agostini et al. First Simultaneous Precision Spectroscopy of pp, 7 Be,and pep Solar Neutrinos with Borexino Phase-II. In: (2017). arXiv: 1707. 09279 [hep-ex].
[3] The Monte Carlo simulation of the Borexino detector. In: Astroparticle Physics (2017).

RECENT RESULTS FROM DAYA BAY NEUTRINO EXPERIMENT

Jilei Xu, on behalf of the Daya Bay collaboration [a]

Institute of High Energy Physics, Chinese Academy of Sciences, 100049, Beijing, China

Abstract. The Daya Bay Reactor Neutrino Experiment was designed to measure the smallest mixing angle θ_{13} in the three-neutrino mixing framework with unprecedented precision. In 2012, Daya Bay measured the first definitive non-zero value of θ_{13}. Then, with more than 4 years' stable data taking, Daya Bay obtained the most precise value of $\sin^2 2\theta_{13}$ and $|\Delta m^2_{ee}|$. In reactor antineutrino terms, the observed flux is consistent with measurements made by previous short baseline experiments, while the observed spectrum differs from prediction with a significance of 4.4σ in the 4-6 MeV energy region. Using our large reactor dataset, the experiment also measured the evolution of the reactor antineutrino flux and found that the IBD yield of ^{235}U is 7.8% lower than predicted. In sterile neutrino terms, there is no hint of a light sterile neutrino from Daya Bay data, giving the most stringent limit for $|\Delta m^2_{41}| < 0.2$ eV^2.

1 Introduction

Neutrinos as a fundamental particle were first discovered in 1956 [1, 2]. With many experimental measurements in recent decades, observation of neutrino mixing and thus non-zero mass is the first evidence of physics beyond the Standard Model of particle physics. Neutrino oscillations can be described using a 3-flavor neutrino mixing matrix, the Pontecorvo-Maki-Nakagawa-Sakata (PMNS) matrix, of the unitary transformation relating the mass and flavor eigenstates [3, 4]. Of these mixing angles, θ_{13} was the least known until measured in 2012 by the Daya Bay reactor anti-neutrino experiment with over 5σ excluded no-oscillation hypothesis [5]. A month later, the RENO experiment released their value with a significance of 4.9σ [6] .

Reactor antineutrino survival probability can be described by

$$P_{\bar{\nu}_e \to \bar{\nu}_e} = 1 - \sin^2 2\theta_{13} \sin^2\left(\Delta m^2_{ee} \frac{L}{4E}\right) - \sin^2 2\theta_{12} \cos^4 \theta_{13} \sin^2\left(\Delta m^2_{21} \frac{L}{4E}\right) \quad (1)$$

where $\sin^2(\Delta m^2_{ee} \frac{L}{4E}) \equiv \cos^2 \theta_{12} \sin^2(\Delta m^2_{31} \frac{L}{4E}) + \sin^2 \theta_{12} \sin^2(\Delta m^2_{32} \frac{L}{4E})$. Since $\Delta m^2_{21} \ll |\Delta m^2_{31}| \approx |\Delta m^2_{32}|$, short distance ($\sim$ km) reactor $\bar{\nu}_e$ oscillation is mainly determined by the $\sin^2 2\theta_{13}$ term. After several years of high quantity

---

[a] E-mail: xujl@ihep.ac.cn

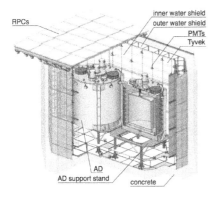

Figure 1: Overview of near site detectors.

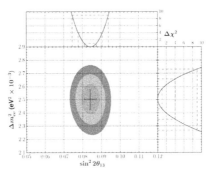

Figure 2: Confidence regions of $\sin^2 2\theta_{13}$ and $|\Delta m^2_{ee}|$.

anti-neutrino data taking, both θ_{13} and Δm^2_{ee} measurement became more and more precise, reached unprecedented precision.

Reactor antineutrinos can be commonly detected via inverse beta decay (IBD) $\bar{\nu}_e + p \rightarrow e^+ + n$. The prompt signal ($e^+$) can be detected over several MeV with a peak at ~ 4 MeV resulting from the convolving of the $\bar{\nu}_e$ flux and IBD cross section. The delay signal is detected from neutron capture on hydrogen (2.2 MeV) in the liquid scintillator (LS) or Gadolinium doped liquid scintillator (GdLS), or on Gadolinium (8 MeV) in the GdLS. Prompt and delay signals are correlated within a time window ~ 30 μs (~ 200 μs) by 0.1% GdLS (LS) to suppress accidental coincidence backgrounds. This paper mainly focuses on recent results from Daya Bay experiment.

2 Daya Bay Reactor Neutrino Experiment

The Daya Bay experiment was built near six reactors with thermal power of 17.4 GW. A description of the experiment can be found elsewhere [7]. Near and far relative measurement is used to effectively cancel the flux uncertainty. Eight identically designed antineutrino detectors (ADs) are employed to decrease detector related errors. The detectors were installed under a mountain to reduce backgrounds caused by cosmic-ray muon flux. Figure 1 shows the Daya Bay near site detector overview. AD is a cylindrical stainless steel vessels (SSV) with diameter and height size 5 m × 5 m. The smart three-zone AD structure were designed with GdLS, LS and mineral oil (MO) from inner to outer. ADs was immersed in octagonal pools filled with ultra-pure water to a depth of 10 m (Fig. 1). Two water shields and an RPC detector compose the veto system.

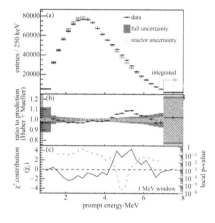
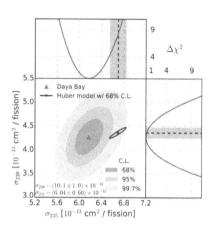

Figure 3: (a) Predicted and measured prompt energy spectra. (b) Ratio of the measured and predicted spectrum. (c) The defined χ^2 distribution of each bin (solid curve) and local p-values for 1 MeV energy windows (dashed curve).

Figure 4: Combined measurement of ^{235}U and ^{239}Pu IBD yields per fission σ_{235} and σ_{239}.

3 Results

3.1 Oscillation results

With a 1230 day dataset, positron spectral analysis and neutron capture on Gd [8], the $\sin^2 2\theta_{13}$ is $0.08421 \pm 0.0027(\text{stat}) \pm 0.0019(\text{syst})$ and $|\Delta m^2_{ee}|$ is $[2.50 \pm 0.006(\text{stat}) \pm 0.006(\text{syst})] \times 10^{-3}\text{eV}^2$. Figure 2 shows the confidence regions of $\sin^2 2\theta_{13}$ and $|\Delta m^2_{ee}|$. They are the most precise measurements in the world. The value of $|\Delta m^2_{32}|$ is consist with reactor and accelerator experiments results.

Daya Bay also has an independent measurement of θ_{13} by neutron capture on hydrogen [9]. The value of $\sin^2 2\theta_{13}$ was 0.071 ± 0.011 by rate analysis.

3.2 Reactor antineutrino flux and spectrum measurement

With 621 days of data, more than 1.2 million IBDs were detected. The measured IBD yield, $(1.53 \pm 0.03) \times 10^{-18}\text{cm}^2/\text{GW}/\text{day}$ or $(5.91\pm0.12) \times 10^{-43}$ cm^2/fission, is consistent with all eight ADs. The measured flux to predicted flux ratio is 0.946 ± 0.020 for the Huber+Mueller model and 0.992 ± 0.021 for the ILL+Vogel model, consistent with the global average of previous short baseline experiments [10]. For reactor spectrum measurement (Fig. 3), a 2.9σ deviation was found in the measured IBD positron energy spectrum compared to the predictions. In particular, an excess of events is seen in the region of 4-6 MeV with a local significance of 4.4σ.

3.3 Reactor antineutrino flux evolution

With 2.2 million IBD candidates from two near sites and detector data spanning effective ^{239}Pu fission fractions F239 from 0.25 to 0.35, an average IBD yield $\bar{\sigma}_f$ of $(5.90\pm0.13) \times 10^{43}$ cm^2/fission and a fuel-dependent variation in the IBD yield, $d\sigma_f/dF_{239}$ of $(-1.86\pm0.18)~10^{-43}$ cm^2/fission were measured [11], consistent with the previous measurement [10].

Based on measured IBD yield variations, two dominant fission parent isotope ^{235}U and ^{239}Pu yields of (6.17 ± 0.17) and $(4.27\pm0.26) \times 10^{-43}$cm^2/fission show a 7.8% discrepancy between the observed and predicted ^{235}U yields (Fig. 4), which suggests that this isotope may be the primary contributor to the reactor antineutrino anomaly.

3.4 Light sterile neutrino search

For light sterile neutrino study, there is no hint of a light sterile neutrino observed from the Daya Bay data giving the most stringent limit for $|\Delta m_{41}^2|$ <0.2 eV^2 [12]. The combine analysis(DayaBay + MINOS + Bugey-3) excludes parameter space allowed by MiniBooNE and LSND for $|\Delta m_{41}^2|$ <0.8 eV^2 [13].

4 Summary

Daya Bay obtained the most precise measurement of $\sin^2 2\theta_{13}$ and $|\Delta m_{ee}^2|$ with 1230 days of data. The measured reactor antineutrino flux results show that the measured flux is consistent with previous short baseline experiments but the spectrum is different from prediction with a local significance of 4.4σ. The IBD yield per fission from two main contributed isotopes, ^{239}Pu and ^{238}U, has been measured and we found that the IBD yield of ^{235}U is 7.8% lower than prediction. There was no hint of a light sterile neutrino observed from Daya Bay data resulting in a stringent limit, $|\Delta m_{41}^2|$ <0.2 eV^2. The experiment is expected to continue running until 2020 and we expect to reduce the uncertainty of the oscillation parameters below 3%.

References

[1] C. L. Cowan *et al.* Science **124** (1956) 103.
[2] F. Reines and C. L. Cowan, Jr., Nature **178** (1956) 446.
[3] B. Pontecorvo, Sov. Phys. JETP **6** (1957) 429 and **26** (1968) 984.
[4] Z. Maki *et al.* Prog. Theor. Phys. **28** (1962) 870.
[5] Daya Bay Collab. (F. An *et al.*), Phys. Rev. Lett. **108** (2012) 171803.
[6] RENO Collaboration, Phys. Rev. Lett.**108** (2012) 191802.
[7] Daya Bay Collab. (F. An et al.), NIM A **811**, (2016) 133.
[8] Daya Bay Collab. (F. An *et al.*), Phys. Rev. D. **95** (2017) 072006.

[9] Daya Bay Collab. (F. An *et al.*), Phys. Rev. D. **93** (2016) 0720.
[10] Daya Bay Collab. (F. An *et al.*), Chinese Physics C, **41(1)** (2017) 013002.
[11] Daya Bay Collab. (F. An *et al.*), Phys. Rev. Lett. **118** (2017) 251801.
[12] Daya Bay Collab. (F. An *et al.*), Phys. Rev. Lett. **117** (2016) 151802.
[13] Daya Bay Collab. (F. An *et al.*), Phys. Rev. Lett. **117** (2016) 151801.

RESULTS FROM THE OPERA EXPERIMENT AT LNGS

Tomokazu Matsuo [a] on behalf of the OPERA Collaboration
Department of Physics, Toho University, Funabashi, Chiba, Japan

Abstract. The OPERA experiment reached its main goal of proving the appearance of tau-neutrinos in the CNGS muon neutrino beam. A sample of 5 candidates was detected with a signal to background ratio of about ten allowing to reject the null hypothesis at 5.1σ. The search has been extended to ν_τ-like interactions failing the kinematic analysis defined in the experiment proposal to obtain a statistically enhanced, low purity, signal sample. Based on the enlarged data sample, the estimation of $\Delta m^2{}_{23}$ in appearance mode is presented. The search for ν_e interactions has been extended over the full data set with a more than twofold increase in statistics with respect to published data. An analysis of $\nu_\mu \to \nu_e$ interactions in the framework of the sterile neutrino model and the results of the study of the annual modulation of the cosmic muon rate are also presented.

1 Introduction

OPERA (Oscillation Project with Emulsion tRacking Apparatus) was a neutrino oscillation experiment designed to detect $\nu_\mu \to \nu_\tau$ oscillations in appearance mode. Its detector was made of two identical super modules, SM1 and SM2, each with a target section and a muon spectrometer. The target section contains ECC bricks, each followed by a doublet of emulsion films (Changeable Sheet, CS). The bricks were arranged in vertical planes ("walls"), interleaved with horizontal and vertical planes of scintillator strips (the Target Tracker, TT). An ECC brick was a sandwich structure of 56 1 mm thick lead plates and 57 nuclear emulsion films. An emulsion film consists of a 205 µm plastic film coated with 44 µm emulsion layers on both sides. The transverse dimensions of each brick were 12.7 cm by 10.2 cm and the thickness along the beam direction was 7.9 cm ($\sim 10 X_0$). About 150,000 such bricks were assembled in the OPERA target sections reaching a total mass of 1.25 kt. A detailed description of the apparatus can be found in Ref. [1].

2 Data taking and processing

The OPERA detector at LNGS was exposed from 2008 to 2012 to the CNGS ν_μ beam. A total exposure corresponding to 17.97×10^{19} POT, resulted in 19,505 neutrino interactions in the target. Neutrino events were classified as "1μ" events if one detected muon was present, or as "0μ" events if without any detected muon. Bricks containing the neutrino interaction were predicted using the TT hits. CS doublets were used to confirm the brick where the neutrino interaction occurred. Tracks measured in the CS were extrapolated to the most

---

[a] E-mail: tomokazu.matsuo@sci.toho-u.ac.jp

downstream film of the brick and followed back until their stopping point which could be either a primary or a secondary vertex. On a secondary vertex a kinematic analysis is performed using the information from multiple scattering in the brick and from the electronic detectors. An event is classified as a ν_τ candidate if it fulfills all the topological and kinematic criteria as set in the OPERA proposal [2].

3 OPERA $\nu_\mu \to \nu_\tau$ oscillation studies

3.1 Background analysis

A tau neutrino charged current (CC) interaction yields a tau lepton which decaying shows, in most cases, a characteristic kink. The following are main sources of background in the search for ν_τ candidate events: large angle muon scattering (> 20 mrad); charmed particle decays if the muon from the primary vertex is missed; re-interaction of a hadron from the primary vertex in the lead target. The numbers of expected background events have been estimated by MC simulations. A dedicated study of hadron interactions was done using an OPERA-like ECC brick exposed to 2, 4 and 10 GeV/c charged pion beams. The data are consistent with the simulation of hadron interactions at 30% level [3]. Furthermore, the probability of nuclear fragments associated with a hadron interaction vertex is well reproduced by the simulation with differences smaller than 10%. In recent times, a "Fine Track Selector" (FTS), having a larger angle of acceptance (up to 72 degrees w.r.t. the perpendicular to the film) [4], has been developed. The FTS allowed to reduce the contribution of hadron interaction background by 40%. Moreover, a re-evaluation of the large angle muon scatterings has been performed and its contribution is reduced considerably [5].

3.2 Golden events

OPERA has observed 5 $\nu_\mu \to \nu_\tau$ events [6–10] under strict selection criteria, so called "golden selection", allowing to confirm $\nu_\mu \to \nu_\tau$ oscillations in appearance mode. The total number of background events is 0.25. The probability of background fluctuation is P $= 1.1 \times 10^{-7}$ (5.1σ significance).

3.3 Silver events

An additional sample of "silver events" was obtained by a less stringent selection than the "golden selection". They are aimed at increasing the statistical sample in order to improve the determination of $\Delta m^2{}_{23}$. With looser selection criteria the hadronic background becomes the dominant one therefore an accurate measurement of hadronic interactions is required. Distributions

Table 1: Silver selection (NEW) and "golden selection" [10].

Variable	$\tau \to 1h$ PREV.	NEW	$\tau \to 3h$ PREV.	NEW	$\tau \to \mu$ PREV.	NEW	$\tau \to e$ PREV.	NEW
z_{dec} (μm)	[44, 2600]	< 2600		< 2600	[44, 2600]	< 2600		< 2600
θ_{kink} (rad)		> 0.02	< 0.5	> 0.02		> 0.02		> 0.02
p_{2ry} (GeV/c)	> 2	> 1	> 3	> 1				
p^T_{2ry} (GeV/c)	> 0.6 (0.3)	> 0.15						
p^T_{miss} (GeV/c)	< 1⋆		< 1⋆					
ϕ_{lH} (rad)	> $\pi/2$⋆		> $\pi/2$⋆					
m, m_{min} (GeV/c^2)			[0.5, 2]		> 0.25	> 0.1	[1, 15]	> 1
								> 0.1

of 2-6 GeV/c π interactions observed in a newly exposed OPERA-like ECC brick are well reproduced by the MC simulation [11]. The fractions of hadron interactions in the "silver event" selection is consistent with the MC simulation at 20% level.

3.4 Results of silver events

Applying the "silver selection" 5 additional ν_τ events were reconstructed (Fig. 1). The expected total numbers of signal and background events ("golden events" + "silver events") are 6.8 ± 1.4 and 2.0 ± 0.5, respectively. These values are consistent with 10 candidate events observed in OPERA. One silver event shows two secondary vertices compatible with short lived particle decays. From the Monte Carlo simulation and Artificial Neural Network analysis, it has been found with a significance of 3.5σ, that the most likely configuration is a ν_τCC interaction with a charm hadron decay [12].

3.5 Δm^2_{23} and the absolute ν_τ cross section

In OPERA, Δm^2_{23} and the ν_τ cross section depend on each other. By fixing the cross section, Δm^2_{23} in appearance mode is $(2.7 \pm 0.6) \times 10^{-3}$ eV^2, consistent with disappearance experiments. The ν_τ cross section is represented by $\sigma_{\nu_\tau} = \sigma^0_{\nu_\tau} EK(E)$. Assuming Δm^2_{23} [13] and maximal mixing $\sin^2 \theta_{23} = 1$, we get $\sigma^0_{\nu_\tau} = \left(8^{+4}_{-3}\right) \times 10^{-39}$cm^{2}GeV^{-1}. This value agrees within 1σ with the standard model value $\sigma^0_{\nu_\tau \text{SM}} = 6.7 \times 10^{-39}$cm^{2}GeV^{-1}.

4 Other results from OPERA

4.1 $\nu_\mu \to \nu_e$ analysis

A search for ν_e interaction events was also performed. The number of observed ν_e events is consistent with what expected from $\nu_\mu \to \nu_e$ oscillations plus the intrinsic ν_e beam component. OPERA carried out also a search for sterile neutrinos excluding the oscillation parameter region shown in Fig. 2.

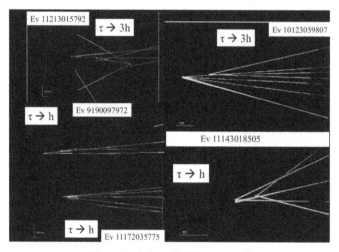

Figure 1: Collected "Silver events".

4.2 Annual modulation of atmospheric muon rate

As the atmospheric density decreases at increasing temperature, the mean free paths of π 's and K 's increase, leading to larger rates of cosmic muons. A study of the annual modulation of atmospheric muon rates in OPERA was performed. The results are consistent with those obtained by other experiments at LNGS.

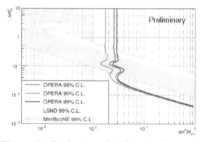

Figure 2: Exclusion plot for sterile neutrino oscillation parameters in the "3+1" neutrino model.

References

[1] R. Acquafredda et al. (OPERA), JINST 4, P04018 (2009).
[2] M. Guler et al., OPERA experimental proposal, CERN-SPSC-2000-028.
[3] H. Ishida et al., PTEP **2014** (2014) 093C01.
[4] T. Fukuda et al., J. Inst. **9** (2014) P12017.
[5] A. Longhin et al., IEEE Trans. Nucl. Sc. **62** (2015) 2216.
[6] N. Agafonova et al. (OPERA), Phys. Lett. B **691** (2010) 138.
[7] N. Agafonova et al. (OPERA), JHEP 11 (2013) 036.
[8] N. Agafonova et al. (OPERA), Phys. Rev. D **89** (2014) 051102(R).
[9] N. Agafonova et al. (OPERA), PTEP **2014** (2014) 101C01.

[10] N. Agafonova *et al.* (OPERA), Phys. Rev. Lett. **115** (2015) 121802.
[11] H. Mizusawa, Master thesis, Toho Univ. (2017).
[12] E. Medinaceli (OPERA), arXiv:1604.08249v2 [hep-ex] (2016).
[13] C. Partrignani *et al.* (Particle Data Group), Chin. Phys. C **40** (2016) 100001.

CROSS-SECTION MEASUREMENTS AT THE T2K EXPERIMENT

Lucie Maret [a] for the T2K Collaboration
Département de Physique Nucléaire et Corpusculaire, Université de Genève,
Geneva, Switzerland

Abstract. T2K (Tokai to Kamioka) is a long-baseline neutrino experiment based in Japan and aimed at studying neutrino oscillations. A muon neutrino beam with a peak energy of 0.6 GeV is produced at the J-PARC facilities and directed towards the near detector at 280 meters and the far detector Super-Kamiokande at 295 kilometers. A precise knowledge of neutrino interactions is essential for an accurate estimation of the oscillation parameters. The results of muon neutrino charged-current interaction measurements at the off-axis detector ND280 are presented with a focus on reactions with no observable pion in the final state.

1 Introduction

The T2K [1] muon neutrino (or anti-neutrino) beam is produced at the J-PARC facility in Tokai and sent towards two near detectors, ND280 and INGRID, and the far detector Super-Kamiokande. The neutrino flux is well constrained by hadron production measurements from the NA61/SHINE experiment [2]. Cross-section uncertainties and nuclear effects are the dominant systematic uncertainties on neutrino oscillation measurements (4–6% uncertainty) [3].

2 The T2K near detector complex

The ND280 near detector is composed of a scintillator-based π^0 detector (P0D), three gas Time Projection Chambers (TPCs) and two Fine Grained Detectors (FGDs) surrounded by an electromagnetic calorimeter (ECal) and a 0.2 T magnet. The TPCs enable 3D tracking, particle identification and determination of momentum and charge. FGD1 and FGD2 are made of finely segmented scintillator bars used as a hydrocarbon target for neutrino interactions and allowing measurements of short tracks. FGD2 also contains water layers alternated with the scintillator layers. The INGRID detector is used to monitor the beam direction and track beam intensity. Cross-sections are measured with either a module made with scintillator bars or modules with iron plates interleaved with the scintillator [4].

3 Cross-section measurements at T2K

The oscillation analysis requires an accurate knowledge of the neutrino energy spectrum. The simplest way to reconstruct the neutrino energy is from the observed muon kinematic variables [7] assuming a stationary target nucleon

---
[a] E-mail: lucie.maret@cern.ch

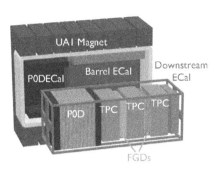

Figure 1: The ND280 off-axis near detector [1]. The neutrino beam comes from the left hand side. One half of the magnet and ECals is not shown on the picture.

and quasi-elastic scattering. However, the contributions from Fermi motion and non-elastic components such as nucleon-nucleon correlations (2p2h) or Final State Interactions (FSI) give rise to biased or unrealistic sensitivity to the oscillation parameters. Thus having a better knowledge of neutrino and nuclear interactions is crucial to study oscillation physics.

At the energy of the T2K neutrino beam the dominant interaction mode is Charged-Current Quasi-Elastic scattering (CCQE), that is $\nu_l + n \to l^- + p$. Because of FSI, different interaction modes can produce the same observable final state particle content (the topology). Moreover, nuclear effects can hide the true interaction mode. Therefore measurements can be made based on the topology in order to reduce the model dependencies, e.g. charged-current reaction with no observable pion in the final state (CC0π).

4 ND280 cross-sections and results

4.1 Muon neutrino CC0π interactions on hydrocarbon with FGD1

ν_μ CC0π interactions on scintillator bars of FGD1 are studied [5], measuring the flux-integrated double-differential cross-section in final-state muon kinematic variables (momentum p_μ and polar angle relative to the incoming neutrino direction $\cos\theta_\mu$). Two analyses are performed with different selections and unfolding methods in order to show the robustness of the results against biases due to model dependencies. The first one uses a binned likelihood fit and control regions to constrain backgrounds, the second one D'Agostini iterative unfolding [6] in restricted phase space.

The results show good agreement between the two analyses. Figure 2 shows some results of the first analysis, compared with the Nieves model with and without nuclear correlations. The models including 2p2h interactions show better agreement with the data.

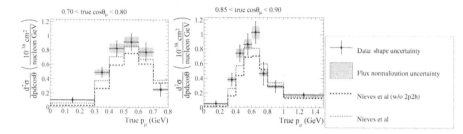

Figure 2: Flux-integrated double-differential ν_μ CC0π cross-section on FGD1 scintillator bars as a function of the true muon momentum for a subset of true muon angle bins. Data (black dots) are compared to the Nieves model prediction without 2p2h contributions (black dashed line) and prediction including them (red dashed line). Including nucleon correlations shows a better agreement with data [5].

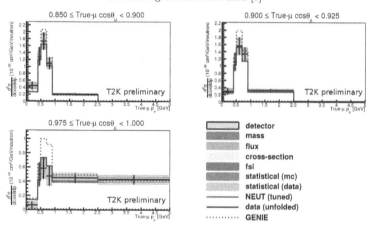

Figure 3: ν_μ CC0π cross-section on water measured with the P0D as a function of true muon momentum for a subset of true muon angle bins. Data (black dots) are compared to NEUT (blue solid line) and GENIE (blue dashed line) predictions. Uncertainties are shown in different colours according to the legend.

4.2 Muon neutrino CC0π interactions on water with P0D

The ν_μ CC0π interactions can also be measured on water with the P0D detector. Water layers can be filled or drained, making it possible to get the interactions on water using a subtraction method. A flux-integrated double-differential cross-section in momentum and angle is extracted using D'Agostini iterative unfolding. Some preliminary results are shown on Fig. 3.

4.3 Muon neutrino CC0π+Np interactions on hydrocarbon with FGD1

Muon kinematics only provide information about neutrino-nucleon scattering assuming a stationary target and elastic scattering. However, measuring proton

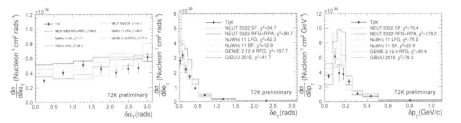

Figure 4: Preliminary results of the ν_μ CC0π cross-section on hydrocarbon as a function of muon and proton kinematics using Single Transverse Variables. Data (black dots) are compared to different MC generator predictions according to the legends.

kinematics gives a new handle on nuclear effects. Performing a binned likelihood fit in the muon angle, proton momentum and angle allows a measurement of the flux-integrated cross section on the FGD1 scintillators. Preliminary results show an excess of the data over the GENIE prediction that does not include 2p2h contributions, as can be seen on Fig. 4.

Another study considers Single Transverse Variables, i.e. the projection of the proton and muon momenta in the transverse plane with respect to the incoming neutrino direction. Measuring the imbalance in transverse-plane kinematics offers an interesting probe of nuclear effects and preliminary results strongly disfavour a simple Fermi-gas nuclear model, see Fig. 4.

5 Conclusion

The T2K detectors allow for a large variety of cross-section measurements. New techniques are developed to improve old results and try new measurements. In particular measurements using proton kinematics provide a very promising tool to probe nuclear effects and separate them from the neutrino cross section. Future analyses are planned to study antineutrino cross sections, as well as increasing the phase space to 4π angular coverage in ND280.

References

[1] K. Abe et al. (T2K Collaboration), Nucl. Instrum. Meth. A 659, 106 (2011).
[2] K. Abe et al. (T2K Collaboration), Phys. Rev. **D 87** 012001 (2013).
[3] K. Abe et al. (T2K Collaboration), Phys. Rev. **D 91** 072010 (2015).
[4] K. Abe et al. (T2K Collaboration), Phys. Rev. **D 90** 052010 (2014), **D 91** 112002 (2015) and **D 93** 072002 (2016).
[5] K. Abe et al. (T2K Collaboration), Phys. Rev. **D 93** 112012 (2016).
[6] G. D'Agostini, Nucl. Instrum. Methods A362, 487 (1995).
[7] O. Lalakulich, U. Mosel and K. Gallmeister, Phys. Rev. **C 86** 054606 (2012).

RECENT RESULTS FROM THE NOvA EXPERIMENT

Jaroslav Zálešák [a]
for the NOvA Collaboration
Institute of Physics of the Czech Academy of Sciences,
Na Slovance 1999/2, Prague, Czech republic

Abstract. The NuMI Off-Axis electron-neutrino Appearance (NOvA) experiment is a second generation, long-baseline, neutrino oscillation experiment. It consists of two finely segmented, liquid scintillator detectors operating at 14 mrad off-axis from the NuMI muon (anti-)neutrino beam. The smaller (300 ton) Near Detector is located underground at Fermilab, and it is used to study the (anti-)neutrino beam spectrum and composition before oscillations; while the 14,000 ton Far Detector, situated on the surface 810 km away, observes the oscillated beam. The measurements of muon neutrino disappearance and electron neutrino appearance potentially allow the extraction of neutrino oscillation unknowns, namely the mass hierarchy, the octant of the largest neutrino mixing angle, and the CP violating phase. In the paper will be presented the current status of the NOvA experiment and presented the latest results of the neutrino oscillation analyses from three years of data taking.

1 NOvA Experiment

The NOvA experiment [1] is a long-baseline neutrino experiment designed to make measurements to determine the neutrino mass hierarchy, neutrino mixing parameters and CP violation in the neutrino sector.

The NOvA experiment formally consists of an upgrade to the Fermilab accelerator complex of NuMI [2] (Neutrinos at the Main Injector) beam line to 700 kW (on Jan 25, 2017 the power record of 715 kW was reached), and two large liquid scintillator based neutrino detectors. The far detector (FD) for the experiment is a 14 kT totally active surface detector that is tuned for the detection of neutrino interactions with an energy around 2 GeV. The near detector (ND) shares an identical design as the far detector but is built to have a 1/4 scale cross section of the far detector. Both detectors lie at an angle of 14 mrad from the focusing axis of Fermilab's NuMI neutrino beam. This off-axis placement causes the neutrino energy spectrum to peak at 2 GeV, which is well-matched to the first oscillation maximum for the NOvA baseline.

The neutrinos in the (predominantly ν_μ) NuMI beam travel from the near detector through the earth to the far detector. During their 810 km journey to the FD, the neutrinos oscillate in their flavor states and we measure the disappearance rate of muon neutrinos (ν_μ) and appearance rate of electron neutrinos (ν_e). From this we can measure the transition probabilities $P(\nu_\mu \to \nu_\mu)$ and $P(\nu_\mu \to \nu_e)$ which are functions of the oscillation parameters [7]. From the measurement of the disappearance probability we can extract information

---
[a] E-mail: zalesak@fzu.cz

about the mixing angle θ_{23} and the neutrino mass difference Δm^2_{32}. From the measurement of the appearance probability we can extract information about the mixing angle θ_{13} and θ_{23}, the CP-violating phase δ_{CP}, and the neutrino mass hierarchy. The mixing angle θ_{13} is now well measured and is fairly large [7], therefore the NOvA concentrates on the phase of CP violation in the neutrino sector δ_{CP}, the hierarchy of the neutrino mass states and determining the octant of θ_{23}, in particular if the angle, and thus ν-e mixing, is maximal or not.

The first measurement of muon-neutrino disappearance and electron-neutrino appearance in the NOvA experiment based on its first years data was produced in 2015, providing solid evidence of neutrino oscillation with the NuMI beam line and some hints on mass-hierarchy and CP [3,4]. This paper describes the neutrino oscillation analyses of the full — Feb, 2014 to May, 2017 — data set, the equivalent to 6.05×10^{20} protons-on-target (POTs) of full-detector exposure of the NOvA detectors in the NuMI beam, as published in [5,6].

2 Detectors and Neutrino Detection

The detectors are fine grained and highly active tracking calorimeters. They consist of plastic (PVC) extrusions (3.9×6.6 cm in cross section and in length of 15.5/3.8 m for FD/ND) filled with liquid-scintillator, with wave-length shifting fibers (WLS) connected to avalanche photodiodes (APDs). The high APD quantum efficiency enables the use of very long scintillator modules, thus significantly reducing the electronics channel count. More information about the NOvA detector can be found in [1].

The NOvA collaboration has also designed and built a highly distributed, synchronized, continuous digitization and readout system that is able to acquire and correlate data from the Fermilab accelerator complex, the near detector at the Fermilab site and the far detector at Ash River, MN. The neutrino beam arrives in 10 μs long pulses that are 1.33 s apart. The NOvA timing system receives messages broadcast by the accelerator systems and the data aggregated over a 550 μs window, containing the 10 μs beam pulse, is written out. The timing side band in these 550 μs data windows, outside of the beam pulse, is used to estimate the cosmogenic background to the oscillation analyses.

Figure 1 shows the simulated signatures of different particle interactions in the detector. It illustrates that we are able to differentiate EM showers (center) from hadronic showers (bottom). With this particle identification we can measure $P(\nu_\mu \to \nu_\mu)$ and $P(\nu_\mu \to \nu_e)$ by looking for a deficit of ν_μ events and an excess of ν_e events, respectively. We can then extract natures parameters based on the standard oscillation equations using a baseline of 810 km and a neutrino energy of 2 GeV.

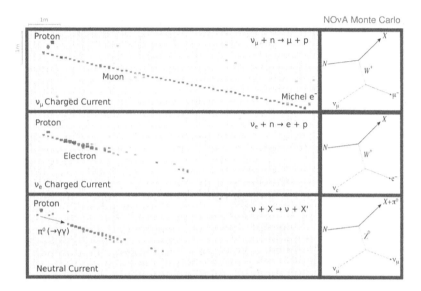

Figure 1: Characteristics of different types of neutrino interactions in the NOvA detectors. ν_μ Charged Current (CC) interactions contain a long muon track, along with some hadronic activity. ν_e CC interactions have an electron shower instead of a track. Neutral Current interactions contain no primary charged lepton in the final state.

3 Muon Neutrino Disappearance

The measured energy spectrum of ν_μ-CC events selected in the ND is used to predict the energy spectrum of events expected in the FD. This extrapolation procedure includes several steps. Firstly the reconstructed energy distribution of events in the ND is unfolded to a true energy spectrum using the simulated migration matrix. Secondly, the ratio of the datas unfolded spectrum to the simulated spectrum in bins of true energy is used as a scale factor to the simulated true energy spectrum of ν_μ-CC events selected in the FD. Then the true energy spectrum is weighted by the oscillation probability computed for a given set of oscillation parameters. Finally, the true energy spectrum is smeared to a reconstructed energy spectrum, again using the simulated migration matrix.

There are 78 ν_μ candidate events in the far detector selected by this process. If there were no oscillations, a projection of the flux seen in the near detector would result in 473 ± 30 such events. In the set of 78 events it is estimated that there are backgrounds of 3.4 neutral current, 0.23 ν_e, 0.27 ν_τ, and 2.7 cosmic-ray-induced events (the blue line in Fig. 2, left). Neutrino oscillations are fit to the resulting spectra, producing a best fit (the red line) of $\Delta m_{32}^2 = (+2.67 \pm 0.11) \times 10^{-3}$ eV^2 and $\sin^2 \theta_{23}$ at the two statistically degenerate values $0.404^{+0.030}_{-0.022}$ and $0.624^{+0.022}_{-0.030}$ (at 68% c.l.) in the normal hierarchy.

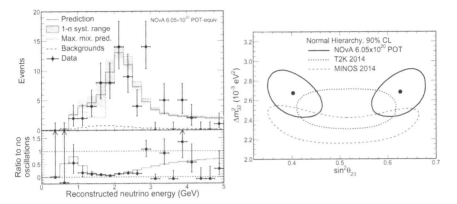

Figure 2: **Left:** A reconstructed ν_μ energy spectrum of data in the nova far detector (black dots with statistical errors) compared to the best oscillation fit prediction with systematic errors (red). Backgrounds are the small blue dashed line, and the prediction at maximal mixing the green dashed line. The ratio of observed data to the unoscillated ν_μ spectrum derived from the near detector is the bottom part of the graph. **Right:** The values of Δm_{32}^2 and $\sin^2\theta_{23}$ in the normal hierarchy allowed at the 90% C.L. for NOvA in black, compared to previous results from MINOS in red and T2K in blue. The plot is taken from [5].

For the inverted hierarchy, $\Delta m_{32}^2 = (-2.72 \pm 0.11) \times 10^{-3}$ eV^2 and $\sin^2\theta_{23} = 0.398^{+0.030}_{-0.022}$ or $0.618^{+0.022}_{-0.030}$. This analysis is mostly insensitive to different values of δ_{CP}, so potential variations therein have been taken as a systematic error. The resulting allowed region is shown in the Fig. 2, right. The analyzed muon neutrino disappearance data exclude maximal mixing at 2.6σ [5].

4 Electron Neutrino Appearance

The selection of events for the ν_e appearance analysis discussed here was optimized to favor oscillation parameter measurement, instead of sensitivity to the appearance of signal events above background as in the first analysis [4]. For this purpose NOvA deployed a new PID framework — the Convolutional Visual Network (CVN) as discussed in [8]. It uses energy calibrated pixels as inputs and the output is a variable that describes the probability to be a ν_e-CC with a range $0 - 1$. In this CVN identifier, convolutional filters are used to automatically extract features from the raw hit map. The output of this neural net is used to classify the event. The CVN approach improves the sensitivity of the result by 30% over traditional energy calibrated reconstruction algorithms, and results in 73% efficiency and 76% purity of ν_e events in the final sample (as determined from MC simulations), while rejecting 97.6% of the neutral current and 99.0% of the ν_μ charged current beam backgrounds.

After reconstruction, 33 ν_e candidates are found at the far detector, of which 8.2 ± 0.8 correspond to the expected background in the absence of ν_e

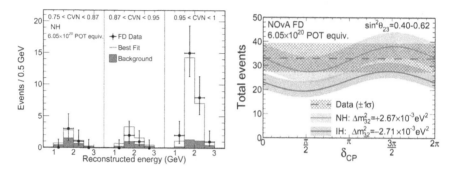

Figure 3: **Left:** The reconstructed energy spectrum of ν_e-CC selected events in the far detector, for three ranges of the CVN event classifier. The black points show the data, while the red histogram shows the spectrum from the best fit. The blue shaded histograms show the background spectrum. **Right:** The total number of events (included background) expected as a function δ_{CP}. The normal hierarchy is shown in blue and the inverted hierarchy is shown in red, resp. The colored shaded regions show the range of values allowed for $\sin^2 \theta_{23} = 0.4 - 0.6$. The number of events selected in the data is shown in gray. The gray band corresponds to the 1σ statistical error.

appearance oscillations. The observed number of events, and the predicted number of events as a function of the hierarchy and δ_{CP} for a range of values of θ_{23} is shown on the right side of Fig. 3. On the left side the energy distributions of FD reconstructed events in three different CVN interval bins are displayed.

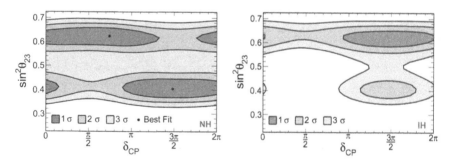

Figure 4: Regions of δ_{CP} versus $\sin^2 \theta_{23}$ parameter space consistent with the observed spectrum of ν_e candidates and the ν_μ disappearance data [5]. The left plot corresponds to normal mass hierarchy ($\Delta m_{32}^2 > 0$) and the right one to inverted hierarchy ($\Delta m_{32}^2 < 0$). The color intensity means the confidence levels (1,2,3 sigmas) at which particular parameter combinations are allowed.

To obtain oscillation parameters, these data are fit simultaneously with the ν_μ data [5], taking the precise value of θ_{13} and its uncertainty from reactor experiments [7]. The global best fit prefers the normal hierarchy, $\delta_{CP} = 1.49\pi$, and $\sin^2\theta_{23} = 0.40$. There is a region of parameter space in the inverted hierarchy, lower octant, around $\delta_{CP} = \pi/2$ that is excluded at the 3σ level. Figure 4 shows the values of $\sin^2\theta_{23}$ and δ_{CP} allowed at 1, 2, and 3 σ for each of the hierarchies for the combined ν_e/ν_μ analysis. For more discussion about the oscillation parameters space refer to [6].

5 Summary

With the 6.05×10^{20} protons on target exposure of the NuMI beam data the NOvA experiment observes both ν_μ disappearance and ν_e appearance oscillations. Muon neutrino disappearance shows to non-maximal mixing with 2.6σ significance. The significance of the ν_e appearance is greater than 8σ, and our data prefer the normal mass hierarchy at low significance. In our analysis, the inverted mass hierarchy for $\delta_{CP} = \pi/2$ is rejected for lower octant. A comparison of an exposure of 9×10^{20} POT neutrino exposure with a similar amount of anti-neutrino data is planned for 2018.

Acknowledgments

I wish to thank the organizers of the 18th Lomonosov Conference on Elementary Particle Physics for an enlightening and entertaining conference, my NOvA colleagues, and Fermilab. Fermilab is operated by Fermi Research Alliance, LLC under Contract No. De-AC02-07CH11359 with the United States Department of Energy. This work was supported by Ministry of Education, Youth and Sports, Czech Republic under grants LM2015068 and LG15047.

References

[1] D. S. Ayres *et al.* [NOvA Collaboration], *The NOvA Technical Design Report*, FERMILAB-DESIGN-2007-01.

[2] P. Adamson *et al.* [MINOS Collaboration], *The NuMI Neutrino Beam*, Nucl. Instrum. Meth. A **806**, 279 (2016) [arXiv:1507.06690 [physics.acc-ph]].

[3] P. Adamson *et al.* [NOvA Collaboration], *First measurement of electron neutrino appearance in NOvA*, Phys. Rev. Lett. **116**, no. 15, 151806 (2016) doi:10.1 [arXiv:1601.05022 [hep-ex]].

[4] P. Adamson *et al.* [NOvA Collaboration], *First measurement of muon-neutrino disappearance in NOvA*, Phys. Rev. **D 93**, no. 5, 051104 (2016) [arXiv:1601.05037 [hep-ex]]

[5] P. Adamson *et al.* [NOvA Collaboration], *Measurement of the neutrino mixing angle θ_{23} in NOvA*, Phys. Rev. Lett. **118**, no. 15, 151802 (2017) [arXiv:1701.05891 [hep-ex]].

[6] P. Adamson *et al.* [NOvA Collaboration], *Constraints on Oscillation Parameters from ν_e Appearance and ν_μ Disappearance in NOvA*, Phys. Rev. Lett. **118**, no. 23, 231801 (2017) [arXiv:1703.03328 [hep-ex]].

[7] C. Patrignani *et al.* [Particle Data Group], *Review of Particle Physics*, Chin. Phys. C **40**, no. 10, 100001 (2016).

[8] A. Aurisano *et al.*, *A Convolutional Neural Network Neutrino Event Classifier*, JINST **11**, no. 09, P09001 (2016) [arXiv:1604.01444 [hep-ex]].

THE KM3NET NEUTRINO TELESCOPE AND THE POTENTIAL OF A NEUTRINO BEAM FROM RUSSIA TO THE MEDITERRANEAN SEA

Dmitry Zaborov [a]
CPPM, Aix-Marseille Université, CNRS/IN2P3, 13288 Marseille, France
for the KM3NeT Collaboration

Abstract. KM3NeT is a new generation neutrino telescope currently under construction at two sites in the Mediterranean Sea. At the Capo Passero site, 100 km off-shore Sicily, Italy, a volume of more than one cubic kilometre of water will be instrumented with optical sensors. This instrument, called ARCA, is optimized for observing cosmic sources of TeV and PeV neutrinos. The other site, 40 km off-shore Toulon, France, will host a much denser array of optical sensors, ORCA. With an energy threshold of a few GeV, ORCA will be capable to determine the neutrino mass hierarchy through precision measurements of atmospheric neutrino oscillations. In this contribution, we review the scientific goals of KM3NeT and the status of its construction. We also discuss the scientific potential of a neutrino beam from Protvino, Russia to ORCA. We show that such an experiment would allow for a measurement of the CP-violating phase in the neutrino mixing matrix. To achieve a sensitivity competitive with that of the other planned long-baseline neutrino experiments such as DUNE and T2HK, an upgrade of the Protvino accelerator complex will be necessary.

1 Introduction

The KM3NeT research infrastructure [1] includes two large underwater neutrino detectors in the Mediterranean Sea: ORCA (Oscillation Research with Cosmics in the Abyss) and ARCA (Astroparticle Research with Cosmics in the Abyss). Both detectors use the same operation principle and instrumentation technique, each optimally configured for a specific energy range. ORCA is optimized for the study of atmospheric neutrino oscillations in the energy range between 3 GeV and 20 GeV, with the primary goal to determine the neutrino mass hierarchy. When completed, ORCA will occupy a volume of 8 Mt of water. ARCA is optimized for the purposes of neutrino astronomy in the TeV–PeV energy range and will occupy a volume of more than 1 km^3 (1 Gt). Both ORCA and ARCA are now under construction. The ORCA site is located 40 km off-shore Toulon, France, 2450 m below the sea level. The ARCA site is 100 km off-shore Capo Passero, Sicily, Italy, 3500 m below the sea level.

This document is organised as follows. Section 2 introduces the KM3NeT detector technology. Sections 3 and 4 introduce the science programmes of ARCA and ORCA, respectively. Section 5 discusses the scientific case for an accelerator neutrino beam from Protvino, Russia, to ORCA. Conclusions are presented in Section 6.

---

[a] E-mail: zaborov@cppm.in2p3.fr

2 KM3NeT technology, operation principle, and current status

ORCA and ARCA detect neutrinos through the detection of the Cherenkov light induced by secondary particles emerging from interactions of the neutrinos with the sea water. Both ORCA and ARCA consist of several thousand Digital Optical Modules (DOMs). Each DOM contains 31 photomultiplier tubes, which detect the Cherenkov light, and associated electronics, all housed in a pressure-resistant glass sphere. The DOMs are arranged in vertical structures, called Detection Units (DU), with 18 DOMs on each DU. The DUs are anchored to the seabed and connected to the shore station via an underwater cable network. In the case of ORCA, the vertical spacing between the DOMs along the DU is 9 m, and the DUs are installed on the seabed with an average spacing of 23 m. When completed, ORCA will comprise 115 DUs. For ARCA, the vertical spacing between the DOMs is 36 m and the horizontal spacing between the DUs is 90 m. Upon completion of the second construction phase, ARCA will comprise 230 DUs grouped in two building blocks of 115 DUs each.

3 Neutrino astronomy with ARCA

The main scientific objective of KM3NeT-ARCA is the detection of high-energy neutrinos of cosmic origin, in particular from sources in our Galaxy [1]. The preferred search strategy is to identify upward-moving muons, which unambiguously indicates neutrino reactions, since only neutrinos can traverse the Earth without being absorbed. The expected angular resolution is better than $0.3°$ at $E > 10$ TeV. It should be noted that most of the potential Galactic sources are in the Southern sky. In this regard, ARCA is complementary to IceCube which is located at the South pole and, therefore, observes mostly the Northern sky [2]. Furthermore, thanks to the excellent optical properties of the deep sea water, the angular resolution of ARCA is significantly better than that of IceCube, providing an unprecendeted sensitivity to point-like sources of neutrino and improved prospects to associate the neutrino sources with specific astronomical objects [1].

The recent observation of a diffuse flux of cosmic neutrinos by IceCube [2] promises an exciting future for ARCA. Indeed, the existence of PeV hadronic accelerators in the Universe has been proved, as well as the feasibility of observing neutrinos originating from such accelerators. However, many questions remain unanswered concerning the origin of the cosmic neutrinos, in particular, whether these neutrinos originate within or outside our Galaxy, what type of astrophysical objects and environments serve as the neutrino production sites, and how particles are accelerated to PeV energies. ARCA will allow to answer these questions by observing the neutrino sources and pinpointing their locations on the sky with an unprecedented accuracy. The ARCA research program also includes studies of the flavour composition of the cosmic neutrino flux, searches for transient neutrino sources, indirect dark matter searches, searches for exotic particles and Lorentz invariance violation. Further information on the ARCA physics case can be found in [1].

The first two ARCA DUs were successfully deployed and connected to the shore station in December 2015 and May 2016, respectively. The recorded data demonstrate a satisfactory operation of all detector subsystems. A measurement of the atmospheric muon flux as a function of depth has been obtained showing a good agreement with expectations.

4 Neutrino mass hierarchy measurement with ORCA

Each of the three flavour eigenstates of neutrino (ν_e, ν_μ, ν_τ) is a superposition of three mass eigenstates (ν_1, ν_2, ν_3) with three different masses (m_1, m_2, m_3) [3]. The neutrino mass ordering (mass hierarchy) is only partially constrained by existing experimental data. More specifically, only the relative ordering of the first and second neutrino mass eigenstates can be fixed ($m_1 < m_2$), based on observations of the Solar neutrino oscillations [4]. It is currently unknown whether the third neutrino mass eigenstate is heavier or lighter than the other two mass eigenstates. Given that the first squared mass splitting $|m_1^2 - m_2^2|$ is much smaller than the the second one $|m_2^2 - m_3^2|$, there are two possible orderings: $m_1 < m_2 < m_3$ (normal mass hierarchy) and $m_3 < m_1 < m_2$ (inverted mass hierarchy). The Standard Model has no particular preference for either type of the mass hierarchy, as it generally assumes neutrinos to be massless. However, the neutrino mass hierarchy may be related to the underlying structure of physics Beyond the Standard Model (BSM), and many BSM theoretical models provide specific predictions for the mass hierarchy [5]. The mass hierarchy is also important for interpreting data from current experiments, as well as for planning of future experiments, for instance on double-beta decay.

When neutrinos propagate in matter, such as inside the Sun or Earth, the flavour transition probabilities are modified due to coherent forward scattering of neutrino on atomic electrons — the so-called Mikheyev-Smirnov-Wolfenstein (MSW) effect [6, 7]. Indeed, contrarily to the other flavours, the ν_e component can undergo charged-current (CC) elastic scattering interactions with the electrons in matter and, consequently, acquire an effective potential $A = \pm\sqrt{2}\,G_F N_e$, where N_e is the electron number density of the medium, G_F is the Fermi constant and the $+(-)$ sign stands for ν_e ($\bar{\nu}_e$). For neutrinos propagating through the Earth, this effective potential leads to a resonant enhancement of ν_e appearance at energies E $\approx$ 3–8 GeV in the case of normal mass hierarchy. In the case of inverted mass hierarchy the resonance occurs instead in the $\bar{\nu}_e$ appearance channel. Hence, observations of the ν_e and/or $\bar{\nu}_e$ appearance at E $\sim$ 5 GeV allow to determine the neutrino mass hierarchy.

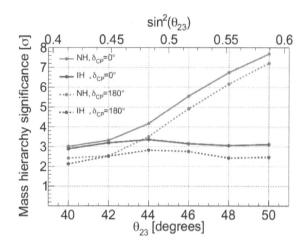

Figure 1: The mass hierarchy sensitivity of ORCA after three years of operation, shown as a function of θ_{23} and for two values of δ_{CP}.

The primary mission of ORCA is to determine the neutrino mass hierarchy (normal vs. inverted). This will be accomplished via precision measurements of the atmospheric neutrino oscillations in the energy range where the MSW effect in the Earth (core, mantle and crust) has the strongest impact on the neutrino oscillation pattern (between 3 GeV and 20 GeV). Thanks to its large volume and low energy threshold (~3 GeV), ORCA will detect about 50000 atmospheric neutrinos every year. Most of these neutrinos will have muon or electron flavour. The relative abundance of muon and electron (anti-)neutrinos will be measured on a statistical basis using the characteristic features of muon neutrino charged current (CC) events, in particular the presence of a long muon track. Studies performed by the KM3NeT Collaboration suggest that at $E_\nu = 5$ GeV the majority of ν_μ CC events detected by ORCA can be correctly identified as muon neutrinos while < 15% of electron CC events are misidentified as muon neutrinos. The neutrino energy resolution of ORCA is about 30% for both ν_μ and ν_e CC events (at $E_\nu = 5$ GeV). Based on latest sensitivity projections, a result with a 3 σ significance on the mass hierarchy is expected after three years of data taking [1]. Under certain conditions (normal mass hierarchy, $\delta_{CP}=0$, $\theta_{23}=50°$), the statistical significance may reach 6 σ already after 3 years (see Fig. 1). ORCA will also provide improved measurements of the atmospheric neutrino oscillation parameters (e.g. θ_{23}) and constraints on non-standard neutrino interactions, as well as sensitivity to astrophysical neutrino sources, dark matter, and other physics phenomena. The detector construction has recently started and is expected to be completed within about 4 yr, providing its first results on the mass hierarchy around 2023.

5 Scientific case for a neutrino beam from Protvino to ORCA

Big Bang cosmology requires that the CP symmetry is violated — at least at some moment during the Big Bang — in order to explain the dominance of normal matter over anti-matter in the present-day Universe [8]. However, the origin of the CP violation remains yet unknown. Indeed, the only source of CP violation experimentally observed so far is a CP-violating phase δ in the mixing of quarks (J. Cronin & V. Fitch, Nobel prize in physics, 1980). However, this CP phase alone is insufficient to explain the present-day abundance of matter in the Universe. Theoretically, the CP symmetry could also be violated by the strong interaction but this has not been experimentally observed, indicating that the effect is negligibly small or strictly zero. The only other known possible source of CP violation is the leptonic CP phase δ_{CP}, associated with neutrino mixing [3]. (Additional CP phases may exist if neutrinos are Majorana particles — these phases do not have observable consequences on neutrino oscillations and will be ignored in the following.)

The fact that all three neutrino mixing angles are different from zero leads to the possibility of probing δ_{CP} in the electron (anti-)neutrino appearance channel using the $\nu_\mu \to \nu_e$ ($\bar{\nu}_\mu \to \bar{\nu}_e$) transition. The most plausible setting for measuring this transition is offered by experiments with accelerator neutrino beams, which provide a clean source of muon neutrinos (ν_μ or $\bar{\nu}_\mu$, depending on the chosen beam polarity). Such a measurement is actively pursued by several long-baseline neutrino experiments, but is so far limited by insufficient statistics. The only significant experimental constraints on δ_{CP} available so far come from the T2K and NOvA experiments [9–11]. However in both cases the statistical significance of the result does not exceed 2 σ. Note that using atmospheric neutrinos ORCA has only a marginal sensitivity to δ_{CP}.

It has recently been proposed that sending an artificial neutrino beam to ORCA would open up a new range of opportunities in neutrino oscillation research [12]. In particular, such an experiment could provide a very sensitive probe of leptonic CP violation (δ_{CP}). A neutrino beam of suitable energy could be produced at the Protvino accelerator facility, located 100 km South of Moscow, Russia. The core component of the accelerator facility is the U-70 synchrotron which accelerates protons up to 70 GeV, allowing for the production of an intense beam of neutrinos or anti-neutrinos with energies up to 7 GeV [13]. With a 2590 km distance between Protvino and ORCA, the first neutrino oscillation maximum will be at 5 GeV, close to the matter resonance energy in the Earth crust (4 GeV) and within an energy range convenient for ORCA as well as for the accelerator. Such an experiment proposal was initially made in Ref. [12]. The proposal is now known as Protvino-to-ORCA experiment, or simply P2O.

The U-70 synchrotron currently operates at a time-averaged beam power of 15 kW. The accelerator power could be increased to $\approx$ 90 kW by means of a

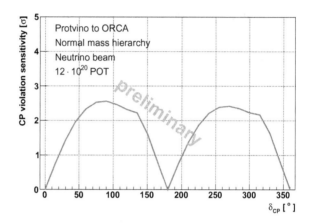

Figure 2: Sensitivity of P2O to the presence of CP violation in the neutrino sector as a function of δ_{CP} after 3 yr of a 450 kW beam, or 15 yr of a 90 kW beam ($\delta_{CP} = 0$ and $180°$ correspond to CP conservation). Normal mass hierarchy.

relatively inexpensive upgrade (new ion injection scheme and optimization of the accelerator cycle from 9 s down to ≈ 4.5 s). An upgrade up to 450 kW could be made possible by a new chain of booster accelerators [14]. A neutrino beam line will need to be constructed to produce a focused neutrino beam in the direction of ORCA. A near beam detector will also be needed, in order to accurately monitor the neutrino beam intensity, energy spectrum, and flavour composition before oscillations.

A preliminary study of the scientific potential of the P2O experiment suggests that the neutrino mass hierarchy could be determined with a 5–10 σ significance after one year of running with a 450 kW beam (or five years with a 90 kW beam). This would provide a solid confirmation of the ≈ 3–5 σ result expected to be obtained by ORCA using atmospheric neutrinos. It was also found that with 3 years (15 years) of the 450 kW (90 kW) beam the P2O experiment could provide a 2–3 σ sensitivity to CP violation (see Fig. 2), along with a better than $20°$ accuracy on the CP phase δ_{CP}. These estimates were obtained using a preliminary data analysis pipeline optimized for the atmospheric neutrino analysis. Potential analysis improvements can be added thanks to the known arrival direction and timing of the neutrino beam. Most of the sensitivity to the mass hierarchy and CP violation comes from the electron (anti-)neutrino appearance channel (see Fig. 3).

It is worth noting that a 90 kW beam will produce about 3000 neutrino events in ORCA every year. For comparison, the DUNE experiment, using a 1 MW beam in combination with a 40 kt detector over a 1300 km baseline, will detect only 1000 neutrino events per year. DUNE is expected to reach a 3 σ sensitivity to CP violation using 15 yr of operation with the beam [15].

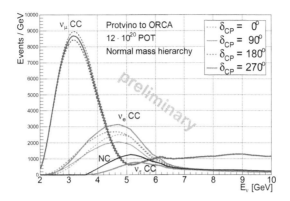

Figure 3: Number of neutrino events expected to be detected by ORCA after 3 years of running with the 450 kW beam, or 15 yr with the 90 kW beam. The case of normal neutrino mass hierarchy has been assumed. The x-axis shows the true neutrino energy. The different line styles correspond to four different values of δ_{CP}, as indicated in the legend. The three groups of colored curves correspond to ν_μ CC (blue), ν_e CC (red) and ν_τ CC (magenta) interactions. The black curve is for neutral current and contains the sum of three neutrino flavors (unaffected by oscillations).

Hence, P2O would be competitive to DUNE, even when using a relatively low intensity beam (90 kW). P2O would also be complementary to DUNE and other long-baseline neutrino experiments, e.g. T2K/T2HK [16], thanks to several unique characteristics of P2O: the highest neutrino event statistics, the longest baseline, and the highest energy of the oscillation maximum.

6 Conclusion

Two giant sea-water Cherenkov detectors, ORCA and ARCA, are now under construction by the KM3NeT Collaboration at two sites in the Mediterranean Sea. ORCA will allow for the study of the atmospheric neutrino oscillations and determine the neutrino mass hierarchy, while ARCA will open a new chapter in high energy neutrino astronomy. Sending an accelerator neutrino beam from Protvino (Russia) to ORCA promises exciting prospects for studies of leptonic CP violation (δ_{CP}), while also providing a robust cross-check on the mass hierarchy. Such an experiment, dubbed P2O, would be complementary to and competitive with other present and future long-baseline experiments, in particular DUNE and T2K/T2HK. The P2O project will require an upgrade of the U-70 accelerator complex at Protvino to serve as a high-intensity neutrino source. Additionally, a near detector would be needed in the vicinity of the accelerator complex for accurate monitoring of the intensity and flavour composition of the neutrino beam.

References

[1] KM3NeT Collaboration: S. Adrián-Martínez et al., KM3NeT 2.0 – Letter of Intent for ARCA and ORCA, J. Phys. G: Nucl. Part. Phys. 43 (2016) 084001.
[2] M. G. Aartsen et al., IceCube Coll., Evidence for high-energy extraterrestrial neutrinos at the IceCube detector, Science 342 (2013) 1242856.
[3] Z. Maki, M. Nakagawa, and S. Sakata, Remarks on the unified model of elementary particles, Prog. Theor. Phys. 28 (1962) 870.
[4] Q. R. Ahmad et al., Direct evidence for neutrino flavor transformation from neutral current interactions in the Sudbury Neutrino Observatory, Phys. Rev. Lett. 89 (2002) 011301.
[5] W. Winter, Neutrino mass hierarchy: Theory and phenomenology, AIP Conf. Proc. 1666 (2015) 120001, Proc. 26th International Conference on Neutrino Physics and Astrophysics (Neutrino 2014).
[6] L. Wolfenstein, Neutrino oscillations in matter, Phys. Rev. D 17 (1978) 2369.
[7] S. P. Mikheev and A. Yu. Smirnov, Resonance amplification of oscillations in matter and spectroscopy of Solar neutrinos, Sov. J. Nucl. Phys. 42 (1985) 913.
[8] Sakharov A. D., Violation of CP invariance, C asymmetry and baryonic asymmetry of the Universe, Pisma Zh. Eksp. Teor. Fiz., 5 (1967) 32.
[9] K. Abe, et al., Measurement of neutrino and antineutrino oscillations by the T2K experiment including a new additional sample of electron neutrino interactions at the far detector, Phys. Rev. D 96 (2017) 092006.
[10] P. Adamson et al., First Measurement of Electron Neutrino Appearance in NOvA, Phys. Rev. Lett. 116 (2016) 151806.
[11] A. Himmel, New neutrino oscillation results from NOvA, CERN seminar, 30 Jan 2018.
[12] J. Brunner, Advances in High Energy Physics, Volume 2013 (2013), Article ID 782538, http://dx.doi.org/10.1155/2013/782538.
[13] V.I. Garkusha, F.N. Novoskoltsev & A.A. Sokolov, Neutrino oscillation research using the U-70 accelerator complex (in Russian), IHEP Preprint 2015-5.
[14] N.E. Tyurin et al., Facility for intense hadron beams (letter of intent), News and Problems of Fundamental Physics 2 (9), 2010, http://exwww.ihep.su/ihep/journal/IHEP-2-2010.pdf.
[15] R. Acciarri, et al., Long-Baseline Neutrino Facility (LBNF) and Deep Underground Neutrino Experiment (DUNE) Conceptual Design Report Volume 1: The LBNF and DUNE Projects, arXiv:1601.05471.
[16] K. Abe, et al., Physics Potential of a Long Baseline Neutrino Oscillation Experiment Using J-PARC Neutrino Beam and Hyper-Kamiokande, Prog. Theor. Exp. Phys. (2015) 053C02.

THE T2HKK EXPERIMENT AND NON-STANDARD INTERACTION

Monojit Ghosh [a] and Osamu Yasuda [b]
Department of Physics, Tokyo Metropolitan University, Hachioji, Tokyo 192-0397, Japan

> *Abstract.* In this work we study the the sensitivity of the T2HKK experiment to probe non-standard interaction in neutrino propagation. As this experiment will be statistically dominated due to its large detector volume and high beam-power, it is expected that the sensitivity will be affected by systematics. This motivates us to study the effect of systematics in probing the non-standard interaction. We also compare our results with the other future proposed experiments i.e., T2HK, HK and DUNE.

1 Introduction

In the standard three flavour framework, the phenomenon of neutrino oscillation where neutrinos evolve from one flavour from another can be parameterized by six oscillation parameters: three mixing angles: θ_{12}, θ_{13} and θ_{23}, two mass squared differences: Δ_{21} ($m_2^2 - m_1^2$) and Δ_{31} ($m_3^2 - m_1^2$) and one phase δ_{CP}. Among them at present the unknowns are: (i) sign of Δ_{31} i.e., $\Delta_{31} > 0$ (normal hierarchy or NH) or $\Delta_{31} < 0$ (inverted hierarchy or IH), (ii) octant of θ_{23} i.e., $\theta_{23} > 45°$ (higher octant or HO) or $\theta_{23} < 45°$ (lower octant or LO) and (iii) δ_{CP}. T2HKK is one of the proposed experiment to determine these unknowns at a higher confidence level. In T2HKK [1], there will be one water cerenkov detector of volume 187 kt in Kamioka and another 187 kt similar detector in Korea. Depending on the location in Korea there are different off-axis (OA) flux possibilities. In this similar context, other future oscillation experiments are T2HK (in which both the detector will be at Kamioka), HK (the atmospheric counterpart of T2HK) and DUNE [2] (the future project in Fermilab). Apart from determining the unknown oscillation parameters, these experiments also give us opportunity to study new physics scenarios: for example non-standard interaction (NSI) which we will discuss in the next section.

2 Non-standard interaction

Existence of NSI implies, the initial and final flavour of the neutrinos during the neutral current (NC) interaction with matter can be different [3]. In this

---

[a] E-mail: mghosh@phys.se.tmu.ac.jp
[b] E-mail: Yasuda@phys.se.tmu.ac.jp

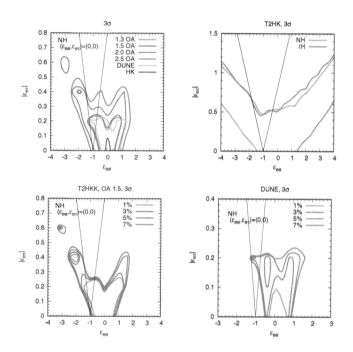

Figure 1: The excluded region in the (ϵ_{ee}, $|\epsilon_{e\tau}|$) plane. The thin solid diagonal straight line stands for the bound $|\tan\beta| \equiv |\epsilon_{e\tau}/(1+\epsilon_{ee})| < 0.8$.

case, the matter term is modified by

$$\mathcal{A} \equiv \sqrt{2} G_F N_e \begin{pmatrix} 1+\epsilon_{ee} & \epsilon_{e\mu} & \epsilon_{e\tau} \\ \epsilon_{\mu e} & \epsilon_{\mu\mu} & \epsilon_{\mu\tau} \\ \epsilon_{\tau e} & \epsilon_{\tau\mu} & \epsilon_{\tau\tau} \end{pmatrix}, \quad (1)$$

where $\epsilon_{\alpha\beta}$ are the NSI parameters. The present 90% bounds of the NSI parameters are given by [4]

$$\begin{pmatrix} |\epsilon_{ee}| < 4 \times 10^0 & |\epsilon_{e\mu}| < 3 \times 10^{-1} & |\epsilon_{e\tau}| < 3 \times 10^0 \\ & |\epsilon_{\mu\mu}| < 7 \times 10^{-2} & |\epsilon_{\mu\tau}| < 3 \times 10^{-1} \\ & & |\epsilon_{\tau\tau}| < 2 \times 10^1 \end{pmatrix}. \quad (2)$$

Thus we understand that the bounds on $\epsilon_{\alpha\mu}$ where $\alpha = e, \mu, \tau$ are stronger than the ϵ_{ee}, $\epsilon_{e\tau}$ and $\epsilon_{\tau\tau}$. One additional bound comes from the high-energy atmospheric data which relates the parameters $\epsilon_{\tau\tau}$ and $\epsilon_{e\tau}$ as [5]

$$\epsilon_{\tau\tau} \simeq \frac{|\epsilon_{e\tau}|^2}{1+\epsilon_{ee}}. \quad (3)$$

Keeping these facts in mind, we perform our analysis with $\epsilon_{\alpha\mu} = 0$. Thus the free parameters are ϵ_{ee}, $|\epsilon_{e\tau}|$ and $\arg(\epsilon_{e\tau}) = \phi_{31}$. In our analysis we have kept the true values of $(\delta_{CP}, \theta_{23})$ fixed at $(-90°, 45°)$.

3 Results

We have done our simulation using GLoBES [6] and MonteCUBES [7]. In Fig. 1, we have given our results in the test ϵ_{ee} - test $|\epsilon_{e\tau}|$ plane. The true value of ϕ_{31} is zero and marginalized in test. For systematic errors, we have considered normalization error which affects the scaling of the events and tilt error which affects the energy dependence of the events. The tilt error is taken as 10% and for the upper panels the normalization errors are taken from Refs. [1, 2]. For the bottom panels, we have given our results for four different values of normalization errors. From the left panel we see that among the different off-axis configurations of T2HKK, the sensitivity of OA 1.3° is best and comparable to DUNE. But the best sensitivity comes from HK. From the right panel we see that compared to T2HKK, the sensitivity of T2HK is quite weak. From the bottom panels we see that the results depend on the systematic uncertainties significantly.

4 Summary

In this work we have studied the sensitivity to NSI for T2HKK and compared our results with T2HK, HK and DUNE. We have also studied the effect of systematic uncertainties. For more details, we refer to [8, 9] on which this work is based upon.

Acknowledgments

This research was partly supported by a Grant-in-Aid for Scientific Research of the Ministry of Education, Science and Culture, under Grants No. 25105009, No. 15K05058, No. 25105001 and No. 15K21734.

References

[1] K. Abe et al. [Hyper-Kamiokande proto- Collaboration], arXiv:1611.06118 [hep-ex].
[2] R. Acciarri et al. [DUNE Collaboration], arXiv:1512.06148 [physics.ins-det].
[3] T. Ohlsson, Rept. Prog. Phys. **76**, 044201 (2013) [arXiv:1209.2710 [hep-ph]].
[4] C. Biggio, M. Blennow and E. Fernandez-Martinez, JHEP **0908**, 090 (2009) [arXiv:0907.0097 [hep-ph]].

[5] A. Friedland and C. Lunardini, Phys. Rev. D **72**, 053009 (2005) doi:10.1103/PhysRevD.72.053009 [hep-ph/0506143].

[6] P. Huber, M. Lindner and W. Winter, Comput. Phys. Commun. **167**, 195 (2005) [hep-ph/0407333].

[7] M. Blennow and E. Fernandez-Martinez, Comput. Phys. Commun. **181**, 227 (2010) [arXiv:0903.3985 [hep-ph]].

[8] S. Fukasawa, M. Ghosh and O. Yasuda, Phys. Rev. D **95**, no. 5, 055005 (2017) [arXiv:1611.06141 [hep-ph]].

[9] M. Ghosh and O. Yasuda, Phys. Rev. D **96**, no. 1, 013001 (2017) [arXiv:1702.06482 [hep-ph]].

STATUS AND PHYSICS POTENTIAL OF THE JUNO EXPERIMENT

Giuseppe Salamanna [a], *on behalf of the JUNO Collaboration*
Roma Tre University and INFN Roma Tre, 00146 Rome, Italy

Abstract. The Jiangmen Underground Neutrino Observatory (JUNO) is an underground 20 kton liquid scintillator detector being built in the south of China and expected to start data taking in 2020. JUNO has a physics programme focused on neutrino properties using electron anti-neutrinos emitted from two near-by nuclear power plants. Its primary aim is to determine the neutrino mass hierarchy from the $\bar{\nu}_e$ oscillation pattern. With an unprecedented relative energy resolution of 3% as target, JUNO will be able to do so with a statistical significance of 3–4 σ within six years of running. It will also measure other oscillation parameters to an accuracy better than 1%. An ambitious experimental programme is in place to develop and optimize the detector and the calibration system, to maximize the light yield and minimize energy biases. JUNO will also be in a good position to study neutrinos from the sun and the earth and from supernova explosions, as well as provide a large acceptance for the search for proton decay. JUNO's physics potential was described and the status of its construction reviewed in my talk at the conference.

1 Introduction

The Jiangmen Underground Neutrino Observatory (JUNO) is a neutrino experiment being built in China, described in [1]. Its primary purpose is to determine the neutrino mass hierarchy (MH) and measure the oscillation parameters using reactor sources. It will consist of a large mass, pure liquid scintillator (LS), placed in an acrylic sphere of 35.4 m of diameter; a system of large-area photomultipliers of new generation (PMT); a veto system. It will be located at a distance of approximately 50 km from two power plants (Yangjiang and Taishan). The two plants are expected to provide an equal thermal power of about 18 GW, but at the start-up of the experiment only 26.6 GW are expected to be available. The baseline is optimized for maximum $\bar{\nu}_e$ disappearance, i.e. a minimum of the survival probability (here in the case of normal neutrino mass hierarchy (NH)):

$$P_{NH}(\bar{\nu}_e \to \bar{\nu}_e) = 1 - \frac{1}{2}\sin^2 2\theta_{13}\left(1 - \cos\frac{L\Delta m^2_{atm}}{2E_\nu}\right)$$
$$- \frac{1}{2}\cos^4\theta_{13}\sin^2 2\theta_{12}\left(1 - \cos\frac{L\delta m^2_{sol}}{2E_\nu}\right)$$
$$+ \frac{1}{2}\sin^2 2\theta_{13}\sin^2\theta_{12}\left(\cos\frac{L}{2E_\nu}(\Delta m^2_{atm} - \delta m^2_{sol}) - \cos\frac{L\delta m^2_{atm}}{2E_\nu}\right), \quad (1)$$

---

[a] E-mail: salaman@fis.uniroma3.it

(the probability for the inverted hierarchy P_{IH} differing only in the coefficient of the last element in the sum, $\cos^2 \theta_{12}$). JUNO will be placed about 700 m below underground, corresponding to about 1900 m.w.e., in a pit dug in the ground afresh. To pursue its main physics goals JUNO will need to attain an unprecedented resolution on the energy of the $\bar{\nu}_e$ produced in the reactors. In order to meet such performance extensive studies have been conducted over the past few years concerning various aspects of engineering, detector design, including optical/light collection in the LS and PMT and response of the read-out electronics, software development and background suppression [2]. In the talk given at the 18th Lomonosov Conference on Elementary Particle Physics, the main physics case of JUNO was described; the status of the detector design optimization and construction was illustrated and some examples of physics measurements possible, beyond the neutrino oscillations, were presented.

2 What drives the detector design....

The experimental signature of the reactor $\bar{\nu}_e$ in the JUNO detector is given by the inverse beta decay (IBD) process $\bar{\nu}_e + p \rightarrow e^+ + n$, where the p and n are a proton from the LS and a neutron, respectively. The resulting signal is given by a visible energy from the positron energy loss and annihilation, plus delayed light at a fixed 2.2 MeV energy from the neutron capture. The main goal of JUNO is to determine the MH with at least a 3 σ significance within the first 6 years of data taking. The correct MH will be extracted by means of a χ^2 fit to the kinetic energy spectrum of the prompt e^+ (T_{e^+}), which is directly related to the $\bar{\nu}_e$ energy ($E_{\bar{\nu}_e} = T_{e^+} + 2 \times m_e + 0.8$ MeV, where m_e is the e^+ mass). From Eqn. 1 one sees that the two hierarchies have a difference in the fine structure of $P_{MH}(\bar{\nu}_e \rightarrow \bar{\nu}_e)$, which is illustrated in Fig. 1, left, and was pointed out in [3]. The correct MH is determined by constructing the $\Delta\chi^2_{min}(NH - IH)$ from the two fits to the experimental reactor data; this can be translated into a statistical significance of the discrimination. It should be noticed that, with such strategy, a residual ambiguity lingers, associated to the correct value of the atmospheric mass difference Δm^2_{atm}; this reduces the final sensitivity of the fitting procedure. JUNO estimates that, in order to reach the desired significance, the most important fogure of performance is the overall resolution on the event-by-event measurement of $E_{\bar{\nu}_e}$. In Fig. 1, right, this relationship appears clearly. It is therefore crucial to design a detector which minimizes the statistical uncertainty from stochastic fluctuations in the scintillation light collection, yet keeping linearity and uniformity of the energy response under control. A 3% overall relative energy resolution will yield, for 6 years of data at 36 GW of reactor thermal power, a median significance of 3.4 (3.5) σ for NH (IH) [1].

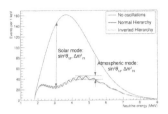 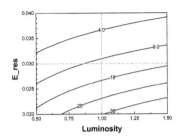

Figure 1: Left: $P_{MH}(\bar{\nu}_e \to \bar{\nu}_e)$ for no oscillation and oscillations under the two MH hypothesis (courtesy: Y.Malyshkin), just for illustration purposes. Right: curves of $\Delta\chi^2_{min}(NH-IH)$ as a function of the energy resolution (y axis) and the "luminosity" of the sample collected with respect to the 6-year benchmark [1]

3 ...and the resulting detector design

To attain this level of precision, the JUNO collaboration has developed a detector made of 3 basic parts, plus the electronics. The central detector is a 20 kton LS target mass, conceived to maximize the photon statistics and minimize the attenuation of the IBD prompt signal. This will be the largest volume of LS to date, composed of a mixture of $> 98\%$ LAB (solvent, 1200 photo-electrons/MeV), PPO (solute) and a less-than-per-cent part of bis-MSB (wavelength shifter). The photons are collected by PMT of two different kinds: about 18000 20 inch PMT, most of which of the micro-channel plate (MCP-PMT) type, will guarantee an extended photo-coverage (75%, as per JUNO requirement) and a high overall detection efficiency (expected by the JUNO specifications to be 27% at $\lambda = 420$ nm); their transit time spread (TTS) being 12 ns. 25000 conventional 3 inch will allow to follow a multi-calorimetric approach, whereby the non-stochastic terms in the energy resolution will be monitored during the calibration runs with known radio-active sources at different energies; the 3 inch PMT will extend the dynamical range in the waveform of large numbers of photo-electrons hitting a localized region and improve time and vertex resolution for muon reconstruction (against cosmic muons). Finally, a veto system will be in place to screen off incoming muons and photons by means of a surrounding water buffer (Cherenkov) and top scintillators. The project of the front-end electronics is also a challenge, because of the many read-out channels, the dark noise rate of the 20 inch PMT and the needed resiliance and efficient heat dissipation of an under-water system. A large effort from the collaboration is being devoted to this task, but it will not be described here.

Such design, especially the large mass and control of the energy scale, will also allow JUNO to perform measurements of the neutrino solar parameters with uncertainties well below 1%, not attained to date.

4 Current status of the detector project

After a careful design phase, the construction of the various elements is underway. About 15000 MCP-PMT were ordered from the NNVT (North Night Vision of Technology CO., LTD, China) [4] manufacturer and are being tested at a dedicated centre in the region around the JUNO site. Further technical details on the design and production of the MCP-PMT was the subject of a dedicated contribution by S. Qian at this conference. Additional 5000 conventional dynode 20 inch PMT were commissioned to Hamamatsu [5], of the type R12860, in order to complement the leading PMT lay-out and provide a faster TTS (3 ns). All the large-area PMT will be equipped with protective masks to reduce propagation of shock waves if one PMT explodes under water pressure: their design has been finalized after extensive pressure tests. The bidding of the 3 inch PMT was completed last spring and the elements ordered from HZC-Photonics. These are custom-made based on the KM3NeT design, with improved TTS for better muon tracking. It is desirable that the LS be purified to a good degree from radioactive isotopes, to reduce the intrinsic background of the detector. The set level are 10^{-15} g/g for ^{238}U and ^{232}Th and 10^{-17} g/g for ^{40}K. The attenuation length (AL) that JUNO requires is greater than 20 m at $\lambda = 430$ nm (for 3 g/l of PPO in LAB). A strategy has been developed by JUNO aimed at obtaining an optimal admixture of solvent and solutes in optical and radio-active terms. The purification will go through four parallel and complementary processes: an Al_2O_3 (alumina) column, a distillation plant, water extraction and gas stripping. A pilot plant has been established in one of the LS halls of the Daya Bay experiment in China to monitor the AL and level of purification, which uses the alumina method. Results are displayed in Fig. 2 and show a good stability and that the desired level of AL has been surpassed.

5 Calibration of energy and vertex position reconstruction

As stressed, a good and understood energy response is paramount for the main aims of the JUNO experimental programme. While the large mass, photocoverage and detection efficiency ensure small statistical fluctuations in the light yield, systematic misestimation from non linearity and non uniformity of the detector response can quickly spoil the precision. Keeping the uncertainty on the energy scale determination at less than 1% over the energy interval is crucial to keep the total $\sigma(E)/E \leq 3\%$. Together with that, reconstructing the

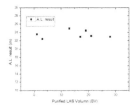

Figure 2: Left: the picture of a 20 inch MCP-PMT (see text). Right: Attenuation length vs level of LAB purification from the Daya Bay purification pilot plant.

position at which the event took place in the LS will play a very useful role in suppressing the backgrounds, mainly related to diffuse radioactivity and cosmic ray muon scattering. To this end JUNO will deploy a redundant calibration system. Complementary methods are envisaged across the detector and for various energy loss mechanisms: an Automatic Calibration Unit (ACU) will place several known radioactive sources along the vertical axis in the sphere; a Cable Loop System (CLS) will move across vertical planes by means of pulleys, while a Guide Tube Calibration System (GTCS) will be in use to probe the outer CD surface. Finally a Remotely Operated under-LS Vehicle (ROV) will provide events of known energy across the whole detector volume. The calibration strategy (in particular number of points in the scan and occurrence of the calibration scans) is being worked out.

6 Other physics possible with JUNO

JUNO's features make it an excellent detector also for other physics topics, including non-reactor neutrino measurements such as solar and supernova neutrino fluxes and geo-neutrino isotopical origin. In many of these case, the main challenge will be to control the intrinsic background activity and the cosmogenic cascades (muons interacting with carbon in the LS and creating Lithium). The latter are expected to be of high occurrence in the relatively shallow JUNO pit (about 250000 per day as opposed to 60 IBD events, tens to thousands of solar neutrinos and about 1 geo-neutrino per day). JUNO will also be complementary to large Cherenkov detectors (e.g. SK, HK) in the search for proton decays. The proton will come from the hydrogen in the LS and should decay to a neutrino and a kaon, with subsequent semi-leptonic meson decays. The initial decay is sub-threshold for Cherenkov light in water but the kaon kinetic energy of about 105 MeV is well visible as scintillation light. Thanks to the peculiar time pattern of the tiered decay, JUNO will be competitive (and complementary) soon after switching on. Additional information on all the programme can be found in [1].

7 Conclusions

The JUNO experiment is on course to start operations within the next few years. Its challenging and multi-faceted physics programme (on and beyond reactor neutrino oscillations) demands a very careful detector design and the use of novel technologies. At the same time, with its unprecedented size and energy resolution, JUNO is poised to have an impact on many areas of neutrino physics. A few selected figures were provided drawn from the many design/technical improvements and tests being put in place to achieve both unprecedented performance and reliability for a large and underwater system.

Acknowledgments

The speaker would like to warmly acknowledge the organizers of the conference for their kindness and the high standard of the physics programme. Also, a heartly thank you for giving me the chance to spend a wonderful time in Moscow and practice the language.

References

[1] F. An *et al.* [JUNO Collaboration], J. Phys. G **43**, 030401 (2016).
[2] T. Adam *et al.* [JUNO Collaboration], arXiv:1508.07166.
[3] S. T. Petcov and M. Piai, Phys. Lett. B **533** 94 (2002) [hep-ph/0112074].
[4] North Night Vision of Technology, China: http://en.nvt.com.cn/.
[5] Hamamatsu Photonics, Japan: http://www.hamamatsu.com.

THE NEUTRINO MASS HIERARCHY FROM OSCILLATION

Luca Stanco [a]
INFN - Padova, Via Marzolo, 8 35131 Padova, Italy

Abstract. The ordering of the neutrino mass eigenstates, also addressed as Mass Hierarchy (MH), is one of the most relevant issues in neutrino physics, currently under investigation by many proposals and experiments. In this short note focus will be given to the different ways to determine MH in the near future.

1 Introduction

The ordering of the neutrino mass eigenstates is one of the most relevant issues, currently under investigation by many proposals and experiments. In the standard scenario and a widely usual convention the three neutrinos ν_1, ν_2 and ν_3 are known to have relative masses measured as $\delta m_{21}^2 = m_2^2 - m_1^2$ (named for historical reasons "solar" mass term) and $|\Delta m_{31}^2| = |m_3^2 - m_1^2| \sim |m_3^2 - m_2^2|$ (called "atmospheric" mass term). The sign of Δm_{31}^2 has not been measured yet, and that allows two different configurations for the mass eigenstates: either $m_1 < m_2 < m_3$ or $m_3 < m_1 < m_2$. That corresponds to have either one or two higher mass states, with huge consequences on the neutrino models [1,2]. The mass ordering is usually identified as normal hierarchy (NH) when $\Delta m_{31}^2 > 0$ and inverted hierarchy (IH) for the case $\Delta m_{23}^2 > 0$. Its importance is enormous to provide inputs for the next studies and experimental proposals, to finally clarify the needs and the tuning of new projects, and to constraint analyses in other fields like cosmology and astrophysics.

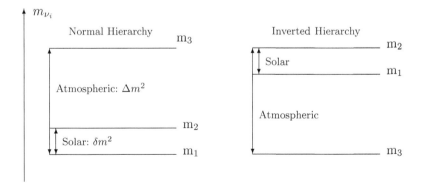

Figure 1: Neutrino mass eigenstates for normal and inverted mass ordering (not to scale).

---

[a] E-mail: luca.stanco@pd.infn.it

In Fig. 1 a cartoon of the two possible configurations for the mass ordering is depicted. In this paper the following notation has been used for the atmospheric mass: $\Delta m^2_{atm} = \Delta m^2_{31}(\text{NH}) = \Delta m^2_{23}(\text{IH})$, for the two different hypotheses, respectively. Δm^2_{atm} is therefore a fundamental physical quantity, which corresponds to the difference of the heaviest squared mass and the lightest neutrino squared-mass.

The achievements of the last two decades brought up a coherent picture, namely the oscillation of three neutrino flavour–states, ν_e, ν_μ and ν_τ, originated by the mixing of the three ν_1, ν_2 and ν_3 mass eigenstates. The issue of the mass ordering has been highly debated in the last decade, but it gained in interest with the discovery of the relatively large value of θ_{13} in 2012. The convolutions between the three mixing angles and the mass parameters are such that measurements of the current experiments may become sensitive to the dependences of the oscillation probabilities to the sign of MH. Surely, the MH determination will be a major point for the next experiments under construction.

2 The MH degeneracies

As far as oscillations are concerned, the dependences on the mass ordering come from the interference between two different effects. In particular, the interference of oscillations driven by $\Delta m^2_{31}(\text{NH})$ or $\Delta m^2_{23}(\text{IH})$ with oscillations driven by another quantity, Q, with a known sign. In vacuum the interference is given by the joint atmospheric and solar oscillations, such that Q corresponds to the solar mass δm^2. For atmospheric and neutrinos from long baseline accelerator the interference is due to the matter effect, Q being the corresponding matter potential, $2\sqrt{2}G_F N_e E$, with obvious meaning for the quantities involved. Moreover, in the three–neutrino framework MH is highly correlated with the neutrino oscillation parameters and the CP phase, δ_{CP}. Specifically, in the neutrino oscillation framework there are three big area of investigation: MH from long baseline accelerators are highly coupled to δ_{CP}, while for the reactor antineutrino (medium baseline) there is no δ_{CP} dependence at first order, in contrast to a strong dependence on the exact value of Δm^2_{atm}. The third area of investigation corresponds to the atmospheric neutrinos, which own a degeneracy both on Δm^2_{atm} and the value of the mixing angle θ_{23}, namely to which octant it belongs.

These correlations correspond to degeneracies that can severely limit the discrimination of the hierarchy, either normal of inverted. If one generally indicates with θ_i the *correlation* parameter ($\theta_1 = \delta_{CP}$, $\theta_2 = \Delta m^2_{atm}$ and $\theta_3 = \theta_{23}$) more solutions may be extracted from the data for MH, e.g. NH($\hat{\theta}_i$) and IH($\hat{\theta}'_i$) with $\hat{\theta}_i \neq \hat{\theta}'_i$. The θ_i parameters are usually evaluated within the standard 3ν

oscillation framework via global fits [3]. Unfortunately, the current uncertainties on θ_i allow several distinct solutions and practically no sensitivity to MH. This is demonstrated in Fig. 2 for the NOvA case and its 2015 data release.

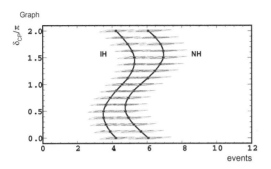

Figure 2: The number of events ($\nu_\mu \to \nu_e$ appearance plus background) as function of δ_{CP} as expected for the 2015 NOvA data analysis from the best fits of the 2017 global fit analysis (GF). The two plain black lines correspond to the expectation from NH (right) and IH (left), respectively. The computation has been performed with the GLoBES package. The two concentric areas for each value of δ_{CP} correspond to the 1 σ and 2 σ contours due to the correlated θ_{23}, θ_{13} uncertainties, σ being estimated by GF. See [4] for more details.

3 A strategy for the MH determination

Given the scenario described in the previous section it is fundamental to control the test statistic that is used in the analysis. The statistical estimator should be robust and should make evidence of the degeneracies and the θ_i dependences. For the MH studies only one estimator has been extensively used so far throughout the several fields of investigation. That is the chi-square difference:

$$\Delta\chi^2 = \chi^2_{min}(\text{IH}) - \chi^2_{min}(\text{NH}), \quad (1)$$

where the two minima are evaluated spanning the uncertainties of the three-neutrino oscillation parameters, namely the solar mass δm^2_{21}, the atmospheric mass $\Delta m^2_{31(23)}$, the CP phase δ_{CP} and the mixing angles θ_{12}, θ_{23}, θ_{13}, as defined by the standard parameterization. On top of that statistical and systematic errors are included in the fitting procedures. The $\Delta\chi^2$ evaluation is based on two distinct hypotheses, NH and IH. For each MH the best solution is found: the χ^2_{min} comes from two different best-fit values for NH and IH, separately, and the $\Delta\chi^2$ is the result of the internal adjustments of the two distinct fits. No real understanding of the weight arising from each single contributions (i.e. the single neutrino oscillation parameters or the statistical/systematic errors) is possible, given to the intrinsic multiple non-linear correlations.

Recently, we suggested a change of perspective: try to identify an estimator that couples NH/IH and decouples the θ_i dependences [1]. As a consequence, each kind of data, long baseline or reactors or atmospheric ones, should be analyzed by different optimized estimators. We already studied a possible new estimator for the accelerator data [4] and a new estimator for reactor antineutrinos [5]. Since these estimators already intrinsically couple NH and IH, it is no more needed to construct an "estimator of the estimators" like the $\Delta\chi^2$. Instead, the two outcomes, one for NH and the other for IH, are directly used to get NH and IH significances.

The change of perspective suggests a pragmatic new strategy in the determination of MH. Once the statistical estimator has been chosen, let us call S, its evaluation over data would simply bring to one of the three following options:

1. both S_{NH} and S_{IH} are *compatible* with data;

2. both S_{NH} and S_{IH} are *incompatible* with data;

3. either S_{NH} or S_{IH} is compatible with data, the other one being incompatible.

The meaning of *compatible* and *incompatible* comes from a long experience in data analysis of experiments in particle physics. Nowadays, it is well accepted that *compatible* means at 95% of C.L., whereas *incompatible* means $\geq 5\,\sigma$. That corresponds to the standard definition of exclusion or observational results [6]. Over the last decades, these choices have been proven to be the right ones by many experimental results. To be more precise, an experimental observation to be conclusive corresponds to the rejection of the background hypothesis at least at $5\,\sigma$. An experimental exclusion limit corresponds to the phase space defined by the set of values of the signal parameter compatible at 95% C.L. with the data themselves, the complementary phase space containing the rejection of the signal at 95% C.L..

When this procedure is applied to the MH determination, a confusing scenario may rise up. The question becomes: the MH determination is a signal or a background rejection? Since NH/IH are mutually exclusive not-nested hypotheses their roles can be interchanged. Then, our proposal is just the above list of options. Specifically, a conclusive experiment, or a global analysis, should provide both a rejection of the wrong hierarchy at $5\,\sigma$ level and a compatibility with the true hierarchy at 95% C.L..

When the analysis should produce a result as in case (1), thus it would be inconclusive. In case (2) probably something wrong were occurring in the analysis procedure (or the 3ν framework is no more appropriate). Case (3) should correspond to the sensitivity with which the experiment/analysis would determine the mass hierarchy. In case (3) and a sensitivity at the level of $5\,\sigma$ the determination of the MH could be finally established.

We also outline that different statistical approaches may be applied, namely a frequentist approach or a Bayesian one. Only when the significance reaches the level of $5\,\sigma$ the different statistical approaches usually give similar results.

4 An estimator for MH at accelerators

For the accelerator basis searches the NOvA experiment is the best placed one [9]. It is foreseen that some information be available after several years of running with data-taking both in neutrino and anti-neutrino modes. Adding measurements on δ_{CP} from few years of T2K exposure will allow to slightly increase the separation between the two options in different portions of δ_{CP} range [7]. Conversely, if MH should be known sometimes in future, T2K would greatly improve its significance on δ_{CP} [8]. The expectation on MH is however not exciting; only a 3-sigma significance could be obtained and only in the most favorable δ_{CP} regions. As a matter of fact the perspectives in the near future for the determination of the neutrino mass ordering with neutrinos from accelerator beams are rather poor, even less favorable than the prospects for the δ_{CP} measurement.

The new technique reported in [4] is based on a new test statistic that properly weights the intrinsic statistical fluctuations of the data and extracts the significances of either NH or IH. The Poisson distributions for n_i observed events, $f_{\text{MH}}(n_i;\mu_{\text{MH}}|\delta_{CP})$ are initially considered, where $\mu_{\text{MH}}(\delta_{CP})$ are the expected number of events as function of δ_{CP}, MH standing for IH or NH. For a specific n the left and right cumulative functions of f_{IH} and f_{NH} are computed and their ratios, q_{MH}, are evaluated. Since for the ν_e appearance at NOvA the number of expected events as function of δ_{CP} is asymmetric towards IH and NH (less events are expected for IH than for NH), the ratios are defined independently for the IH and the NH cases:

$$q_{\text{IH}}(n,\delta_{CP}) = \frac{\sum_{n_i^{\text{IH}} \geq n} f_{\text{IH}}(n_i^{\text{IH}};\mu_{\text{IH}}|\delta_{CP})}{\sum_{n_i^{\text{NH}} \geq n} f_{\text{NH}}(n_i^{\text{NH}};\mu_{\text{NH}}|\delta_{CP})},$$

$$q_{\text{NH}}(n,\delta_{CP}) = \frac{\sum_{n_i^{\text{NH}} \leq n} f_{\text{NH}}(n_i^{\text{NH}};\mu_{\text{NH}}|\delta_{CP})}{\sum_{n_i^{\text{IH}} \leq n} f_{\text{IH}}(n_i^{\text{IH}};\mu_{\text{IH}}|\delta_{CP})}.$$

q_{IH} and q_{NH} are two discretized random variables comprised to the $[0,1]$ interval. As n goes to zero q_{IH} goes to one, while when n increases q_{IH} asymptotically tends to zero. q_{NH} behaves the other way around towards n.

The probability mass functions, $P(n)$, of each q_{MH} have been computed via toy Monte Carlo simulations based either on f_{IH} (test of IH against NH) or f_{NH} (test of NH against IH). They are further compared to the real number of observed data n_D. By evaluating the p–value probabilities for n_D the significance is finally computed.

With the new method an averaged increase of 0.5 σ with respect to the standard $\Delta\chi^2_{min}$ is obtained [4]. Worth to note that the increase is not constant but it depends on the discrimination threshold n_D and δ_{CP}: the gain of the new method in terms of the number of sigma's strongly raises with n_D and "favorable" regions of δ_{CP}. As demonstrated in the appendix of [4] the new method is generally better than $\Delta\chi^2_{min}$ for many reasons: it deals with the full probability distributions, it profits of the intrinsic fluctuations of the data and, most relevant, it answers the right question (to disprove one MH option). In fact the new q estimator focusses on the possibility to reject the wrong hierarchy, disregarding the other one. Therefore, once one option is selected (e.g. rejection of IH) it does not provide any evaluation on the other option (rejection of NH). Instead, the $\Delta\chi^2_{min}$ method treats the two options in a symmetric way with the disadvantage of mixing up the information.

This new procedure becomes asymptotically equivalent to the $\Delta\chi^2$ one when the luminosity increases. Nevertheless, it allows to achieve a similar level of significance with about a factor three less data.

5 An estimator for MH at reactors

The determination of the mass hierarchy with reactor neutrinos is a very challenging task. Both an exceedingly high energy-resolution and a large mass detectors are required. The JUNO experiment has been proposed and it is currently under construction trying to match these two requirements [10]. The foreseen achievement on MH is nevertheless limited. A significance around 4σ, after 6 years of exposure, at the full reactor power of 36 GW, is predicted for the median sensitivity.

A new technique that would provide a robust 5σ measurement in less than six years of running was recently proposed in [5]. It is based on the introduction of a new statistical estimator, F, which revises the approach followed in the last ten years based on the $\Delta\chi^2$ estimator and an effective parameterization of the neutrino masses. The effective parameterization [11] was very valuable in boosting the studies and the proposals for a large reactor neutrino experiment at medium baseline (30-50 km). It predicted the possible determination of MH without any degeneracy, even taking into account the rather large uncertainty on Δm^2_{atm} at that time (larger than 40%). However, the effective parameterization reduces the information above a certain neutrino energy threshold. For example, at JUNO the discrimination between NH and IH vanishes for $E_\nu > 4 - 5$ MeV.

Instead, using the F estimator the two mass orderings could be discriminated at the price of allowing for two different values of Δm^2_{atm}. This degeneracy on Δm^2_{atm} (around 12×10^{-5} eV^2) can in any case be measured at an unprecedented accuracy of much less than 1%, i.e. 10^{-5} eV^2, within the same analysis.

The key picture is shown in Fig. 3. It demonstrates that, whatever be the algebraic construction and the implicit assumptions of the F-test, for each real/simulated data sample (x-axis) the F technique identifies two main possibilities (y-axis): the true MH associated to the true Δm^2_{atm} and a wrong MH solution with a Δm^2_{atm} shifted of 12×10^{-5} eV^2 with respect to the true value. Each of the two solutions own a Δm^2_{atm} resolution of about 0.3%.

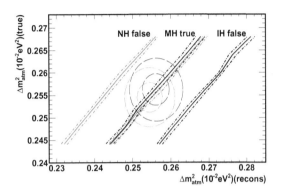

Figure 3: Δm^2_{atm}(true) vs Δm^2_{atm}(recons) is drawn, Δm^2_{atm}(recons) being obtained by applying the F estimator. The continuous lines correspond to the central values, the dashed ones to the $\pm \sigma$ bands. Black (red) curves corresponds to the NH (IH) generation. The central circles correspond to the 68% and 95% C.L. contours of the current Δm^2_{atm} uncertainties from the global fits for NH and IH. See [5] for more details.

It is worth to add that in [5] evaluation and inclusion of systematic errors and backgrounds have been performed, the most relevant among them being the addition of the two remote reactor plants 250 km away. Baselines of each contributing reactor core and its spatial resolution have been taken into account. Possible results after two years of running and the foreseen initially-reduced available reactor power have been studied, too.

Last but not least, using the F estimator, the estimated significance grows linearly with the size of the data sample, contrary to the $\Delta\chi^2$ outcome that is asymptotically limited.

6 Conclusions

A major enterprise of the neutrino community is the future determination of the neutrino mass ordering. Unfortunately, it appears to be a challenging task for any framework would be used. The atmospheric neutrino framework, as well as the cosmological framework, were not discussed in this note. Nevertheless their sensitivities on the MH determinations seem either rather poor or with severe concerns, respectively. The same occurs to the framework of the accelerator baseline and the reactor neutrinos. Therefore, it is mandatory to evaluate whether new tools of analysis can overcome these limits. We reported about the recent techniques developed for the two latter frameworks. They are encouraging and could provide more robust and significant/complementary results than the standard technique based on the $\Delta\chi^2$ estimator, and in a shorter time.

Acknowledgments

It is a pleasure to thank the organizers, in particular A. Studenikin, for their kind invitation, the warm hospitality in Moscow and the overall very stimulating series of presentations at this 18th Lomonosov conference. I would like to acknowledge discussions about the issues illustrated in this brief note with my colleagues S. Dusini, A. Lokhov, G. Salamanna C. Sirignano and M. Tenti. The help of F. Sawy in some lateral work about the reactor neutrino study is also acknowledged, as well as deep discussions with S. Petcov.

References

[1] L. Stanco, Rev. in Phys. **1** (2016) 90.
[2] C. Patrignani et al. (Particle Data Group), Chin. Phys. C, 40, 100001 (2016) and 2017 update.
[3] P. F. de Salas et al., arXiv:1708.01186, and I. Esteban et al. of the nu-fit group, www.nu-fit.org, report the most recent results. See also S. Goswami at this conference.
[4] L. Stanco, S. Dusini and M. Tenti, Phys. Rev. D **95**, 053002 (2017) [arXiv: 1606.09454v3].
[5] L. Stanco, G. Salamanna, A. Lokhov, F.H. Sawy and C. Sirignano., arXiv:1707.07651v3.
[6] G. Cowan et al., Eur. Phys. Jou. C **71** (2011) 1554 [arXiv:1007.1727v3].
[7] T. Nagaya and R. K. Plunkett, New J. Phys. **18**, 015009 (2016) [arXiv: 1507.08134].
[8] K. Abe et al. (T2K collaboration), arXiv:1607.08004.
[9] R.B. Patterson, Ann. Rev. Nucl. Part. Sci. **65** 177 (2015) [arXiv:1506. 07917].
[10] F. An et al. (JUNO collaboration), J. Phys. G **43**, no. 3, 030401 (2016).
[11] A de Gouvea, J. Jenjins and B. Kayser, PRD 71, (2005) 113009;
H. Nunokawa, S. Parke and R. Zukanovich Funchal, PRD72, (2005) 013009

NEUTRINO MASS ORDERING AND NEUTRINOLESS DOUBLE-BETA DECAYS

Shun Zhou [a]

Institute of High Energy Physics, Chinese Academy of Sciences, Beijing 100049, China
Center for High Energy Physics, Peking University, Beijing 100871, China

Abstract. In this talk, the importance of neutrinoless double-beta ($0\nu\beta\beta$) decays in understanding the origin of tiny neutrino masses has been emphasized. In addition, the Bayesian approach is applied to the study of whether it is possible to determine neutrino mass ordering in future ^{76}Ge-based $0\nu\beta\beta$ experiments.

1 Introduction

In the past two decades, a number of elegant neutrino oscillation experiments have firmly established that neutrinos are massive and the lepton flavors are significantly mixed [1]. However, it remains to be understood how neutrinos acquire their tiny masses. To this end, it is extremely important to first find the answer to a fundamental question initially raised by Majorana in 1937 [2]: Are neutrinos their own antiparticles?

As first pointed out by Furry in 1939, the neutrinoless double-beta ($0\nu\beta\beta$) decays $A(Z,N) \to A(Z+2, N+2) + 2e^-$, where Z and N denote respectively the numbers of protons and neutrons in the even-even nuclei $A(Z,N)$, can take place when massive neutrinos are their own antiparticles, namely, Majorana fermions [3]. For massive Dirac neutrinos, only the lepton-number-conserving double-beta decays $A(Z,N) \to A(Z+2, N+2) + 2e^- + 2\overline{\nu}_e$ can happen [4]. The primary motivations for the experimental searches for $0\nu\beta\beta$ decays are the following. First, the discovery of such a kind of decays will unambiguously prove that the lepton number is not conserved in nature and neutrinos are Majorana particles [5–7]. Second, the demonstration of the Majorana nature of massive neutrinos will pave the way to the ultimate understanding of small neutrino masses and the new physics beyond the standard model [8]. Finally, the $0\nu\beta\beta$ decay experiments are currently the most promising place to constrain or probe the Majorana-type CP violating phases, appearing only in the lepton-number-violating processes.

In assumption of Majorana neutrinos, the half life of an even-even nuclear isotope in the $0\nu\beta\beta$ decay mode is given by the following formula [8]

$$\left(T_{1/2}^{0\nu}\right)^{-1} = G_{0\nu} |\mathcal{M}_{0\nu}|^2 m_{\beta\beta}^2 , \qquad (1)$$

where $G_{0\nu}$ denotes the phase-space factor, $\mathcal{M}_{0\nu}$ the relevant nuclear matrix element (NME) and $m_{\beta\beta}$ the effective neutrino mass. The phase-space factor

---
[a] E-mail: zhoush@ihep.ac.cn

can in general be calculated accurately, while the uncertainties associated with the NME can be as large as a factor of a few. In the standard parametrization of leptonic flavor mixing matrix, the effective neutrino mass can be expressed in terms of neutrino masses m_i (for $i = 1, 2, 3$) and flavor mixing parameters

$$m_{\beta\beta} \equiv \left| m_1 \cos^2\theta_{12} \cos^2\theta_{13} e^{2i\phi_1} + m_2 \sin^2\theta_{12} \cos^2\theta_{13} e^{2i\phi_2} + m_3 \sin^2\theta_{13} \right| , \quad (2)$$

where θ_{12} and θ_{13} are two of three flavor mixing angles, while ϕ_1 and ϕ_2 are the Majorana-type CP-violating phases. Notice that we have assumed that the dominant contribution to the $0\nu\beta\beta$ decays comes from the exchange of light Majorana neutrinos. In the case of normal neutrino mass ordering (NO) with $m_1 < m_2 < m_3$, the effective neutrino mass in Eq. (2) could be exactly vanishing, leading to no $0\nu\beta\beta$ decays even for Majorana neutrinos. On the other hand, if neutrinos take the inverted mass ordering (IO) with $m_3 < m_1 < m_2$, the effective neutrino mass has a lower bound $m_{\beta\beta} > 15$ meV, which has been the target for the next-generation $0\nu\beta\beta$ decay experiments [8].

An immediate question is then whether it is possible to determine neutrino mass ordering in future $0\nu\beta\beta$ decay experiments. If the $0\nu\beta\beta$ decays are observed and the extracted value of $m_{\beta\beta}$ is lying in the right region as predicted by Eq. (2) in the IO case, one may simply conclude that neutrino mass ordering is inverted. In this talk, we will explain why the conclusion is not that solid as expected, utilizing the Bayesian approach to discriminate between neutrino mass orderings in the future large ^{76}Ge-based experiments.

2 Bayesian Analysis

First, let us briefly introduce the Bayesian approach for statistical analyses. See, e.g., Ref. [9], for more details. In this approach, the probability can be interpreted as the degree of belief in a certain hypothesis $\mathcal{H}_i$ (for $i = 1, ..., r$), given the data set $\mathcal{D}$. Note that the hypotheses $\mathcal{H}_i$ are mutually exclusive, and only one of them is actually true. According to the Bayes' theorem, the posterior probability of the hypothesis $\mathcal{H}_i$, denoted as $\Pr(\mathcal{H}_i|\mathcal{D})$, is given by

$$\Pr(\mathcal{H}_i|\mathcal{D}) = \Pr(\mathcal{D}|\mathcal{H}_i) \Pr(\mathcal{H}_i)/\Pr(\mathcal{D}) . \quad (3)$$

In Eq. (3), $\Pr(\mathcal{H}_i)$ stands for the prior probability of $\mathcal{H}_i$, reflecting our prior degree of belief in such a hypothesis. $\Pr(\mathcal{D}|\mathcal{H}_i)$ is the probability of obtaining the data $\mathcal{D}$, assuming $\mathcal{H}_i$ to be true, and is called the evidence $\mathcal{Z}_i$ of the hypothesis $\mathcal{H}_i$. The overall probability of observing the data $\mathcal{D}$ is denoted by $\Pr(\mathcal{D})$, and it is equal to $\sum_{i=1}^{r} \Pr(\mathcal{D}|\mathcal{H}_i)\Pr(\mathcal{H}_i)$, because of the normalization condition $\sum_{i=1}^{r} \Pr(\mathcal{H}_i|\mathcal{D}) = 1$.

A typical application of the Bayesian approach is to make the model selection. With the help of Eq. (3), one can compute the posterior odds of two competing

Table 1: The Jeffreys scale used for the statistical interpretation of Bayes factors, posterior odds and model probabilities [9].

$|\ln(\text{odds})|$	Odds	Probability	Interpretation
< 1.0	≤ 3 : 1	≤ 75.0%	Inconclusive
1.0	≃ 3 : 1	≃ 75.0%	Weak evidence
2.5	≃ 12 : 1	≃ 92.3%	Moderate evidence
5.0	≃ 150 : 1	≃ 99.3%	Strong evidence

hypotheses by taking the ratio of their posterior probabilities, namely,

$$\frac{\Pr(\mathcal{H}_i|\mathcal{D})}{\Pr(\mathcal{H}_j|\mathcal{D})} = \frac{\mathcal{Z}_i}{\mathcal{Z}_j} \frac{\Pr(\mathcal{H}_i)}{\Pr(\mathcal{H}_j)}, \quad (4)$$

where the ratio of evidences $\mathcal{B} \equiv \mathcal{Z}_i/\mathcal{Z}_j$ is termed Bayes factor. In the case of no prior preference for any hypothesis (i.e., equal prior probabilities), the posterior odds is then directly reflected by the Bayes factor. To interpret the value of this posterior odds or the Bayes factor, one often adopts the Kass-Raftery or Jeffreys scale. In Table 1 we list the Jeffreys scale and will implement them to interpret the results as in Ref. [10]. Now suppose that there exists a single hypothesis $\mathcal{H}$ that describes the data well, and this hypothesis is specified by a set of free parameters Θ. the evidence $\mathcal{Z}$ is found to be

$$\mathcal{Z} = \int \Pr(\mathcal{D}|\Theta,\mathcal{H}) \Pr(\Theta|\mathcal{H}) \, \mathrm{d}^N\Theta = \int \mathcal{L}(\Theta) \, \pi(\Theta) \, \mathrm{d}^N\Theta, \quad (5)$$

where N is the dimension of parameter space, while $\Pr(\mathcal{D}|\Theta,\mathcal{H})$ and $\Pr(\Theta|\mathcal{H})$ are also denoted as the likelihood function $\mathcal{L}(\Theta)$ and the prior probability distribution $\pi(\Theta)$, respectively.

3 Discriminating Neutrino Mass Orderings

Then, we apply the Bayesian approach to the discrimination between the NO and IO hypothesis. Regarding the future experiments, we choose the combination of GERDA-II and Majorana Demonstrator as an example, and also the even larger ^{76}Ge-based detector with an exposure of 10^4 kg yr and a background index of $B = 10^{-4}$. The ^{76}Ge-based detectors are implemented since a very simple model of backgrounds is required and a high energy resolution can be achieved. For the details of numerical simulations, one should be referred to Ref. [10]. The main results are summarized in Fig. 1, and some comments are in order.

The exposure at least 500 (or 2500) kg yr needs to be accumulated to reach a weak (or moderate) evidence for NO. According to the Jeffreys scale in Table 1,

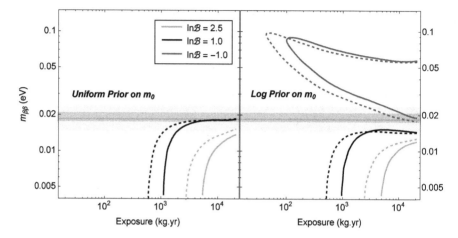

Figure 1: The contour plots for the Bayes factor $\ln \mathcal{B} \equiv \ln \left(\mathcal{Z}_{\rm NO}/\mathcal{Z}_{\rm IO} \right)$ in the plane of the effective neutrino mass and the exposure [10], where the dark and light bands are the 1σ and 3σ regions for $m_{\beta\beta}$. In addition, the thick solid (dashed) curves correspond to the small NME $\mathcal{M}_{0\nu} = 4.6$ (the large NME $\mathcal{M}_{0\nu} = 5.8$).

the weak and moderate evidence should be understood as a degree of belief of 75.0% and 92.3%, respectively. As shown in Fig. 1, even for the future detector with an exposure of 10^4 kg yr and the true value of $m_{\beta\beta} > 15$ meV, namely, above the lower bound in the IO case, it is impossible to significantly distinguish between NO and IO within the $0\nu\beta\beta$ experiments.

Acknowledgments

I thank Professor Alexander Studenikin for kind invitation to the Lomonosov Conference, and Dr. Jue Zhang for enjoyable collaboration.

References

[1] C. Patrignani et al., Chin. Phys. C **40**, 100001 (2016).
[2] E. Majorana, Nuovo Cim. **14**, 171 (1937).
[3] W. H. Furry, Phys. Rev. **56**, 1184 (1939).
[4] M. Goeppert-Mayer, Phys. Rev. **48**, 512 (1935).
[5] J. Schechter and J. W. F. Valle, Phys. Rev. D **25**, 2951 (1982).
[6] M. Duerr, M. Lindner and A. Merle, JHEP **1106**, 091 (2011).
[7] J. H. Liu, J. Zhang and S. Zhou, Phys. Lett. B **760**, 571 (2016).
[8] S. M. Bilenky and C. Giunti, Int. J. Mod. Phys. A **30**, 1530001 (2015).
[9] D. S. Sivia and J. Skilling, *"Data Analysis: a Bayesian Tutorial"* (Oxford University Press, 2006).
[10] J. Zhang and S. Zhou, Phys. Rev. D **93**, 016008 (2016).

STATUS AND PROSPECTS OF THE SEARCH FOR NEUTRINOLESS DOUBLE BETA DECAY OF ^{76}Ge

Karl Tasso Knöpfle for the GERDA collaboration [a]
Max-Planck-Institut für Kernphysik, Heidelberg, Germany

Abstract. This paper presents a review of the search for neutrinoless double beta decay of ^{76}Ge with emphasis on the recent results of the GERDA experiment. It includes an appraisal of fifty years of research on this topic as well as an outlook.

1 Introduction

Fifty years ago Fiorini and collaborators published the first paper on the study of neutrinoless double beta $(0\nu\beta\beta)$ decay of the ^{76}Ge isotope, ^{76}Ge$\rightarrow^{76}$Se$+2e^-$ [1]. Physics motivation was the search for lepton number violation (LNV), technical innovation the use of a Ge(Li) crystal acting both as source and high-resolution detector. With a lead-shielded Ge(Li) coaxial detector of 90 g of natural germanium — containing 7.8% of ^{76}Ge — Fiorini et al. observed $\sim 10^{-2}$ counts /(keV·hr) around the transition energy $Q_{\beta\beta} = 2039$ keV, but no $0\nu\beta\beta$ signal, i.e. a sharp peak at $Q_{\beta\beta}$, and inferred after 712 hours of data accumulation a lower half-life limit of $T^{0\nu}_{1/2} > 3 \cdot 10^{20}$ yr (68% CL). The improvement

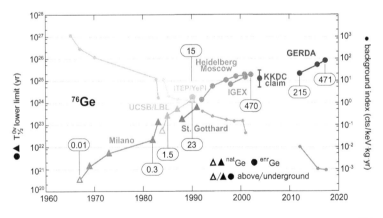

Figure 1: History of $T^{0\nu}_{1/2}$ lower limits (68% CL, 90% CL since 1991) and BIs from ^{76}Ge $0\nu\beta\beta$ studies, see refs. [1–9] and refs. therein. Framed numbers show exposures in mol(^{76}Ge)·yr.

of the $T^{0\nu}_{1/2}$ limit in the following decades by 5 orders of magnitude (Fig. 1) is due to a reduction of the background index (BI) — commonly quoted as count rate at $Q_{\beta\beta}$ normalized to energy bin and detector mass M — and an increase of source strength, or exposure $\mathcal{E} = $M·T, with T being the run time. The

---
[a] E-mail: ktkno@mpi-hd.mpg.de

former became possible by better shielding, including the reduction of cosmic rays by going underground, the latter by progress in the fabrication of larger Ge crystals. Both trends experienced a big boost with the introduction of Ge detectors made from Ge material enriched up to 90% in ^{76}Ge [5].

This article reports mainly on the GERDA experiment which is expected to break the $T^{0\nu}_{1/2} = 10^{26}$ yr frontier in 2018, and on current preparations to reach $T^{0\nu}_{1/2} \geq 10^{27}$ yr sensitivity. These efforts reflect the intensified interest in $0\nu\beta\beta$ decay caused by the discovery that neutrinos have mass, and the theorem that the observation of $0\nu\beta\beta$ decay would establish neutrinos to have a Majorana neutrino mass component, as predicted by various extensions of the Standard Model of particle physics (SM). Among the many models for LNV that of light Majorana neutrino exchange relates $T^{0\nu}_{1/2}$ to an effective Majorana neutrino mass providing access to the absolute neutrino mass scale and its hierarchy [10].

2 Present experiments

The GERDA and MAJORANA collaborations are currently operating a total of 65.3 kg of enriched high purity Ge detectors underground. The main difference of the experiments is the shielding against external radiation, a common feature that they discriminate $0\nu\beta\beta$ decays from background events by their different topology: while the two electrons of a $0\nu\beta\beta$ decay would release their energy within a small detector volume of a few mm^3, background events most likely deposit energy in several locations of the detector, on its surface, or in adjacent detectors. Background events can thus be identified by detector-detector anti-coincidences and, in particular, by the analysis of the time profile of the detector signal. For the latter task, both collaborations have independently developed small read-out electrode high purity Ge detectors, p-type Broad Energy (BEGe [11]) or point contact (PPC) detectors [12], which exhibit not only better pulse shape discrimination (PSD) than the traditional coaxial detectors but also superior energy resolution, < 3 keV full width at half maximum (FWHM) at $Q_{\beta\beta}$. To prevent any bias GERDA has adopted, first in the field, the concept of 'blind analysis': events within $Q_{\beta\beta} \pm 25$ keV are cached until all analysis procedures and cuts are finalized [13].

2.1 The GERDA experiment

The Germanium Detector Array (GERDA) experiment [14] is located at the INFN Laboratori Nazionali del Gran Sasso (LNGS) below a rock overburden of ~3500 m water equivalent. The innovative feature of GERDA is the operation of bare Ge detectors in an ultra-pure cryogenic liquid, liquid argon (LAr), that serves both as cooling and shielding medium. The LAr cryostat is enclosed by a tank filled with ultra-pure water (Fig. 2) as additional shield against external radiation and as medium for a Cherenkov veto system against muons. A clean

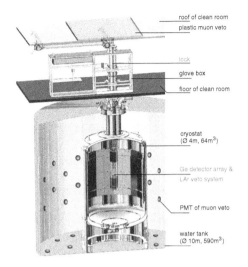 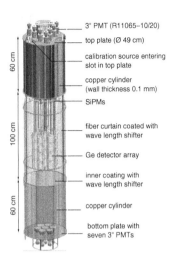

Figure 2: The GERDA experimental setup (left) and the Phase II LAr veto system (right) [15].

room on top of cryostat and water tank houses a glove box and lock for assembly and deployment of the Ge detectors.

In Phase I of GERDA (Nov 11–May 13) the detector array consisted of 15.6 kg of refurbished semi-coaxial Ge detectors from the former HDM [6] and IGEX [7] collaborations and of 3.0 kg of BEGe detectors, all enriched to 86% in ^{76}Ge. The background goal of BI = 10^{-2} cts/(keV·kg·yr) was reached. With $\mathcal{E} = 21.6$ kg·yr no $0\nu\beta\beta$ signal was observed and a new 90% CL limit of $T_{1/2}^{0\nu} > 2.1 \cdot 10^{25}$ yr was set (median sensitivity $2.4 \cdot 10^{25}$ yr) that excluded the claim of observation by part of the HDM collaboration [8] with 99% probability [13].

Phase II of GERDA started in Dec 2015 with the aim to improve the half-life sensitivity beyond 10^{26} yr. To achieve this at the exposure of ~100 kg·yr within reasonable time, e.g. 3 years of running, the detector mass has been doubled by augmenting the enriched BEGe detector mass to 20 kg (30 pcs). Simultaneously, the BI had to be reduced by a factor 10 to 10^{-3} cts/(keV·kg·yr) to stay within the 'background-free' regime, i.e. a mean expected background of <1 in the region of interest (ROI), $Q_{\beta\beta}$ ±0.5·FWHM. This warrants the sensitivity to scale about linearly with exposure $\mathcal{E}$ while at larger backgrounds it scales as $\sqrt{\mathcal{E}}$ (Fig 4, left). The background reduction is achieved by various improvements [15]: by the superior PSD performance of the BEGe detectors, by new low-mass cables and detector mounts of improved radio-purity, and by the instrumentation of the LAr surrounding the detector array with light sensitive sensors. The latter allows the detection of LAr scintillation light using photomultiplier tubes (PMTs) and a fiber curtain read out by SiPMs (Fig. 2, right) thus creating an effective LAr veto system against background events.

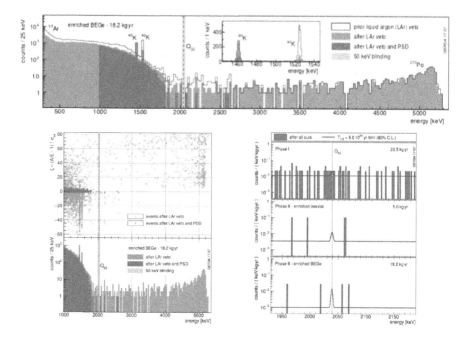

Figure 3: Top: Phase II energy spectra obtained with the BEGe detectors after indicated cuts. The energy region of the ^{40}K and ^{42}K lines is shown enlarged in the inset. The vertical band indicates the blinded region about $Q_{\beta\beta}$. Prominent features are below 500 keV the tail of the ^{39}Ar β spectrum, between 600–1600 keV the broad structure from $2\nu\beta\beta$ decays, individual γ lines between 400–2650 keV, and α structures above 2650 keV, predominantly due to ^{210}Po. Bottom: Scatter plot of the PSD parameter $\zeta = (A/E - 1)/\sigma_{A/E}$ vs energy for all Phase II BEGe detectors, and the corresponding energy spectra after indicated cuts (left). Energy spectra in the analysis window with 2 keV binning obtained in Phase I and II after all cuts. The blue lines show a fit of a constant background and a hypothetical $0\nu\beta\beta$ signal corresponding to the 90% CL limit of $T_{1/2}^{0\nu} > 8 \cdot 10^{25}$ yr [9] (right).

Figure 3 shows the energy spectra obtained with the BEGe detectors after $\mathcal{E} = 18.2$ kg·yr and indicated cuts. In the analysis window between 1930 and 2190 keV, the spectrum is composed of degraded α particles, Compton scattered γ's from ^{208}Tl and ^{214}Bi decays, and ^{42}K β decays at the detector surface. The power of the LAr veto is illustrated by the ^{40}K and ^{42}K lines at 1461 keV and 1525 keV. Following electron capture, the line of the 1461 keV transition in ^{40}Ar is unaffected by the LAr veto since no energy is deposited in the LAr. On the other hand, the 1525 keV transition follows a β decay which deposits up to 2 MeV in the LAr; thus the corresponding line can be suppressed by $\sim 80\%$.

The PSD for BEGe detectors is based on a single parameter determined by the ratio A/E where A is the maximum of the detector current pulse and E the total energy [11]. Figure 3 also shows a scatter plot of the PSD parameter $\zeta = (A/E - 1)/\sigma_{A/E}$ and its projection on the energy scale. Accepted $\beta\beta$

candidates at $\xi = 0$, mostly $2\nu\beta\beta$ decays, are shown in red; their survival fraction is $(85 \pm 2)\%$. The two K peaks and Compton scattered events located at $\xi < 0$ are easily cut, like the α events at $\xi > 0$.

After all cuts a BI of $1.0^{+0.6}_{-0.4} \cdot 10^{-3}$ cts/(keV·kg·yr) is deduced confirming a former result from lower exposure [16]. GERDA is thus the first experiment in the field that will stay background-free up to its design exposure. Assuming in the analysis window a flat background and a Gaussian signal at $Q_{\beta\beta}$ (and excluding 10 keV wide intervals around two known γ lines), a combined unbinned profile likelihood fit to the Phase I and II spectra after all cuts (Fig. 3, bottom right) yields a half-life limit for $0\nu\beta\beta$ decay of $T^{0\nu}_{1/2} > 8 \cdot 10^{25}$ yr (90% CL, med. sensitivity $5.8 \cdot 10^{25}$ yr). For light Majorana neutrino exchange this limit converts to an upper limit of the effective Majorana neutrino mass of 0.12 - 0.26 eV using $g_A = 1.27$ and nuclear matrix elements (NMEs) from 2.8–6.1 [9].

2.2 The MAJORANA DEMONSTRATOR

The MAJORANA DEMONSTRATOR (MJD) [17] is operating at the Sanford Underground Research Facility, South Dakota. Its goal is to prove the design for a 1 ton-scale experiment as to background level and modular design. The MJD contains 35 PPC detectors (29.7 kg) enriched up to 88% in ^{76}Ge. The detectors are mounted in two vacuum cryostats within a traditional graded passive Cu/Pb shield with the inner layer consisting of ultraclean electroformed copper. The full detector array is running since August 2016.

With 2.5(1) keV FWHM resolution at $Q_{\beta\beta}$, the MJD has achieved the best energy resolution of any $0\nu\beta\beta$ decay experiment [18], and its sub-keV threshold allowed to perform sensitive tests of physics beyond the SM [19]. The lowest background runs ($\mathcal{E} = 5.24$ kg·yr) yield a BI of $1.6^{+1.2}_{-1.0} \cdot 10^{-3}$ cts/(keV·kg·yr), being compatible with the projected background level of 3 cts/(ROI·t·yr) based on the material assay. At $\mathcal{E} = 10$ kg·yr no $0\nu\beta\beta$ signal candidate has been observed which implies a lower limit of $T^{0\nu}_{1/2} > 1.9 \cdot 10^{25}$ yr (90% CL) [18].

3 The next generation of experiments

Already in 2005 the GERDA and MAJORANA collaborations recognized that the ton-scale experiment which is needed to cover the inverted mass hierarchy would call for a world-wide effort. Hence they signed a MoU on open exchange of knowledge and technologies and declared their intentions to merge for a ton-scale experiment combining the best features of both experiments. The recently formed LEGEND (Large Enriched Germanium Experiment for Neutrinoless Double Beta Decay) collaboration is the result of these joint efforts including new members. It proposes a staged approach [20]: the operation of up to 200 kg of detectors in the existing GERDA infrastructure at LNGS, and in a next step the use of 1000 kg of detectors in a new installation at a location to be still determined. Fig. 4 shows the envisaged $T^{0\nu}_{1/2}$ limit setting sensitivities.

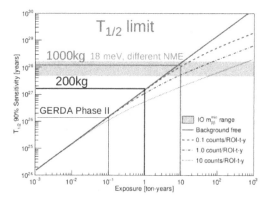 

Figure 4: Left: $T_{1/2}^{0\nu}$ sensitivities (90% CL) of GERDA Phase II and LEGEND-200/1000 vs exposure; the horizontal band indicates the lower bound of the inverted neutrino mass ordering (IO) [21]. Right: Tentative baseline design of a cryostat for LEGEND-1000 [20].

For LEGEND-200 the GERDA lock and cryostat piping will be modified to accommodate up to 200 kg of detectors. The major challenge will be the reduction of background to 0.6 cts/(FWHM·t·yr), i.e. by a factor 5 compared to GERDA; this is needed to stay 'background-free' up to the design exposure of 1 t·yr where the sensitivity of 10^{27} yr is reached. Envisaged improvements include: (i) the use of PPC detectors of larger mass, 1.5 - 2 kg, which becomes possible with the novel inverted coaxial PPC detector type [22]; one of these devices will substitute 2-3 BEGe detectors, keeping the excellent PSD properties while reducing the number of detector holders, cables and electronic channels by the same factor; (ii) better LAr scintillation light collection with a denser fiber curtain for increased ^{214}Bi rejection; (iii) a new design of the electronic readout chain for better noise reduction such that PSD will become more effective; and (iv) the overall reduction of radio-impurities close to the detector array by using low-mass MAJORANA DEMONSTRATOR style components. Preparations for LEGEND-200 are in progress: the PSD power of prototype inverted coaxial detectors has been verified [23], the purchase of Ge material enriched in ^{76}Ge has started, and the cryostat piping has been redesigned.

The LEGEND-1000 goal to stay 'background-free' for 10^{28} yr sensitivity up to the exposure of 10 t·yr requires a BI of less than 0.1 cts/(FWHM·t·yr), a factor of 30 lower than achieved in GERDA. Several options for reaching this goal are under study, and the experience gained with LEGEND-200 will help thereby. An initial baseline design (Fig. 4, right) shows the main cryostat volume separated by thin copper walls from 4 smaller volumes of $\sim 3\,\mathrm{m}^3$ each. Each volume carries on top a shutter and lock, like in GERDA, such that 4 payloads with up to 250 kg detectors can be deployed separately. The $3\,\mathrm{m}^3$ volumes might be filled with depleted LAr eliminating the background due to

^{42}Ar/^{42}K β decays. An important R&D effort will be the minimization of all construction materials which do not scintillate and hence cannot be used to veto background events.

4 Conclusion

The GERDA Phase II upgrade has achieved the desired background goal of 10^{-3} cts/(keV·kg·yr). GERDA will thus run 'background-free' up to its design exposure of 100 kg·yr reaching in 2018 a $T^{0\nu}_{1/2}$ sensitivity beyond 10^{26} yr. The MAJORANA DEMONSTRATOR is expected to exhibit a similar performance. ^{76}Ge experiments have shown the best energy resolution and the lowest background in the ROI of any isotope for $0\nu\beta\beta$ searches [24]. This is motivation for future experiments with 200 kg of ^{76}Ge and more. The newly formed LEGEND collaboration plans a ton-scale ^{76}Ge experiment for probing half-lifes $T^{0\nu}_{1/2}$ up to 10^{28} yr. Preparations for the first stage, LEGEND-200, are in progress.

References

[1] E. Fiorini et al., Phys. Lett. **25B** (1967) 602.
[2] E. Bellotti et al. (Milano), Phys. Lett. **146B** (1984) 450.
[3] D.O. Caldwell et al., Nucl. Phys. **B 13** (1990) 547.
[4] D. Reusser et al. (St. Gotthard), Phys.Rev. **D 45** (1992) 2548.
[5] A.A. Vasenko et al., Mod. Phys. Lett. **5** (1990) 1299.
[6] H.V. Klapdor-Kleingrothaus (HDM), Eur. Phys. J. **A 12** (2001) 147.
[7] C.E. Aalseth et al. (IGEX), Phys. Rev. **D 65** (2002) 198.
[8] H.V. Klapdor-Kleingrothaus et al. (KKDC), Phys. Lett. **B 586** (2004) 198.
[9] M. Agostini et al. (GERDA), TAUP 2017 Proc., preprint arXiv:1710.07776.
[10] H. Päs, W. Rodejohan, New J. Phys. **17** (2015) 115010 and refs. therein.
[11] M. Agostini et al. (GERDA), Eur. Phys. J. **C 75** (2015) 39.
[12] P.S. Barbeau, J.I. Collar, O. Tench, JCAP **2007** (2007) 009.
[13] M. Agostini et al. (GERDA), Phys. Rev. Lett.**111** (2013) 122503.
[14] K.-H. Ackermann et al. (GERDA), Eur. Phys. J. **C 73** (2013) 2330.
[15] M. Agostini et al. (GERDA), submitted to EPJC, arXiv:1711.01452.
[16] M. Agostini et al. (GERDA), Nature **544** (2017) 47.
[17] N. Abgrall et al. (MAJORANA), Adv. High Energy Phys. (2014) 365432.
[18] C.E. Aalseth et al. (MAJORANA), preprint arXiv:1710.11608.
[19] N. Abgrall et al. (MAJORANA), Phys. Rev. Lett. **118** (2017) 161801.
[20] N. Abgrall et al. (LEGEND), AIP Conf. Proc. **1894** (2017) 020027.
[21] J. Detwiler, private communication.
[22] R.J. Cooper et al., Nucl. Instr. Meth. **A 665** (2011) 25.
[23] A. Domula et al., to appear in Nucl. Instr. Meth., arXiv:1711.01433.
[24] M. Agostini, G. Benato, J.A. Detwiler, Phys. Rev. **D 96** (2017) 053001.

THE CUORE EXPERIMENT AT LNGS

Davide Chiesa [a] on behalf of the CUORE collaboration
Dipartimento di Fisica, Università di Milano-Bicocca, Milano I-20126 - Italy
INFN - Sezione di Milano Bicocca, Milano I-20126 - Italy

Abstract. The Cryogenic Underground Observatory for Rare Events (CUORE) is the first bolometric experiment searching for neutrinoless double beta decay that has been able to reach the 1-ton scale. The detector consists of an array of 988 TeO_2 crystals arranged in a cylindrical compact structure of 19 towers. The construction of the experiment and, in particular, the installation of all towers in the cryostat was completed in August 2016, followed by the cooldown to base temperature at the beginning of 2017. The CUORE detector is now operational and has been taking science data since Spring 2017. We present here the initial performance of the detector and the preliminary results from the first detector run.

CUORE (Cryogenic Underground Observatory for Rare Events) is an experiment that searches the neutrinoless double beta decay ($0\nu\beta\beta$) of ^{130}Te [1]. This nuclear transition, if observed, would imply that lepton number is not conserved and that neutrinos are Majorana particles [2].

The CUORE detector (Fig. 1) is composed by 988 TeO_2 bolometers operated at a temperature of ~ 10 mK in a cryostat installed at *Laboratori Nazionali del Gran Sasso* (LNGS) in Italy. Each bolometer is a cubic TeO_2 crystal with 5 cm side and 750 g average mass, equipped with a neutron transmutation doped (NTD) thermistor and a silicon heather. The TeO_2 crystals are made with natural tellurium, that includes ^{130}Te with 34.2% isotopic abundance. The 988 bolometers are arranged in a closely packed array of 19 towers, each consisting of 13 floors of 4 crystals. With a total detector mass around 740 kg of TeO_2 (206 kg of ^{130}Te), CUORE is the first ton-scale cryogenic detector for the search of $0\nu\beta\beta$ decay.

The CUORE detector measures the energy released inside the bolometers by particle interaction. The experimental signature of the $0\nu\beta\beta$ decay is a peak of events at the Q-value of the transition ($Q_{\beta\beta}$=2528 keV for ^{130}Te). Since the $0\nu\beta\beta$ decay is an extremely rare decay, the challenge of CUORE is to maximize its experimental sensitivity. For this purpose, the CUORE detector is designed to reach good energy resolution and low background rate in the $0\nu\beta\beta$ Region of Interest (ROI). With an energy resolution of 5 keV at $Q_{\beta\beta}$ and a background rate of 10^{-2}counts/(keV·kg·y) in the ROI, CUORE is expected to reach an experimental sensitivity of $T^{0\nu}_{1/2} > 9 \times 10^{25}$yr (90% C.L.) with 5 yr of live time [3].

The materials used to build the CUORE detector and its cryostat were selected with the aim of minimizing the radioactive contaminations that can

---

[a] E-mail: davide.chiesa@mib.infn.it

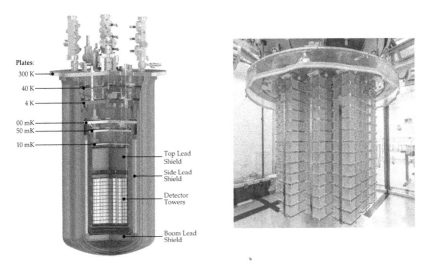

Figure 1: An illustration of the CUORE cryostat (left) and a picture of the CUORE detector (right).

produce background events in the ROI. Moreover, the detector components were produced, handled and cleaned according to specific protocols, and the detector installation was performed in a radon free environment to avoid any possible recontamination. The background due to muons is strongly suppressed thanks to the experiment location at LNGS, with ∼1400 of overlying rock. The environmental background due to external γ-rays and neutrons is shielded by different layers of copper, lead, borated polyethylene and boric acid, that are included in the cryostat itself or surround it. To evaluate the expected background in the ROI, we developed a detailed Monte Carlo simulation that exploits the information about the contaminations of materials obtained through radio-assay screening campaigns and bolometric measurements [4, 5], proving that the goal of a background rate of 10^{-2} counts/(keV·kg·y) is within the reach.

The installation of the 19 towers was successfully completed in Summer 2016, obtaining 984 functioning bolometers out of 988. The cryostat interfaces and radiation shields were assembled in the following months. At the beginning of 2017, we started the detector pre-operation phase, to optimize the signal readout and the working points of the bolometers. In May 2017, we collected three weeks of physics data bracketed by two calibration periods. The detector stability has been much improved compared to Cuoricino and CUORE-0 predecessor experiments [6, 7] and, for the first analysis, we acquired an exposure of 38.1 kg·yr of TeO_2 (10.6 kg·yr of ^{130}Te). During calibration measurements, 12 strings populated with ^{232}Th sources were temporarily deployed inside the detector region. Six γ-lines of ^{232}Th decay chain (from 239 keV to 2615 keV) are then used to perform the energy calibration of the detected pulses.

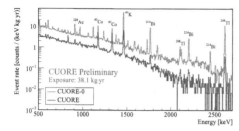 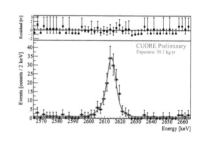

Figure 2: Left: comparison of physics spectra in the gamma region measured with CUORE and CUORE-0, with prominent γ-lines labelled. Right: detail of the 2615 keV line.

The physics spectrum (Fig. 2, left) is built after applying a series of selection criteria aimed at improving the experimental sensitivity. First, we remove periods of low quality data — caused by noisy laboratory conditions — and we reject pile-up events. Then, we select only signals consistent with a proper template waveform (pulse shape analysis) in order to identify real particle events. Finally, we exclude events that simultaneously trigger more than one crystal, to reduce the background due to events depositing energy in multiple crystals. We evaluate an overall detection efficiency of (55.3±3.0)%, which includes the (88.35±0.09)% probability that a $0\nu\beta\beta$ decay is fully contained in a single crystal and the (62.6±3.4)% probability that a physics event is not discarded when the selection criteria are applied.

We evaluate the detector energy resolution near the ROI by fitting the 2615 keV line in the physics spectrum. The armonic mean of the detector FWHM resolutions is 7.9±0.6 keV (Fig. 2, right).

To estimate the background in the ROI and the $0\nu\beta\beta$ decay rate ($\Gamma_{0\nu}$), we perform an Unbinned Extended Maximum Likelihood fit in the [2465–2575] keV range (Fig. 3, left) with the same procedure used for CUORE-0 [8]. The best-fit values are $0.98^{+0.17}_{-0.15} \times 10^{-2}$counts/(keV·kg·y) for the background rate, and $(-0.03^{+0.07}_{-0.04}(\text{stat.})\pm 0.01(\text{syst.}))\times 10^{-24}\text{yr}^{-1}$ for $\Gamma_{0\nu}$. We find no evidence for the $0\nu\beta\beta$ decay of ^{130}Te and we can only calculate an upper limit of $\Gamma_{0\nu}$, by integrating the profile likelihood in the physical region ($\Gamma_{0\nu} \geq 0$). This corresponds to a half-life lower limit of $T^{0\nu}_{1/2} > 4.5 \times 10^{24}$yr (90% C.L.).

Finally, we combine the first results from CUORE with those obtained from CUORE-0 [7] and Cuoricino [6] with 9.8 kg·yr and 19.8 kg·yr exposure of ^{130}Te, respectively (Fig. 3, right). The half-life lower limit obtained by combining the profile negative-log-likelihood curves of the three experiments is $T^{0\nu}_{1/2} >$ 6.1 × 10^{24}yr (90% C.L.).

In summary, CUORE is the first ton-scale cryogenic detector array in operation and with three week of physics data we were able to set the most stringent limit on the $0\nu\beta\beta$ half-life of ^{130}Te, surpassing the previous one obtained by combining the results from CUORE-0 and Cuoricino experiments. From the

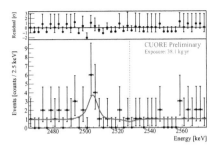 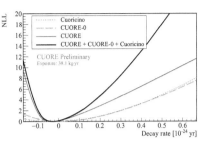

Figure 3: Left: Fit of the CUORE spectrum in the ROI. The peak near 2506 keV is attributed to ^{60}Co. Right: Profile negative-log-likelihood curves for CUORE, CUORE-0, Cuoricino, and their combination.

analysis of the first dataset, we got important information about detector performances, that allow to confirm the expected sensitivity of CUORE for the search of the $0\nu\beta\beta$ decay of ^{130}Te.
Note: A new science run was carried out during August 2017. The new dataset and the one described in this paper were re-processed with a slightly improved analysis procedure. The results were submitted for publication to PRL [9].

Acknowledgments

The CUORE Collaboration thanks the directors and staff of the Laboratori Nazionali del Gran Sasso and the technical staff of our laboratories. This work was supported by the Istituto Nazionale di Fisica Nucleare (INFN); the National Science Foundation; the Alfred P. Sloan Foundation; the University of Wisconsin Foundation; and Yale University. This material is also based upon work supported by the US Department of Energy (DOE) Office of Science; and by the DOE Office of Science, Office of Nuclear Physics. This research used resources of the National Energy Research Scientific Computing Center (NERSC).

References

[1] D. R. Artusa *et al.*, Advances in High Energy Physics **2015** (2015) 879871.
[2] E. Majorana, Nuovo Cimento **14** (1937) 171.
[3] C. Alduino *et al.*, European Physical Journal **C 77** (2017) 532.
[4] C. Alduino *et al.*, European Physical Journal **C 77** (2017) 543.
[5] C. Alduino *et al.*, European Physical Journal **C 77** (2017) 13.
[6] E. Andreotti *et al.*, Astropart. Phys. **34** (2011) 822.
[7] K. Alfonso *et al.*, Phys. Rev. Lett. **115** (2015) 102502.
[8] C. Alduino *et al.*, Phys. Rev. **C 93** (2016) 045503.
[9] C. Alduino *et al.*, arxiv:1710.07988 (2017), submitted to Phys. Rev. Lett.

PROBING THE MAJORANA NEUTRINO NATURE AT THE CURRENT AND FUTURE COLLIDERS

Rachik Soualah [a]
Department of Applied Physics and Astronomy, University of Sharjah, UAE
The Abdus Salam International Centre for Theoretical Physics, Strada Costiera 11,
I-34014, Trieste, Italy
INFN, Sezione di Trieste, Gruppo collegato di Udine, Italy
Salah Nasri [b]
Department of physics, United Arab Emirates University, Al-Ain, UAE
Safa Naseem [c]
Department of Applied Physics and Astronomy, University of Sharjah, UAE

Abstract. We present the phenomenology of class of neutrino mass models as an extension of the Standard Model (SM) with charged singlet scalars and three right handed neutrinos. Probing this class of radiative neutrino mass models provides simultaneous explanations for neutrino masses and Dark Matter. The neutrino mass in these models is generated radiatively at three-loop, the lightest right handed neutrino is a good dark matter candidate, and the electroweak phase transition strongly first order as required for baryogenesis. In the presented works, comprehensive physics analyses at lepton-lepton, lepton-hadron as well as proton-proton colliders (such as the ILC, FCC-ee, FCC-eh, LHC and FCC-hh) were carried-out for detecting any possible significant signatures of one of the most striking extensions of the SM that can shed some light on Dark Matter (DM) nature. Along this line, we study different physics processes like: $e^-e^-(e^+) \to \ell_\alpha \ell_\beta + \not{E}$ ($\alpha, \beta = e, \mu$) and $e^-p \to \ell^\pm \ell^\pm \ell^\mp + \not{E} + jet$ ($\ell = e, \mu$) at center of mass energies $\sqrt{s}$=250 GeV, 350 GeV, 500 GeV, 1 TeV in addition to $pp \to \ell^\pm \ell^\mp + \not{E}$ with 8 and 14 TeV LHC energies.

1 Introduction

The Standard Model of particle physics is very successful in describing nature around the electroweak scale. However, many questions remained unanswered such as neutrino masses and mixing, the nature of Dark Matter, and the origin of baryon asymmetry of the universe. There is no unique way to introduce the neutrino masses in the SM. Therefore many extensions beyond the Standard Model were introduced to tackle these questions. Among them, one of the well motivated models KNT model [1]; which has been generalised later on where the singlets Z_2 odd singlet charged S_2 and the RH Majorana neutrinos N_i are promoted to triplets, quintuplet, septuplets, and in a scale invariant framework [2]. The phenomenology of these models has been investigated in [4] and [6]. This paper is providing an overview of our previous studies of any possible physics signatures that can be produced from the neutrino sector at ILC, LHC and FCC.

---

[a] E-mail: rsoualah@sharjah.ac.ae
[b] E-mail: snasri@uaeu.ac.ae
[c] E-mail: U00039313@sharjah.ac.ae

2 The model and its phenomenological studies

Within this class of models, the SM was extended with two electrically charged singlet fields under $SU(2)_L$ scalars and three right-handed (RH) neutrino, N_i where a Z_2 symmetry was imposed to forbid the Dirac neutrino mass terms at tree-level. Once the electroweak symmetry is broken, neutrino masses are generated at three loops, and the lightest RH neutrino, N_1, could be a candidate for dark matter. Some generalisations of this model have been proposed where one of the two singlet charged scalars and N_i are replaced by $SU(2)_L$ triplet, or by $SU(2)_L$ quintuplet fields. The scalar spectrum of this model has a pair of charged scalar particles which can be produced at colliders via their couplings to photon and the SM higgs boson. The signatures of charged scalars as well as other multi-leptons topologies at the LHC and the future high energy colliders has been studied and given more interests by many authors. The full theoretical discussion and its details can be found at [4, 6] where it was shown that this class of models should satisfy a set of constraints before selecting any considered benchmark [4, 5]. After that we perform many studies by taking into account the considered benchmark. Fig. 1 illustrates the significance versus luminosity for the $e^-e^+ \to e\mu$ process, without and with polarised beams for the considered CM energies. Clear signal enhancement can be seen when using polarised beams. Another physics signal has been investigated at the LHC-pp collisions. After studying the signal to background and taking into account the kinematical cuts, the production cross section of S^+S^- pair production at $\sqrt{s}=$ 8 TeV and 14 TeV vs the charged scalar mass M_S has been obtained (Fig. 2) in addition to the significance vs the luminosity at $\sqrt{s}=$ 8 TeV and 14 TeV for the different signature topologies [4]. Table 1 provides the signal and background cross section yields for S_{100} [6]. In addition, we investigate another promising channel $e^-e^- \to \ell_\alpha \ell_\beta + \not{E}$ $(\alpha, \beta = e, \mu)$ [7] (see Fig. 3 for

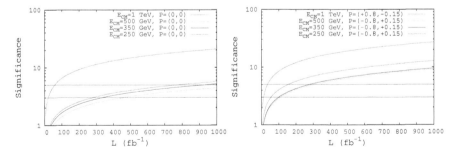

Figure 1: The significance versus luminosity for the $e^-e^+ \to e\mu$ process, without (left) and with (right) polarised beams for the considered CM energies.

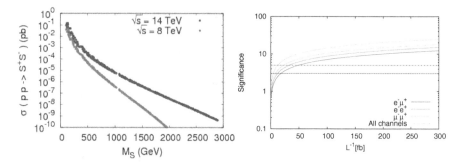

Figure 2: The production σ (in fb units) of S^+S^- pair production at $\sqrt{s}=$ 8 TeV (in magenta) and 14 TeV (in blue) vs the charged scalar mass M_S (left). The production of σ (in fb units) The significance vs the luminosity at $\sqrt{(s)}=$ 14 TeV (right) for the different signatures. The horizontal dashed lines represents the S=3 and S=5 significance values, respectively (right).

Table 1: The cross section of the total expected signals (σ^M) and the corresponding background (σ^B) are used to estimate the significance S_{100} at 14 TeV) with the recorded integrated luminosity value $L = 100 \ fb^{-1}$.

Process	$\sigma^M(fb)$	$\sigma^B(fb)$	$(\sigma^M - \sigma^B)/\sigma^B$	S_{100}
$pp \to e^-\mu^+ + \not{E}$	1.253	0.459	1.7	7.093
$pp \to e^-e^+ + \not{E}$	44.45	38.65	0.150	8.699
$pp \to \mu^-\mu^+ + \not{E}$	65.27	56.86	0.148	10.409

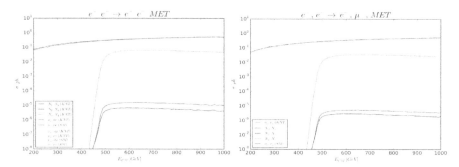

Figure 3: The production of σ (in fb units) The significance vs the luminosity at $\sqrt{(s)}=$ 14 TeV (right) for the different signatures. The horizontal dashed lines represents the S = 3 and S = 5 significance values, respectively.

the production cross section) as well $e^-p \to \ell^\pm \ell^\pm \ell^\mp + \displaystyle{\not}E + jet$ ($\ell = e, \mu$) via the production of singlet charged scalar $S^\pm$ at the expected FCC-ee, FC-eh, LHeC and HL-LHC different collider designs that would very possibly enable higher luminosity values.

3 Conclusions

We have investigated the possibility of testing the singlet charged scalars effect as well as multi-lepton final states in a class of neutrino mass models that could probe directly or indirectly the Majorana neutrino nature at e^-e^-, e^-e^+, e^-p and pp colliders. Very promising results are expected to tackle the neutrino sector in the hope to understand more the neutrino masses and Dark Matter.

Acknowledgments

We would like to thank the organizers of the 18th Lomonosov Conference on Elementary Particle Physics for their invitation. S. Naseem would like to thank the Center of Advanced Materials Research (University of Sharjah) for supporting her work at earlier stages.

References

[1] L.M. Krauss, S. Nasri and M. Trodden, Phys. Rev. D 67, 085002 (2003).
[2] A. Ahriche, C. S. Chen, K. L. McDonald and S. Nasri, Phys. Rev. D 90 (2014) 015024 [arXiv:1404.2696 [hep-ph]]. A. Ahriche, K. L. McDonald and S. Nasri, JHEP 1410, 167 (2014) [arXiv:1404.5917 [hep-ph]]. A. Ahriche, K. L. McDonald, S. Nasri and T. Toma, Phys. Lett. B 746, 430 (2015) [arXiv:1504.05755 [hep-ph]]. A. Ahriche, K. L. McDonald and S. Nasri, arXiv:1508.02607 [hep-ph].
[3] A. Ahriche and S. Nasri, JCAP 1307, 035 (2013) [arXiv:1304.2055].
[4] A. Ahriche, S. Nasri and R. Soualah, Phys. Rev. D 89, 095010 (2014) [arXiv:1403.5694 [hep-ph]].
[5] A. Ahriche, K. L. McDonald and S. Nasri, Phys. Rev. D 92, 095020 (2015) [arXiv:1508.05881 [hep-ph]].
[6] C. Guella, D. Cherigui, A. Ahriche, S. Nasri, R. Soualah, Phys. Rev. D 93, 095022 (2016) [arXiv:1601.04342 [hep-ph]].
[7] R. Soualah, S. Nasri and S. Naseem, *ongoing work*.

STUDY OF THE ν_μ CHARGED CURRENT QUASIELASTIC-LIKE INTERACTIONS IN THE NOvA NEAR DETECTOR

Luchuk Stanislav
Institute for Nuclear Research, Moscow, Russia

Abstract. We report on the status of the measurement of the semi-inclusive charged current quasielastic-like (CCQE-like) cross-section of ν_μ interactions in the NOvA near detector. As a first step, we have developed selection criteria, including a multivariate selector, for events with one reconstructed track in the final state. We present the efficiency and purity of this selection based on studies with simulated neutrino interactions in the NOvA near detector.

1 NOvA experiment

The NOvA experiment is a long-baseline neutrino oscillation experiment. NOvA is designed to determine Θ_{23} octant, the neutrino mass hierarchy and find is there CP violation in the lepton sector. The experiment uses two detectors (Near and Far) which are functionally identical. Both detectors that are at distance of 810 km between each other are placed 14.8 milliradians off the NuMI beam axis. The Near Detector is 3.9 m × 3.9 m × 16 m in size and made of horizontally and vertically oriented 3.8 m × 3.9 cm × 6.6 cm PVC cells filled with liquid scintillator [1]. Each cell contain wave length shifting fibre for light collection produced by charged particles. Fully active and fine grained NOvA Near Detector allows to reconstruct calorimetric neutrino energy with good accuracy [2] and study neutrino interactions with high precision. Off-axis beam has a narrow energy range with a peak at 2 GeV. In this energy range(0.5-3.5 GeV) quasielastic(QE) $\nu_\mu + A \to \mu + p + B$, meson exchange current(MEC) $\nu_\mu + A \to \mu + 2N + B$ and resonance(RES) $\nu_\mu + A \to \mu + \pi + p + B$ processes are dominated. The signature of these interactions is one, two or three reconstructed tracks. We study CCQE-like interactions, selecting events without pion in the final states(so called CC0pi events). The topology of CCQE-like events includes one, two or three reconstructed non pion tracks. The CCQE interactions, as well as, MEC and RES interactions give the main contribution to this type of events. The resonance interactions give contribution to CCQE-like events when pion is absorbed in nuclei. In this paper we regard the topology with only one muon track events.

2 Signal selection

Analysis of the cross-section requires correct reconstruction of muon and neutrino energies. Therefore we select reconstructed muon track that is fully

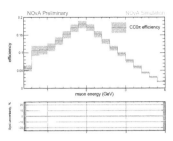 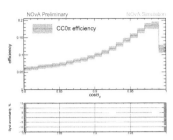

Figure 1: The efficiency of CCQE-like event selection as function of measured kinematic variables: muon energy and muon angle.

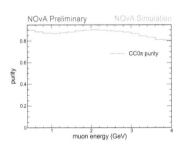 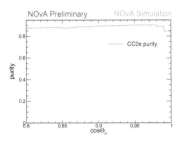

Figure 2: The purity of CCQE-like event selection as function of measured kinematic variables: muon energy and muon angle.

contained in the detector, applying containment cut. To exclude the contamination of muons produced by neutrino interactions outside the Near detector and determine the detector fiducial volume we use the fiducial volume cut. To reject the events with pion in the final state we apply the Michel electrons cut. The identification of the muon track is based on the multivariate analysis which uses particle energy loses, particle scattering, track length and hadronic contamination of the track. For selection of CCQE-like events we apply multivariate analysis with following variables:

- visible energy outside of muon track

- difference between two reconstructed neutrino energies E_ν^{cal} and E^{QE}:
  a) reconstructed neutrino energy: $E_\nu^{cal} = E_\mu + E_{Had}$, where E_μ and E_{Had} are muon and hadronic energy correspondingly
  b) kinematic CCQE neutrino energy: $E_\nu^{QE} = [mE_b - E_b/2 - m_\mu^2/2 - E_\mu(E_b - m)]/(p_\mu cos\Theta - E_\mu + m - E_b)$

Applying this method we selected about 600000 CCQE-like events produced in the Near detector.

The selection quality of one track CCQE-like events is characterized by two main figures: efficiency and purity. The efficiency is determined as ratio of the number of selected true signal events to the number of true signal events in the fiducial volume. The purity is determined as ratio of the number of selected true signal events to the number of all selected events. The preliminary systematic uncertainties of the efficiency due to simulation of the neutrino interactions are estimated. Preliminary results are following: the averaged efficiency is about $0.1142^{+0.0081}_{-0.0098}$ and the averaged purity is about 0.8858. The efficiency and purity of CCQE-like event selection as function of muon energy and muon angle are shown in Figs. 1 and 2. The purity depends weakly as a function of muon energy and muon angle.

3 Conclusion

Fully active and fine grained NOvA Near Detector allows to reconstruct calorimetric neutrino energy with good accuracy and study neutrino interactions with high precision. The applying of the calorimetric hadronic energy in the multivariate algorithm allows to select CCQE-like event with high purity. The preliminary CCQE-like selection purity is about 89%, preliminary efficiency is about 11%.

Acknowledgments

We are grateful for the help and contribution of the NOvA collaboration.

References

[1] The NOvA Technical Design Report - NOvA Collaboration (Ayres, D.S. et al.) FERMILAB-DESIGN-2007-01.
[2] P. Adamson et al., Phys. Rev. Lett. **118** (2017) 151802.

SEARCH FOR SUPERNOVA NEUTRINO BURSTS WITH THE LVD EXPERIMENT

N.Yu. Agafonova [a] and V.V. Ashikhmin [b] on behalf of the LVD Collaboration
Institute for Nuclear Research of RAS, 117312, Moscow

Abstract. The Large Volume Detector in the LNGS is a 1 kton liquid scintillator neutrino observatory well suited to study low energy neutrinos from gravitational stellar collapses. The detector is sensitive to core-collapse supernovae via neutrino burst detection with 100 % efficiency over the Galaxy. We establish the following upper limit on the rate of core collapse and failed supernova explosions out to distances of 25 kpc: 0.092 yr^{-1} at 90% c.l.

1 Introduction

The detection of neutrinos from the optically bright supernova in the Large Magellanic Cloud, SN1987A led to important inferences on the physics of core collapse supernovae. Two scintillation detectors, LSD [1] and BUST [2], and two Cherenkov detectors, Kamiokande-II [3] and IMB [4], were operational on February 23, 1987. The importance of the discovery and the established connection between neutrino observation and the most powerful events occurring in the Galaxy called for a new generation of larger detectors.

We present the results of the search for supernova neutrino bursts in data taken by the Large Volume Detector [5] in 25 years of operation. LVD is a large-mass long-term neutrino experiment located in the INFN Gran Sasso National Laboratory in Italy.

2 The Large Volume Detector

The Large Volume Detector (LVD), operating since June 1992, is located underground at a depth of 3600 m w.e. The experiment consists of an array of 840 scintillator counters, 1.5 m^3 each, viewed from the top by three photomultipliers (PMTs) and arranged in a modular geometry [6]. This modularity allows LVD to achieve a very high duty cycle, that is essential in the search of unpredictable sporadic events. On final configuration the LVD has about 1000 ton of scintillator $C_n H_{2n}$, n=9.6 and 1000 ton of iron.

Neutrinos can be detected in LVD through charged current and neutral current interactions on proton, Carbon nuclei and electrons of the liquid scintillator as well as the Iron nuclei, which is included in the design of counters, portatanks and bearing support of the set-up. Given the relevance of the IBD reaction, the trigger has been optimized for the detection of both products of

---

[a] E-mail: Agafonova@inr.ru
[b] E-mail: vole4ka86@mail.ru

this interaction, namely the positron and the neutron. Each PMT is thus discriminated at two different threshold levels, the higher one, $E_H = 4$ MeV (from 2006 up to now), being also the main trigger condition for the detector array. The lower one ($E_L = 0.5$ MeV) is in turn active only in a 1 ms time-window following the trigger, allowing the detection of (n, p)-captures. One millisecond after the trigger, all memory buffers are read out, independently in the three towers. The mean trigger rate is $\sim 0.013\ s^{-1}t^{-1}$.

3 Event Selection

The method used in LVD to search for neutrino bursts from gravitational stellar collapses essentially consists in searching, in the time series of single counter signals (events), for a sequence (cluster) whose probability of being simulated by fluctuations of the counting rate is very low.

The higher the event frequency, the higher is the probability of a "background-cluster", given by accidental coincidences. At the trigger level, the bulk of events in LVD is due to natural radioactivity products both from the rock surrounding the detector and from the material that constitutes the detector itself and to atmospheric muons. The set of cuts aims at reducing such a background while isolating signals potentially due to neutrinos. The first condition functions as a filter to remove events triggered in malfunctioning counters. The second and third conditions reject cosmic-ray muons and most of the radioactive background. The fourth one refines the rejection of defective counters, through the analysis of the time series of the events. As we will show below, after the background reduction, the counting rate is decreased by a factor of about 400.

4 Search for Neutrino Bursts

The aim is the search for significant clusters of events that could be indicative of neutrino bursts. To search for supernova neutrino bursts, we analyze the time series of the selected events, i.e. triggers in the 10-100 MeV energy range ($f_{bk} = 0.03$Hz), and look for clusters. While to provide the SNEWS, the on-line network of running neutrino detectors [6], with a prompt alert we use in the burst search method (on-line mode) a fixed time window (20 s) [7], in this analysis (off-line mode) we consider different burst durations up to 100s as discussed in detail in [8]. As discussed in [7], [8], the latter method is less model dependent than the former at a cost of a more complex procedure, which is not feasible on-line when the clusters selection has to be quite fast.

In both cases the selection is essentially a two-step process. In the first step, we analyze the entire time series to search for clusters of events. The rationale of the search is that every nth event could be the first of a possible neutrino burst. As we do not know *a priori* the duration of the burst, we consider all clusters formed by the nth event and its successive ones. The duration of each cluster is given by the time difference δt between the first event and the last one of each sequence. The advantage of the described analysis is that it is unbiased with respect to the duration of the possible neutrino burst. The second step of the process consists in determining if one or more among the detected clusters are neutrino bursts candidates. To this aim, we associate to each of them (characterised by multiplicity m_i and duration δt_i) a quantity that we call imitation frequency F^i_{im}. This represents the frequency with which background fluctuations can produce, by chance, clusters with multiplicity m mi and duration δt_i. As shown in [8], this quantity, which depends on $(m_i, \delta t_i)$, on the instantaneous background frequency, f^i_{bk} and on the maximum cluster duration chosen for the analysis, δt_{max} (100 s), can be written as: $F^i_{im} = f^2_{bk} \delta t_{max} \sum_{k>m_i-2} P(k, f^i_{bk}\delta t_i)$, where $P(k, f^i_{bk}\delta t_i)$ is the Poisson probability to have k events in the time window δt_i and being f^i_{bk} the background frequency.

The introduction of the imitation frequency has a double advantage. From the viewpoint of the search for neutrino bursts, it allows us to define a priori the statistical significance of each cluster in terms of frequency. Also, it allows us to monitor the performance of the search algorithm and the stability of the detector as a function of the imitation frequency threshold (see [8]).

5 Conclusion

None of the observed clusters passes the 0.01 year^{-1} threshold: 6 clusters have a $F_{im} < 1$ year^{-1} and they have been individually checked in terms of energy spectra and low energy signals that may be the signature of the IBD interactions. They are fully compatible with chance coincidence among background signals. We conclude that no evidence of neutrino burst signal is found and taking into account the total live-time of LVD, we fix the corresponding upper limit to the rate of Gravitational stellar collapse out to 25 kpc of 0.1 per year at 90% c.l.

Acknowledgment

This work was supported by the Russian Foundation for Basic Research (project No. 15-02-01056 a) and the program of investigations of the Presidium of Russian Academy of Sciences High Energy Physics and Neutrino Astrophysics.

References

[1] M. Aglietta et al., EPL **3** (1987) 1315.
[2] E. N. Alekseev et al., JETPL **45** (1987) 589.
[3] K. Hirata et al., PhRvL **58** (1987) 1490.

[4] R. M. Bionta et al. (IMB collaboration), PhRvL **58** (1987) 1494.
[5] M. Aglietta et al., NcimA, **105** (1992) 1793.
[6] P. Antonioli et al., NJPh, **6** (2004) 114.
[7] N. Yu. Agafonova et al., APh, **28** (2008) 516.
[8] N. Yu. Agafonova et al., ApJ, **802** (2015) 47.

COSMOLOGICAL RELIC NEUTRINO DETECTION WITH THE PTOLEMY EXPERIMENT

Alfredo G. Cocco [a]

Istituto Nazionale di Fisica Nucleare - Sezione di Napoli
Via Cintia 80126 Napoli, Italy

Abstract. The PTOLEMY experiment aim at the direct detection of the Cosmological Relic Neutrino Background (CNB) by the use of a Tritium target. Neutrino produced in the early stage of the Big Bang are predicted to have thermally decoupled from other forms of matter at approximately 1 second after the Big Bang; they represent the oldest detectable Big Bang relics and as such they carry an invaluable content of information about the genesis and evolution of our Universe. Despite their very small energy they present a sizable interaction cross section on nuclei that decay through beta decay, as it was pointed out by recent studies. In particular Tritium is among the nuclei having the most favorable detection conditions. We will report here the status of construction of the PTOLEMY proof-of-principle prototype being installed at Laboratori Nazionali del GranSasso (Italy) in order to demonstrate the key aspects of the detection technique. Outline plan for the realization of the full experimental setup needed in order to detect CNB induced events will be reported.

1 Introduction

According to the Big Bang model light neutrinos produced in the early universe are predicted to have thermally decoupled from other forms of matter at approximately 1 second after the Big Bang. The light neutrinos, believed to be stable particles, have cooled to a temperature of 1.9 K (1.7×10^{-4} eV) in the present day and are predicted to have an average number of density of approximately 56/cm^3 per lepton flavor. The number density prediction is based on the annihilation rate of e^+e^- into three flavors of neutrinos through the weak neutral current interaction in the dense, high temperature conditions of the early universe.

At present, the direct measurements on the lightest neutrino mass eigenstate give results in the order of the eV. Oscillation experiments only provide a lower limit on the mass of (at least) one neutrino mass eigenstate of the order of 0.05 eV, while direct measurements of the electron energy spectrum in ^{3}H decay give $m_\nu < 2$ eV [1]. A large improvement in this respect will be provided by the KATRIN experiment, whose expected sensitivity is 0.2 eV. On the other hand, cosmological data from Cosmic Microwave Background (CMB) anisotropies and the large scale structure power spectrum provide an independent bound on the sum of neutrino masses which, depending on the particular model adopted and the number of free parameters, lies in the range $0.3 \div 2$ eV; see e.g. [2]. As we will argue in the following, if m_ν is of the order of 0.1 eV, the

---
[a] E-mail: alfredo.cocco@na.infn.it

PTOLEMY experiment may represent the unique way to detect Cosmological Relic Neutrino Background.

The idea of using Neutrino Capture on Beta decaying nuclei to measure the cosmological relic neutrino background predicted in the framework of the hot big bang model was recently pointed out in [3, 4]; in these papers authors make an extensive study of the threshold-less reaction, of its cross section and on the effects of the recently discovered neutrino oscillations on the possible detection of very low energy neutrino. They found that a Tritium target mass of 100 grams is predicted to produce approximately 10 events/year from relic neutrino capture with the relic neutrino density modeled as a uniform Fermi-Dirac number density throughout space. The local neutrino density may be enhanced in galactic clusters by factors that range from 1–100 depending on the neutrino mass [5]. The uncertainty on the neutrino capture cross-section on tritium is constrained to the sub-percent level, and among the possible nuclei unstable to β-decay the tritium nucleus is considered optimal based on the product of the capture cross section and the half-life of 12.3 years uniquely coming from the β-decay process.

The PTOLEMY experiment [6] aim at the detection of cosmological relic neutrino background using a 100 grams tritium target and leading edge technology in order to achieve an unprecedented discrimination power of signal events with respect to the overwhelming amount of electrons produced in the β-decay. Similarly, PTOLEMY is capable of reaching the highest sensitivity in the search for keV-scale dark matter candidates in the form of sterile neutrinos with an admixture of electron flavor at the level of $|U_{e4}|^2$ of $10^{-4} \div 10^{-6}$, depending on the sterile neutrino mass. Moreover, a side project born from the studies on Tritium loaded graphene target may provide unique results on directionality of Dark Matter in the MeV mass scale [10].

2 The PTOLEMY experiment

The PTOLEMY experiment conceptual design is described in Fig. 1; the electrons produced in the beta decay and neutrino capture process of the 100 g Tritium target are focused to the bore of a MAC-E filter where a first stage of background event rejection is performed. An ideal cut on the last eV of electron beta spectrum would give an event rate reduction of about 1.55×10^{-13}. The surviving electrons are then accelerated by an electric field and focused into an RF tracking device able to provide both energy measurement and time-of-flight tag for the downstream micro calorimeter. The electron is then de-accelerated to about 100 eV where a cryogenic micro-calorimeter can measure the energy with an accuracy of about 0.1 eV. The PTOLEMY experiment is based on an atomic thickness tritium target. The large surface area required for relic neutrino detector can be factorized into isolated tritium target areas using a

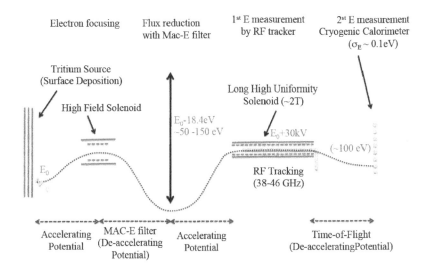

Figure 1: Schematic description of the PTOLEMY prototype experiment.

method of magnetic ducts and E×B filtering. Half of the 4π solid angle of the isotropically emitted electrons from a planar Tritium target are accelerated with a precision voltage reference into a high magnetic field region at the opening of the MAC-E filter. Graphene substrates are suitable to hold monoatomic Tritium layers through chemical absorption. The binding energy of tritium on graphene is much less than the rotational excitation levels of the T_2 molecule [7] and the differential density of final states of the ^{3}He is expected to be reduced by an order of magnitude. By binding the tritium to individual layers of graphene, the high conductivity of graphene is also expected to eliminate the effect of charging and voltage reference shifting.

MAC-E filter in PTOLEMY acts as an integrating high-pass filter. The use of a detector with precision calorimetry removes the need for a sharp cutoff; the main role of the MAC-E filter is to remove enough of the spectrum below the endpoint so that the detector may function without being swamped by large signal rates. The new filter concept will be based on the E×B filtering: electrons from the target enter at a fixed reference voltage into one end of an E·B bottle; as they bounce back and forth in the bottle, they trade kinetic energy for potential energy as they slowly drift vertically in the voltage potential and also drift into lower B field regions. This new approach should allow to realize compact filters that could fit in a high packing factor design.

Tritium decay endpoint electrons have energy of about 18.6 keV, corresponding to a total velocity of approximately $\beta = 0.26$. For B = 1.9 T, the value of the cyclotron frequency is of approximately 46 GHz. An endpoint electron moving transversely to the magnetic field will radiate approximately $P_{tot} = 3 \times 10^{-14}$

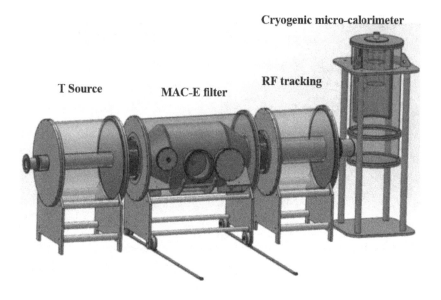

Figure 2: The PTOLEMY prototype setup.

W of coherent RF power. Single electron measurement in RF trackers has been recently proven as a viable technology in Project-8 prototypes [8]. Thread electron trajectories through an array of such antennas with wide bandwidth can be used to identify cyclotron RF signal in transit times of order 0.2 msec. The timing resolution is expected to be of the order of 10 ns, depending on micro-calorimeter response.

Electron calorimeter should have an energy resolution sufficient to resolve the neutrino mass. In case of the Tritium decay endpoint decelerated to the $10 \div 100$ eV energy window the electron can be stopped with a very thin absorber (much thinner then the one used in X-ray measurements) thus reducing significantly the detector heat capacity. This has also a beneficial effect on the detector time response reaching bandwidths of about 1 MHz to record 10 kHz of electrons hitting the individual sensors. Recent development of TES cryogenic micro calorimeters may prove the feasibility of having a resolution well below 0.1 eV in the measure of the electron energy; resolution of about 0.05 eV for infrared photons have just be achieved [9] and tests are ongoing to demonstrate the same level of accuracy in low energy electrons operating the sensor below 100 mK. The design for PTOLEMY incorporates magnetic shielding for the TES and the microwave-readout massive SQUID multiplexer (MMSM) in order to readout all the calorimeter sensors.

The construction of the PTOLEMY prototype is completed and the detector is ready to be test commissioned at Princeton University Campus in the first months of the 2018. The detector will be then dismounted and moved to the Laboratori Nazionali del GranSasso in Italy, in order to profit from the very low environmental background to perform a complete test program with electron tracking from the graphene tritiated target through the MAC-E filter to the endpoint TES. The experimental challenges for the PTOLEMY proof-of-principle will be then sort-out:

1. prove reduced molecular smearing in the CNB target to below 0.05 eV

2. achieve an electron energy measurement resolution of 0.05 eV to separate the CNB signal from the decay spectrum,

3. demonstrate high radio-purity in the Graphene target and low background rate in the CNB signal region (concurrent with a physics program in the MeV Dark Matter searches [10])

4. demonstrate intrinsic triggering capability for selecting endpoint electrons

5. design and simulate a scalable target mass with high acceptance kinematic filtering

After the proof-of-principle phase is successfully established with the prototype setup, the PTOLEMY Collaboration foresee a second stage of development to conduct the validation of prototypes demonstrating the scalability concepts for the Graphene target and instrumentation toward the realization of the full scale experiment.

References

[1] C. Patrignani *et al.* (Particle Data Group), *Chinese Physics* C **40**, 100001 (2016).
[2] J. Lesgourgues and S. Pastor, *Physic Reports* **429**, 307 (2006).
[3] A.G. Cocco, G. Mangano and M. Messina, *Journal of Cosmology and Astroparticle Physisics* **06**, 015 (2007).
[4] A.G. Cocco, G. Mangano and M. Messina, *Physical Review* D **79**, 053009 (2009).
[5] A. Ringwald and Y.Y.Y. Wong, *Journal of Cosmology and Astroparticle Physics* **12**, 005 (2004).
[6] S. Betts *et al.*, [arXiv:1307.4738v2].
[7] C. Lin *et al.*, *Nano Letters* **15**, 903 (2015).
[8] D.M. Asner *et al.*, *Physical Review Letters* 114 (2015) 162501.
[9] C. Portesi *et al.*, *IEEE Transactions on Applied Superconductivity* 25 **3** (2015).
[10] Y. Hochberg, *et al.*, *Physics Letters* **B772** (2017) 239.

EFFECTS OF BSMS ON θ_{23} DETERMINATION

C.R. Das [a]
*Bogoliubov Laboratory of Theoretical Physics, Joint Institute for Nuclear Research,
Joliot-Curie 6, 141980 Dubna, Moscow region, Russian Federation*
Jukka Maalampi [b]
*University of Jyvaskyla, Department of Physics, P.O. Box 35,
FI-40014 University of Jyvaskyla, Finland*
João Pulido [c]
*Centro de Física Teórica das Partículas (CFTP), Instituto Superior Tcnico,
Av. Rovisco Pais, P-1049-001 Lisboa, Portugal*
Sampsa Vihonen [d]
*University of Jyvaskyla, Department of Physics, P.O. Box 35,
FI-40014 University of Jyvaskyla, Finland*

Abstract. We investigate the prospects for determining the octant of θ_{23} in the future long baseline oscillation experiments. We present our results as contour plots on the $(\theta_{23} - 45°, \delta)$–plane, where δ is the CP phase, showing the true values of θ_{23} for which the octant can be experimentally determined at $3\,\sigma$, $2\,\sigma$ and $1\,\sigma$ confidence level.

1 The causatum

The recent data indicate that the neutrino mixing angle θ_{23} deviates from the maximal-mixing value of $45°$, showing two nearly degenerate solutions, one in the lower octant (LO) ($\theta_{23} < 45°$) and one in the higher octant (HO) ($\theta_{23} > 45°$). Please check arXiv:1708.05182 for detail analysis.

We have presented the sensitivity to the determination of the θ_{23} octant ($\theta_{23} \leq \pi/4$ or $\theta_{23} \geq \pi/4$) in DUNE in four different scenarios. On the one hand, we have updated the $1\,\sigma$, $2\,\sigma$ and $3\,\sigma$ confidence level contours for the Standard Model, where oscillations are constituted between three active neutrinos. On the other hand, we have also given these contours for three different scenarios where the octant sensitivity is interfered by sterile neutrinos and other potential sources for physics beyond Standard Model. We analyzed these scenarios by parametrizing the new physics with the methods that were originally introduced in Refs. [1] and [2] to describe non-unitarity of the light neutrino mixing matrix.

We found that the non-unitarity of the mixing matrix caused the sensitivity θ_{23} octant to decrease from the Standard Model case. Nevertheless, due to the strictness of the existing bounds for the non-unitarity parameters α_{ij}, $i,j = 1,2,3$ derived in Ref. [3] and for $\alpha_{ll'}$, $l, l' = e, \mu, \tau$ derived in Ref. [2] the observed drop in the octant sensitivity was found to be very small. The worsening of

---

[a] E-mail: das@theor.jinr.ru
[b] E-mail: jukka.maalampi@jyu.fi
[c] E-mail: pulido@cftp.ist.utl.pt
[d] E-mail: sampsa.p.vihonen@student.jyu.fi

the octant sensitivity due to sterile neutrino was found larger than this. The sensitivity was calculated in this case using the bounds on $\alpha_{ll'}$ given in Ref. [2]. The worsening of the sensitivity was found to be less than $1°$ in each octant.

Table 1: Bounds on non-unitary parameters in $\alpha_{ll'}$ representation, taken from [2]. In this scenario the constraints would correspond to mixing with a light sterile neutrino in two mass scales: $\Delta m_{41}^2 \sim 0.1 - 1\,\mathrm{eV}^2$ (left column) and $\Delta m_{41}^2 \geq 100\,\mathrm{eV}^2$ (right column). The constraints are presented at a 95% confidence level.

Parameter	$\Delta m_{41}^2 \sim 0.1 - 1\,\mathrm{eV}^2$	$\Delta m_{41}^2 \geq 100\,\mathrm{eV}^2$
α_{ee}	1.0×10^{-2}	2.4×10^{-2}
$\alpha_{\mu\mu}$	1.4×10^{-2}	2.2×10^{-2}
$\alpha_{\tau\tau}$	1.0×10^{-1}	1.0×10^{-1}
$|\alpha_{\mu e}|$	1.7×10^{-2}	2.5×10^{-2}
$|\alpha_{\tau e}|$	4.5×10^{-2}	6.9×10^{-2}
$|\alpha_{\tau\mu}|$	5.3×10^{-2}	1.2×10^{-2}

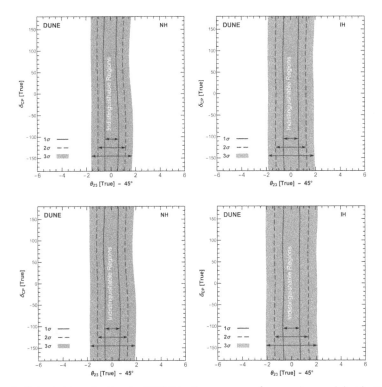

Figure 1: Octant determination in DUNE in the presence of non-unitary mixing in top two and light sterile mixing with $0 < \Delta m_{41}^2 < 1\,\mathrm{eV}^2$ or $\Delta m_{41} \geq 100\,\mathrm{eV}^2$ in bottom two figures.

We found the decrease in sensitivity due to the light sterile neutrino to be substantially less significant than that reported in Ref. [4] where the impact of a sterile neutrino with mixing angles $\theta_{14} = \theta_{24} = 9°$ and $\theta_{34} = 0°$ was considered in the determination of the θ_{23} octant in DUNE. Evidence of this sensitivity decrease, can be seen from the comparison between our Fig. 1 with Fig. 3 of Ref. [4]. When converted to the non-unitarity formalism (see the appendix of Ref. [1]), this kind of sterile neutrino would imply non-unitarity whose parameter values lie close to the existing bounds we presented for $0 < \Delta m^2_{41} < 1\,\text{eV}^2$ in Table 1. On the other hand, our investigation takes into account all possibilities for light sterile neutrinos, whereas the authors of Ref. [4] consider a specific model. Thus our results are in this respect more general, therefore statistically favoured by comparison and hence the difference between the two sets. If the model (disfavoured by roughly 95% CL) of Ref. [4] is realized in nature, then the ability of DUNE to tell the θ_{23} octancy is deteriorated.

We also tested how the octant sensitivity changed when the new physics parameters α_{ij} were left unconstrained. This type of simulation corresponds to a new physics scenario, where sterile neutrinos are associated with other new physics effects, not taken into account in Refs. [2] and [3] when deriving the bounds for the non-unitary and light sterile mixing effects. An example of this could be non-standard interactions involved in the neutrino propagation. Our simulations showed that in the worst case the octant could be determined at $3\,\sigma$ CL or better for $\theta_{23} \lesssim 41.0°$ and $\theta_{23} \gtrsim 48.5°$ for the normal hierarchy to be compared with the bounds $\theta_{23} \lesssim 43.5°$ and $\theta_{23} \gtrsim 44.5°$ of the standard case.

2 The succinct

We found that non-unitarity of the neutrino mixing matrix or the possible existence of light sterile neutrinos affect only mildly the sensitivity of DUNE to determine the octant of θ_{23}. This is in contrast with the determination of the CP violation, where the presence of sterile neutrinos could jeopardize the sensitivity [2,3,5].

References

[1] F.J. Escrihuela, D.V. Forero, O.G. Miranda, M. Tortola and J.W.F. Valle, Phys. Rev. **D 92** (2015) 053009 [Erratum: Phys. Rev. **D 93** (2016) 119905], [arXiv: 1503.08879].

[2] M. Blennow, P. Coloma, E. Fernandez-Martinez, J. Hernandez-Garcia and J. Lopez-Pavon, JHEP **04** (2017) 153, [arXiv: 1609.08637].

[3] F.J. Escrihuela, D.V. Forero, O.G. Miranda, M. Tortola and J.W.F. Valle, New J. Phys. **19** (2017) 093005, [arXiv: 1612.07377].

[4] S.K. Agarwalla, S.S. Chatterjee and A. Palazzo, Phys. Rev. Lett. **118** (2017) 031804, [arXiv: 1605.04299].
[5] S-F. Ge, P. Pasquini, M. Tortola and J.W.F. Valle, Phys. Rev. **D 95** (2017) 033005, [arXiv: 1605.01670].

THE AGING BEHAVES AND THE SMALL BATCH TEST OF THE 20" MCP-PMTs

Yao Zhu^{a,b}, Feng Gao^{a,c,d}, Guorui Huang^e, Yuekun Heng^{a,c}, Dong Li^e, Huilin Liu^f, Shulin Liu ^{a,c}, Weihua Li^f, Zhe Ning^{a,c}, Ming Qi^g,
Sen Qian^{a,c*}, Lin Ren^e, Jianning Sun^e, Shuguang Si^e, Jinshou Tian^f, Xingchao Wang^e, Yifang Wang^{a,c}, Yonglin Wei^f, Liwei Xin^e, Tianchi Zhao^a,
on behalf of the MCP-PMT workgroup*
a.Institute of High Energy Physics, Chinese Academy of Sciences, 100049 Beijing, China
b.Harbin Institute of Technology, 150001 Harbin, China
c.State Key Laboratory of Particle Detection and Electronics, 100049 Beijing, China
d.University of Chinese Academy of Sciences, 100049 Beijing, China
e.Nanjing, North Night Vision Tech. Ltd., 211106 NanJing, China
f.Xian Institute of Optics and Precision Mechanics, Chinese Academy of Sciences, 710068 Xian, China
g. Department of Physics, Nanjing Unversity, 210093 Nanjing, China

Abstract. An new concept of large area photomultiplier based on MCPs was conceived for JUNO by the scientists in IHEP in 2009, and with the collaborative work of the MCP-PMT collaboration in China, 20 inch prototypes were produced in 2015 and its characteristic was tested in the lab. Especially the aging behaves tested in 400 days was discussed in this manuscript. At the end, also introduced the small batch test result about 1000 pics 20 inch MCP-PMT prototypes.

1 Introduction

Photomultiplier tubes are (PMTs) widely used in various detection and imaging field for their high detection efficiency and fast time response. With large area photocathode and high detection efficiency, the 20-inch PMT was used in the large-scale neutrino experiments [1], which require the high-energy resolution and light coverage area.

Microchannel plate (MCP) is used in the small PMTs as the electron multiplier [2], which greatly improved the time resolution of PMT. In the prototypes, MCP assemblies are used as the electron amplifiers. When a primary photoelectron comes from the photocathode of the PMT, it will impinge on the inner wall of the micro channel of the MCP to generate the secondary electrons. Being accelerated by the electric field created by the voltage applied across both sides of the MCP, these secondary electrons bombard the channel wall again to produce additional secondary electrons. The electrons output from the first MCP enter the second MCP and hit the channel wall again, thus producing more

* Corresponding author. Institute of High Energy Physics, CAS, Beijing, 100049, China. Tel.: +8610 8823 6760. E-mail address: qians@ihep.ac.cn.
* The MCP-PMT work group: Microchannel-Plate-Based Large Area Photomultiplier Collaboration.

secondary electrons collected by the anode. Due to the short multiplication path of the MCP system, the MCP-PMT has fast time response [3].

In 2009, researchers in the Institute of High Energy Physics (IHEP) have conceived a new concept of large PMT with MCPs [4], for the Jiangmen Underground Neutrino Observatory (JUNO) [5]. And the North Night Vision Tech. Ltd. (NNVT) in Nanjing, China, one of the members of the MCP-PMT collaboration group has produced this type of 20-inch MCP-PMT in 2015 [6]. A photograph of the 20 inch MCP-PMTs is shown in Figure.1.

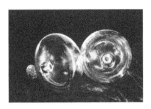

Figure 1: A photograph of 20 inch MCP-PMT.

2 The performance of the 20 inch MCP-PMT prototype

After a series of joined work by the IHEP and NNVT people, the new 20-inch MCP-PMT prototype with high detection efficiency (HDE) was produce at the middle of 2015. This HDE 20-inch MCP-PMT met the requirements for JUNO, and got the 75% order of the JUNO PMT contract.

The performance of this type of 20-inch PMTs was carefully studied and measured. Such as the I-V curves, the quantum efficiency (QE) vs. wavelength, the mapping the QE uniformity, the SPE charge spectra, ranges of gain linearity and dynamic range, timing characteristics of output signals et.al are shown in the Table 1. Noise characteristics and after pulse properties were studied at gain $1.0*10^7$.

3 The aging behaves of the 20 inch MCP-PMT

JUNO experiment requires the PMTs running in the feeble light sources about 20 years. Therefore, the study of the aging characterization for PMT is of great important. As shown in Fig. 1, a test system was setup for the aging test of the MCP-PMT in our lab [7]. Single-photo mode was obtained by modifying the pulsed light source LDs and an online labview program was used to modify the driving voltage of LD.

In order to maintain the stability of the large area PMT input light intensity from the LD, a small PMT with the type of XP2020 was uesd to monitor the changes of the intensity of LD's lighting. It was put opposite to the tested PMT

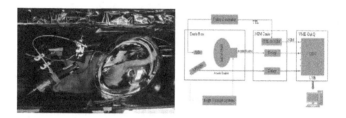

Figure 2: Experiment setup.

to test the reflecting light. Any changes to the XP2020 MPE spectrum could be seen as the changes to the intensity of LD lighting.

The PMT detected about 1000 photoelectrons under strong light condition. In the experiment, parameters were tested daily. Its convenient to switch the light condition between a single photoelectron state and a 1000 photoelectron state.

The Gain of the MCP-PMT prototype is associated with the lattices of the inner hole of MCP. The aging of MCP can be described by measuring the signal photoelectron output charge. The results of aging of the 20 inch MCP-PMT is shown in Fig. 2. When the output charge of the MCP was about 40C, the gain only decresced to 70%.

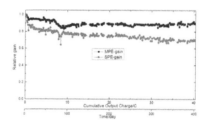

Figure 3: Aging result.

4 Mass production and batch test result

The PMT mass production line was operated in 2016 and 30 pic PMTs could be produced per day. Also the batch test system has been designed and finished in 2017. At present, NNVT has already started shipping the MCP-PMT from Nanjing to JUNO test site. All parameters of the PMTs have been tested in the NNVT based on batch test system. Performance parameters can be obtained from the database. The first batch of tests had 336 pic MCP-PMTs. Table 1 shows the MCP-PMT performance parameters for the PMTs in the mass production by batch test.

Compared with the prototype, some parameters have changed. The QE and DE have been improved, but the TTS, DR and FT are larger. These parameters

Table 1: MCP-PMT performance parameters in the mass production by batch test.

PMT Parameters	Data in Contract	Prototype	336 Mass Production	1000 Mass Production
QE@410nm(%)	26.5	26	29.53	29.2
QE Uniformity(%)	15	10	8.1	7.8
SPE-P(V)	2.8	5.6	8.08	7.1
SPE-ER(%)	40	41	31.1	33.1
Gain	26.5	26	29.53	29.2
HV(V)	2800	1780	1781	1767
DE(%)	24	26	28.95	29.3
DR(KHz)	30	30	33.9	36.9
TTS(ns)	15	12	19.2	19.5
APR(%)	26.5	2.5	1.16	0.85
Linearity<10%(pe)	1000	1000	1183	1160
RT(ns)	2	1.2	1.43	1.41
FT(ns)	12	10.2	24.5	24.5

are worse due to the change of the collection, the photocathode and technics of MCP. In addition, the electron from the transmission photocathode and the reflection photocathode do not match when arriving in the anode.

Acknowledgments

The MCP-PMT development project has been partially supported by the Strategic Priority Research Program of the Chinese Academy of Sciences (Grant No. XDA10010200 and No. XDA10010400) and the National Natural Science Foundation of China (Grant No.11175198 and No.11475209 and No.11675205).

References

[1] Abe K et al. Letter of intent: The hyper-kamiokande experiment — detector design and physics potential —. *Physics*, 2011.
[2] Yang Yuzhen et al. Single electron counting using a dual mcp assembly.
[3] M. Yu Barnyakov and A. V Mironov. Photocathode aging in mcp pmt. *Journal of Instrumentation*, 6(12):C12026, 2011.
[4] Wang Yifang et al. A new design of large area mcp-pmt for the next generation neutrino experiment. *Nuclear Inst & Methods in Physics Research A*, 695:113C117, 12 2012.
[5] Li Yufeng et al. Unambiguous determination of the neutrino mass hierarchy using reactor neutrinos. *Physical Review D*, 88(1):57–61, 2013.
[6] Chang Yaping et al. The r&d of the 20 in. mcpcpmts for juno. *Nuclear Instruments & Methods in Physics Research*, 824:143–144, 2016.
[7] Wang Wenwen et al. The study of the aging behavior on large area mcp-pmt. *Nuclear Science and Techniques*, 27, 03 2015.

ICARUS: FROM CNGS TO BOOSTER BEAM

Alberto Guglielmi [a]
INFN Sezione di Padova, via Marzolo 8, 35131 Padova, Italy

Abstract. The ICARUS T600 liquid argon detector underwent an intensive overhauling after the successful run at LNGS; it was transported to Fermilab in view of a sensitive search for sterile neutrinos at Booster beam in conjunction with SBND and MicroBooNE detectors. ICARUS data taking is expected to start in late 2018.

1 Introduction

Cherenkov radiation detection has been so-far one of the key choices for exploring neutrino's properties with > 10 kton mass water and ice detectors. As an alternative, the Liquid Argon Time Projection Chamber (LAr-TPC) which allows for accurately identifying and reconstructing each ionizing track in complex topology neutrino events, has been proposed in 1977 [1] and successfully developed during the last two decades. With the continuing effort of the ICARUS Collaboration and INFN support, the LAr-TPC technology has been taken to full maturity with the T600 detector, the largest LAr-TPC ever built 760 t of LAr mass, installed at LNGS underground Laboratories and exposed to CNGS ν_μ's from CERN in 2010–2012 [2].

ICARUS performed a sensitive search for an additional fourth sterile neutrino state driving neutrino oscillations at a small distance with a mass at ~ 1 eV [3], as hinted by the pioneering LSND experiment and by the anomalies observed at accelerators, reactors and with intense radioactive sources [4].

Global fits to all experimental data incorporating also the new constraints from MINOS combined with the Daya Bay and Bugey-3 experiment and from IceCube [5], would indicate $0.8 \div 2$ eV^2 Δm^2 range for a sterile neutrino mass. However recent results appear to disfavor this interpretation [6], making the sterile neutrino scenario inconclusive and calling for a definitive experiment. Among the new proposed searches for sterile neutrinos, a conclusive clarification on such phenomena is expected from the new SBN experiment [7] in preparation at FNAL Booster Neutrino Beam (BNB), based on the search for spectral differences in both electron and muon-like channels in three LAr-TPCs, SBND, MicroBooNE and ICARUS T600 at different baselines.

2 ICARUS T600 LAr-TPC run at Gran Sasso

ICARUS T600 consists of two identical modules filled with 760 t of ultra-pure liquid Argon (~ 476 t active mass) [2] each one housing on the long sides two TPC chambers separated by a central common cathode with 1.5 m drift length.

---

[a] on behalf on the ICARUS Collaboration, e-mail: alberto.guglielmi@pd.infn.it

A 500 V/cm uniform electric field allows drifting ionization electrons produced by charged particles along their path to 3 parallel readout wire planes facing the drift volume, 54000 wires in total, 3 mm pitch and plane spacing, oriented at 0^0, $\pm 60^0$ with respect to the horizontal direction. The measurements of 3 independent projections, result in a full 3D reconstruction of any ionizing event with a $\sim$ mm scale resolution. The electron charge signal detected on the last Collection wire plane, proportional to the deposited energy, allows for a calorimetric measurement of the particle energy. The TPCs are complemented by a photo-multipliers system installed behind the wire planes, coated with a TPB (tetraphenyl butadiene) wavelength shifter, to detect the VUV scintillation light emitted at 128 nm by ionizing particles. The relative time of each ionization signal, combined with ~ 1.6 mm/μs electron drift velocity, provides the position of the track along the drift coordinate.

To prevent the absorption of the drifting electrons by electronegative impurities which would result in a reduction of the signal, the LAr was kept at an exceptionally high purity level, < 50 ppt O_2 residual impurity corresponding to an electron lifetime in excess of 7 ms, by new industrial LAr purification methods developed to continuously filter both liquid and gas argon [2]. Further improvements in the recirculation system allowed to reach 16 ms lifetime (20 ppt O_2 LAr contamination) [8] demonstrating the effectiveness of single-phase LAr-TPC technique and paving the way to huge LAr-TPC detectors.

The long ICARUS run at LNGS demonstrated its excellent performance as tracking device, ~ 1 mm^3 spatial resolution, and homogenous calorimeter measuring the contained e.m. showers with $\sigma_E/E \simeq 0.03/\sqrt{(E(GeV))}$ resolution. Momentum of escaping muons has been determined via Multiple Coulomb Scattering (MCS) by statistically analyzing track deflection angles, which are inversely proportional to momentum it-self. The measurement algorithm has been validated in the few GeV/c range on muons from CNGS ν_μCC interactiong in the upstream rock and stopping in the LAr volume by comparing the MCS momentum p_{MCS} with calorimetric measurement p_{CAL} (Fig. 1). After correcting for the $\sim$ cm observed non-planarity of the TPC cathode plane ($\sim 1\%$ distortion of the drift field in its proximity) the MCS and calorimetric estimates of momentum agree on average within the errors, confirming the viability of the measurement. A $\sim 15\%$ average resolution in the 1-4 GeV/c range has been determined, depending on the muon momentum and track length [9].

Exposed to the CNGS beam the T600 collected with a livetime $> 93\%$ 2650 neutrino interactions, i.e. ~ 3.4 ν interactions and 12 beam related μ's per 10^{17} p.o.t. on average, consistent within 6% with the MC predictions. The detector demonstrated a remarkable e/γ separation and particle identification exploiting the measurement of dE/dx versus range. The capabilities to reconstruct the neutrino interaction vertex, to identify and measure e.m. showers generated by primary electrons and to accurately measure invariant mass of photon pairs

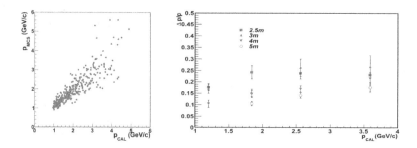

Figure 1: Left: stopping muon momentum p_{MCS} obtained by MCS compared to the corresponding p_{CAL} by calorimetry before the cathode planarity correction. Right: Some small underestimation, due to the non-perfect planarity of the TPC cathode, appears for p>3.5 GeV/c. momentum resolution as a function of muon momentum for different track lenghts.

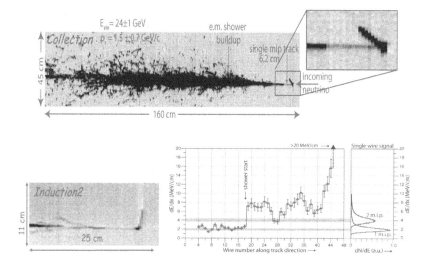

Figure 2: ν_e CC CNGS event in Collection and Induction 2 views of ICARUS (top) with the corresponding evolution of ionization density dE/dx in the first wires with the marked shower onset (bottom). The leading electron is identified by the 2.1 MeV m.i.p. signal in the first wires from the interaction vertex. The dE/dx expectations for 1 and 2 m.i.p. are also shown.

for the π^0 recognition, allowed to reject by a factor $\sim 10^4$ the NC background in the study of $\nu_\mu \to \nu_e$ transitions. Globally 7 electron-like events have been observed (see Fig. 4) to be compared with the 8.5 ± 1.1 expected from the $\sim 1\%$ intrinsic beam component and standard 3-flavor oscillations, providing the limit on $\nu_\mu \to \nu_e$ oscillations $P < 3.86\ 10^{-3}$ at 90 % C.L. (Fig. 4 left). The ICARUS result [3], confirmed by OPERA [10], allowed to define a narrow parameter region at $\sim \Delta m^2 \sim 1 eV^2$ to be definitely investigated.

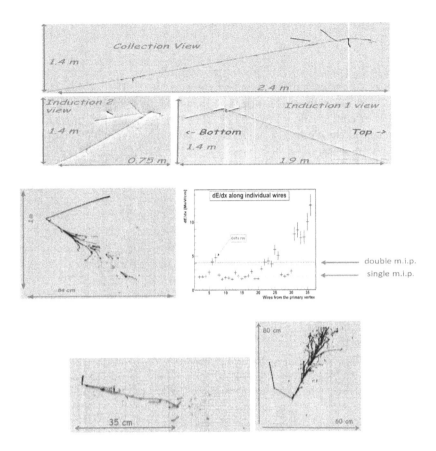

Figure 3: Atmospheric neutrino event candidates: (Top) a 2 GeV upward going ν_μ CC showing a 4 m escaping muon with $p = 1.8 \pm 0.3$ GeV/c as measured by MCS, a $E_{DEP} = 250$ MeV proton and two low energy π's well reconstructed in all three views; (Center) a downward-going quasi-elastic ν_e CC event, $E_{DEP} = 0.9$ GeV in Collection view (left) with the leading electron identified by the ionization density dE/dx signal in the first ~ 20 wires before the shower onset; (Bottom) a downward-going quasi-elastic ν_eCC, $E_{DEP} = 240$ MeV with a short proton (left) and a $E_{DEP} = 1.9$ GeV ν_eCC (right); the single primary electron developing in an e.m. shower is clearly identified.

Atmospheric neutrinos collected by ICARUS at LNGS, ~ 0.73 kt year exposure, are being analyzed. Multi-prong neutrino event candidates with $E_{dep} > 20$ MeV are selected by an automatic procedure to reconstruct 3D tracks and interaction vertices reducing to 0.5 % of events undergoing further visual analysis. In 1/3 exposure 7 muon-like and 8 electron-like candidates have been identified so far (see Fig. 3), demonstrating that the automatic search for ν_eCC in sub-GeV range at FNAL is feasible. Moreover, a similar analysis could also address nucleon decay searches with an efficiency of $\sim 80\%$ in channels involving kaons.

3 The SBN program for sterile neutrino search

In the SBN short-baseline experiment [7] at FNAL the SBND, MicroBooNE and ICARUS-T600 LAr-TPC's are installed at shallow depth, protected by 3 m concrete overburden from neutral cosmics, in the BNB neutrino beam at 110 m, 470 m and 600 m distance from target respectively. While, in the deep underground conditions of the LNGS laboratory, a single prompt trigger has always ensured the timing of the main image of the event which determine the track position along the drift coordinate, at the FNAL shallow depth location $\sim$ 11 additional cosmic muons are expected in the T600 in the 1 ms drift time, introducing some difficulties for the event track reconstruction. A 4π, segmented cosmics tagging system of plastic scintillation slabs read out by silicon PMTs surrounding the T600 will allow, in conjunction with the T600 internal PMT system, to unambiguously identify cosmic rays entering the detector.

This multiple LAr-TPC configuration will allow for a very sensitive search for $\nu_\mu \to \nu_e$ appearance signals by comparing the neutrino interactions at the different distances, covering in three year data taking (6.6 10^{20} p.o.t.) the LSND 99 % C.L. allowed region at $\sim 5\sigma$ C.L. (Fig. 4 left). In addition the high correlations between the near detector and the MicroBooNE/ICARUS event samples combined with the huge statistics at the near site will make the SBN program the most sensitive ν_μ disappearance experiment at $\Delta m^2 \sim 1$ eV^2 (Fig. 4 right). The disappearance measurement is a critical aspect of the SBN program which is needed to confirm a signal, if seen in ν_e appearance, as oscillations since the transition probabilities are related through a common mixing matrix. By collecting at the same time the ν_μ and ν_e event samples in the same experiment, correlations between the samples can be well understood and common systematics measured. The simultaneous analysis of ν_e CC and ν_μ CC events will be a very powerful way to explore oscillations and disentangle the effects of ν_μ disappearance and ν_e appearance in this mass-splitting range.

The ICARUS physics outreach will be enhanced with the study of neutrino cross-sections and interaction topologies at energies relevant to the Long Baseline Neutrino Facility (LBNF) program with the multi-kt DUNE LAr-TPC detector [11], exploiting the off-axis neutrinos from the NuMI. The T600 will collect, in parallel to SBN, an event statistics comparable to the one from the Booster beam, in the $0 \div 3$ GeV energy range with an enriched component of ν_es from the dominant three body decay of secondary K. A detailed analysis of these events will be highly beneficial for the future LBNF/DUNE LAr program, allowing to study very precisely the detection efficiencies and kinematical cuts in all the neutrino channels and event topologies.

To face the new shallow depth operation at Fermilab the ICARUS detector underwent an intensive overhauling at CERN in the framework of the WA104/NP01 project before being shipped to US, introducing some technology developments while maintaining the already achieved performance. The TPC

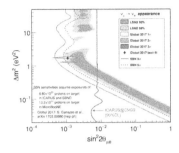 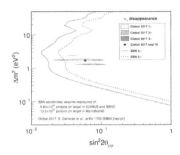

Figure 4: Sensitivities of the SBN experiment: $\nu_\mu \to \nu_e$ appearance sensitivity, compared with LSND allowed region and current global best fit [5] (left). The 90% C.L. limit obtained by ICARUS at CNGS is also shown. On the right, sensitivity in the ν_μ disappearance channel.

chambers, with improved planarity cathodes, have been installed in new cold vessels which will be surrounded by a new purely passive thermal insulation with a new cryogenics and LAr purification systems. The inner light detection system has been upgraded with 360 new 8″ photomultipliers, TPB coated, installed behind the wire planes ($\sim$ 5% of photo-cathode coverage). This will allow a more precise event timing and localization, exploiting also the BNB bunched beam structure at FNAL ($\sim$ 2 ns width bunches every 19 ns) to reject out-of-bunch cosmics. The T600 will be equipped with a new upgraded "warm" TPC read-out electronics. Improvements concern:

- serial 12 bit ADCs, one per channel, 400 ns synchronous sampling on the whole detector in place of the multiplexed ones used at LNGS;
- a serial bus architecture with optical links for a Gbit/s data transmission;
- a new compact design to host both analogue and digital electronics directly on ad-hoc detector feed-through flanges acting as electronic backplane. The digital part is fully contained in a single high performance FPGA in each board handling the ADC signals.

The adoption of the same new preamp in the front-end for both Induction and Collection wires with a $\sim 1.5\mu s$ faster shaping time of analogue signal to match the electron transit time in the wire plane spacing, will remove undershoot in the preamp response as well as the low frequency noise while maintaining the same S/N $\geq$ 10 ratio. The throughput of the read-out system has been improved up to 10 Hz by replacing the VME and the single board sequential access mode inherent to shared bus architecture, with a modern switched I/O where transactions are carried over optical Gigabit/s serial links.

The performance of the new electronic chain has been tested with a 50 liter LAr-TPC prototype exposed to cosmic-rays (Fig. 5). The optimized preamp

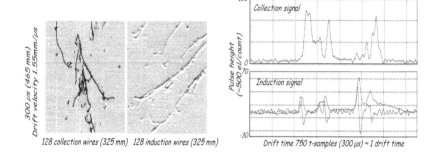

Figure 5: A shower event developing along the drift direction in the 50 liters LAr-TPC equipped with the new read-out electronics (left). The very stable baseline without any undershoot allows for a better hit position definition (right). The bipolar signal in Induction allows to extract with a dedicated algorithm (e.g. a running sum in green) the energy deposition in Induction view also.

architecture resulted in an unprecedented image sharpness with a better hit separation without signal undershoot even in crowed and complex events both in Collection and Inductions views. The use of dedicated algorithms for the bipolar signals from Induction wires will allow measuring the energy deposition in the Induction view also, with a $\Delta E/E \sim 27\%$ on the single wire hit, with a significant increase of the electron neutrino detection efficiency.

The two T600 modules have been shipped to Fermilab for the installation in the new pit in far position. According to the schedule, the ICARUS activation, commissioning and the start of data taking will follow in 2018.

References

[1] C. Rubbia, CERN Report 77-8 (1977).
[2] C. Rubbia et al. [ICARUS Coll.], JINST **6** (2011) P07011.
[3] M. Antonello et al. [ICARUS Coll.], EPJ C **C 73** (2013) 2599.
[4] A. Aguilar et al. [LSND Coll.], Phys. Rev. **D 64** (2001) 112007; G. Mention et al., Phys. Rev. **D 83** (2011) 073006; C. Giunti et al., Phys. Rev. **D 83** (2011) 065504.
[5] S. Gariazzo et al., JHEP **06** (2017) 135.
[6] P.A.R. Ade et al., A&A **594** (2016) A13; M.G. Aartsen et al., PRL **117** (2016) 071801; Y. Ko et al., PRL **118** (2017) 121802.
[7] R. Acciarri et al., arXiv:1503.01520 (2015).
[8] M. Antonello et al. [ICARUS Coll.], JINST **9** (2014) P12006.
[9] M. Antonello et al. [ICARUS Coll.], JINST **12** (2017) P04010.
[10] N. Agafonova et al. [OPERA Coll.], JHEP **1307** (2013) 004.
[11] R. Acciarri et al., arXiv:1601.05471 (2016).

PURSUIT FOR OPTIMAL BASELINE FOR MATTER NONSTANDARD INTERACTIONS IN LONG BASELINE NEUTRINO OSCILLATION EXPERIMENTS

Timo Kärkkäinen [a]
Department of Physics, University of Helsinki, FI-00014 Helsinki, Finland

Abstract. We investigate the prospects for probing the strength of the possible matter nonstandard neutrino interactions (mNSI) in long baseline neutrino oscillation experiments and the interference of the leptonic CP angle δ_{CP} with the constraining of the mNSI couplings. The interference is found to be strong in the case of the $\nu_e \leftrightarrow \nu_\mu$ and $\nu_e \leftrightarrow \nu_\tau$ transitions but not significant in the other cases.

1 Introduction

After the confirmation of atmospheric and solar oscillations, the neutrino physicist community has fitted the standard three flavor neutrino oscillations (NO) framework to the continuously accumulating data. As the mixing parameters are determined more and more precisely, it is clear that neutrino flavor transition can be very well interpreted as a purely oscillatory phenomenon. However, there may be subleading contributions to flavor transition. In that scenario, the future datasets will turn out to be unfittable to the NO framework, and additional degrees of freedom must be introduced.

2 Formalism

The possible deviations of standard NO scheme may be parametrized by nonstandard interactions (NSI). Consider the following Lagrangians:

$$\mathcal{L}_{\text{NSI}}^{\text{CC}} = -2\sqrt{2}G_F \varepsilon_{\alpha\beta}^{ff',C} (\overline{\nu}_\alpha \gamma^\mu P_L \ell_\beta)(\overline{f}\gamma^\mu P_C f'),\ f \neq f', \\ \mathcal{L}_{\text{NSI}}^{\text{NC}} = -2\sqrt{2}G_F \varepsilon_{\alpha\beta}^{f,C} (\overline{\nu}_\alpha \gamma^\mu P_L \nu_\beta)(\overline{f}\gamma^\mu P_C f). \quad (1)$$

In the most general case, f and f' are charged leptons or quarks, G_F is the Fermi coupling constant, α and β are flavor labels, and C is the chiral label, P_L and P_R being the chiral projection operators. The NSI parameters $\varepsilon_{\alpha\beta}^{ff',C}$ and $\varepsilon_{\alpha\beta}^{f,C}$ are dimensionless complex numbers. The charged current Lagrangian $\mathcal{L}_{\text{NSI}}^{\text{CC}}$ is relevant for the NSI effects in the neutrino creation and detection processes and the neutral current Lagrangian $\mathcal{L}_{\text{NSI}}^{\text{NC}}$ is relevant for the NSI matter (mNSI) effects. The effective low-energy NSI Lagrangians (1) are assumed to follow from some unspecified beyond-the-standard-model (BSM) theory after integrating out heavy degrees of freedom.

---

[a] E-mail: timo.j.karkkainen@helsinki.fi

mNSI matrix elements are gained by summing over chirality and fermion states (N_f and N_e are the fermion f and electron number densities),

$$\varepsilon_{\alpha\beta}^m = \sum_{f,C} \varepsilon_{\alpha\beta}^{fC} \frac{N_f}{N_e}, \qquad (2)$$

Description of NO in matter is given by standard interaction (SI) Hamiltonian

$$H_{\rm SI} = \frac{1}{2E_\nu} \left[U {\rm diag}(m_1^2, m_2^2, m_3^2) U^\dagger + \begin{pmatrix} V_{\rm CC} + V_{\rm NC} & 0 & 0 \\ 0 & V_{\rm NC} & 0 \\ 0 & 0 & V_{\rm NC} \end{pmatrix} \right], \qquad (3)$$

where E_ν is neutrino energy, $V_{\rm CC} = \sqrt{2} G_F E_\nu N_e$ and $V_{\rm NC} = -\frac{\sqrt{2}}{2} G_F E_\nu N_n$ are the matter potentials. In the case of mNSI, NO probability is calculated with NSI Hamiltonian, which is the SI case with an extra term:

$$P_{\alpha\beta} = \left| \langle \nu_\beta | e^{-i(H_{\rm SI} + V_{\rm CC} \varepsilon^m / 2E_\nu) L} | \nu_\alpha \rangle \right|^2. \qquad (4)$$

Note that we get the SI case at the $\varepsilon \to 0$ limit.

3 Numerical analysis and discussion

We study how the future NO experiments would constrain various mNSI parameters, setting the baseline free and letting the NO parameters vary within their 3σ limits. We consider both mass hierarchies but only higher θ_{23} octant.

We use two benchmark setups, both designed for a future long baseline neutrino experiment with double-phase liquid argon detector. The first benchmark (**SPS**) utilized the LBNO setup with 20 kt detector and beam optimization at 2288 km [1]. Our second benchmark (**DUNE**) utilized the DUNE setup with 40 kt detector and beam optimization at 1300 km. The analysis is done by using the GLoBES simulation software [2].

Because the neutrino fluxes are optimized at a certain baseline, once we perform the simulation with a different baseline, we reoptimize the flux assuming $L/E =$ constant, by

$$E_{\rm new} = \frac{L_{\rm new}}{L_{\rm old}} E_{\rm old}, \quad \phi_{\rm new}(E_{\rm new}) = \phi_{\rm old}(E_{\rm old}). \qquad (5)$$

We determine the upper bounds for $\varepsilon_{\alpha\beta}^m$ by evaluating the NSI discovery potential. The non-observation of NSI then allows to set new limits for $\varepsilon_{\alpha\beta}^m$. For each δ_{CP} value, 90 % CL contour is found and the results merged in a contour band. The bands in $(L, \varepsilon_{\alpha\beta}^m)$-plane are plotted in Fig. 1. As expected, the discovery potential is increased when baseline increases, even though the effect shrinks when the baseline is long to begin with. Changing the hierarchy produces slightly tighter constraints.

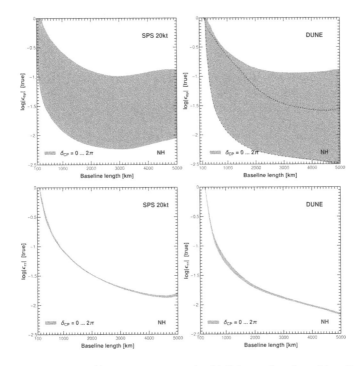

Figure 1: Upper plots: 90 % CL discovery reach of $|\varepsilon_{e\mu}^m|$ as a function of baseline length. Band thickness results from the ambiguity of δ_{CP}, which visibly interferes with constraining mNSI. Dashed line in DUNE plot represents the case $\delta_{CP} = 0$ and dot-dashed line the case $\delta_{CP} = \pi/2$. Lower plots: same, but for $|\varepsilon_{\tau\tau}^m|$. Band is thick also for $|\varepsilon_{e\tau}^m|$, but not for other parameters.

Acknowledgments

The author expresses his gratitude to the Magnus Ehrnrooth foundation for financial support and to the organizers of 18th Lomonosov Conference on Elementary Particle Physics for their invitation and hospitability.

References

[1] K. Huitu, T. Kärkkäinen, J. Maalampi, S. Vihonen, Phys. Rev. D **93**, 053016 (2016).
[2] P. Huber, M. Lindner, and W. Winter, Comput. Phys. Commun. **167**, 195 (2005), P. Huber, J. Kopp, M. Lindner, M. Rolinec, and W. Winter, Comput. Phys. Commun. **177**, 432 (2007), J. Kopp, Int. J. Mod. Phys. **C19**, 523 (2008), J. Kopp, M. Lindner, T. Ota, and J. Sato, Phys. Rev. **D77**, 013007 (2008).

NEUTRINO PHYSICS AND LEPTONIC WEAK BASIS INVARIANTS

M. N. Rebelo [a]

Centro de Física Teórica de Partículas – CFTP and Dept de Física, Instituto Superior Técnico – IST, Universidade de Lisboa (UL), Av. Rovisco Pais, P-1049-001 Lisboa, Portugal

Abstract. In this talk we present a powerful tool applied to the study of Leptonic Physics. This tool is based on the construction of Weak Basis invariant relations associated to different properties of leptonic models. The rationale behind these constructions is the fact that fermion mass matrices related though weak basis transformations look different but lead to the same physics. Such invariants can be built, for instance, with the aim to test leptonic models for different types of CP violation. These invariants are also relevant beyond such tests and have been applied to the study of implications from zero textures appearing in the leptonic mass matrices. In this case an important question is, how can a flavour model corresponding to a set of texture zeros be recognised, when written in a different weak basis, where the zeros are not explicitly present. Another important application is the construction of invariants sensitive to the neutrino mass ordering and the θ_{23} octant.

1 Introduction

One of the major puzzles in Particle Physics is the origin of fermion masses, mixing and CP violation, the so-called Flavour Puzzle. It is by now established that at least two of the three active light neutrinos have non-zero masses. In the Standard Model neutrinos are strictly massless. Accounting for neutrino masses requires physics beyond the Standard Model and has profound phenomenological implications. Neutrino Physics is at present an important field of research with many different experiments taking data and future new facilities and upgrades being planned. Among the fundamental open questions in this field are whether or not neutrinos are Dirac or Majorana particles, whether or not there is CP violation in the leptonic sector, what is the absolute neutrino mass scale and what is the mass ordering, meaning, what is the sign of $m_3^2 - m_1^2$? Is the leptonic mixing matrix, U_{PMNS} unitary? Are there sterile neutrinos? Do neutrinos have nonstandard interactions?

Attempts at solving the Flavour Puzzle are often based on the use of symmetries or else of special textures for the mass matrices. The fact that fermion mass matrices related through weak basis transformations look different while leading to the same physics raises the question of how to recognise the same model written in different bases where the symmetry (or the special texture) is not apparent. A special technique based on the use of weak basis (WB) invariants has been developed to tackle this problem. In this talk special WB

---

[a] E-mail: rebelo@tecnico.ulisboa.pt

invariant conditions adapted to answer several different specific questions concerning leptonic physics are presented. The same technique has been applied to the quark sector. The use of Higgs basis invariants based on the same rationale are extremely useful in the study of the scalar potential for multi-Higgs models.

2 Building of weak basis invariant conditions

The technique employed in this talk was developed for the fist time by the authors of Ref. [1] applied to the study of the CP properties of the quark sector in the Standard Model. After spontaneous symmetry breaking quark masses are generated and can be written as:

$$\mathcal{L}_m(\text{quarks}) = -\overline{u_L^0}\, m_u\, u_R^0 - \overline{d_L^0}\, m_d\, d_R^0 + \text{h.c.} \tag{1}$$

still in a weak basis where, by definition of weak basis, the charged current is diagonal. The charged current has the form:

$$\mathcal{L}_W(\text{quarks}) = -\frac{g}{\sqrt{2}}\, W_\mu^+\, \overline{u_L^0}\, \gamma^\mu\, d_L^0 + \text{h.c.} \tag{2}$$

Weak basis transformations are given by:

$$d_L^0 \longrightarrow U' d_L^0, \quad u_L^0 \longrightarrow U' u_L^0, \quad d_R^0 \longrightarrow W' d_R^0, \quad u_R^0 \longrightarrow V' r_R^0 \tag{3}$$

with U', V' W' arbitrary unitary matrices, whereas the most general CP transformation for fermion fields in a weak basis allows for the combination of the definition of the CP transformation of a single fermion with a weak basis transformation [2], this is so because the fermionic CP transformation must be defined taking into account the part of the Lagrangian that conserves CP, to wit, the fermion gauge couplings. Pure gauge theories together with fermions do not violate CP [3]. The most general CP transformation allowed by these couplings is then:

$$\begin{aligned}
\text{CP} u_L^0 (\text{CP})^\dagger &= U' \gamma^0 C\, \overline{u_L^0}^T; \quad \text{CP} u_R^0 (\text{CP})^\dagger = V' \gamma^0 C\, \overline{u_R^0}^T \\
\text{CP} d_L^0 (\text{CP})^\dagger &= U' \gamma^0 C\, \overline{d_L^0}^T; \quad \text{CP} d_R^0 (\text{CP})^\dagger = W' \gamma^0 C\, \overline{d_R^0}^T
\end{aligned} \tag{4}$$

In order for the Lagrangian to be CP invariant the following relations have to be verified:

$$U'^\dagger m_u V' = m_u^* \tag{5}$$
$$U'^\dagger m_d W' = m_d^* \tag{6}$$

this means that there must be unitary matrices U', V' W' that obey these relations. For real mass matrices these relations are trivially satisfied with identity matrices. Looking for such unitary matrices in more general cases is not simple, however by combining these relations in such a way as to produce similarity transformations it is possible to obtain simple conditions expressed only in terms of the mass matrices by applying traces and determinants. In this way one may obtain a simple condition [1]:

$$\text{tr}\,[H_u, H_d]^3 = 0 \tag{7}$$

which is a necessary and sufficient condition for CP conservation in the SM. For three generations this condition is equivalent to

$$\det[H_u, H_d] = 0 \tag{8}$$

which was obtained in Ref. [4] for the particular weak basis where the quark mass matrices are Hermitian. However this choice of a particular basis is not necessary since both conditions are weak basis invariant.

Reference [1] was the starting point for the development of an extremely powerful technique to test for CP violation in many different scenarios.

2.1 Testing for CP violation, leptonic sector, low energies

At low energies, assuming that the lepton number is violated, one can write an effective Majorana neutrino mass matrix and the leptonic mass terms are then of the form:

$$\mathcal{L}_{\text{mass}} = -\frac{1}{2}\nu_L^{0\,T} C^{-1} m_\nu \nu_L^0 - \overline{\ell_L^0} m_\ell \ell_R^0 + \text{h.c.}\,, \tag{9}$$

the charged currents have a form similar to the one of Eq. (2) written in terms of leptons. The WB transformations in the leptonic sector are given by:

$$\nu_L^0 \to V\nu_L^0, \quad \ell_L^0 \to V\ell_L^0, \quad \ell_R^0 \to W\ell_R^0 \tag{10}$$

with V and W unitary 3×3 matrices. Leptonic mixing and CP violation in the leptonic sector is described by the Pontecorvo-Maki-Nakagawa-Sakata (PMNS), U_{PMNS} matrix. Using the standard parametrisation [5] this matrix can be written:

$$U_{PMNS} = \begin{pmatrix} c_{12}c_{13} & s_{12}c_{13} & s_{13}e^{-i\delta} \\ -s_{12}c_{23} - c_{12}s_{23}s_{13}e^{i\delta} & c_{12}c_{23} - s_{12}s_{23}s_{13}e^{i\delta} & s_{23}c_{13} \\ s_{12}s_{23} - c_{12}c_{23}s_{13}e^{i\delta} & -c_{12}s_{23} - s_{12}c_{23}s_{13}e^{i\delta} & c_{23}c_{13} \end{pmatrix} \cdot P$$

$$P = \text{diag}\,(1, e^{i\alpha_{21}}, e^{i\alpha_{31}}) \tag{11}$$

where $c_{ij} \equiv \cos\theta_{ij}$ and $s_{ij} \equiv \sin\theta_{ij}$. There are three CP violating phases, δ, α_{21}, and α_{31} in Eq. (11). In Ref. [6] a set of necessary and sufficient WB invariant conditions for CP invariance were derived, valid in the case of three generations for nonzero and nondegenerate masses:

$$\text{Im Tr } [h_\ell \cdot m_\nu^* \cdot m_\nu \cdot m_\nu^* \cdot h_\ell^* \cdot m_\nu] = 0 \quad (12)$$
$$\text{Im Tr } [h_\ell \cdot (m_\nu^* \cdot m_\nu)^2 \cdot (m_\nu^* \cdot h_\ell^* \cdot m_\nu)] = 0 \quad (13)$$
$$\text{Im Tr}[h_\ell \cdot (m_\nu^* \cdot m_\nu)^2 \cdot (m_\nu^* \cdot h_\ell^* \cdot m_\nu)(m_\nu^* \cdot m_\nu)] = 0 \quad (14)$$
$$\text{Im Tr}[(m_\nu \cdot h_\ell \cdot m_\nu^*) + (h_\ell^* \cdot m_\nu \cdot m_\nu^*)] = 0 \quad (15)$$

In Ref. [7] a minimal set of necessary and sufficient conditions for CP invariance was given:

$$\text{Tr } [m_\nu^* \cdot m_\nu^T, \ h_\ell]^3 = 0 \quad (16)$$
$$\text{Tr } [m_\nu \cdot h_\ell \cdot m_\nu^*, h_\ell^*]^3 = 0 \quad (17)$$
$$\text{ImTr } (h_\ell \cdot m_\nu^* \cdot m_\nu \cdot m_\nu^* \cdot h_\ell^* \cdot m_\nu) = 0 \quad (18)$$

Equation (16) is similar to Eq. (7) which was derived for the quark sector, this invariant is only sensitive to the Dirac-type phase, δ. The other two invariants are sensitive both to Dirac and Majorana type phases. The invariant of Eq. (17) was first derived in Ref [8] applied to the study of CP violation in the context of three degenerate neutrinos which still allows for Majorana-type CP violation. The invariant of Eq. (18) coincides with Eq. (12) and was derived before in Ref. [6], applied to the case of two generations, which also allows for Majorana-type CP violation.

2.2 Testing for CP violation, leptonic sector, Leptogenesis

In the minimal seesaw framework [9–13], one introduces righthand neutrino fields which are singlets of $SU(2) \times U(1)$. The most general leptonic mass term may then be written as:

$$\mathcal{L}_m = -[\overline{\nu_L^0} m_D \nu_R^0 + \frac{1}{2} \nu_R^{0T} C M_R \nu_R^0 + \overline{l_L^0} m_l l_R^0] + h.c. \quad (19)$$

Let us assume that three righthanded neutrinos are introduced, this does not need to be the case, one needs at least two. The scale of M_R can be much higher than the electroweak scale and in this case the seesaw mechanism operates. In the context of seesaw, it is possible to generate a lepton number asymmetry through the decays of the heavy Majorana neutrinos. This is the so-called Leptogenesis mechanism [14–16] and requires CP violation at high energies. In the general seesaw framework it is not possible to establish a connection between leptonic CP violation at low energies and CP violation at high energies [17,18]. Such a relation can only be established in the context of a flavour theory.

For Leptogenesis in the single flavour approximation, i.e., in the case when washout effects are not sensitive to the different flavours of charged leptons into which the heavy neutrino decays, the possibility of having Leptogenesis can be probed by means of the following invariant conditions [17]:

$$I_1 \equiv \mathrm{Im Tr}[m_D m_D^\dagger M_R^\dagger M_R M_R^* m_D^T m_D^* M_R] = 0 \qquad (20)$$

$$I_2 \equiv \mathrm{Im Tr}[m_D m_D^\dagger (M_R^\dagger M_R)^2 M_R^* m_D^T m_D^* M_R] = 0 \qquad (21)$$

$$I_3 \equiv \mathrm{Im Tr}[m_D m_D^\dagger (M_R^\dagger M_R)^2 M_R^* m_D^T m_D^* M_R M_R^\dagger M_R] = 0 \qquad (22)$$

The first one of these invariants was discussed in Ref. [19]. For a detailed discussion of these invariant conditions see Ref. [17]. In the case of flavoured Leptogenesis additional CP-odd weak basis invariant conditions are required. A simple choice of additional invariant conditions is obtained by replacing $m_D m_D^\dagger$ in the above equations by $m_D h_l m_D^\dagger$, where $h_l = m_l m_l^\dagger$ [17].

2.3 Beyond tests for CP violation. Texture zeros

The imposition of texture zeros in the Yukawa couplings allows to establish a connection between low energy physics and physics at high energies, for instance Leptogenesis. In Ref. [20] we addressed the question of how to recognise flavour models corresponding to a set of texture zeros when written in a different weak basis where the zeros are not explicitly present. We considered texture zeros in m_D in the seesaw framework, appearing in the weak basis where M_R and m_l are diagonal and real and we found invariants that vanish for different classes of textures. One such example is:

$$I = \mathrm{Tr}[m_D M_R^\dagger M_R m_D^\dagger, h_l]^3 \qquad (23)$$

Implications for two texture zeros in m_D in the case of two righthanded neutrinos were studied in [21]. In Ref [22] we classified all allowed four zero textures in m_D with three righthanded neutrinos and we showed that in general CP may be violated both at low and high energies. Furthermore, in all these cases the parameters relevant for Leptogenesis can be fully specified in terms of light and heavy neutrino masses and low energy leptonic mixing.

2.4 Beyond tests for CP violation. The octant of θ_{23} together with the neutrino mass ordering

In Ref. [23] we have built weak basis invariants that provide a clear indication of whether a particular lepton flavour model leads to normal or inverted hierarchy for neutrino masses and what is the octant of θ_{23}, of the standard parametrisation of U_{PMNS}. It was shown in Ref. [23] that the sign of the invariant:

$$\tilde{I}_1 \equiv Tr[H_\ell \ H_\nu] - \frac{1}{3} Tr[H_\ell] Tr[H_\nu] \qquad (24)$$

indicates the ordering of the neutrino masses and that the invariant:

$$\tilde{I}_2 \equiv Tr[H_\ell]\, Tr[H_\ell^2\, H_\nu] - Tr[H_\ell^2]Tr[H_\ell\, H_\nu] \qquad (25)$$

is sensitive to the θ_{23} octant. In Table 1 we show how to combine the information provided by these two invariants. In Ref. [23] these invariants were applied

Table 1: Combination of the two invariants. NO stands for normal ordering, IO for inverted ordering.

	$\tilde{I}_2 > 0$	$\tilde{I}_2 < 0$
$\tilde{I}_1 > 0$	NO, $\theta_{23} < \pi/4$	NO, $\theta_{23} > \pi/4$
$\tilde{I}_1 < 0$	IO, $\theta_{23} > \pi/4$	IO, $\theta_{23} < \pi/4$

to specific Ansätze studied in the literature [24]. For a different strategy see [25].

3 Conclusions

This talk presents a very powerful tool based on the derivation of weak basis invariant conditions that are extremely useful for model building. Such conditions have been widely used and derived in the literature in different contexts by many different authors. Ref. [26] provides a long list of references for conditions relevant to the study CP violation both in the quark and in the leptonic sector, as well as in the Higgs sector, in several extensions of the Standard Model.

Acknowledgments

The author thanks the Organisers of the 18th Lomonosov Conference for the very fruitful scientific meeting and the warm hospitality. This work was partially supported by Fundação para a Ciência e a Tecnologia (FCT, Portugal) through the projects CERN/FIS-NUC/0010/2015, and CFTP-FCT Unit 777 (UID/FIS/00777/2013) which are partially funded through POCTI (FEDER), COMPETE, QREN and EU.

References

[1] J. Bernabeu, G. C. Branco and M. Gronau, Phys. Lett. B **169** (1986) 243.
[2] G. Ecker, W. Grimus and W. Konetschny, Nucl. Phys. B **191** (1981) 465.
[3] W. Grimus and M. N. Rebelo, Phys. Rept. **281** (1997) 239.

[4] C. Jarlskog, Phys. Rev. Lett. **55** (1985) 1039.
[5] C. Patrignani et al. [Particle Data Group], Chin. Phys. C **40** (2016) no.10, 100001.
[6] G. C. Branco, L. Lavoura and M. N. Rebelo, Phys. Lett. B **180** (1986) 264.
[7] H. K. Dreiner, J. S. Kim, O. Lebedev and M. Thormeier, Phys. Rev. D **76** (2007) 015006.
[8] G. C. Branco, M. N. Rebelo and J. I. Silva-Marcos, Phys. Rev. Lett. **82** (1999) 683.
[9] P. Minkowski, Phys. Lett. **67B** (1977) 421.
[10] T. Yanagida, Conf. Proc. C **7902131** (1979) 95.
[11] S. L. Glashow in M. Levy, J. L. Basdevant, D. Speiser, J. Weyers, R. Gastmans and M. Jacob, "Quarks And Leptons. Proceedings, Summer Institute, Cargese, France, July 9-29, 1979," NATO Sci. Ser. B **61** (1980) pp.1.
[12] M. Gell-Mann, P. Ramond and R. Slansky, Conf. Proc. C **790927** (1979) 315.
[13] R. N. Mohapatra and G. Senjanovic, Phys. Rev. Lett. **44** (1980) 912.
[14] M. Fukugita and T. Yanagida, Phys. Lett. B **174** (1986) 45.
[15] For a review on Leptogenesis see: S. Davidson, E. Nardi and Y. Nir, Phys. Rept. **466** (2008) 105 and references therein.
[16] See G. C. Branco, R. G. Felipe and F. R. Joaquim, Rev. Mod. Phys. **84** (2012) 515 for a review on leptonic CP violation and references therein.
[17] G. C. Branco, T. Morozumi, B. M. Nobre and M. N. Rebelo, Nucl. Phys. B **617** (2001) 475.
[18] M. N. Rebelo, Phys. Rev. D **67** (2003) 013008.
[19] A. Pilaftsis, Phys. Rev. D **56** (1997) 5431.
[20] G. C. Branco, M. N. Rebelo and J. I. Silva-Marcos, Phys. Lett. B **633** (2006) 345.
[21] A. Ibarra and G. G. Ross, Phys. Lett. B **591** (2004) 285.
[22] G. C. Branco, D. Emmanuel-Costa, M. N. Rebelo and P. Roy, Phys. Rev. D **77** (2008) 053011.
[23] G. C. Branco, M. N. Rebelo and J. I. Silva-Marcos, JHEP **1711** (2017) 001.
[24] P. H. Frampton, S. L. Glashow and D. Marfatia, Phys. Lett. B **536** (2002) 79.
[25] L. M. Cebola, D. Emmanuel-Costa and R. G. Felipe, Phys. Rev. D **92** (2015) no.2, 025005
[26] G. C. Branco, L. Lavoura and J. P. Silva, *"CP Violation,"* Int. Ser. Monogr. Phys. **103** Oxford University Press (1999).

NONSTANDARD NEUTRINO INTERACTIONS IN A DENSE MAGNETIZED MEDIUM

A. V. Borisov [a]

Faculty of Physics, Moscow State University, 119991 Moscow, Russia

P. E. Sizin [b]

National University of Science and Technology MISiS, 119049 Moscow, Russia

Abstract. We calculate the neutrino luminosity of a degenerate electron gas in a strong magnetic field under the conditions of the neutron star crust via plasmon decay to a neutrino pair due to nonstandard interaction induced by a neutrino electric millicharge. We obtain the relative upper bounds on the millicharge by comparison with the standard electroweak mechanism of the plasmon decay as well as with one via a neutrino magnetic moment.

Currently, great attention is devoted to investigation of nonstandard neutrino interactions (NSNI) for searching for new physics beyond the Standard Model (SM) (see, e.g., [1] and references therein). Among various NSNI, neutrino electromagnetic interactions (NEMI) perform a very important role [2]. In the SM, NEMI are generated by radiative corrections due to strictly zero neutrino electric charge. But there are many ways to extend the SM for introducing particles of small (not necessarily rational in the units of the positron charge e) electric charges (millicharges). A minimal extension is to remain with the SM particle content and allow the neutrinos to have millicharges $e_\nu = q_\nu e$ ($|q_\nu| \ll 1$). The SM hypercharge operator can be redefined in a special way, and as a result neutrinos acquire millicharges without destroying the SM anomaly cancellation (for a review, see [2–4]). The present upper bounds on the neutrino millicharge (NMC) cover a wide range [5]:

$$10^{-4} \lesssim |q_\nu| \lesssim 10^{-15}. \quad (1)$$

The best model-independent astrophysical limit was obtained from the SN 1987A neutrino signal measurements [6]:

$$|q_\nu| \lesssim 2 \times 10^{-17}. \quad (2)$$

Analysis of the influence of the NMC on the rotation of a magnetized star during a core-collapse supernova explosion gives a stronger bound [7]:

$$|q_\nu| \lesssim 1.3 \times 10^{-19}. \quad (3)$$

But the strongest bound on the NMC was derived from electric charge conservation in the neutron beta decay and the neutrality of matter [8]:

$$|q_\nu| \lesssim 3 \times 10^{-21}, \quad (4)$$

---

[a] E-mail: borisov@phys.msu.ru
[b] E-mail: mstranger@list.ru

while a detailed analysis gives [2]: $q_\nu = (-0.6 \pm 3.2) \times 10^{-21}$.

In the present report, we investigate a contribution of the NMC to the neutrino luminosity due to plasmon decay to a neutrino pair $\gamma \to \nu\bar{\nu}$ in a strongly magnetized degenerate electron gas under conditions of the neutron star crust [9, 10]. Comparing this contribution to the luminosity and the SM one [11], we obtain relative upper bounds on $|q_\nu|$. The analogous method has been used for deriving bounds on the neutrino magnetic moment from comparative analysis of contributions of electromagnetic and weak mechanisms to the neutrino luminosity via neutrino-pair photoproduction on an electron [12] and plasmon decay to a neutrino pair [13] under the same conditions.

We have calculated the neutrino luminosity taking into account the SM contribution [11] and one via the NMC (we use the system of units where $\hbar = c = k_B = 1$, $\alpha = e^2/4\pi \simeq 1/137$):

$$Q = \frac{G_F^2 \omega_p^4 T^5}{48\pi^4 \alpha} \left[\zeta(3) \left(\langle g_V^2 \rangle - \langle g_A^2 \rangle + \langle g_V \rangle F q_\nu + \frac{3}{2} F^2 q_\nu^2 \right) \left(\frac{\omega_p}{T} \right)^2 \right.$$
$$\left. + 8\zeta(5) \langle g_A^2 \rangle \right] = Q_{SM} + Q_{el}^{(1)} + Q_{el}^{(2)}, \quad (5)$$

where $Q_{SM} = Q(q_\nu = 0)$ is the SM luminosity, $Q_{el}^{(2)} \sim q_\nu^2$ is the NMC contribution, and $Q_{el}^{(1)} \sim q_\nu$ is a result of the interference between the SM and NMC contributions to the amplitude of the plasmon decay. In Eq. (5), G_F is the Fermi coupling constant, the effective (after summation over the neutrino flavors) vector and axial couplings are $\langle g_V^2 \rangle \simeq 0.929$, $\langle g_V \rangle \simeq 0.886$, $\langle g_A^2 \rangle = 3/4$; $F = 4\sqrt{2}\pi\alpha/(G_F \omega_p^2)$; $\omega_p = [(2\alpha/\pi)(p_F/\varepsilon_F)H/H_0]^{1/2} m$ is the plasmon frequency [14]; T, ε_F and p_F are the temperature, Fermi energy and momentum of the degenerate electron gas; H is the magnetic field strength, $H_0 = m^2/e \simeq 4.41 \times 10^{13}$ G, m and $-e$ are the electron mass and charge. We note that Eq. (5) is an asymptotic formula derived under conditions

$$T \ll \varepsilon_F - m, \; H > (p_F/m)^2 H_0/2, \; \omega_p/T \ll 1, \quad (6)$$

forcing relativistic electrons ($p_F/m \gg 1$) to occupy only the lowest Landau level in the magnetic field with $p_F = 2\pi^2 n_e/(eH)$, where n_e is the electron concentration.

Assuming $Q_{el}^{(1)} + Q_{el}^{(2)} < Q_{SM}$, from Eq. (5), we obtain the constraint

$$\left(q_\nu^2 + 1.46 \times 10^{-14} H_{13} q_\nu \right)^{1/2} < 2.39 \times 10^{-14} H_{13}^{1/2} T_8, \quad (7)$$

where $H_{13} = H/(10^{13}$ G$)$, $T_8 = T/(10^8$ K$)$. From (7), taking the numerics suitable for the conditions (6), $H_{13} = 40$ and $T_8 = 35$, the upper bounds on the NMC has been derived: $|q_\nu| \lesssim 5.0$ (5.6) $\times 10^{-12}$ for $q_\nu > 0$ (< 0), which are somewhat weaker than one derived in [15] from the reactor experiment on antineutrino scattering: $|q_\nu| \lesssim 1.5 \times 10^{-12}$.

The stronger bound on $|q_\nu|$ can be obtained from the requirement that the NMC luminosity $Q_{el}^{(2)}$ be less that one due to plasmon decay via the neutrino magnetic moment μ_ν [13]. It gives the limit

$$|q_\nu| < 6.62 \times 10^{-3} \hat{\mu}_\nu H_{13}^{1/2}, \qquad (8)$$

where $\hat{\mu}_\nu = \mu_\nu/\mu_B$ with μ_B being the Bohr magneton. Using the above value $H_{13} = 40$ and the recent bound $\hat{\mu}_\nu \lesssim 1.4 \times 10^{-13}$ set by the Borexino Collaboration [16], we obtain $|q_\nu| \lesssim 5.9 \times 10^{-15}$. This limit is stronger than one derived in [8] from the red giant cooling due to plasmon decay: $|q_\nu| \lesssim 2 \times 10^{-14}$.

The most severe (but academic) constraint on the NMC can be derived, if we insert the SM value $\hat{\mu}_\nu = 3.2 \times 10^{-19}(m_\nu/1 \text{ eV})$ in (8) (see [2,17]): $|q_\nu| < 2.1 \times 10^{-21}(m_\nu/1 \text{ eV})H_{13}^{1/2}$. For $H_{13} = 40$ and the neutrino mass $m_\nu \simeq 0.2$ eV (the cosmological bound on the sum of light neutrino masses [5]), we reproduce the bound (4).

In conclusion, we have obtained the relative bounds on the neutrino millicharge that are in agreement with the known ones [2,5].

References

[1] A. N. Khan and D. W. McKay, JHEP **1707**, 143 (2017).
[2] C. Giunti and A. Studenikin, Rev. Mod. Phys. **87**, 531 (2015).
[3] R. Foot, G. C. Joshi, R. R. Volkas, Mod. Phys. Lett. A **5**, 2721 (1990).
[4] R. Foot, H. Lew, and R. R. Volkas, J. Phys. G **19**, 361 (1993).
[5] C. Patrignani et al. (Particle Data Group), Chin. Phys. C **40**, 100001 (2016).
[6] G. Barbiellini and G. Cocconi, Nature **329**, 21 (1987).
[7] A. I. Studenikin and I. V. Tokarev, Nucl. Phys. B **884**, 396 (2014).
[8] G. G. Raffelt, Phys. Rep. **320**, 319 (1999).
[9] R. C. Duncan and C. Thompson, Astrophys. J. Lett. **392**, L9 (1992).
[10] P. Haensel, A. Y. Potekhin, and D. G. Yakovlev, *Neutron Stars 1. Equation of State and Structure* (Springer, New York, 2007).
[11] M. V. Chistyakov and D. A. Rumyantsev, JETP **107**, 533 (2008).
[12] A. V. Borisov, B. K. Kerimov, and P. E. Sizin, Phys. Atom. Nucl. **75**, 1305 (2012).
[13] A. V. Borisov and P. E. Sizin, Moscow Univ. Phys. Bull. **68**, 114 (2013).
[14] A. E. Shabad, Tr. FIAN **192**, 5 (1988).
[15] A. I. Studenikin, Europhys. Lett. **107**, 21001 (2014).
[16] J. Barranco, D. Delepine, M. Napsuciale, and A. Yebra, arXiv:1704.01549 [hep-ph].
[17] R. E. Shrock, Nucl. Phys. B **206**, 359 (1982).

NON-CONSERVATION OF THE LEPTON CURRENT AND ASYMMERTY OF RELIC NEUTRINOS

V.B. Semikoz [a] and Maxim Dvornikov [b]

Pushkov Institute of Terrestrial magnetism, Ionosphere and Radiowave Propagation (IZMIRAN) of the Russian Academy of Sciences, Russia, 108840 Moscow, Troitsk, Kaluzhskoe High Way, 4

Abstract. The neutrino asymmetry in the early universe plasma, $n_\nu - n_{\bar\nu}$, is calculated after the electroweak phase transition (EWPT). Using the neutrino Boltzmann equation, modified by the Berry curvature term in the momentum space, we establish the violation of the macroscopic neutrino current in plasma after EWPT.

1 Adler-Bell-Jackiw anomaly in early universe and Berry curvature

One can expect that after EWPT there appears at $T < T_{\rm EWPT}$ an initial neutrino asymmetry $n_{\nu_a} - n_{\bar\nu_a} = \xi_{\nu_a} T^3/6$, where $\xi_{\nu_a} = \mu_{\nu_a}/T \neq 0$ is the asymmetry parameter and μ_{ν_a} is the neutrino chemical potential. The concrete initial neutrino asymmetry in a hot plasma at $\mathcal{O}({\rm MeV}) < T < T_{\rm EWPT}$ is unknown except of the primordial nucleosynthesis (upper) bound on it, $|\xi_{\nu_a}| < 0.07$ at $T_{\rm BBN} = 0.1\,{\rm MeV}$ [1], accounting for neutrino oscillations with equivalent $\xi_{\nu_e} \sim \xi_{\nu_\mu} \sim \xi_{\nu_\tau}$ at $T \sim \mathcal{O}({\rm MeV})$. Let us try to find other ways to put a bound on relic neutrino asymmetries before the BBN time.

Since a neutrino has the zero electric charge, there is no triangle anomaly for it after EWPT, in contrast to charged fermions in QED. Let us remind that the Adler's triangle anomaly for chiral (massless) charged fermions,

$$\partial_t n_{\rm R,L} + (\nabla \cdot {\bf j}_{\rm R,L}) = \pm \frac{\alpha_{\rm em}}{\pi} ({\bf E} \cdot {\bf B}), \qquad \alpha_{em} = \frac{1}{137} \qquad (1)$$

provides the instability of a seed magnetic field in a relativistic plasma, e.g., in the case of the hot plasma of early universe in the presence of a difference between chemical potentials of right-handed and left-handed charged particles, $\mu_5 = (\mu_{\rm R} - \mu_{\rm L})/2 \neq 0$, which is diminishing because of the helicity flip when the fermion mass is accounted for [2].

Note that, recently, the anomaly in Eq. (1) was reproduced independently through the Boltzmann kinetic equation accounting for the Berry curvature in the momentum space in Ref. [3] (see Eq. (20) therein). Below we consider the chiral matter of massless neutrinos after EWPT described by the kinetic Boltzmann equation and obtain the analogues of the anomaly in Eq. (1) accounting for the Berry curvature resulting in the nonconservation of the four-current for the $\nu\bar\nu$ gas.

---

[a] E-mail: semikoz@yandex.ru
[b] E-mail: maxdvo@izmiran.ru

1.1 Berry curvature

The Berry curvature,

$$\mathbf{\Omega}_\mathbf{k}^\pm = \nabla_\mathbf{k} \times \mathbf{a}_\mathbf{k}^\pm = \pm \frac{\hat{\mathbf{k}}}{2k^2}, \quad \mathbf{n} \equiv \hat{\mathbf{k}} = \frac{\mathbf{k}}{k}, \quad n^2 = 1, \qquad (2)$$

where the upper sign stays for $\bar{\nu}_a$ and the lower one for ν_a, is given by the Berry connection $\mathbf{a}_\mathbf{k}^\pm$ which enters as the additional term in the action for chiral right-handed antineutrino and left-handed neutrino in the presence of electroweak interaction with $e^\pm$ plasma (compare for charged particles in [3],

$$S = \int dt \left[(\mathbf{k} - G_F \sqrt{2} c_V^a \delta \mathbf{j}^{(e)}) \dot{\mathbf{x}} - (\varepsilon_\mathbf{k} - G_F \sqrt{2} c_V^a \delta n^{(e)}) - \mathbf{a}_\mathbf{k}^\pm \cdot \dot{\mathbf{k}} \right]. \qquad (3)$$

To get Boltzmann equation $\partial_t f + \dot{\mathbf{x}} \partial_\mathbf{x} f + \dot{\mathbf{k}} \partial_\mathbf{k} f = J_{coll}$ we write down the corresponding equations of motion. Taking into account the Berry curvature in Eq. (2), these equations have the form (see the case of charged particles in Refs. [3, 4]):

$$\dot{\mathbf{x}} = \frac{\partial \varepsilon_\mathbf{k}}{\partial \mathbf{k}} - (\dot{\mathbf{k}} \times \mathbf{\Omega}_\mathbf{k}^\pm),$$

$$\dot{\mathbf{k}} = G_F \sqrt{2} c_V^a \left[-\frac{\partial \delta \mathbf{j}^{(e)}}{\partial t} - \nabla \delta n^{(e)} + \dot{\mathbf{x}} \times (\nabla \times \delta \mathbf{j}^{(e)}) \right]. \qquad (4)$$

Here $G_F = 1.17 \times 10^{-5} \text{GeV}^{-2}$ is the Fermi constant, $c_V^{(a)} = 2\xi \pm 0.5$ is the vector coupling constant for $\nu_a e$ interactions (upper sign stays for electron neutrinos), $\xi = \sin^2 \theta_W = 0.23$ is the Weinberg parameter in SM, $\delta n^{(e)}(\mathbf{x}, t) = n_e(\mathbf{x}, t) - n_{\bar{e}}(\mathbf{x}, t)$ is the asymmetry of the electron number density in $e^- e^+$ plasma, and $\delta \mathbf{j}^{(e)}(\mathbf{x}, t) = \mathbf{j}_e(\mathbf{x}, t) - \mathbf{j}_{\bar{e}}(\mathbf{x}, t)$ is the asymmetry of the electron three-current density.

2 Neutrino asymmetry generation accounting for the Berry curvature

2.1 Without a Berry curvature

Let us remind that the Boltzmann equation for neutrinos (antineutrinos) in unpolarized matter at the temperature $T \ll T_{\text{EWPT}}$ and without a Berry curvature has the form [5, 6], which results from the action in Eq. (3), when for $\mathbf{\Omega}_\mathbf{k} = 0$ and the velocity becomes a usual unit velocity of a massless particle, $\dot{\mathbf{x}} = \partial \varepsilon_\mathbf{k} / \partial \mathbf{k} = \mathbf{n}$:

$$\frac{\partial f^{(\nu_a, \bar{\nu}_a)}(\mathbf{k}, \mathbf{x}, t)}{\partial t} + \mathbf{n} \frac{\partial f^{(\nu_a, \bar{\nu}_a)}(\mathbf{k}, \mathbf{x}, t)}{\partial \mathbf{x}} \pm \left[\mathbf{E}_e(\mathbf{x}, t) + \mathbf{n} \times \mathbf{B}_e(\mathbf{x}, t) \right]$$

$$\times \frac{\partial f^{(\nu_a,\bar{\nu}_a)}(\mathbf{k},\mathbf{x},t)}{\partial \mathbf{k}} = J^{(\nu_a,\bar{\nu}_a)}(\mathbf{k},\mathbf{x},t). \quad (5)$$

Here for massless ν_a ($\bar{\nu}_a$) one substitutes the upper (lower) sign for the third (force) term given by the weak νe interactions in the Fermi approximation:

$$\mathbf{E}_e(\mathbf{x},t) = G_F\sqrt{2}c_V^a\left[-\nabla\delta n^{(e)}(\mathbf{x},t) - \frac{\partial \delta \mathbf{j}^{(e)}(\mathbf{x},t)}{\partial t}\right],$$
$$\mathbf{B}_e(\mathbf{x},t) = G_F\sqrt{2}c_V^a\nabla \times \delta\mathbf{j}^{(e)}(\mathbf{x},t). \quad (6)$$

Let us stress that neglecting the Berry curvature, we have the spectrum $k_0 = \varepsilon_\mathbf{k} = k$, for which the four-current

$$j_\mu^{(\nu_a,\bar{\nu}_a)}(\mathbf{x},t) = (n_{\nu_a,\bar{\nu}_a}(\mathbf{x},t), \mathbf{j}_{\nu_a,\bar{\nu}_a}(\mathbf{x},t)) = \int \frac{d^3k}{(2\pi)^3}\frac{k_\mu}{\varepsilon_k}f^{(\nu_a,\bar{\nu}_a)}(\mathbf{k},\mathbf{x},t), \quad (7)$$

is conserved as it follows from integration of Eq. (5) over momentum $\mathbf{k}$,

$$\frac{\partial j_\mu^{(\nu_a,\bar{\nu}_a)}(\mathbf{x},t)}{\partial x_\mu} = \frac{\partial n^{(\nu_a,\bar{\nu}_a)}(\mathbf{x},t)}{\partial t} + \frac{\partial \mathbf{j}^{(\nu_a,\bar{\nu}_a)}(\mathbf{x},t)}{\partial \mathbf{x}} = 0, \quad (8)$$

Note that the collision integral being integrated over $\mathbf{k}$ vanishes both for elastic and inelastic neutrino (antineutrino) scattering, $\int d^3k J^{(\nu_a,\bar{\nu}_a)}(\mathbf{k},\mathbf{x},t) = 0$.

2.2 Accounting for the Berry curvature

Now let us turn to the case of the Berry curvature in Eq. (2) considering, e.g., a generalization of the neutrino kinetic Eq. (5), see details in Ref. [7]. In full analogy with the approach in Refs. [3,4], one can obtain in that case instead of the neutrino asymmetry conservation, $d[n^{(\nu_a)}(\mathbf{x},t)) - n^{(\bar{\nu}_a)}(\mathbf{x},t))]/dt = 0$, as given by the current difference in Eq. (8) integrated over d^3x, its generation due to weak interactions via the effective electromagnetic fields $\mathbf{E}_e, \mathbf{B}_e$ given by Eq. (6), see details in [7],

$$\frac{d}{dt}(n_{\nu_a} - n_{\bar{\nu}_a}) = -\frac{1}{4\pi^2}\int\frac{d^3x}{V}(\mathbf{E}_e\cdot\mathbf{B}_e) = \frac{A^2}{8\pi^2}\int k^4\frac{\partial}{\partial t}h(k,t)dk. \quad (9)$$

Here $A = G_F\sqrt{2}c_V^a/e$, $e = \sqrt{4\pi\alpha_{em}} = 0.3 > 0$ are constants, $h(k,t)$ is a spectrum of the magnetic helicity density $h(t) = V^{-1}\int d^3x(\mathbf{A}\cdot\mathbf{B}) = \int dk h(k,t)$ for standard Maxwellian fields, $\mathbf{B} = \nabla \times \mathbf{A}$.

2.3 Monochromatic helicity spectrum

To demonstrate the possibility for generation of the neutrino asymmetry, driven by the electroweak interaction with e^-e^+ plasma, we consider a maximally helical seed magnetic field and adopt monochromatic spectrum of the magnetic helicity $h(k,t) = h(t)\delta(k - k_0)$, where $k_0 = r_D^{-1}$, $r_D = v_T/\omega_p = \omega_p^{-1}$ is the Debye radius in relativistic plasma, $T \gg m_e$, $\omega_p = \sqrt{4\pi\alpha_{em}n_e/3T}$ is the plasma frequency. Then we obtain from Eq. (9) the new conservation law for neutrinos, which is similar to the well-known one for charged particles, $d[(n_{eR} - n_{eL}) + \alpha_{em}h(t)/\pi]/dt = 0$ coming from the Adler anomaly in Eq. (1),

$$\frac{d}{dt}\left[(n_{\nu_a} - n_{\bar{\nu}_a}) - \frac{\alpha_{ind}^a}{2\pi}h(t)\right] = 0, \qquad (10)$$

where $\alpha_{ind}^a = \left[e_{ind}^{(\nu_a)}\right]^2/4\pi$ is the effective electromagnetic constant and $e_{ind}^{(\nu_a)}$ is the induced charge of a neutrino in plasma. For a Dirac neutrino, $e_{ind}^{(\nu_a)}$ was found in Refs. [8,9],

$$e_{ind}^{(\nu_a)} = -\frac{G_F c_V^a (1-\lambda)}{\sqrt{2}er_D^2}, \qquad (11)$$

where $\lambda = \mp 1$ is the helicity of a neutrino and the lower sign stays for a sterile particle. From the conservation law in Eq. (10) one obtains for the neutrino asymmetry $n_{\nu_a} - n_{\bar{\nu}_a} = T^3 \xi_{\nu_a}/6$ at temperature $T \ll T_0 \ll T_{EWPT}$,

$$\xi_{\nu_a}(T) = -0.712(c_V^a)^2 \times 10^{-15} \left(\frac{T_0}{m_p}\right)^4 \left(\frac{T_0}{T}\right)^3. \qquad (12)$$

For the initial temperature $T_0 = 10\,\text{GeV}$ still acceptable in the Fermi approximation for weak interactions, $T_0 \ll T_{EWPT} = 100\,\text{GeV}$, one gets at $T = O(\text{MeV})$ a huge asymmetry $\xi_{\nu_a} = -9.12(c_V^a)^2$ that contradicts the upper limit $|\xi_{\nu_e}| < 0.07$, $c_V^e = 0.96$ [1]. Requiring that the magnitude of ξ_{ν_e} in Eq. (12) does not exceed the limit in Ref. [1], we put here the upper bound for the initial temperature, $T_0 < 5\,\text{GeV}$. Thus, we reveal a new way how to generate the relic neutrino asymmetry.

References

[1] A. S. Dolgov et al., Nucl. Phys. B **632** (2002) 363.
[2] A. Boyarsky, J. Fröhlich, O. Ruchayskiy, Phys. Rev. Lett. **108** (2012) 031301.
[3] D. T. Son, N. Yamamoto, Phys. Rev D **87** (2013) 085016.
[4] N. Yamamoto, Phys. Rev. D **93** (2016) 065017.
[5] L. O. Silva et al., Phys. Rev. Lett. **83** (1999) 2703.
[6] V. N. Oraevsky, V. B. Semikoz, Astropart. Phys. **18** (2002) 261.
[7] M.S. Dvornikov, V.B. Semikoz, J. Exp. Theor. Phys. **124** (2017) 731.
[8] V. N. Oraevsky, V. B. Semikoz, Physica A **142** (1987) 135.
[9] J. Nieves, P. Pal, Phys. Rev. D **49** (1994) 1398.

BREAKINGS OF THE NEUTRINO μ-τ REFLECTION SYMMETRY

Zhen-hua Zhao [a]

Department of Physics, Liaoning Normal University, Dalian 116029, China

Abstract. The neutrino μ-τ reflection symmetry has been attracting a lot of interest as it predicts the interesting results $\theta_{23} = \pi/4$ and $\delta = \pm\pi/2$. But it is reasonable to consider its breaking either for some theoretical considerations or on the basis of experimental results. We perform a systematic study for the possible symmetry-breaking patterns and their implications for the mixing parameters. And the general treatment is applied to the specific symmetry breaking arising from the renormalization group equation effects for illustration.

1 Introduction

The neutrino mixing is described by the 3×3 unitary matrix

$$U = P_\phi \begin{pmatrix} c_{12}c_{13} & s_{12}c_{13} & s_{13}e^{-i\delta} \\ -s_{12}c_{23} - c_{12}s_{23}s_{13}e^{i\delta} & c_{12}c_{23} - s_{12}s_{23}s_{13}e^{i\delta} & s_{23}c_{13} \\ s_{12}s_{23} - c_{12}c_{23}s_{13}e^{i\delta} & -c_{12}s_{23} - s_{12}c_{23}s_{13}e^{i\delta} & c_{23}c_{13} \end{pmatrix} P_\nu , \quad (1)$$

with $P_\phi = \text{Diag}(e^{i\phi_1}, e^{i\phi_2}, e^{i\phi_3})$ and $P_\nu = \text{Diag}(e^{i\rho}, e^{i\sigma}, 1)$. The three mixing angles θ_{ij} (for $ij = 12, 13, 23$) and two neutrino mass-squared differences $\Delta m_{ij}^2 = m_i^2 - m_j^2$ (for $ij = 21, 31$) have been measured to a good accuracy [1]. Interestingly, θ_{23} is close to the special value $45°$ while the best-fit result for δ is around the special value $270°$ ($261° \pm 55°$ for NH and $277° \pm 43°$ for IH) [1]. But the sign of Δm_{31}^2 remains undetermined, thereby allowing for two possible mass orderings $m_1 < m_2 < m_3$ (the normal hierarchy or NH) or $m_3 < m_1 < m_2$ (the inverted hierarchy or IH). And the absolute neutrino mass scale (the lightest neutrino mass) is not known either.

How to understand the neutrino mixing pattern (particularly the special values of mixing parameters) poses an interesting question. Flavor symmetries might have played an important role in shaping the observed neutrino mixing pattern [2]. In this connection, the μ-τ reflection symmetry [3] may serve as a unique example: In the basis of the charged-lepton mass matrix being diagonal, the neutrino mass matrix M_ν features

$$M_{e\mu} = M_{e\tau}^* , \quad M_{\mu\mu} = M_{\tau\tau}^* , \quad M_{ee} \text{ and } M_{\mu\tau} \text{ being real} . \quad (2)$$

Such a symmetry leads us to the following special mixing parameters [4]

$$\phi_1 = 0 , \quad \phi_2 = -\phi_3 , \quad \theta_{23} = \pi/4 , \quad \delta = \pm\pi/2 , \quad \rho, \sigma = 0 \text{ or } \pi/2 . \quad (3)$$

---
[a] E-mail: zhzhao@itp.ac.cn

Because of these interesting predictions, this symmetry has been attracting a lot of interest recently.

However, the μ-τ reflection symmetry can hardly remain as an exact one. On the one hand, a recent NOvA result ($\theta_{23} = 39.5° \pm 1.7°$ or $52.1° \pm 1.7°$ in the NH case) disfavors $\theta_{23} = 45°$ with 2.6σ significance [5]. On the other hand, flavor symmetries are generally implemented at a superhigh energy scale and so the renormalization group equation (RGE) effects may provide a source for the symmetry breaking. We thus perform a systematic study for the possible symmetry-breaking patterns and their implications for the mixing parameters. For illustration, the general treatment will be applied to the specific symmetry breaking arising from the RGE effects.

2 Breakings of the μ-τ reflection symmetry

Above all, let us define some dimensionless parameters to measure the breaking strength of μ-τ reflection symmetry

$$\epsilon_1 = \frac{M_{e\mu} - M_{e\tau}^*}{M_{e\mu} + M_{e\tau}^*}, \quad \epsilon_2 = \frac{M_{\mu\mu} - M_{\tau\tau}^*}{M_{\mu\mu} + M_{\tau\tau}^*}, \quad \epsilon_3 = \frac{\text{Im}(M_{ee})}{\text{Re}(M_{ee})}, \quad \epsilon_4 = \frac{\text{Im}(M_{\mu\tau})}{\text{Re}(M_{\mu\tau})}, \quad (4)$$

which correspond to the four symmetry conditions in Eq. (2) one by one. These parameters should be small (say $|\epsilon_{1,2,3,4}| \leq 0.1$) in order to keep the symmetry as an approximate one. We point out that two of Im($\epsilon_{1,2}$) and $\epsilon_{3,4}$ can be made vanishing by rephasing the neutrino fields. Therefore, we take $\epsilon_{3,4} = 0$ in the following discussions without loss of generality.

When the μ-τ reflection symmetry is slightly broken, the deviations of mixing parameters from the special values given by Eq. (3)

$$\Delta\phi_1 = \phi_1' - 0, \quad \Delta\phi = (\phi_2' + \phi_3')/2 - 0, \quad \Delta\theta = \theta_{23}' - \pi/4,$$
$$\Delta\delta = \delta' - \delta, \quad \Delta\rho = \rho' - \rho, \quad \Delta\sigma = \sigma' - \sigma, \quad (5)$$

are expected to be some small quantities. These mixing-parameter deviations are directly controlled by the symmetry-breaking parameters

$$m_3 s_{13}^2 \Delta\delta + \overline{m}_1 c_{12}^2 \Delta\rho + \overline{m}_2 s_{12}^2 \Delta\sigma = (m_3 s_{13}^2 - m_{11})\Delta\phi_1,$$
$$2m_{12}\bar{s}_{13}\Delta\theta - m_{11}s_{13}^2\Delta\delta - \overline{m}_1 s_{12}^2 \Delta\rho - \overline{m}_2 c_{12}^2 \Delta\sigma = (m_{22} - m_3)\Delta\phi,$$
$$[m_{12} + \mathrm{i}(m_{11} + m_3)\bar{s}_{13}]\Delta\theta - (m_{11} - m_3)\bar{s}_{13}\Delta\delta - 2\overline{m}_1 c_{12}(\mathrm{i}s_{12} + c_{12}\bar{s}_{13})\Delta\rho$$
$$+2\overline{m}_2 s_{12}(\mathrm{i}c_{12} - s_{12}\bar{s}_{13})\Delta\sigma = [m_{12} - \mathrm{i}(m_{11} + m_3)\bar{s}_{13}](\mathrm{i}\Delta\phi_1 + \mathrm{i}\Delta\phi - \epsilon_1),$$
$$2(m_{22} - m_3)\Delta\theta - 2(m_{12} - \mathrm{i}m_{11}\bar{s}_{13})\bar{s}_{13}\Delta\delta - 2\overline{m}_1 s_{12}(\mathrm{i}s_{12} + 2c_{12}\bar{s}_{13})\Delta\rho$$
$$-2\overline{m}_2 c_{12}(\mathrm{i}c_{12} - 2s_{12}\bar{s}_{13})\Delta\sigma = (m_{22} + m_3 - 2\mathrm{i}m_{12}\bar{s}_{13})(2\mathrm{i}\Delta\phi - \epsilon_2). \quad (6)$$

In the above equation, the definitions

$$m_{11} = \overline{m}_1 c_{12}^2 + \overline{m}_2 s_{12}^2, \quad m_{12} = (\overline{m}_1 - \overline{m}_2)c_{12}s_{12}, \quad m_{22} = \overline{m}_1 s_{12}^2 + \overline{m}_2 c_{12}^2,$$
$$\overline{m}_1 = m_1 \exp[2\mathrm{i}\rho], \quad \overline{m}_2 = m_2 \exp[2\mathrm{i}\sigma], \quad \bar{s}_{13} = -\mathrm{i}s_{13}\exp[\mathrm{i}\delta], \quad (7)$$

have been used.

By solving Eq. (7) one will obtain $\Delta\theta$, $\Delta\delta$, $\Delta\rho$ and $\Delta\sigma$ as linear functions of $R_{1,2} = \text{Re}(\epsilon_{1,2})$ and $I_{1,2} = \text{Im}(\epsilon_{1,2})$, which can be parameterized as

$$\Delta\theta = c_{r1}^\theta R_1 + c_{i1}^\theta I_1 + c_{r2}^\theta R_2 + c_{i2}^\theta I_2 \,, \quad \Delta\delta = c_{r1}^\delta R_1 + c_{i1}^\delta I_1 + c_{r2}^\delta R_2 + c_{i2}^\delta I_2 \,,$$
$$\Delta\rho = c_{r1}^\rho R_1 + c_{i1}^\rho I_1 + c_{r2}^\rho R_2 + c_{i2}^\rho I_2 \,, \quad \Delta\sigma = c_{r1}^\sigma R_1 + c_{i1}^\sigma I_1 + c_{r2}^\sigma R_2 + c_{i2}^\sigma I_2 \,. \quad (8)$$

The coefficients measure the sensitive strengths of mixing-parameter deviations to the symmetry-breaking parameters. For example, c_{r1}^θ measures the sensitive strength of $\Delta\theta$ to R_1. The contribution of a given R_1 to $\Delta\theta$ is expressed as $c_{r1}^\theta R_1$. In consideration of $|\epsilon_{1,2}| \leq 0.1$, the coefficients must have magnitudes $\geq \mathcal{O}(1)$ to induce some sizeable (say $0.1 \simeq 6°$) mixing-parameter deviations. If one coefficient is much greater than 1, then even a tiny symmetry-breaking parameter can give rise to some sizable mixing-parameter deviation. But if one coefficient is much smaller than 1, then the resulting mixing-parameter deviation will be negligibly small.

It is found that the values of the coefficients are strongly dependent on the neutrino mass spectrum and the values of ρ and σ. So in Fig. 1 we choose to present the coefficients associated with R_1 (for the coefficients associated with I_1, R_2 and I_2, see Ref. [6]) against the lightest neutrino mass (m_1 in the NH case or m_3 in the IH case) for various combinations of ρ and σ. The black, red, green and blue colors are assigned to the coefficients for $\Delta\theta$, $\Delta\delta$, $\Delta\rho$ and $\Delta\sigma$, respectively. In order to save space, the absolute value of a coefficient will be shown in the dashed line if it is negative. By contrast, the full line will be used when the coefficients are positive. In the calculations we have specified $\delta = -\pi/2$.

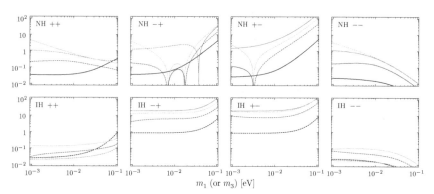

Figure 1: The coefficients associated with R_1 (c_{r1}^θ in black, c_{r1}^δ in red, c_{r1}^ρ in green and c_{r1}^σ in blue) against the lightest neutrino mass for various combinations of ρ and σ. The signs $++, -+, +-$ and $++$ respectively stand for $[\rho, \sigma] = [0,0], [\pi/2, 0], [0, \pi/2]$ and $[\pi/2, \pi/2]$.

From these numerical results one finds that: (1) For $\rho \neq \sigma$ (or $[\rho, \sigma] = [\pi/2, \pi/2]$), the coefficients get greatly enhanced (or suppressed) when the neutrino masses are quasi-degenerate. (2) $\Delta\theta$ is most sensitive to R_2 while $\Delta\delta$,

$\Delta\rho$ and $\Delta\sigma$ to all the symmetry-breaking parameters. In magnitude, the coefficients for $\Delta\delta$, $\Delta\rho$ and $\Delta\sigma$ (which can even obtain some magnitudes around 100 when the neutrino masses are quasi-degenerate and $\rho \neq \sigma$) are generally much greater than those for $\Delta\theta$. To be specific: (1) $|c_{r2}^{\theta}|$ takes values of $\mathcal{O}(1)$ in most cases. But it decreases to $\mathcal{O}(0.1)$ in the case of $m_3 \ll m_1 \simeq m_2$ combined with $\rho \neq \sigma$. $|c_{r1}^{\theta}|$ can also reach $\mathcal{O}(1)$ in the case of IH combined with $\rho \neq \sigma$. $|c_{i1}^{\theta}|$ and $|c_{i2}^{\theta}|$ are well below $\mathcal{O}(0.1)$. (2) $|c_{r1}^{\delta}|$ (except in the case of IH combined with $\rho = \sigma$), $|c_{i1}^{\delta}|$ and $|c_{i2}^{\delta}|$ generally have values of $\mathcal{O}(1)$ or greater. $|c_{r2}^{\delta}|$ can be significant only in the case of IH combined with $\rho \neq \sigma$. (3) The coefficients for $\Delta\rho$ obtain magnitudes of $\mathcal{O}(1)$ or greater in most cases, with the exceptions: $|c_{r1}^{\rho}|$ and $|c_{r2}^{\rho}|$ are substantially suppressed in the case of IH combined with $\rho = \sigma$. The coefficients for $\Delta\sigma$ almost share the same properties as their counterparts for $\Delta\rho$ except that their magnitudes are somewhat smaller.

Finally, we point out that the symmetry breaking triggered by the RGE effects is characterized by $\epsilon_2 = 2\epsilon_1 = \Delta_\tau$ and $\epsilon_{3,4} = 0$ [6]. Here Δ_τ is the only symmetry-breaking parameter and takes a value $0.042 \left(\tan\beta/50\right)^2$ for a flavor-symmetry scale being 10^{13} GeV in the MSSM (while the RGE-induced symmetry breaking is negligibly small in the SM). Obviously, one will have $\Delta\eta = c_\tau^\eta \Delta_\tau$ with $c_\tau^\eta = c_{r1}^\eta/2 + c_{r2}^\eta$ (for $\eta = \theta, \delta, \rho$ and σ).

Acknowledgments

I am grateful to Prof. Studenikin and the organizing committee for their warm hospitality at Moscow State University where the 18th Lomonosov Conference on Elementary Particle Physics was held. This work is supported in part by the National Natural Science Foundation of China under grant No. 11605081.

References

[1] I. Esteban, M. C. Gonzalez-Garcia, M. Maltoni, I. Martinez-Soler, T. Schwetz, JHEP **01** (2017) 087.
[2] For a review with extensive references, see S. F. King, C. Luhn, Rept. Prog. Phys. **76** (2013) 056201.
[3] P. F. Harrison, W. G. Scott, Phys. Lett. **B 547** (2002) 219. For a review with extensive references, see Z. Z. Xing, Z. H. Zhao, Rept. Prog. Phys. **79** (2016) 076201.
[4] W. Grimus, L. Lavoura, Phys. Lett. **B 579** (2004) 113.
[5] P. Adamson et al. (NOvA Collaboration), Phys. Rev. Lett. **118** (2017) 151802.
[6] Z. H. Zhao, JHEP **1709** (2017) 023.

NEUTRINO CLUSTERING IN THE MILKY WAY

Stefano Gariazzo [a]

Instituto de Física Corpuscular (CSIC-Universitat de València)
Parc Científic UV, C/ Catedrático José Beltrán, 2
E-46980 Paterna (Valencia), Spain

Abstract. The Cosmic Neutrino Background is a prediction of the standard cosmological model, but it has been never observed directly. In the experiments with the aim of detecting relic CNB neutrinos, currently under development, the expected event rate depends on the local density of relic neutrinos. Since massive neutrinos can be attracted by the gravitational potential of our galaxy and cluster around it, a local overdensity of cosmic neutrinos should exist. Considering the minimal masses guaranteed by neutrino oscillations, we review the computation of the local density of relic neutrinos and we present realistic prospects for a PTOLEMY-like experiment.

1 Introduction

The standard cosmological model predicts the existence of a relic population of neutrinos produced in the early Universe, which is usually referred to as the Cosmic Neutrino Background (CνB). These neutrinos have nowadays a distribution very close to a Fermi-Dirac with an effective temperature of about 1.9 K, or 0.17 meV. This is small enough to say that at least two neutrinos over three are non-relativistic today, since neutrino oscillation experiments tell us that the second lightest neutrino must have a mass of at least ~ 8 meV [1].

Despite being the second most copious species in the Universe after the photons of the Cosmic Microwave Background, with a mean number density of $\sim 330\,\mathrm{cm}^{-3}$, relic neutrinos are extremely difficult to detect, due to their very small energy. The most interesting method that we can exploit for their direct detection is the mechanism of neutrino capture (NC) in β-decaying nuclei, acting through the process $\nu + n \to p + e^-$ [2]. The interaction of a relic neutrino with a detector atom forces the production of an electron that has an energy above the endpoint of the standard β-decay by twice the neutrino mass: this can be visible in the experiment if the energy resolution is sufficient to distinguish the peak due to NC from the standard β-decay events (see e.g. [3]).

The PTOLEMY experiment [4], currently under development, aims at detecting relic neutrinos with a mass above ~ 150 meV, as the expected energy resolution of ~ 100 meV allows. The event rate from NC in the PTOLEMY detector, built with 100 g of atomic tritium, is expected to be of order ten per year, if the mean number density of the CνB is considered.

Even if neutrinos are very light, however, they are expected to cluster around our galaxy, thanks to the gravitational potential of the matter which forms it. An increased local number density of relic neutrinos would correspond to an

---
[a] E-mail: gariazzo.ific.uv.es

increased event rate in the detector. In the following we will show the results on the neutrino clustering in the Milky Way (MW), published in Ref. [5], which are based on the N-one-body simulation technique firstly presented in [6], and the corresponding prospects for the event rate in a PTOLEMY-like experiment.

2 N-one-body simulations and the Milky Way

We compute the clustering of relic neutrinos in the MW using the N-one-body technique [6], which consists in independently evolving the trajectories of a high number N of neutrinos of mass m_ν in the gravitational potential of the MW, from some early time until today, sampling all the possible initial conditions (neutrino position and momentum). We assume initial homogeneity and spherical symmetry. The final positions of these test particles are then employed to reconstruct the relic neutrino distribution in the MW today. The ratio between the local number density at Earth, $n(m_\nu)$, and the mean number density, n_0, gives the clustering factor $f_c(m_\nu) = n(m_\nu)/n_0$, which enters the calculation of the event rate (see next section). In order to compute the neutrino trajectories, we must adopt a description for the MW content and its time evolution. We use results from the literature to describe the profiles and the evolution of the MW content (dark matter, baryons) as follows.

For the dark matter, we assume two possible descriptions for the halo: the Navarro-Frenk-White (NFW) and the Einasto (EIN) profiles, whose parametrizations are detailed in Ref. [5]. The parameters which enter the NFW and EIN profiles are determined using the astrophysical data from Ref. [7] on the dark matter density. The time dependence of the profiles is computed using the standard evolution of the universe assuming the ΛCDM model, the evolution of virial quantities and the results of N-body simulations as given in Ref. [8].

The baryon content of the MW is described using the profiles for the five components proposed in Ref. [9]: stars, warm and cold dust, atomic and molecular hydrogen gas. The time evolution of the total baryon profile is approximated as a global renormalization constant, which we obtain from N-body simulations of MW-sized objects [10]. For more details, we refer to Ref. [5].

3 Clustering factors and PTOLEMY prospects

We firstly show the results obtained considering neutrinos with nearly minimal masses. Assuming that the heaviest mass eigenstate has a mass of $\sim$60 meV, we run our N-one-body simulation and reconstruct the profile of the neutrino halo using different assumptions on the MW content, as shown in Figure 1, where we plot the results obtained using the NFW (EIN) dark matter profiles in the left (right) panel, alone and in combination with the MW baryons. We can see comparing the two plots that the local neutrino density can be up to

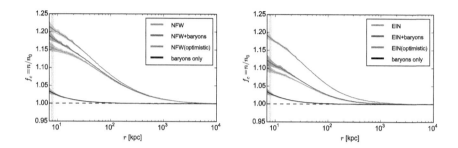

Figure 1: Profiles of the neutrino halo in the MW, for a neutrino with a mass of 60 meV and different parametrizations of the MW contents. From Ref. [5].

20% larger than the mean neutrino density. The most relevant source of error is represented by the MW structure, since the results can significantly change when different dark matter or baryon distributions are considered.

The rate of relic neutrino events expected in a PTOLEMY-like experiment can be computed using [3]:

$$\Gamma_{C\nu B} = \sum_{i=1}^{3} |U_{ei}|^2 \, f_c(m_i) \, [n_{i,0}(\nu_{h_R}) + n_{i,0}(\nu_{h_L})] \, N_T \, \bar{\sigma} \,, \qquad (1)$$

where U_{ei} is the matrix element that encodes the mixing between the neutrino mass eigenstate i and the electron neutrino flavour, $n_{i,0}(\nu_{h_{R(L)}})$ is the mean number density of right (left) helical neutrinos, N_T is the number of hydrogen nuclei in the detector and $\bar{\sigma} \simeq 3.834 \times 10^{-45}$ cm^{-2}. Since it contains the mixing matrix elements, Eq. (1) tells us that the event rate depends on the neutrino mass ordering, when clustered neutrinos are considered. In the case of normal mass ordering, for which the mixing between the electron neutrino and the heaviest mass eigenstate is the smallest, the enhanced local neutrino density has no impact on the expected event rate. On the other hand, if the ordering is inverted, the situation is opposite: the U_{e1} and U_{e2} terms are large and the increase in the event rate is directly proportional to the clustering factor.

The planned resolution for PTOLEMY, unfortunately, will not allow a detection of 60 meV neutrinos. For this reason we have also analysed the case of neutrinos with a mass of 150 meV, that should be the minimum mass detectable by the experiment. Considering this larger value of m_ν, neutrinos are practically degenerate in mass and the event rate is not influenced by the mass ordering. We get a clustering factor between 1.7 and 2.9, as depicted in Fig. 2, which corresponds to an increase of the event rate by the same factor. A precise determination of the event rate at PTOLEMY, in this case, would allow us to put constraints on the structure of our galaxy.

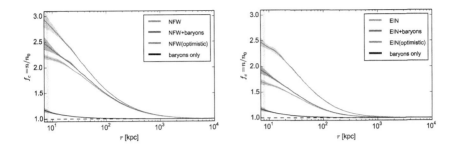

Figure 2: The same as Fig. 1, but for a neutrino mass of 150 meV. From Ref. [5].

Acknowledgments

Work supported by the Spanish grants FPA2014-58183-P and SEV-2014-0398 (MINECO), and by the PROMETEOII/2014/084 (Generalitat Valenciana).

References

[1] P. F. de Salas et al. Status of neutrino oscillations 2017. Arxiv:1708.01186. 2017.
[2] S. Weinberg. Universal Neutrino Degeneracy. *Phys. Rev.*, 128:1457–1473, 1962.
[3] A. J. Long et al. Detecting non-relativistic cosmic neutrinos by capture on tritium: phenomenology and physics potential. *JCAP*, 1408:038, 2014.
[4] S. Betts et al. Development of a Relic Neutrino Detection Experiment at PTOLEMY: Princeton Tritium Observatory for Light, Early-Universe, Massive-Neutrino Yield. Arxiv:1307.4738. 2013.
[5] P. F. de Salas et al. Calculation of the local density of relic neutrinos. *JCAP*, 09:034, 2017.
[6] A. Ringwald and Y. Y. Y. Wong. Gravitational clustering of relic neutrinos and implications for their detection. *JCAP*, 0412:005, 2004.
[7] M. Pato and F. Iocco. The Dark Matter Profile of the Milky Way: a Non-parametric Reconstruction. *Astrophys. J.*, 803(1):L3, 2015.
[8] A. A. Dutton and A. V. Macciò. Cold dark matter haloes in the Planck era: evolution of structural parameters for Einasto and NFW profiles. *Mon. Not. Roy. Astron. Soc.*, 441(4):3359–3374, 2014.
[9] A. Misiriotis et al. The distribution of the ISM in the Milky Way A three-dimensional large-scale model. *Astron. Astrophys.*, 459:113, 2006.
[10] F. Marinacci et al. The formation of disc galaxies in high resolution moving-mesh cosmological simulations. *Mon. Not. Roy. Astron. Soc.*, 437(2):1750–1775, 2014.

HIGH ENERGY NEUTRINOS AND DARK MATTER

Arman Esmaili [a]

Departamento de Física, Pontifícia Universidade Católica do Rio de Janeiro,
C. P. 38071, 22452-970, Rio de Janeiro, Brazil

Abstract. We discuss the possibility to interpret the observed high energy neutrino events at the IceCube detector by the PeV-scale decaying Dark Matter (DM). We consider both the energy and angular distributions of the observed events and confront them with the expected distributions in the decaying DM scenario. We consider the scenario that all the events are originating from the decay of DM; and also the scenario where both the decaying DM and the conventional power-law astrophysical flux contribute to the observed flux. We will show that the observed distributions mildly prefer the decaying DM scenario.

1 Introduction

After 4 years of data taking, the IceCube neutrino telescope at the south pole has observed 54 high energy starting events with the deposited energies from 20 TeV to 2 PeV [1-4]. Although the angular and energy distributions of these events strongly favor an extragalactic origin, still the precise source(s) of these neutrinos is (are) unknown. We will discuss the possibility of interpreting these events by PeV-scale decaying DM [5-12]. We consider the recent 4-year IceCube HESE data, which contains a combination of muon track as well as shower events. In particular, we perform a simultaneous likelihood analysis of the topology and energy distributions, including also the hemisphere of the events. We consider three possible scenarios as the origin of IceCube neutrinos: isotropic unbroken power-law flux, decaying DM, or a combination of both. We allow all signal parameters to vary, both the normalization and the spectral index for the astrophysical flux, as well as the DM mass, its decay lifetime and branching ratio for several two-body decay channels. In this way, we explore the parameter space of several combinations of fluxes from an astrophysical source and DM decays that gives the best fit to the observed event spectrum.

2 Neutrinos from dark matter decay

From the DM decay, the neutrino flux has two contributions: 1) an extragalactic component which originates from the decay of DM particles in halos at all redshifts, and thus it is isotropic; 2) a galactic contribution that comes from DM decays within our galactic halo and it is anisotropic, i.e., it follows the galactic morphology. The total flux is the sum of these two components (taking into account the neutrino oscillation). The spectrum of neutrino from DM decay at the point of production has been calculated using the event generator PYTHIA

---

[a] E-mail: arman@puc-rio.br

8.2 [13], which includes the weak gauge bosons radiation corrections. The details of this calculation are presented at [6,11].

3 Results

First we consider the scenario where both the DM decay and the conventional power-law astrophysical flux contribute to the observed neutrino flux at the IceCube detector. The decaying DM contribution is described in the previous section, while for the astrophysical flux we consider the following parametrization:

$$\left.\frac{d\Phi_{\text{astro},\nu_\alpha}}{dE_\nu}\right|_\oplus = \phi_{\text{astro}} \left(\frac{E_\nu}{100 \text{ TeV}}\right)^{-\gamma}, \qquad (1)$$

where γ is the spectral index and ϕ_{astro} is the flux normalization (per flavor). In what follows we consider this flux to be mainly originate from the usual hadronic production mechanism at neutrino sources, which results in the canonical homogeneous flavor ratios at Earth, $(1 : 1 : 1)_\oplus$, for both neutrinos and antineutrinos. Therefore, the combined flux from an astrophysical power-law and single-channel DM decays can be represented in terms of four fundamental parameters: the spectral index γ and normalization ϕ_{astro} of the astrophysical flux and the lifetime τ_{DM} and DM mass m_{DM} governing the neutrino flux from DM decays into channel c,

$$\frac{d\Phi^c}{dE_\nu}(E_\nu; \tau_{\text{DM}}, m_{\text{DM}}, \phi_{\text{astro}}, \gamma) = \frac{d\Phi^c_{\text{DM}}}{dE_\nu}(E_\nu; \tau_{\text{DM}}, m_{\text{DM}}) + \frac{d\Phi_{\text{astro}}}{dE_\nu}(E_\nu; \phi_{\text{astro}}, \gamma). \qquad (2)$$

We perform an unbinned extended maximum likelihood analysis, using the EM-equivalent deposited energy, event topology and hemisphere of origin of the events in the 4-year HESE sample. We calculate the best-fit values of the parameters and the 2D contour plots using the likelihood analysis.

The obtained best-fit for the DM mass for quark channels span more than an order of magnitude (from $m_{\text{DM}} \sim 500$ TeV for decays into $u\bar{u}$ to $m_{\text{DM}} \sim 11$ PeV for decays into $t\bar{t}$), while decays into the gauge and Higgs bosons, charged leptons, and neutrinos best-fit the data with DM masses in a narrower range, $m_{\text{DM}} \sim 4$–8 PeV. The former tend to better explain the low-energy excess in the sample, whereas the latter help to explain the PeV events (gauge boson channels also partly contribute to events at ~ 100 TeV). Except for DM decays into $u\bar{u}$ and $b\bar{b}$, the best fit for the astrophysical index points to a very soft spectrum ($\gamma > 3$), hard to explain with standard acceleration mechanisms. However, for the few cases with harder astrophysical flux ($\gamma < 2.5$), the corresponding DM lifetime is inevitably too low ($\tau_{\text{DM}} \lesssim 10^{27}$ s), and in tension with constraints from gamma-ray observations. Tables of the best-fit values for all the parameters in Eq. (2) for different decay channels c can be found in [11].

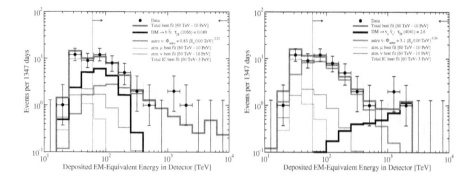

Figure 1: Event spectra in the IceCube detector for the decay channels: DM → $b\bar{b}$ (left panel) and DM → $\nu_e \bar{\nu}_e$ (right panel).

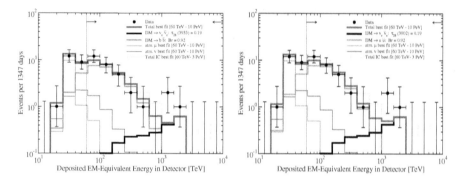

Figure 2: Event spectra in the IceCube detector in the scenario where the PeV-scale dark matter decay into two channels. The left panel is for DM → $\{92\% \, b\bar{b}, \, 8\% \, \nu_e \bar{\nu}_e\}$ and the right panel is for DM → $\{92\% \, u\bar{u}, \, 8\% \, \nu_e \bar{\nu}_e\}$, where the numbers depict the branching ratios.

We show in Fig. 1 the event spectra corresponding to the best-fit parameters for the single-channel DM decays plus astrophysical signal, for two representative DM decay channels: DM → $b\bar{b}$ and DM → $\nu_e \bar{\nu}_e$. Figure 2 shows the event spectra for DM decaying into two channels with the branching ratio indicated in the caption.

4 Conclusions

To conclude, in the scenario where the observed data is the result of a combination of events from DM decay and astrophysical power-law spectrum, we provide limits on the DM lifetime and indicate how future IceCube data could pin down the properties of the DM particle. We have also investigated the scenario where DM decays via two distinct channels, instead of restricting it to just one. In order to avoid increasing the number of parameters involved in the fit, we have left out the astrophysical flux from this scenario altogether. We have shown that if DM decays to a combination of soft and hard channels, the corresponding events can, by themselves, explain the IceCube dataset to a degree as good as, and indeed slightly better than, that obtained from combining single-channel DM decays with an astrophysical power-law flux. As intuitively expected, the best-fit $m_{\rm DM}$ in this case is found to be in the $\sim$ few PeV range, ensuring that DM decays into the hard channel explains the multi-PeV data, while the softer DM decay channel fill up the events at sub-PeV energies.

We conclude that the current 1347-day IceCube dataset prefers a fit involving multiple-component flux to a single power-law flux. In both of the scenarios that we have considered, we find excellent fits to the data and determine the preferred DM masses and lifetimes consistent with these fits. If future data from IceCube strengthens the trend of disfavoring an astrophysical-only power-law flux, a multi-component fit might likely be required to explain the observations.

References

[1] M. G. Aartsen *et al.* [IceCube Collaboration], Phys. Rev. Lett. **111**, 021103 (2013) [arXiv:1304.5356 [astro-ph.HE]].
[2] M. G. Aartsen *et al.* [IceCube Collaboration], Science **342**, 1242856 (2013) [arXiv:1311.5238 [astro-ph.HE]].
[3] M. G. Aartsen *et al.* [IceCube Collaboration], Phys. Rev. Lett. **113**, 101101 (2014) [arXiv:1405.5303 [astro-ph.HE]].
[4] IceCube collaboration, M. G. Aartsen et al., Proceedings of the 34th International Cosmic Ray Conference (ICRC 2015), 2015.
[5] A. Esmaili and P. D. Serpico, JCAP **1311**, 054 (2013) [arXiv:1308.1105 [hep-ph]].
[6] A. Esmaili, S. K. Kang and P. D. Serpico, JCAP **1412**, no. 12, 054 (2014) [arXiv:1410.5979 [hep-ph]].
[7] A. Esmaili and P. D. Serpico, JCAP **1510**, no. 10, 014 (2015) [arXiv:1505.06486 [hep-ph]].
[8] A. Esmaili, A. Ibarra and O. L. G. Peres, JCAP **1211**, 034 (2012) [arXiv:1205.5281 [hep-ph]].
[9] A. Esmaili Taklimi and P. Serpico, PoS DSU **2015**, 047 (2016).
[10] A. Esmaili, A. Palladino and F. Vissani, EPJ Web Conf. **116**, 11002 (2016).
[11] A. Bhattacharya, A. Esmaili, S. Palomares-Ruiz and I. Sarcevic, JCAP **1707**, no. 07, 027 (2017) [arXiv:1706.05746 [hep-ph]].
[12] A. Bhattacharya, M. H. Reno and I. Sarcevic, JHEP **1406**, 110 (2014) [arXiv:1403.1862 [hep-ph]].
[13] T. Sjstrand *et al.*, Comput. Phys. Commun. **191**, 159 (2015) [arXiv:1410.3012 [hep-ph]].

A SCALE-INVARIANT RADIATIVE NEUTRINO MASS GENERATION AND DARK MATTER

Salah Nasri [a]
Department of Physics, College of Science, United Arab Emirates University, PO Box 15551, Al Ain, United Arab Emirates

Abstract. I will discuss a minimal scale-invariant scotogenic model for neutrino mass and show that viable electroweak symmetry breaking can occur while simultaneously generating one-loop neutrino masses and the dark matter relic abundance.

1 The Scale Invariant Scotogenic Model Model

The scotogenic model is a simple framework that offers an explanation for the origin of the smallness of neutrino mass and the nature of dark matter [1]. Motivated by its simplicity, and our aim of understanding the origin of the electroweak scale, we investigate a minimal-scale invariant (SI) implementation of the scotogenic model [2, 3]. This is obtained by extending the standard model (SM) with three generations of gauge of gauge-singlet fermions, $N_{iR} \sim (1,1,0)$, where $i = 1, 2, 3$, labels generations, a second SM-like scalar doublet, $S \sim (1,2,1)$, and a singlet scalar $\phi \sim (1,1,0)$. A Z_2 symmetry with action $\{N_{iR}, S\} \to \{-N_{iR}, -S\}$ is imposed on the model, with all other fields being Z_2 even. The scalar ϕ, as well as the SM fields, transform trivially under the Z_2 symmetry. The scalar ϕ plays the dual role of sourcing lepton number violation to allow neutrino masses and triggering electroweak symmetry (EW) breaking.

With this field content, the most general Lagrangian consistent with both the SI and Z_2 symmetries contains the terms

$$\mathcal{L} \supset i\overline{N_{iR}}\gamma^\mu \partial_\mu N_{iR} + \frac{1}{2}(\partial^\mu \phi)^2 + |D^\mu S|^2 - \frac{y_i}{2} \phi \overline{N^c_{iR}} N_{iR}$$
$$- g_{i\alpha}\overline{N_{iR}} L_\alpha S - V(\phi, S, H), \quad (1)$$

where $L_\alpha \sim (1,2,-1)$ denotes the SM lepton doublets, with generations labeled by Greek letters, $\alpha, \beta = e, \mu, \tau$. We denote the SM scalar doublet as $H \sim (1,2,1)$ and $V(\phi, S, H)$ is the most-general scalar potential consistent with the symmetries of the model. Note that the SI symmetry precludes any dimensionfull parameters in the model, forbidding bare Majorana mass terms for the fermions N_i.

1.1 Symmetry breaking

In the absence of dimensionful parameters, the scalar potential contains only quartic interactions:

[a] E-mail: snasri@uaeu.ac.ae

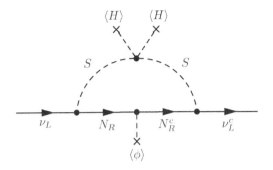

Figure 1: One-loop diagram for neutrino mass in the scale-invariant scotogenic model.

$$V(\phi, S, H) = \lambda_H |H|^4 + \frac{\lambda_\phi}{4}\phi^4 + \frac{\lambda_S}{2}|S|^4 + \frac{\lambda_{\phi H}}{2}\phi^2|H|^2 + \frac{\lambda_{\phi S}}{2}\phi^2$$
$$+ |S|^2 + \lambda_3|H|^2|S|^2 + \lambda_4|H^\dagger S|^2 + \frac{\lambda_5}{2}(S^\dagger H)^2 + \text{H.c.} \quad (2)$$

where λ_5 can be taken real without loss of generality. The desired VEV pattern has $\langle S \rangle = 0$, to preserve the Z_2 symmetry, with $\langle H \rangle \neq 0$ and $\langle \phi \rangle \neq 0$ to break both the SI and EW symmetries. The loop corrections to the potential involving SM fields are dominated by top-quark loops, and so to ensure viable symmetry breaking (equivalently, to give a positively-valued dilaton mass), loop corrections from the scalar S must be large.

1.2 Neutrino mass and lepton flavor violation

The combined terms in explicitly break lepton number symmetry. Neutrinos therefore acquire mass radiatively at the one-loop level, as shown in Fig. 1. Calculating the mass diagram gives

$$(\mathcal{M}_\nu)_{\alpha\beta} = \sum_i \frac{g_{i\alpha} g_{i\beta} M_i}{16\pi^2} \left\{ \frac{M_{S^0}^2}{M_{S^0}^2 - M_i^2} \ln \frac{M_{S^0}^2}{M_i^2} - \frac{M_A^2}{M_A^2 - M_i^2} \ln \frac{M_A^2}{M_i^2} \right\}. \quad (3)$$

In the limit that $M_{S^0}^2 \approx M_A^2 \equiv M_0^2$, this simplifies to

$$(\mathcal{M}_\nu)_{\alpha\beta} \simeq \sum_i \frac{g_{i\alpha} g_{i\beta} \lambda_5 v^2}{16\pi^2} \frac{M_i}{M_0^2 - M_i^2} \left\{ 1 - \frac{M_i^2}{M_0^2 - M_i^2} \ln \frac{M_0^2}{M_i^2} \right\}. \quad (4)$$

Note that the Z_2 symmetry prevents mixing between SM neutrino and the exotics N_i.

1.3 Dark matter

Since the gauge singlet fermions are odd under Z_2, the lightest among them, N_1, is stable, and hence is a candidate for dark matter (DM). In our model, the different DM annihilation channels are: (1) charged leptons and neutrinos LFV final states $\ell_\alpha^- \ell_\beta^+$ and $\nu_\alpha \bar\nu_\beta$, (2) SM fermions and gauge bosons $b\bar b$, $t\bar t$, W^+W^-, ZZ and the scalars SS final states, and (3) the Higgs dilaton final states $h_i h_k$. The first class channels are $h_{1,2}$-mediated s-channel processes, the second class ones are S-mediated t-channel processes while the third class channels are both Higgs s- and t-channels processes.

2 Results and Conclusion

We performed a numerical scan of the parameter space to determine whether radiative electroweak symmetry breaking is compatible with radiative neutrino mass and singlet neutrino dark matter. We also enforced the constraints from LEP (OPAL) on a light Higgs [4] and the experimental bound on the branching fraction of the LFV processes [5] and of the Higgs invisible decay [6]. In Figure 2-left we plot the contribution of each channel relative to the total cross section at freeze-out, σ_X/σ_{tot}, versus the DM mass. Figure 2-right shows the

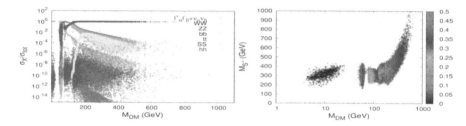

Figure 2: Left: The cross-section ratio σ_X/σ_{tot} at freeze-out versus the DM mass. Here X denotes lepton pairs, gauge bosons, heavy quarks and scalars. Right: The charged scalar masses M_{S^+} versus the DM mass. The palette shows the DM Yukawa coupling.

mass of the charged scalar, M_{S^+}, versus the DM mass. In the lighter DM mass range, $M_{DM} \leq \mathcal{O}(100)$ GeV, one notices that the charged scalar mass should not exceed 450 GeV, while for larger values of M_{DM} one can have M_{S^+} at the TeV scale. Such light charged scalars may be of phenomenological interest as they can be within reach of collider experiments. A severe constraints on model parameters come from dark direct detection searches, in particular the recent limits from LUX experiment [7]. Figure 3 shows that the surviving benchmark sets have $M_{DM} \leq 10$ GeV or $M_{DM} \geq 20$ GeV, while benchmarks with intermediate MDM values are excluded. The viable parameter space typically requires a lighter dilaton mass, $M_{h_2} \leq 10$ GeV, with all benchmarks with $M_{h_2} \geq 50$ GeV excluded.

In conclusion, we investigated a minimal SI scotogenic model and demonstrated the existence of viable parameter space in which one obtains radiative electroweak symmetry breaking, one-loop neutrino masses and a good

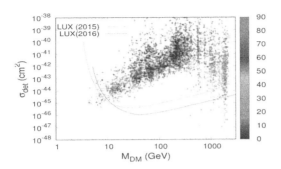

Figure 3: The direct detection cross section versus the DM mass. The dashed line shows the the recent constraints from LUX, while the palette gives the mass for the neutral beyond-SM scalar (dilaton), M_{h_2}, in units of GeV.

DM candidate. The model predicts a new scalar with $\mathcal{O}(\text{GeV})$ mass, which plays the dual roles of triggering electroweak symmetry breaking and sourcing lepton number symmetry violation. A viable parameter space was found for $M_{DM} \leq 10$ GeV and $M_{DM} \geq 200$ GeV, while intermediate values for M_{DM} appear excluded.

Acknowledgments

I would like to thank Alexander I. Studenikin for inviting me to give this talk at the 18th Lomonosov Conference on Elementary Particle Physics in Moscow.

References

[1] E. Ma, Phys. Rev. D **73**, 077301 (2006).
[2] A. Ahriche, A. Manning, K. L. McDonald and S. Nasri, Phys. Rev. D **94**, no. 5, 053005 (2016).
[3] A. Ahriche, K. L. McDonald and S. Nasri, JHEP **1606**, 182 (2016).
[4] G. Abbiendi *et al.* [OPAL Collaboration], Eur. Phys. J. C **27**, 483 (2003).
[5] J. Adam et al. [MEG Collaboration], arXiv:1303.0754 [hep-ex].
[6] P. Bechtle, S. Heinemeyer, O. Stl, T. Stefaniak and G. Weiglein, JHEP **1411**, 039 (2014).
[7] D. S. Akerib *et al.* [LUX Collaboration], Phys. Rev. Lett. **116**, no. 16, 161301 (2016).

THE COHERENT WEAK CHARGE OF MATTER

Alejandro Segarra [a]
*Departament de Física Teòrica and IFIC, Universitat de València – CSIC,
C/Dr. Moliner 50, E-46100 Burjassot (Spain)*

Abstract. We study the long-range force arising between two aggregates of ordinary matter due to a neutrino-pair exchange, in the limit of zero neutrino mass. Even if matter is neutral of electric charge, it is charged for this weak force. The interaction is described in terms of a coherent charge, which we call the weak flavor charge of aggregated matter. For each one of the interacting aggregates, this charge depends on the neutrino flavor as $Q_W^{\nu_e} = 2Z - N$, $Q_W^{\nu_\mu} = Q_W^{\nu_\tau} = -N$, where Z is the number of protons and N the number of neutrons. $Q_W^{\nu_e}$ depends explicitly on Z because of the charged current contribution to $\nu_e e$ elastic scattering, while the N term in the three charges comes from the universal neutral current contribution. The effective potential describing this force is repulsive and decreases as r^{-5}. Due to its specific behavior on (Z, N) and r, this interaction is distinguishable from both gravitation and residual electromagnetic forces. As neutrinos are massive and mixed, this potential is valid for $r \lesssim 1/m_\nu$, distances beyond which a Yukawa-like attenuation kicks in.

1 Introduction

The study of the origin of neutrino mass is one of the directions in which we can expect finding new Physics, even though its small value ($m_\nu \lesssim 1$ eV [1,2]) makes it hard to observe experimentally. As well as determining the absolute mass of the neutrino, there's still a more fundamental question about their nature unanswered: their finite mass could be explained through a Dirac mass term (implying there is a conserved total lepton number L distinguishing neutrinos from antineutrinos, which are described by 4-component Dirac spinors) or through a Majorana one (implying that neutrinos are self-conjugate of all charges, described by 2 independent degrees of freedom).

In any case, the fact that their masses are very low stands, and we discuss here another property of neutrinos as mediators of a new force. As is well known, the processes represented in Quantum Field Theory by the exchange of a massless particle give raise to long-range interactions. An easy example is the scattering of two particles mediated by a photon, which—at tree level— describes Coulomb scattering. Our objective in this work is the application of these ideas to a process mediated by neutrinos. According to the Electroweak Lagrangian, the lowest-order process is a neutrino-pair exchange, which—since neutrinos are nearly massless—describes an interaction of long range.

---

[a]E-mail: alejandro.segarra@uv.es

2 Long-range weak interaction between aggregate matter

We are interested in obtaining the interaction potential due to a neutrino-pair exchange between two matter aggregates, say A and B. In doing so, we will not impose any restriction on the internal structure of the aggregates—whatever they are, we only ask them to be neutral of electric charge. Therefore, for each aggregate, its composition is specified by two numbers: Z will represent the number of protons and the number of electrons, which must be the same, and N will represent the number of neutrons.

The picture is now clear. Elastic interactions of matter constituents with neutrinos is through either W or Z exchange, as well as the aggregate structure, determine the 1-loop neutrino-pair exchange $AB \to AB$ elastic interaction, as shown in Fig. 1.

The whole interaction potential between the two aggregates is given by the Fourier Transform of the Feynman amplitude in Fig. 1. Since we are only interested in the long-range part of the potential, a few simplifications can be performed. Through rewriting the amplitude as an unsubtracted dispersion relation, we find the long-range behavior is fully determined by its absorptive part. In turn, the absorptive part is determined, after unitarity-cutting the diagram in the t-channel, by a simple tree-level $A\nu \to A\nu$ amplitude. This tree-level calculation is straightforward—in the process, we only kept the dominant contributions, neglecting both incoherent and relativistic corrections.

The analysis described above leads [3] to the interaction potential

$$V(r) = \frac{G_F^2}{8\pi^3} \left[(2Z_A - N_A)(2Z_B - N_B) + 2N_A N_B\right] \frac{1}{r^5}. \quad (1)$$

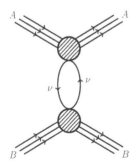
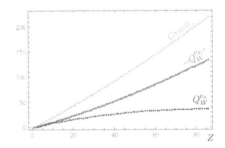

Figure 1: Effective neutrino-pair exchange interaction between two aggregates of matter. The blobs in the vertices represent any structure the aggregates could have.

Figure 2: Weak flavor charges of the elements specified by the atomic number Z, compared to their mass $\approx Z + N$. The isotope chosen is the one in which the (Z, N) pair lies in the valley of nuclear stability.

3 Coherent weak flavor charges

A carefoul reading of Eq. (1) shows the standard structure of an interaction potential. By defining, for each of the aggregates, their weak flavor charges as $Q_W^{\nu_e} = 2Z - N$ and $Q_W^{\nu_\mu} = Q_W^{\nu_\tau} = -N$, one gets the usual structure $V =$ (coupling)^2 × (product of charges) × (power law).

Indeed, this shows that, within the Standard Model, matter is charged! The values of the weak charges of all stable atoms are represented in Fig. 2. At this point, we remark that the unique (Z, N) dependence of these charges makes them scale with the size of the system in a different way than gravitation. Also, the fact that the charges of all elements have the same sign (for each flavor) implies that this interaction is always repulsive. These two properties may become crucial in disentangling this weak interaction from gravitation experimentally.

4 Prospects

The long-range potential obtained in this work, Eq. (1), is valid and of interest for distances between nanometers and microns. The short-distance limit comes from the requirement of having neutral (of electric charge) systems of aggregate matter, while the long-distance limit is imposed by a non-vanishing value of the absolute mass of the neutrino—indeed, the range of this interaction for neutrinos of $m \sim 0.1$ eV is of the order of $R \sim 1/m_\nu \sim 1\mu$m. In this region, the effective potential will become of Yukawa type instead of the pure inverse power law.

The neutrino mass dependence of the effective potential in the long-range behavior opens novel directions in the study of the most interesting pending questions on neutrino properties: absolute neutrino mass (from the range), flavor dependence and mixing (from the weak charges in the interaction) and, hopefully, with two neutrino exchange, the exploration of the most crucial open problem in neutrino physics: whether neutrinos are Dirac or Majorana particles.

Acknowledgments

The author acknowledges the Spanish MECD financial support through the FPU14/04678 grant.

References

[1] J. Bernabéu, Nucl. Phys. Proc. Suppl. **A 28** (1992) 503.
[2] C. Patrignani et al. (PDG), Chin. Phys. C **40** (2016) 100001.
[3] A. Segarra, arXiv:1606.05087 [hep-ph].

SIGNATURES OF NEUTRINO MAGNETIC MOMENT IN COLLECTIVE OSCILLATIONS OF SUPERNOVA NEUTRINOS

Oleg Kharlanov [a] and Pavel Shustov [b]
Faculty of Physics, Moscow State University, 119991 Moscow, Russia

Abstract. We reanalyze a recently claimed ultrahigh sensitivity of collective Majorana neutrino oscillations to a neutrino transition magnetic moment and discuss its observability based on numerical simulations and analytical stability analysis.

1 Introduction

Compact astrophysical objects, in particular, protoneutron stars born in supernova explosions, provide a scene to test physics at its extreme, in particular, physics beyond the Standard Model (BSM) [1]. Luckily, core-collapse supernovae typically deposit the lion's share of the explosion energy into neutrinos, which lets one probe exotic properties of this elusive particle. Moreover, thirty years since the first observation of supernova neutrinos from SN1987A have substantially improved the experimental techniques, e.g., the JUNO detector that should start operating in a couple of years will be able to collect over 5000 neutrino+antineutrino events for an explosion in the Milky Way, also offering an unprecedented energy resolution $3\%/\sqrt{E/\text{MeV}}$ [2]. In this connection, it its very interesting to study the effect of BSM neutrino physics near a protoneutron star on the neutrino energy spectra observed at infinity.

In the present paper, we study the effect of a tiny transition magnetic moment μ of a Majorana neutrino on the neutrino collective oscillations taking place near the neutrinosphere of a protoneutron star. Our study is inspired by two recent papers [3], whose authors claimed an ultrahigh sensitivity of the neutrino spectra to μ, at least at the level of $10^{-20}\mu_B$ (cf. the present experimental constraints $\mu \lesssim 10^{-15}\mu_B$ or even weaker [4]). We rederive the equation for flavor evolution of neutrinos, analyze the development of instabilities within the single-angle scheme [5], and conclude that the latter are in fact strongly suppressed, so that the sensitivities to the neutrino magnetic moment turn out to be orders of magnitude lower than those reported in Ref. [3].

2 The evolution equation for collective oscillations and its stability

Our analysis is based on the Standard Model Lagrangian after the breakdown of the electroweak symmetry, in which the neutrino-neutrino interaction term reads

$$\mathcal{L}^{(4)} = -\frac{G_F}{\sqrt{2}} : (\bar{\nu}_a \gamma_L^\mu \nu_a)^2 :, \quad a = 1, 2, \; \gamma_L^\mu \equiv \gamma^\mu \frac{1-\gamma_5}{2}, \quad (1)$$

---
[a] E-mail: kharlanov@physics.msu.ru
[b] E-mail: pi.shustov@physics.msu.ru

where $\nu_a(x)$ are the neutrino fields with Majorana masses m_a, G_F is the Fermi constant. As usual within studies of collective oscillations, we introduce the 4×4 density matrix describing mass + helicity states of neutrinos

$$\rho_{a\alpha,b\beta}(\boldsymbol{p}) \equiv \langle \hat{a}^\dagger_{b\beta \boldsymbol{p}} \hat{a}_{a\alpha \boldsymbol{p}} \rangle, \quad a,b=1,2,\ \alpha,\beta=\pm, \qquad (2)$$

where $\hat{a}^\dagger_{a\alpha \boldsymbol{p}}$ is the neutrino annihilation operator. A Heisenberg equation for the density matrix can be derived in the mean-field regime

$$\frac{\partial \rho_{a\alpha,b\beta}(\boldsymbol{p})}{\partial t} = \mathrm{i} \langle \Phi | [\hat{H}, \hat{a}^\dagger_{b\beta \boldsymbol{p}} \hat{a}_{a\alpha \boldsymbol{p}}] | \Phi \rangle \approx \mathrm{i}[h(\boldsymbol{p}), \rho(\boldsymbol{p})]_{a\alpha,b\beta}, \qquad (3)$$

where, in the flavor basis, the matrix Hamiltonian for collective oscillations

$$h(\boldsymbol{p}) = \frac{\Delta m^2}{4|\boldsymbol{p}|} \begin{pmatrix} \mathbb{M} & 0 \\ 0 & \mathbb{M} \end{pmatrix} + G_F \sqrt{2} \begin{pmatrix} \mathbb{V} & 0 \\ 0 & -\mathbb{V} \end{pmatrix} - \mathrm{i} B_\perp \begin{pmatrix} 0 & \mathfrak{m} \\ \mathfrak{m} & 0 \end{pmatrix} + h_{\text{self}}, \qquad (4)$$

$$h_{\text{self}}(\boldsymbol{p}) = G_F \sqrt{2} \int \mathrm{d}^3 q (1 - \hat{p} \cdot \hat{q}) \Big\{ \mathrm{tr}(\rho(\boldsymbol{q})G)G + [\rho(\boldsymbol{q}) - \rho^{\mathrm{cT}}(\boldsymbol{q})]^\times \Big\}, \qquad (5)$$

$$\mathbb{M} \equiv \begin{pmatrix} -\cos 2\theta & -\sin 2\theta \\ -\sin 2\theta & \cos 2\theta \end{pmatrix}, \ \mathbb{V} \equiv \begin{pmatrix} n_e - \frac{n_n}{2} & 0 \\ 0 & -\frac{n_n}{2} \end{pmatrix}, \mathfrak{m} \equiv \begin{pmatrix} 0 & \mu \\ -\mu & 0 \end{pmatrix}, \qquad (6)$$

where $\Xi^c \equiv C\Xi C$, $\Xi^\times \equiv (\Xi + G\Xi G)/2$, $G \equiv \begin{pmatrix} 1 & 0 \\ 0 & -1 \end{pmatrix}$, $C \equiv \begin{pmatrix} 0 & 1 \\ 1 & 0 \end{pmatrix}$; upper/lower (left/right) 2×2 blocks of ρ describe neutrino/antineutrino states, respectively; $n_{e,n}$ are densities of electrons and neutrons, $B_\perp$ is the component of the magnetic field transversal to the neutrino momentum $\boldsymbol{p}$; Δm^2 and θ are the mass and mixing parameters. Within the single-angle scheme [5],

$$\mathrm{i}\partial_r \rho_E(r) = [h_{0E}(r) + h_{\text{AMM}}(r) + h_{\text{self}}(r), \rho_E(r)], \qquad (7)$$

$$h_{\text{self}} = G_F \sqrt{2} n_\nu(r) \int \mathrm{d}E \Big\{ \mathrm{tr}(\rho_E(r)G)G + [\rho_E(r) - \rho_E^{\mathrm{cT}}(r)]^\times \Big\}, \qquad (8)$$

where $n_\nu(r)$ is the effective neutrino number density including the geometric factor and $h_{0E}(r) + h_{\text{AMM}}(r)$ contain the three non-collective terms in (4). Our result differs from [3] in the collective term $[\rho_E - \rho_E^{\mathrm{cT}}]^\times$, where they have simply $G(\rho_E - \rho_E^{\mathrm{cT}})G$ instead.

Linear stability analysis comparing the block-diagonal $\mu = 0$ solution $\rho_E^{(0)}(r)$ with a $\mu \to 0$ solution $\rho_E = \rho_E^{(0)}(r) + \delta\rho_E(r) + \mathcal{O}(\mu^2)$ leads to an equation

$$\mathrm{i}\partial_r \delta\rho_E = [h_{0E} + h_{\text{self}}, \delta\rho_E] + [\delta h_{\text{self}}, \rho_E^{(0)}] + [h_{\text{AMM}}, \rho_E^{(0)}]. \qquad (9)$$

It is easy to demonstrate that, due to the structure of the self-interaction, the same linear equation but with the opposite inhomogeneity $-[h_{\text{AMM}}, \rho_E^{(0)}]$ is valid for $G\delta\rho_E(r)G$, thus, $G\delta\rho_E G = -\delta\rho_E$. For such a matrix, in the above equation, δh_{self} vanishes and, in the linear order in μ, $\|\delta\rho_E(r)\| = \mathcal{O}(r\|[h_{\text{AMM}}, \rho_E^{(0)}]\|) = \mathcal{O}(\mu B r \rho_E^{(0)})$, i.e., the instabilities do not develop in the linear approximation, unlike, however, the case of Hamiltonian from Ref. [3].

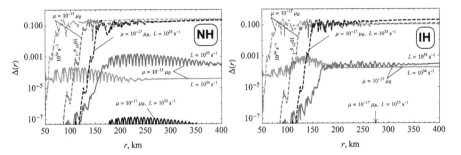

Figure 1: Spectral differences (10) for collective neutrino oscillations by Eq. (7) (solid lines) and Ref. [3] (dashed lines). The simulation parameters are those from [3], but with $\theta = 9°$.

3 The numerical simulation and the magnitude of the effect

We have performed a numerical solution of Eq. (7) comparing the (anti)neutrino spectra $s_f(E;r)$ with the $\mu = 0$ case, in terms of the spectral difference

$$\Delta(r) = \frac{\sum_f \int |s_f(E;r) - s_f^{(0)}(E;r)| \mathrm{d}E}{\sum_f \int s_f^{(0)}(E;r) \mathrm{d}E}, \qquad (10)$$

and plotted them in Fig. 1 for the Hamiltonian of Eq. (7) and the one from Ref. [3] for different neutrino luminosities L and magnetic moments μ. One easily observes that instabilities of (7) are strongly suppressed, and the resulting sensitivities hardly exceed $10^{-15}\mu_B$ for the luminosities shown, while the Hamiltonian from Ref. [3] leads to an avalanche-like growth of magnetic moment-induced perturbations to a figure of the order of unity.

Acknowledgments

The reported study was supported by Supercomputing Center of Moscow State University [6].

References

[1] G. G. Raffelt, *"Stars as Laboratories for Fundamental Physics"* (University of Chicago Press, Chicago, 1996).
[2] Y. F. Li, Int. J. Mod. Phys. Conf. Ser. **31** (2014), 1460300.
[3] A. de Gouvêa and S. Shalgar, JCAP **10 (2012)**, 027; **04 (2013)**, 018.
[4] A. Studenikin, J. Phys. Conf. Ser. **718** (2016), 062076.

[5] H. Duan, G. M. Fuller, and Y. Zh. Qian, Annu. Rev. Nucl. Part. Sci. **60** (2010), 569.

[6] V. Sadovnichy *et al.* in *"Contemporary High Performance Computing: From Petascale toward Exascale"*, Ed. by J. S. Vetter, CRC Press, USA, 2013, p. 283.

A DECAY OF AN ULTRA-HIGH-ENERGY NEUTRINO $\nu_e \to e^- W^+$ IN AN EXTREMELY HIGH MAGNETIC FIELD

Alexander Kuznetsov [a], Alexander Okrugin [b] and Anastasiya Shitova [c]
*Division of Theoretical Physics, P. G. Demidov Yaroslavl State University,
Sovietskaya 14, 150000 Yaroslavl, Russia*

Abstract. An intense electromagnetic field makes possible the process of the neutrino decay into electron and W boson that is kinematically forbidden in a vacuum. The exact formula for the probability of ultra-high-energy neutrino decay in a superstrong magnetic field is derived. Possible manifestations of the process in the conditions of the Early Universe are briefly discussed.

Extreme physical conditions actively influence the run of quantum processes, thus allowing or enhancing the transitions that are forbidden or strongly suppressed in vacuum. The external field effects should be taken into account on the basis of exact solutions of the field theory equations for a charged particle in an external electromagnetic field, but not on the basis of perturbation theory.

In the present work we discuss the field with a strength essentially greater than the so-called Schwinger limit $B_e = m_e^2/e \simeq 4.4 \times 10^{13}$ G, which is the quantizing field strength for an electron. Another limit that can be naturally introduced regarding the conditions of the Early Universe [1, 2], is defined by the mass of the gauge boson m_W and can serve as a boundary of applicability of Standard Model: $B_W = m_W^2/e \simeq 1.1 \times 10^{24}$ G. As it was shown in Ref. [3], the radiation corrections act to prevent the instability of the electroweak vacuum in strong fields. A class of models of the Early Universe where such strong fields $\sim B_W$ could exist, exploit an idea of the domain structure of the Universe. Namely, in a non-equilibrium primordial electroweak phase transition, the low temperature phase starts to form, due to quantum fluctuations, simultaneously and independently in many parts of the system, and domains arise separated by domain walls. An attractive hypothesis was claimed in Ref. [4] that domain walls were ferromagnetic that allowed to generate magnetic field strong enough to be a possible candidate of a primordial magnetic field leading to presently observable galactic fields. The idea was lately supported and discussed in Ref. [5–7], but received several contradict arguments [8, 9]. The authors of Ref. [10] claim that the submitted objections could only slightly decrease the effect but do not nullify it. It was suggested [6] that the effective surface tension of the domain walls can be made vanishingly small due to a peculiar magnetic condensation induced by fermion zero modes localized on the wall. As a consequence, the domain wall acquires a non-zero magnetic

---

[a] E-mail: avkuzn@uniyar.ac.ru
[b] E-mail: okrugin@uniyar.ac.ru
[c] E-mail: pick@mail.ru

field perpendicular to the wall, and it becomes almost invisible as far as the gravitational effects are concerned.

A problem of studying possible influence of an active environment on the neutrino properties is quite important. Probably, the most interesting process which is kinematically forbidden in a vacuum and is possible in a strong magnetic field, is the decay of a massless ultra-high-energy neutrino into W^+ boson and a charged lepton, $\nu_\ell \to \ell^- W^+$ ($\ell = e, \mu, \tau$). The calculation of the width of the process $\nu_e \to e^- W^+$ in an external magnetic field has a long history started by Ref. [11], a detailed list of references can be found e.g. in Ref. [12].

In our research, we consider extremely strong magnetic fields $\sim B_W$, the details of calculation techniques can be found e.g. in Ref. [13].

The probability of the decay $\nu_e \to e^- W^+$ can be presented in the form:

$$W = \frac{G_F\, m_W^2\, \beta}{\pi\sqrt{2}\, E} \sum_{k=0}^{k_{max}} \sum_{n=0}^{n_{max}(k)} \frac{\exp(-\chi)}{\sqrt{D(\chi)}} \Big\{ (\chi - \mu - k + 1/2 + n) \qquad (1)$$

$$\times \left[\frac{k!}{n!} \chi^{n-k} \left(L_k^{n-k}(\chi)\right)^2 + \frac{(k-2)!}{(n-1)!} \chi^{n-k+1} \left(L_{k-2}^{n-k+1}(\chi)\right)^2 \right]$$

$$- 4(-1)^{k+n} \chi\, L_{k-1}^{n-k+1}(\chi)\, L_{n-1}^{k-n}(\chi) \Big\} \theta\left(\sqrt{\chi} - \sqrt{\mu + k - 1/2} - \sqrt{n}\right),$$

where k and n are the numbers of the Landau levels for the W boson and the electron correspondingly, $L_n^s(x)$ are the associated Laguerre polynomials, the dimensionless variables are introduced: $\chi \equiv p_\perp^2/(2\beta)$, $\mu \equiv m_W^2/(2\beta)$, and

$$D(\chi) = (\chi - \mu - k + 1/2 + n)^2 - 4n\chi. \qquad (2)$$

The theta-function in Eq. (1) allows to define the limits on summation over n and k, $n_{max}(k) = \lfloor (\sqrt{\chi} - \sqrt{\mu + k - 1/2})^2 \rfloor$, and $k_{max} = \lfloor \chi - \mu + 1/2 \rfloor$, where $\lfloor x \rfloor$ means for $x > 0$, as in our case, an integer part of x. As a consequence there are kinematically allowed Landau states for creating W boson and electron originated from the threshold for neutrino energy.

The structure of the derived expression (1) indicates the existence of its abrupt increase (and decrease of the mean free path) at the energy values corresponding to the "turning on" the new Landau states. It is worth to note that the similar saw-tooth profile arises also in the problem of electron-positron pair creation by a photon in strong magnetic fields [14,15]. Thus, for some specific values of parameters, the neutrino mean free path is significantly lower than it is predicted by the asymptotic formula [12]. That corresponds to generation of a great amount of electrons and W^+ bosons, and the subsequent W^+ boson decay can influence the picture of the baryosynthesis in the Early Universe.

It would be important that the process width (1) essentially depends on $p_\perp = E \sin\theta$, where θ is an angle between the neutrino momentum and the field direction. Suppose that the hypothesis of magnetic field generation in domain walls is correct. Then neutrinos moving almost perpendicular to the domain wall (i.e. almost parallel to the field direction) would fly it right through. On the other hand, neutrinos propagating almost parallel to the wall (in the way of the maximum magnetic field influence) with specific energies would predominantly decay, and would generate a great amount of W bosons in the thickness of the wall. The subsequent decay of the W boson by dominant quark channels could lead to some overabundance of hadron matter and antimatter inside domain walls to outside.

If the lepton-antilepton asymmetry, induced by the CP violation in the lepton sector, has arisen before the electroweak phase transition, leading to overabundance of neutrinos over antineutrinos in the Universe, the considered mechanism would provide an overabundance of W^+ and e^- over W^- and e^+ inside domain walls. The W^+ bosons would finally transform to protons while the W^- bosons would dominantly transform to antiprotons. The subsequent formation of the baryon asymmetry would paint the final picture of the Universe.

Acknowledgments

The research is financed by the grant of the Russian Science Foundation (Project No. 15-12-10039).

References

[1] T. Vachaspati, Phys. Lett. **B 265** (1991) 258.
[2] D. Grasso, H. R. Rubinstein, Phys. Rep. **348** (2001) 163.
[3] V. V. Skalozub, Phys. At. Nucl. **77** (2014) 901.
[4] A. Iwazaki, Phys. Rev. D **56**, 2435 (1997).
[5] P. Cea and L. Tedesco, Phys. Lett. B **425**, 345 (1998).
[6] P. Cea and L. Tedesco, Phys. Lett. B **450**, 61 (1999).
[7] M. M. Forbes and A. R. Zhitnitsky, Phys. Rev. Lett. **85**, 5268 (2000).
[8] M. B. Voloshin, Phys. Lett. B **491**, 311 (2000).
[9] M. B. Voloshin, Phys. Rev. D **63**, 125012 (2001).
[10] L. Campanelli, P. Cea, G. L. Fogli and L. Tedesco, JCAP **0603**, 005 (2006).
[11] A. V. Borisov, V. Ch. Zhukovskii, A. V. Kurilin, A. I. Ternov, Yad. Fiz. **41** (1985) 743 [Sov. J. Nucl. Phys. **41** (1985) 473].
[12] A. V. Kuznetsov, N. V. Mikheev, A. V. Serghienko, Phys. Lett. **B 690** (2010) 386.
[13] A. V. Kuznetsov, A. A. Okrugin, A. M. Shitova, Int. J. Mod. Phys. **A 30** (2015) 1550140.
[14] J. K. Daugherty and A. K. Harding, Astrophys. J. **273**, 761 (1983).
[15] V. N. Baier and V. M. Katkov, Phys. Rev. D **75**, 073009 (2007).

Physics at Accelerators and Studies in SM and Beyond

ELECTROWEAK PRECISION MEASUREMENTS IN ATLAS

Evgeny Soldatov [a]
On behalf of the ATLAS Collaboration
National Research Nuclear University "Moscow Engineering Physics Institute",
115409, Kashirskoe shosse, 31, Moscow, Russia

Abstract. The ATLAS collaboration has performed detailed integrated and differential cross-section measurements of heavy bosons and di-boson pairs production in fully-leptonic and semi-leptonic final states at the centre-of-mass energies of 8 and 13 TeV for pp collisions provided by the LHC. These measurements represent stringent tests of the electroweak sector of the Standard Model and provide a model-independent means to search for the new physics at the TeV scale. The results are compared to theory predictions at NLO (and NNLO) and provide constraints on the new physics, by setting limits on anomalous gauge couplings.

1 Introduction

Precision measurements in the electroweak sector are very important to probe the Standard Model (SM) and search for any possible deviations, so-called "new physics". Electroweak theory can provide very high precision of theoretical predictions (NLO, NNLO and higher) for comparison to experimental data. These comparisons allow to examine the higher order QCD and QED effects, to understand the irreducible backgrounds for the Higgs measurements and new resonances searches and to search for the "new physics" via anomalous triple/quartic gauge couplings (aTGC/aQGC).

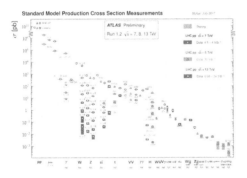

Figure 1: Summary of total and fiducial Standard Model production cross-section measurements by ATLAS, corrected for leptonic branching fractions, compared to the corresponding theoretical expectations [2].

---

[a] E-mail: EYSoldatov@mephi.ru

ATLAS experiment [1] is performing a lot of studies (Fig. 1) in most of the possible production channels of the SM: from well-known single W/Z production to the brand new and very rare di-boson and tri-boson productions.

This report contains an overview of recent precision electroweak measurements of the inclusive Z and W boson productions, as well as the multi-boson productions (VV, VVV, where $V=W, Z, \gamma$) by using data from the first ATLAS data-taking period (Run 1) and the first data from the Run 2.

2 Single W and Z bosons production

Single W and Z bosons serve as standard candles at the LHC experiments.

The inclusive measurement of single W/Z bosons production (leptonic decays) [3] was performed by using 4.5 fb^{-1} of 7 TeV data. Measured cross-section values are in a good agreement with the predictions. Dominant uncertainties for the measurements come from signal modelling and main multi-jet background.

Single W/Z bosons can be produced through the various mechanisms, one of the rarest and most interesting is the electroweak (EWK) production via vector-boson fusion (VBF). In this case, single boson production associates by two high-energetic jets. The main problem of identification of this process is its separation from the usual QCD production, which has almost the same final state, but much higher probability. For VBF W Run 1 measurement [4], which used full 2011 and 2012 sets of data, it was done using requirements on jet variables such as the invariant mass of di-jet system, the absolute difference in rapidity between two jets and ΔR between jet and the closest lepton. Integrated and differential cross-sections measurement demonstrates the agreement with predictions, Fig. 2. An observed significance of the measurement is more than 5σ. Dominant sources of uncertainty for the measurement are: jet energy scale and resolution, the knowledge of the main multi-jet background. This channel is also sensitive to aTGCs via charged vertices $WWZ/WW\gamma$.

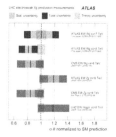
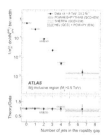

Figure 2: Measurements of the cross section times branching fractions of electroweak production of a single W, Z, or Higgs boson at high di-jet invariant mass, divided by the SM predictions (left) [4]; Unfolded normalized distribution of the N_{jets} in the rapidity interval bounded by the two highest-p_{T} jets in the fiducial region with $M_{jj} > 0.5$ TeV (right) [4].

VBF Z production measurement was done by using 3.2 fb^{-1} of 13 TeV data. [5]. Cross-section measurement demonstrates agreement with predictions.

3 Di-boson production

Associated di-boson production basically serves as the main instrument to study the gauge boson couplings and self-couplings.

ATLAS recently performed 13 TeV measurements of inclusive WW (leptonic) [6], WZ (leptonic) [7] and $ZZ(\to 4\ell)$ [8] productions. Main sources of uncertainties on the measurements are: the lepton identification and reconstruction efficiencies, the knowledge of the background from lepton misidentification. Observed cross sections are in the fair agreement with the NLO theory predictions. And the agreement is much better with the NNLO theory predictions. In addition, WZ and ZZ production measurements were used to set the limits on aTGCs using the vertex functions and effective field theory (EFT) approaches.

Diboson production is also possible via electroweak mechanisms, including vector-boson scattering (VBS). Recent study of the electroweak $Z\gamma$ ($\to \ell\ell\gamma / \nu\bar{\nu}\gamma$) production [9] at 8 TeV is one of the first VBS measurements in ATLAS. Dominant uncertainties here come from statistics, calibration of the jet energy scale and (for $\nu\bar{\nu}\gamma$ channel) from the $W\gamma$ and Z+jets backgrounds estimation. Cross-section measurement in $\ell\ell\gamma$ channel demonstrates the significance of about 2σ. Both channels were used to set the limits on aQGC EFT parameters.

4 Conclusions

An overview of the recent ATLAS precision electroweak results was presented. The amount of data taken by the ATLAS detector allows to perform precise cross-section measurements. The sensitivity of such measurements are so high, that it is necessary to calculate theory predictions with the NNLO corrections.

Also it allows to perform the first measurements of the rare electroweak processes. The most challenging task is the recognition of the electroweak production component from the QCD one. Simple rectangular cuts approach is used for this purpose in the considered analyses. However machine learning techniques were also tested and can improve the results in future.

We gratefully acknowledge the financial support from Russian Science Foundation grant No.17-72-10021.

References

[1] ATLAS Collaboration, JINST **3** (2008) S08003.
[2] https://atlas.web.cern.ch/Atlas/GROUPS/PHYSICS/CombinedSummary Plots/SM/.

[3] ATLAS Collaboration, Eur. Phys. J. C **77** (2017) 367.
[4] ATLAS Collaboration, Eur. Phys. J. C **77** (2017) 474.
[5] ATLAS Collaboration, Phys. Lett. B **775** (2017) 206.
[6] ATLAS Collaboration, Phys. Lett. B **773** (2017) 354.
[7] ATLAS Collaboration, ATLAS-CONF-2016-043.
[8] ATLAS Collaboration, arXiv:1709.07703 [hep-ex].
[9] ATLAS Collaboration, JHEP **07** (2017) 107.

SEARCHES FOR SUPERSYMMERTY WITH THE ATLAS DETECTOR

Manfredi Ronzani [a] (On behalf of the ATLAS Collaboration)
Physikalisches Institut, Albert-Ludwigs-Universität Freiburg (DE)

Abstract. Despite the absence of experimental evidence, weak scale supersymmetry remains one of the best motivated and studied Standard Model extensions. This paper summarises recent ATLAS results for searches for supersymmetric (SUSY) particles. Weak and strong production in both R-parity conserving and R-parity violating SUSY scenarios are considered. The searches used a dataset collected by the ATLAS detector in 2015 and 2016 from the proton-proton collisions at $\sqrt{s} =$ 13 TeV at the LHC, and involved final states including jets, missing transverse momentum, light leptons as well as long-lived particle signatures.

1 Introduction and Motivation

Supersymmetry (SUSY) [1] is a theoretically favored candidate for physics beyond the Standard Model (SM). Many SUSY models predict sparticles that could be accessible at the LHC and detected by the ATLAS experiment [2]. The MSSM (Minimal Supersymmetric Standard Model) predicts a new bosonic (fermionic) partner for each SM fundamental fermion (boson). The SUSY partners of the Higgs boson are the higgsinos. In the MSSM, two scalar Higgs doublets along with their higgsinos are necessary. This extension provides an elegant solution for several of the SM shortcomings, such as the gauge hierarchy problem. In SUSY models where R-parity[b] conservation is assumed (RPC), sparticles are produced only in pairs and the Lightest Supersymmetric Particle (LSP) is stable and weakly interacting. The LSP is supposed to be undetected resulting in a signature of missing transverse momentum (E_T^{miss}). If the R-parity is violated (RPV), the LSP can decay to light quarks via the RPV Yukawa couplings, resulting in a different phenomenology.

2 Inclusive Searches for SUSY Strong Production

Searches for strong production of squarks ($\tilde{q}$) and gluinos ($\tilde{g}$) are favored at the LHC due to the large cross sections. Simplified models are considered with direct to the LSP or via intermediate chargino. The signature can include various multiplicities of jets coming from the hadronization of the produced particles, E_T^{miss}, and leptons in the final state. Efficient tagging algorithms can be employed to identify jets as coming from b-hadrons.

---

[a] E-mail: manfredi.ronzani@cern.ch
[b] R-parity, $R = (-1)^{3(B-L)+2s}$, with s the spin of the particle, is introduced in SUSY models with a minimal particle content in order to reconcile the possible baryonic and leptonic number violation with the strong constraints from the non-observation of these processes.

The *effective mass* m_{eff} [c] is the main discriminant variable used in the signal region (SR) definition of the inclusive analyses with 0 or 1 lepton in the final state [3,4]. In the inclusive analysis using multiple b-jets [5], the number of b-jets and the total jet mass M_J^Σ are used. More complex methods as the *recursive jigsaw* (RJR) [6] technique [d] are employed in a parallel complementary search with no leptons in the final state [3].

No significant excess is observed above the SM expectations. A simplified model with intermediate decay of squarks and gluinos via chargino is considered by both the 0- and 1-lepton analysis. The resulting exclusion limits are displayed in Fig. 1. Improvement in the region $x \to 1$ (left), with a large mass splitting between $\tilde{\chi}_1^\pm$ and $\tilde{\chi}_1^0$, are achieved by specialized boosted SRs [e].

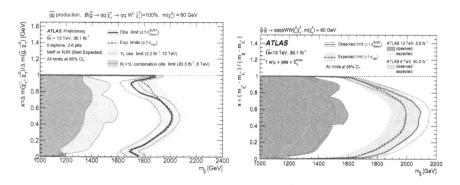

Figure 1: Exclusion limits for a simplified model scenario with direct production of gluino pairs with decoupled squarks and a fixed LSP mass $m(\tilde{\chi}_1^0) = 60$ GeV, from the 0-lepton [3] (left) and 1-lepton [4] (right) analysis.

3 Searches for Third Generation SUSY particles

Scenarios where the supersymmetric partners of the third generation SM quarks (b,t) are the lightest colored SUSY particles are motivated by naturalness considerations. The SUSY partner of the top quark, the stop ($\tilde{t}$), arrange for the cancellation of potentially large top-quark loop corrections in the mass of the Higgs boson.

Stops and sbottoms can be produced directly by strong interactions and decay directly to their SM partners and neutralinos or via intermediate charginos. Simplified models with a stop produced indirectly from gluino decays are investigated by the analysis with 0 leptons in the final state [7].

---

[c] The scalar sum of the p_T of the leading n-jets in the n-jets channel, the E_T^{miss}, and the p_T of the lepton (if present).
[d] The RJR exploits combinations of observables reconstructed in different reference frames.
[e] Boosted SRs use reclustered large radius jets in order to reconstruct W boson mass.

A summary of the limits on simplified models with direct stop production can be seen in Fig. 2. The limits exclude stop quarks with masses up to 1000 GeV for small neutralino masses.

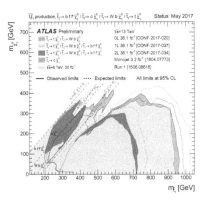

Figure 2: Summary of exclusion limits of the dedicated ATLAS searches for top squark (stop) pair production for simplified model scenarios with four different stop decay modes [8].

4 Searches for SUSY Electroweak Production

In a scenario where colored sparticles have masses above the reach of the LHC, the electroweak (EW) production could be the only one accessible. Cross sections for SUSY EW production is generally much lower than in the strong case. Searches with 2–3 leptons in final states with or without the presence of jets focus on RPC models, while the RPV case is investigated in events with 4 leptons in final state. Simplified models explore the direct production of $\tilde{\chi}^{\pm}_{1,2}, \tilde{\chi}^0_2$ or $\tilde{l}\tilde{l}$ pairs [9]. A search involving the intermediate production of tau sleptons, giving final states with hadronically decaying taus, is also performed [10].

The best limits in the $((\tilde{\chi}^{\pm}_1, \tilde{\chi}^0_2, \tilde{\chi}^0_3), \tilde{\chi}^0_1)$ plane are obtained in channels using sleptons, excluding masses up to ~ 1100 GeV [11].

5 Long-Lived Particles

Several BSM models predict SUSY particles to be produced with long lifetimes, in the ps to ns range.

In the squark-mediated gluino decay case, depending on the scale of the squark mass, the gluino lifetime is of the order of picosends or longer, which is above the hadronisation time scale. The gluino forms a bound color singlet state with SM particles known as an R-hadron, whose decay vertices can be reconstructed in the ATLAS inner detector. A search for displaced vertices sensitive to long lived particle decays occurring in the range of (1 − 100) mm from the reconstructed primary vertex is performed [12].

An other common scenario includes a chargino which is nearly degenerate with a neutralino, called the "pure wino-LSP" model. The decays of the long-lived chargino produces a pion and a LSP, where the pion has typically very low-momentum and it is often not reconstructed, giving a signature of a disappeared track, which is investigated by a dedicated search [13]. Figure 3 shows the results of the displaced vertices (left) and the disappearing track (right) analyses, in terms of the lifetime of the sparticle considered.

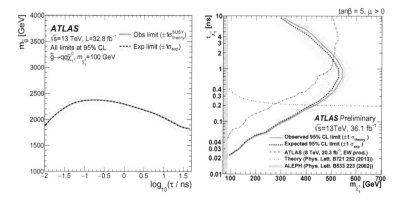

Figure 3: Left: Limits of the ATLAS search for displaced vertices on the gluino mass $m(\tilde{g})$ as a function of the gluino lifetime for a fixed $m(\tilde{\chi}_1^0) = 100$ GeV [12]. Right: Limits of the ATLAS search for disappearing tracks in terms of the chargino lifetimes and $m(\tilde{\chi}_1^0)$ [13].

6 Conclusions

A selection of the most recent searches for SUSY particles at the ATLAS detector is presented. No significant excess above SM expectations has been found. Limits on specific models widely extend the results at $\sqrt{s} = 8$ TeV.

References

[1] A. Salam and J. A. Strathdee, Phys. Lett. **B 51** (1974) 353.
[2] ATLAS Collaboration, JINST 3 S08003 (2008).
[3] ATLAS Collaboration, ATLAS-CONF-2017-022.
[4] ATLAS Collaboration, arXiv:1708.08232, CERN-EP-2017-140.
[5] ATLAS Collaboration, arXiv:1711.01901, CERN-EP-2017-182.
[6] Jackson, Paul et al., Phys. Rev. **D 95** (2017) 3.

[7] ATLAS Collaboration, arXiv:1709.04183, CERN-PH-2017-162.
[8] https://atlas.web.cern.ch/Atlas/GROUPS/PHYSICS/
CombinedSummaryPlots/SUSY/ATLAS_SUSY_Stop_tLSP/ATLAS_SUSY_
Stop_tLSP_201705.pdf.
[9] ATLAS Collaboration, ATLAS-CONF-2017-039.
[10] ATLAS Collaboration, arXiv:1708.07875, CERN-PH-2017-173.
[11] https://atlas.web.cern.ch/Atlas/GROUPS/PHYSICS/
CombinedSummaryPlots/SUSY/ATLAS_SUSY_EWSummary/ATLAS_SUSY_
EWSummary.png.
[12] ATLAS Collaboration, arXiv:1710.04901, CERN-EP-2017-202.
[13] ATLAS Collaboration, ATLAS-CONF-2017-017.

HADRONIC RESONANCE PRODUCTION WITH ALICE AT THE LHC

Sergey Kiselev [a] for the ALICE collaboration
Institute for Theoretical and Experimental Physics, 117259 Moscow, Russia

Abstract. We present recent results on short-lived hadronic resonances obtained by the ALICE experiment in pp, p-Pb and Pb-Pb collisions at LHC energies, including the most recent measurements of $\Lambda(1520)$ and $\Xi(1530)^0$ resonances.

Hadronic resonance production plays an important role both in elementary proton-proton and in heavy-ion collisions. In heavy-ion collisions, since the lifetimes of short-lived resonances are comparable with the lifetime of the late hadronic phase, regeneration and rescattering effects become important and resonance ratios to longer lived particles can be used to estimate the time interval between the chemical and kinetic freeze-out [1]. The measurements in pp and p-Pb collisions constitute a reference for nuclear collisions and provide information for tuning event generators inspired by Quantum Chromodynamics.

Recent results on short-lived mesonic $K^*(892)^0$, $\phi(1020)$ and baryonic $\Lambda(1520)$, $\Xi(1530)^0$ resonances (hereafter K^{*0}, ϕ, Λ^*, Ξ^{*0}) obtained by the ALICE experiment are presented. The K^{*0} and ϕ have been measured in pp collisions at $\sqrt{s}$ = 13 TeV and in Pb-Pb collisions at $\sqrt{s_{NN}}$ = 5.02 TeV (results for the K^{*0} and ϕ in pp at $\sqrt{s}$ = 7 TeV, p-Pb at $\sqrt{s_{NN}}$ = 5.02 TeV and Pb-Pb at $\sqrt{s_{NN}}$ = 2.76 TeV published in [2], [3] and [4,5], respectively). The Λ^* has been measured in pp collisions at $\sqrt{s}$ = 7 TeV, in p-Pb collisions at $\sqrt{s_{NN}}$ = 5.02 TeV and in Pb-Pb collisions at $\sqrt{s_{NN}}$ = 2.76 TeV. The Ξ^{*0} has been measured in Pb-Pb collisions at $\sqrt{s_{NN}}$ = 2.76 TeV (results for the $\Sigma^{*\pm}$ and Ξ^{*0} in pp at $\sqrt{s}$ = 7 TeV and p-Pb at $\sqrt{s_{NN}}$ = 5.02 TeV published in [6] and [7], respectively).

The resonances are reconstructed in their hadronic decay channels and have very different lifetimes.

	K^{*0}	ϕ	Λ^*	Ξ^{*0}
decay channel (B.R.)	$K\pi(0.67)$	$KK(0.49)$	$pK(0.22)$	$\Xi\pi(0.67)$
lifetime (fm/c)	4.2	46.2	12.6	21.7

Figure 1 (left) shows the mean transverse momentum $\langle p_T \rangle$ of K^{*0}, ϕ and stable hadrons in Pb-Pb collisions at $\sqrt{s_{NN}}$ = 5.02 TeV. The $\langle p_T \rangle$ of resonances exhibits similar increasing trend with multiplicity as other hadrons. Moreover, mass ordering is observed in central Pb-Pb collisions. The K^{*0}, p and ϕ, which have similar masses, are observed to have similar $\langle p_T \rangle$ values, as expected if their spectral shape is dominated by radial flow. The results for Pb-Pb collisions at 5.02 TeV confirm the results in Pb-Pb collisions at 2.76 TeV [4].

---
[a] E-mail: Sergey.Kiselev@cern.ch

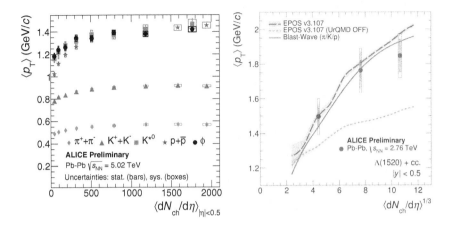

Figure 1: (color online) (left) The mean transverse momentum of K^{*0}, ϕ mesons and stable hadrons as a function of the mean charged particle multiplicity at mid-rapidity in Pb-Pb collisions at $\sqrt{s_{NN}} = 5.02$ TeV. (right) The mean transverse momentum of Λ^* as a function of the mean charged particle multiplicity at mid-rapidity in Pb-Pb collisions at $\sqrt{s_{NN}} = 2.76$ TeV. The measurements are also compared to EPOS3 [8] and Blast-Wave [9] predictions.

Figure 1 (right) presents the $\langle p_T \rangle$ of Λ^* in Pb-Pb collisions at $\sqrt{s_{NN}} = 2.76$ TeV. The results are in agreement with the prediction from the EPOS3 generator with UrQMD [8], which includes a modeling of re-scattering and regeneration in the hadronic phase. The results are also in agreement with the average momentum extracted from the Blast-Wave model [9] with parameters obtained from the simultaneous fit to pion, kaon, and (anti)proton p_T distributions [10].

Figure 2 (left) shows the particle yield ratios K^{*0}/K and ϕ/K in Pb-Pb collisions at $\sqrt{s_{NN}} = 5.02$ TeV. Results for Pb-Pb collisions at $\sqrt{s_{NN}} = 2.76$ TeV [5] and p-Pb collisions at $\sqrt{s_{NN}} = 5.02$ TeV [3] are also shown. The K^{*0}/K ratio shows a significant suppression going from p-Pb and peripheral Pb-Pb collisions to most central Pb-Pb collisions. This suppression is consistent with rescattering of K^{*0} daughters in the hadronic phase of central collisions as the dominant effect and confirms the trend observed in Pb-Pb at 2.76 TeV [4]. The ϕ/K ratio is nearly flat. This suggests that rescattering effects are not important for ϕ, which has 10 times longer lifetime than K^{*0} and decays mainly after the kinetic freeze-out.

Figure 2 (right) presents the ratio K^{*0}/K in pp collisions at $\sqrt{s} = 13$ TeV. Results for pp at $\sqrt{s} = 7$ TeV [11] and p-Pb at $\sqrt{s_{NN}} = 5.02$ TeV [3] are also shown. There is a hint of decrease of the ratio with increasing multiplicity. The values of the ratio are consistent for similar multiplicities across collision systems (pp, p-Pb) and energy (7, 13 TeV). The decrease of the ratio might be

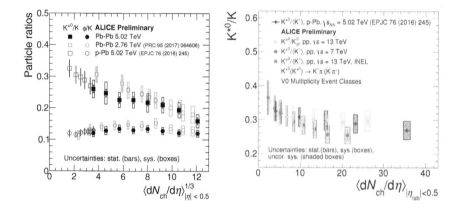

Figure 2: (color online) Particle yield ratios as a function of the mean charged particle multiplicity at mid-rapidity. (left) K^{*0}/K and ϕ/K in Pb-Pb collisions at $\sqrt{s_{NN}} = 5.02$ TeV. Results for Pb-Pb collisions at $\sqrt{s_{NN}} = 2.76$ TeV [5] and p-Pb collisions at $\sqrt{s_{NN}} = 5.02$ TeV [3] are also shown. (right) K^{*0}/K in pp collisions at $\sqrt{s} = 13$ TeV. Results for pp collisions at $\sqrt{s} = 7$ TeV [11] and p-Pb collisions at $\sqrt{s_{NN}} = 5.02$ TeV [3] are also shown.

an indication of a hadron-gas phase with non-zero lifetime in high-multiplicity pp and p-Pb collisions.

Figure 3 (left) shows the particle yield ratio Λ^*/Λ for various collision systems. The Λ^*/Λ ratio demonstrates a significant suppression going from pp, p-Pb and peripheral Pb-Pb collisions to most central Pb-Pb collisions. The suppression confirms the trend seen by STAR at $\sqrt{s_{NN}} = 200$ GeV [15]. Although predictions of the EPOS3 model with UrQMD overestimate the data, the trend of the suppression is qualitatively reproduced. In Pb-Pb collisions at $\sqrt{s_{NN}} = 2.76$ TeV the Λ^*/Λ ratio suppression is similar to the behavior observed for the ρ^0/π [16] and K^{*0}/K [5] ratios.

Figure 3 (right) presents the particle yield ratio Ξ^{*0}/Ξ in Pb-Pb collisions at $\sqrt{s_{NN}} = 2.76$ TeV. Results for pp collisions at $\sqrt{s} = 7$ TeV [6] and p-Pb collisions at $\sqrt{s_{NN}} = 5.02$ TeV [3] are also shown. There is a hint of suppression in central Pb-Pb collisions with respect to pp and p-Pb collisions, but systematics are to be reduced in peripheral Pb-Pb collisions before making any conclusive statement. EPOS3 with UrQMD overestimates the data and predicts only a slight decrease of the Ξ^{*0}/Ξ ratio.

Thermal model predictions overestimate all particle ratios under study in central Pb-Pb collisions, except the ϕ/K ratio.

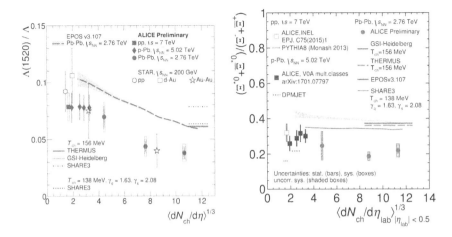

Figure 3: (color online) Particle yield ratios as a function of the mean charged particle multiplicity at mid-rapidity. (left) Λ^*/Λ for various collision systems. The measurements are also compared to model predictions: EPOS3 [8], THERMUS [12], GSI-Helderberg [13], SHARE [14]. STAR data from [15]. (bottom left) Ξ^{*0}/Ξ in Pb-Pb collisions at $\sqrt{s_{NN}} = 2.76$ TeV. Results for pp collisions at $\sqrt{s} = 7$ TeV [6] and p-Pb collisions at $\sqrt{s_{NN}} = 5.02$ TeV [3] are also shown. The measurements are also compared to model predictions: PYTHIA8 [17] and DPMJET [18].

References

[1] G. Torrieri and J. Rafelski, Phys. Lett. **B 509** (2001) 239.
[2] B. Abelev et al., (ALICE Collaboration), Eur. Phys. J. **C 72** (2012) 2183.
[3] J. Adam et al., (ALICE Collaboration), Eur. Phys. J. **C 76** (2016) 245.
[4] B. Abelev et al., (ALICE Collaboration), Phys. Rev. **C 91** (2015) 024609.
[5] J. Adam et al., (ALICE Collaboration), Phys. Rev. **C 95** (2017) 064606.
[6] B. Abelev et al., (ALICE Collaboration), Eur. Phys. J. **C 75** (2015) 1.
[7] D. Adamova et al., (ALICE Collaboration), Eur. Phys. J. **C 77** (2017) 389.
[8] A.G. Knospe et al., Phys. Rev. **C 93** (2016) 014911.
[9] E. Schnedermann et al., Phys. Rev. **C 48** (1993) 2462.
[10] B. Abelev et al., (ALICE Collaboration), Phys. Rev. **C 88** (2013) 044910.
[11] A.G. Knospe (for the ALICE Collaboration), J. Phys. Conf. Series **779** (2017) 012072.
[12] S. Wheaton et al., Comput. Phys. Commun. **180** (2009) 84.
[13] J. Stachel et al., J. Phys.: Conf. Ser. **509** (2014) 012019.
[14] M. Petran et al., Comput. Phys. Commun. 185 (2014) 2056.
[15] B. Abelev et al., (STAR Collaboration), Phys. Rev. C **78** (2008) 044906.

[16] V.G. Riabov (for the ALICE Collaboration), J. Phys. Conf. Series **778** (2017) 012054.
[17] T. Sjöstrand et al., Comput. Phys. Commun. **178** (2008) 852.
[18] S. Roesler, R. Engel and J. Ranft, Conference Proceedings, MC2000, Lisbon, Portugal, October 23-26 (2000) 1033, arXiv:0012252v1 [hep-ph].

SEARCH FOR EXOTIC CHARMONIUM-LIKE STATES AT COMPASS

Andrei Gridin [a b c]

Joint Institute for Nuclear Research Joliot-Curie, 6, Dubna, Moscow region, Russia, 141980

Abstract. Exotic charmonium-like states have been the main objective of various experiments over the last 15 years, but their nature is still unknown. Exclusive leptoproduction can be a new instrument to study their nature. COMPASS [1], a fixed target experiment at CERN, analyzed the full set of the data collected with a muon beam between 2002 and 2011, covering the range from 7 GeV to 19 GeV in the center-of-mass energy of the photon-nucleon system. The results obtained for the production of the $Z_c^\pm(3900)$ and $X(3872)$ are given as well as future perspectives.

1 Introduction

The existence of states built from more than three quarks was predicted by the quark model, while still there are not any experimental observations of such systems. In the early 2000s a number of such "exotic" candidates, named X, Y, Z states, have been discovered in the charmonium and bottomonium sectors. These states were discovered in several production channels like direct production in e^+e^- collisions, direct production in hadron interractions, decays of B-mesons. Photoproduction as a production mechanism has not been studied at all while it has been predicted since 2005 [2].

The $X(3872)$ was the first exotic charmonium-like state observed by the Belle collaboration in 2003 [3]. An explanation of its nature is that $X(3872)$ has a $c\bar{c}$ core and it is a superposition of several states like D^+D^{*-}, $J/\psi\rho$, $J/\psi\omega$, etc. [4]. But during its long history the nature of $X(3872)$ still remains unknown.

The $Z_c^\pm(3900)$ was discovered by BESIII [5] and Belle [6] experiments in 2013. The most probable explanation is that $Z_c^\pm(3900)$ is a tetraquark state because of decay channel $Z_c^\pm(3900) \to J/\psi\pi^\pm$.

The COMPASS experiment at CERN performed the search for exclusive leptoproduction of the $Z_c^\pm(3900)$ [7] in

$$\mu^+ N \to \mu^+ N' Z_c^\pm(3900) \to \mu^+ N' J/\psi \pi^\pm \to \mu^+ \mu^+ \mu^- \pi^\pm N' \quad (1.1)$$

and $X(3872)$ [8] in

$$\mu^+ N \to \mu^+ X(3872) \pi^\pm N' \to \mu^+ (J/\psi \pi^+ \pi^-) \pi^\pm N' \to \mu^+ (\mu^+ \mu^- \pi^+ \pi^-) \pi^\pm N' \quad (1.2)$$

---

[a] E-mail: andrei.gridin@cern.ch
[b] Speaker
[c] On behalf of the COMPASS collaboration

2 Leptoproduction of $Z_c^{\pm}(3900)$ and $X(3872)$

The COMPASS experimentis located at the M2 beam line of the Super Proton Synchrotron at CERN. The experiment uses a muon beam with energies from 160 GeV up to 200 GeV and hadron beam with the energy of 190 GeV. Using muon beam, COMPASS can produce a virtual photon which behaves like a J/ψ and interracts with the target nucleon via virtual pion or pomeron exchange.

This leptoproduction mechanism was used for studying $Z_c^{\pm}(3900)$ and $X(3872)$ photoproduction.

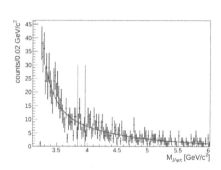

Figure 1: $J/\psi\pi^{\pm}$ mass spectrum. The searching range is shown by the vertical lines, the curve represents the background fitting.

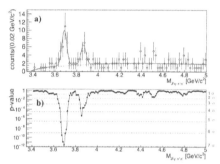

Figure 2: (a) $J/\psi\pi^+\pi^-$ mass spectrum. The curve represents the fit of assumed signals and background. (b) The probability to obtain the observed or a larger number of events due to statistical fluctuation of the Poissonian background.

Figure 1 shows the mass spectrum of the $J/\psi\pi^{\pm}$. This spectrum does not contain any statistically significant peaks in the range $3.84\text{ GeV}/c^2 < \text{M}_{J/\psi\pi^{\pm}} < 3.96\text{ GeV}/c^2$. The upper limit for the number of $Z_c^{\pm}(3900)$, produced in the reaction 1.1, in this range was estimated to be $N_{Z_c}^{UL} = 15.1$ events. The estimation of the upper limit of the $Z_c^{\pm}(3900)$ production rate was performed by the comparison with the production rate of exclusively produced J/ψ in reaction

$$\gamma N \to J/\psi N \qquad (2.1)$$

$$BR(Z_c^{\pm}(3900) \to J/\psi\pi^{\pm}) \cdot \sigma_{\gamma N \to Z_c^{\pm}(3900)N}\Big|_{\sqrt{s_{\gamma N}}=13.8\,GeV} < 52\,pb,\ CL = 90\%. \qquad (2.2)$$

The $J/\psi\pi^+\pi^-$ invariant mass spectrum for the $J/\psi\pi^+\pi^-\pi^{\pm}$ final state of the reaction 1.2 is shown in Fig. 2(a). Two possible combinations of the $\pi^+\pi^-$ are taken into account. Two peaks below 4 GeV/c^2 are assigned to the production and decay of $\psi(2S)$ and $X(3872)$. The statistical significance of the signals is 6.9σ and 4.5σ, respectively (Fig. 2(b)). There are $N_{\psi(2S)} = 24.2 \pm 6.5$ and $N_{X(3872)} = 13.2 \pm 5.2$ events of $\psi(2S)$ and $X(3872)$ and their masses $M_{\psi(2S)} = 3683.7 \pm 6.5 MeV/c^2$ and $M_{X(3872)} = 3860.4 \pm 10.0 MeV/c^2$.

Using the same technique of normalization as for $Z_c^{\pm}(3900)$, the cross-section of the reaction $\gamma N \to X(3872)\pi N$ multiplied by the branching fraction for the decay $X(3872) \to J/\psi \pi^+ \pi^-$ was calculated:

$$\sigma_{\gamma N \to X(3872)\pi N'} \cdot BR(X(3872) \to J/\psi \pi \pi) = 71 \pm 28(stat) \pm 39(syst)\, pb. \quad (2.3)$$

The COMPASS result for the $\pi^+\pi^-$ mass spectrum from the decay of the $\psi(2S)$ is in a good agreement with previous observations [10–13] while the shape of the $\pi^+\pi^-$ mass distribution for $X(3872)$ is different. For $X(3872)$, the $\pi^+\pi^-$ mass spectrum, measured by COMPASS, is in agreement with a three-body phase-space decay, while the other experiments found a dominance of the $X(3872) \to J/\psi \rho^0$ decay mode. Additional studies made by COMPASS have shown that the observed difference cannot be explained by acceptance effects. The possibility to obtain the observed two-pion spectrum from the decay $X(3872) \to J/\psi \omega \to J/\psi \pi^+ \pi^- \pi^0$, where the π^0 has been lost, was also investigated and excluded.

Also the reaction $\gamma N \to X(3872) N$ was studied. The upper limit for the production rate of $X(3872)$ in this reaction is

$$\sigma_{\gamma N \to X(3872) N'} \cdot BR(X(3872) \to J/\psi \pi \pi) < 2.9\, pb, \; CL = 90\%. \quad (2.4)$$

3 COMPASS: new possibilities

For the GPD program [1], the COMPASS setup in 2016–2017 has several features which provide wide opportunities for the search for events of leptoproduction in other channels, especially with π^0 and γ in the final state.

The COMPASS calorimeter system is extended with the third electromagnetic calorimeter ECAL0. It will provide a much better selection of exclusive events. The search for $Z_c^0(3900)$ in its decay channel to $J/\psi \pi^0$ will also be possible.

The target region was upgraded with the recoil-proton detector CAMERA, which surrounds the 2.5 m liquid hydrogen target. Absence of neutrons in the target will allow to use only the $J/\psi \pi^+$ final state for searching for $Z_c^+(3900)$ while the $J/\psi \pi^-$ channel will be suppressed. It would be possible to use the CAMERA detector for reconstruction of a recoil proton in the reactions with neutral exchange and as a veto for the reactions with positive charge exchange.

4 Summary

The COMPASS experiment performed the search for exclusive events of leptoproduction of $Z_c^\pm(3900)$ and $X(3872)$ using reactions 1.1 and 1.2. The production rate of these states was estimated. The COMPASS result for the dipion mass spectrum for $X(3872)$ is in tension with previous observations. It could be an indication that the $X(3872)$ signal could contains a component with quantum numbers different from 1^{++}. The study of exotic charmonia photoproduction also could be carried out using high intensity beams like in CLAS [14] and GlueX [15].

References

[1] (COMPASS Collaboration), COMPASS-II proposal, SPSC-2010-014/P-340.
[2] Bing An Li, Phys. Lett. B605, 306 (2005).
[3] S. K. Choi, et al., (Belle Collaboration), Phys. Rev. Lett. 91 (2003), 262001.
[4] S. Takeuchi, K.Shimizu and M.Takizawa, PTEP2014(2014)123D01.
[5] M. Ablikim, et al.(BESIII Collaboration), Phys. Rev. Lett. 110 (2013), 252001.
[6] Z. Q. Liu, et al.(Belle Collaboration), Phys. Rev. Lett. 110 (2013), 252002.
[7] C. Adolph et al. (COMPASS Collaboration), Phys. Lett. B 742 (2015) 330 [arXiv:1407.6186[hep-ex]].
[8] M. Aghasyan et al.(COMPASS Collaboration) CERN-EP/2017-165 [arXiv:1707.01796[hep-ex]].
[9] R. Barate et al.(NA14 Collaboration), Z.Phys. C33 (1987), 505.
[10] S. K. Choi, et al., (Belle Collaboration), Phys. Rev. D84 (2011), 052004.
[11] A. Abulencia, et al., (CDF Collaboration), Phys. Rev. Lett. 96 (2006), 102002.
[12] S. Chatrchyan, et al., (CMS Collaboration), JHEP 04 (2013) 154.
[13] M. Aaboud et al. (ATLAS Collaboration), JHEP 1701 (2017) 117.
[14] B. A. Mecking et al. (CLAS Collaboration), Nucl. Instrum. Meth. A503 (2003), 513.
[15] H. Al Ghoul et al. (GlueX Collaboration), AIP Conf. Proc. 1735, 020001.

OVERVIEW OF THE CEPC PROJECT

Xin Shi [a]

Institute of High Energy Physics, Beijing 100049, People's Republic of China
State Key Laboratory of Particle Detection and Electronics, Beijing 100049, Hefei 230026, People's Republic of China

Abstract. The Circular Electron-Positron Collider (CEPC) project is the next generation collider program proposed by the high energy physics community in China. Since the release of the preCDR (Conceptual Design Report), significant progress has been made for both accelerator and detector design towards the CDR. In addition, status and major development of the program are presented.

1 Introduction

The BEPCII program is the only running collider experiment in China currently, and will likely to complete its mission around 2020s. The HEP community in China proposed a possible accelerator based particle physics program in China after the BEPCII — the CEPC-SppC (Circular Electron-Positron Collider and Super proton-proton Collider).

The CEPC will be an e^+e^- Higgs (Z) factory with the center-of-mass energy around 240GeV and luminosity about $2 \times 10^{34} cm^{-2}s^{-1}$ for two interaction points (IP). It will produce one million Higgs particles in 10 years and produce 10^{10} Z bosons per year while running at the Z-pole. These data will be used to precisely measure the Higgs boson (and the Z boson), in which a 1% or better precision of Higgs is expected.

Using the same tunnel, the CEPC is upgradable to pp collision with $E_{cm} \sim 50 - 100$ TeV, with electron-proton (ep) and heavy ion (HI) collision options. Running at the energy frontier, the SppC will be a discovery machine for new physics beyond standard model (BSM).

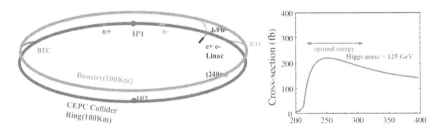

Figure 1: (left) Overview of the CEPC project. (right) Production cross section of $e^+e^- \to ZH$ as a function of E_{cm} [GeV].

---

[a] E-mail: shixin@ihep.ac.cn

The schedule of CEPC started with the pre-studies to identify design issues and R&D items from 2013 to 2015. As a milestone, the pre-CDR was released in early 2015 [1]. The R&D for engineering design is planned to begin from 2016 to 2022, which includes international collaboration and site study. The construction is scheduled from 2022 to 2030, which includes seeking approval, site decision, construction during 14th 5-year plan and commissioning. Finally, the data taking will begin from 2030 to 2040, before the LHC program ends. The schedule is possibly runs con-currently with the ILC program.

The baseline design of the CEPC Conceptual Design Report (CDR) are: circumference of 100km with E_{cm} = 240GeV, power per beam ≤ 30MW. The design luminosity is $\sim 2 \times 10^{34} cm^{-2} s^{-1}$ for E_{cm} = 240GeV and $\sim 1 \times 10^{34} cm^{-2} s^{-1}$ for E_{cm} = 91GeV. There are two layouts for the accelerator ring: double ring as the default and the advanced local double ring as an option. Two independent detectors will be located on the ring. The CEPC will be based on mature technologies and covers both Higgs and Z physics program. It also has the potential to upgrade to high energy pp collision as well as the synchrotron light source.

2 The Conceptual Design Report

The Conceptual Design Report (CDR) is composed of the CEPC accelerator design, the detector design and physics performance. The CDR drafts are expected by the end of 2017, while reviews and finalization will be done in the mid of 2018.

The layout and hardware satisfy both the Z and Higgs programs, with two ring schemes towards CDR. The baseline is the Fully Partial Double ring, as shown on the left of Fig. 2. It has better performance for Higgs and Z without bottleneck problems, but with higher cost. An alternative option is the Advanced Partial Double Ring (right figure of Fig. 2), which has a lower cost and could reach the fundamental requirement for Higgs and Z luminosities, under the condition that sawtooth and beam loading effects will be solved.

Parameters for CEPC double ring shows co-existence of Z/H programs are possible. Reconfiguration of CEPC can lead to much better luminosity at the Z pole. The CEPC detector is more compact that HCAL reduced from 48 layers in preCDR to 40 layers. It also accommodates to double ring geometry and MDI (machine detector interface) design implemented. Based on the preliminary study, no visible impact is observed on physics performance. For example, the di-jet mass resolution of $H \to gg$ events only changed from 4.7GeV to 4.9GeV.

3 Status and Major Development

The main CEPC ring super conducting RF hardware R&D is benefit from the ILC development and will in turn contribute back to the ILC construction. To

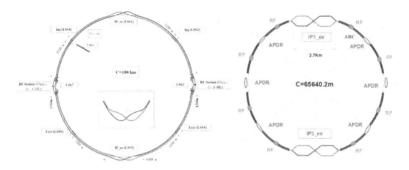

Figure 2: Layout of CEPC-SppC. (left) Fully Partial Double Ring as baseline design. (right) Advanced Partial Double Ring as alternative design.

support the R&D, a new SRF facility will be built in the Platform of Advanced Photon Source Technology (PAPS) complex in Huairou Science Park, Beijing. The ground breaking of 4500 m^2 SRF lab is already taken on May 31, 2017 and will be done by 2019. The whole PAPS complex is funded by the city of Beijing. The construction period is from May 2017 to June 2020. It includes RF system and cryogenic systems, magnet technology, and beam test, etc.

The CEPC SRF R&D and test at the PAPS facility includes advanced superconducting RF technology R&D, preparation, diagnostics and test tools for high performance cavity, nitrogen-doping & infusion; Nb_3Sn thin film for high Q and high gradient, high resolution optical inspection, temperature and X-ray mapping, second sound quench detection, defects local grinding. The PAPS can also host the test facilities for key components of SRF accelerator, very high power variable input coupler with low heat load, high power HOM(Higher-order modulation) coupler and absorber, components horizontal test with tuner and LLRF in low magnetic field. Common cutting-edge research with ILC and SCLF (Shanghai XFEL) and possible breakthroughs in Fe-pnictides superconducting cavity. The development of high performance cryomodule prototypes for CEPC and demonstration of mass-production capability are being done at the PAPS facility as well. Within five years, two small test cryomodules and two full scale prototype cryomodules will be build and tested at PAPS.

The funding for CEPC is encouraging. Started with the IHEP initial fund for three years (2015–2017), CEPC gained increasing support by NSFC with 5 projects in 2015 and 7 projects in 2016. It also got significant support from the Ministry of Science and Technology (MOST) and Chinese Academy of Sciences (CAS) talent program fund by Beijing city. In 2017, additional funding request will be sent to MOST and other agencies as well. The overall funding needs to carrying out key CEPC design and R&D should be fully met by the end of 2018.

Many sites have been explored to host the CEPC. For example, Qinghuangdao in Hebei province has been completed in preCDR. Shenshan in Guangdong province was completed in August 2016 and Huangling in Shannxi province signed contract to exploration company in January 2017. More sites are under investigation.

The international advisory committee (IAC) meetings of CEPC have been held once a year since 2015. The IAC has been impressed with the amount of work done by the CEPC team for the significant progress on many fronts including accelerator R&D, detector R&D, simulation and theory. To make the CEPC study more international, the CEPC workshop [2] will be held in IHEP in November 2017 to encourage global contributions.

For the first time, the HEP division of the Chinese Physical Society reached a consensus in August, 2016 that placed CEPC as the top priority accelerator based program for the future.

4 Summary

The CEPC CDR is progressing very well. The design and R&D needs are largely met with various sources of funding and support. People are working hard on the key items and to build a stronger CEPC team with international collaboration and participation. For the future, economic HTS magnet program is being explored in China with a carefully constructed consortium. Infrastructure, experience and engineering proficiency gained through current projects (light source, CSNS, etc.) is helpful for the CEPC. Upon successfully completing the R&D program, we expect to make the case to the national government for building the CEPC around 5 years later.

Acknowledgments

We thank all the people working hard to support the CEPC project. We also acknowledge the organizers of Lomonosov conference to give us this opportunity to report the progress of CEPC. X. Shi thanks "The Thousand Talents Plan" in China to provide funding for this program.

References

[1] The CEPC-SppC Study Group, *"CEPC-SPPC preCDR Vol I and II"*, (2015).

[2] International Workshop on High Energy Circular Electron Positron Collider 2017: http://cepcws17.ihep.ac.cn.

RECENT RESULTS ON DIFFRACTION AT HERA

Sergey Levonian [a]
DESY, Notkestraße 85, 22607 Hamburg, Germany

Abstract. Four new measurements are presented from the area of diffractive and exclusive production at HERA. Isolated photons are studied in diffractive photoproduction, while open charm cross section and cross-section ratio $\sigma_{\psi(2S)}/\sigma_{J/\psi(1S)}$ are measured in diffractive deep-inelastic scattering (DIS) regime. Finally, exclusive ρ^0 meson photoproduction associated with a leading neutron is investigated for the first time at collider experiments.

1 Introduction

Diffraction is an important and challenging part of physics landscape at the electron-proton collider HERA. It represents a complicated interplay of soft and hard phenomena, thus providing rich opportunities for QCD studies. Diffractive ep scattering at high photon virtualities Q^2 for the first time allows a partonic content of the Pomeron, a central object in diffractive physics deeply related to QCD vacuum, to be probed. Using these diffractive parton densities (DPDF) to predict various final states in ep and also in pp diffraction relies on collinear factorisation theorem [1] and hence provides crucial tests of major concepts and overall consistency of QCD picture of diffraction in high energy particle collisions.

Below, four new measurements are presented, performed by the H1 and ZEUS collaborations, which benefit from large integrated luminosity, or from the detector upgrades implemented for HERA2 running period (2004–2007).

2 Isolated photons in diffractive photoproduction

Prompt photons in diffractive events are interesting in two respects. First, they are direct messengers from hard subprocess which, unlike jets, are not affected by fragmentation. In addition, photons couple to charged partons only and hence they probe quark content of the Pomeron in diffractive processes. ZEUS Collaboration studied this process [2] using two data samples of 91 pb^{-1} (HERA1) and 374 pb^{-1} (HERA2) respectively. High transverse energy photons $4 < E_T^{\gamma} < 15$ GeV are selected in central pseudo-rapidity range $-0.7 < \eta^{\gamma} < 0.9$ in diffractive photoproduction: $Q^2 < 1$ GeV^2, $0.2 < y < 0.7$, $x_{I\!P} < 0.03$, where $x_{I\!P}$ is the fraction of the proton energy taken by the Pomeron. In $\sim 85\%$ of such events an isolated photon is accompanied by a jet with $E_T^{jet} > 4$ GeV. A template fit of the energy weighted cluster width is employed to statistically separate prompt photons and background originating mainly from π^0 and η decays (see Fig. 1a). The data are corrected to the hadron level and compared

---

[a] E-mail: levonian@mail.desy.de

with theory as provided by the Rapgap MC program [3], normalised to the data. In Fig. 1b the cross section is given as a function of $z_{I\!P}$, a fraction of the Pomeron momentum transferred to the hard subprocess. The data show prominent peak at largest $z_{I\!P}$ which is not expected and not reproduced by the Rapgap, containing so called *resolved Pomeron* contributions only. All other distributions are well described by this model. These events at $z_{I\!P} \approx 1$ are interpreted as a first evidence of *direct Pomeron* contribution.

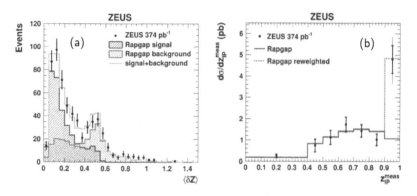

Figure 1: (a) The distribution of the energy-weighted cluster width for isolated photon candidates. (b) Differential cross section as a function of $z_{I\!P}$ for diffractive events containing an isolated photon and a jet, compared to normalised prediction from Rapgap MC.

3 D^* in diffractive DIS

Contrary to prompt photons, charm in diffractive ep scattering is produced predominantly via boson-gluon fusion, and thus is strongly sensitive to the gluon content of the Pomeron. D^* mesons reconstructed via so called 'golden' decay mode $D^{*+} \to D^0 \pi^+_{slow} \to (K^-\pi^+)\pi^+_{slow} + C.C.$ provide a convenient way of tagging open charm production. H1 Collaboration has used 281 pb^{-1} data sample to measure D^* cross section in diffractive DIS [4] for $p_{T,D^*} < 1.5$ GeV and $-1.5 < \eta_{D^*} < 1.5$ in the range of $5 < Q^2 < 100$ GeV^2 and $0.02 < y < 0.65$. Diffractive events are selected using Large Rapidity Gap signature and correspond to the following phase space: $x_{I\!P} < 0.03$, $|t| < 1$ GeV^2 and proton dissociative mass $M_Y < 1.6$ GeV.

In Fig. 2 the measured differential cross sections are confronted with the NLO QCD calculations [5] which use $\mu_r^2 = \mu_f^2 = m_c^2 + Q^2$ scale choice and DPDF set [6] as determined in H1 inclusive diffraction. For $c \to D^*$ fragmentation the Kartvelishvili parametrisation is used with the parameters fixed by the H1 inclusive D^* measurements. NLO QCD describes the data fairly well both in

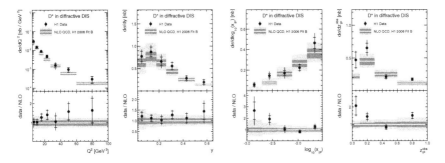

Figure 2: D^* meson cross sections as a function of the Q^2, the inelasticity y, and the diffractive variables $x_{I\!P}$ and $z_{I\!P}^{obs}$ as compared to NLO QCD prediction.

shapes and in absolute normalisation. This supports the universality of charm fragmentation and the validity of QCD factorisation ansatz in diffractive DIS.

4 Ratio $\sigma_{\psi(2S)}/\sigma_{J/\psi(1S)}$ in exclusive DIS

An important testing ground for theoretical concepts and quantitative calculations in diffraction is provided by the study of exclusive vector meson production at HERA $e+p \to e+V+Y$. The exclusive deep inelastic electroproduction of $\psi(2S)$ and $J/\psi(1S)$ has been studied with the ZEUS detector in the kinematic range $2 < Q^2 < 80$ GeV^2, $30 < W < 210$ GeV and $|t| < 1$ GeV^2 [7]. Since these two charmonium states have same quark content, but different radial distributions of the wave functions (w.f.), their cross section ratio allows QCD predictions of the w.f. dependence on the $\bar{c}c$-proton cross section to be tested. In particular, a suppression of $\psi(2S)$ relative to $J/\psi(1S)$ is expected due to the node in the $\psi(2S)$ w.f. leading to the destructive interference of the contributions from small and large $\bar{c}c$ dipoles to the production amplitude.

Since the reactions studied are statistically limited all available data have been used, amounting to 468 pb^{-1}. The final sample contains ∼2500 $J/\psi(1S)$ and ∼190 $\psi(2S)$ events. After correcting for the detector acceptance, efficiency and the branching ratios, the cross section ratio $R = \sigma_{\psi(2S)}/\sigma_{J/\psi(1S)}$ was determined in bins of Q^2, W and $|t|$ with statistical precision of ∼20% and systematic uncertainty ∼10%. While as a function of W and $|t|$ the values of R are found to be compatible with a constant, the Q^2 dependence shows a positive slope with the significance of ∼2.5σ.

In Fig. 3 the results are compared to the previous H1 measurements [8] and confronted with a set of QCD models. One can see, that all models correctly predict the suppression strength at $Q^2 = 0$ and qualitatively reproduce the rise of R with Q^2, however with a large spread in its magnitude. Thus, the

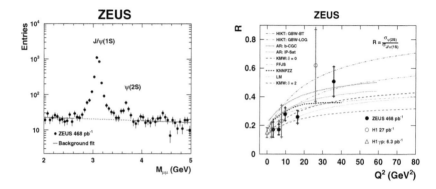

Figure 3: Two-muon mass distribution, $M_{\mu\mu}$, for exclusive dimuon events (left panel) and the cross-section ratio $R = \sigma_{\psi(2S)}/\sigma_{J/\psi(1S)}$ as a function of Q^2 (right panel). HERA measurements are compared to different QCD model predictions (see ref. in [7]).

experimental data, albeit with relatively large uncertainties, show a definite discriminating power allowing to disfavour some extreme model predictions.

5 Exclusive ρ^0 photoproduction with a leading neutron

Measurements of leading baryon production in high energy particle collisions provide important inputs for the theoretical understanding of strong interactions in the soft, non-perturbative regime. The aim of the analysis presented here is to investigate exclusive ρ^0 production on virtual pions in the photoproduction regime at HERA and to extract for the first time experimentally the quasi-elastic $\gamma\pi \to \rho^0\pi$ cross section. At small transverse momenta squared $t \to 0$ dominant contribution comes from one-pion-exchange (OPE) diagram shown in Fig. 4a.

The analysis is based on $\sim$ 6600 events, containing only two charged pions from ρ^0 decay and a leading neutron with energy $E_n > 120$ GeV, and nothing else above noise level in the detector. This ensures the exclusivity and limits the dissociative background to the range $M_Y < 1.6$ GeV. The sample corresponds to an integrated luminosity of 1.16 pb^{-1}, collected by a special minimum bias track trigger in the years 2006–2007 at $\sqrt{s_{ep}} = 319$ GeV. Diffractive dissociation background is modelled by the Diffvm program [10] and statistically separated from the signal, $\gamma p \to \rho^0 n\pi^+$, on the basis of the neutron energy distribution shape. It amounts to $(34\pm5)\%$ and gives the dominant experimental uncertainty to the cross section measurement. Further details of the analysis can be found in [9].

The cross sections for the reaction $\gamma p \to \rho^0 n\pi^+$ are measured for two ranges of the neutron transverse momentum, $p_{T,n}$: $p_{T,n} < 0.69 E_n$, corresponding to

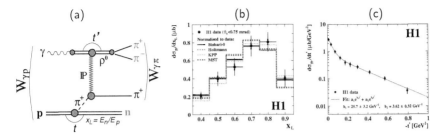

Figure 4: (a) OPE diagram for exclusive photoproduction of ρ^0 mesons associated with leading neutrons at HERA. (b) Differential cross section $\mathrm{d}\sigma_{\gamma p}/\mathrm{d}x_L$ in the range $20 < W_{\gamma p} < 100$ GeV compared to the predictions based on different versions of pion fluxes (see ref. in [9]). (c) Differential cross section $\mathrm{d}\sigma_{\gamma p}/\mathrm{d}t'$ of ρ^0 mesons fitted with the sum of two exponential functions. The inner error bars in (b,c) represent statistical and uncorrelated systematic uncertainties added in quadrature and the outer error bars are total uncertainties, excluding an overall normalisation error of 4.4%.

the full angular acceptance of the neutron calorimeter, $\theta_n < 0.75$ mrad, and for so called OPE safe range, $p_{T,n} < 0.2$ GeV. In Fig. 4b the data are compared to the predictions, based on different models for the pion flux. The shape of x_L distribution is described by most of the models, although some of them can be ruled out already at that level [9]. The pion flux models compatible with the data in shape of the x_L distribution are used to extract the photon-pion cross section at $\langle W_{\gamma\pi}\rangle = 24$ GeV from $\mathrm{d}\sigma/\mathrm{d}x_L$ in the OPE approximation:

$$\sigma(\gamma\pi^+ \to \rho^0\pi^+) = 2.33 \pm 0.34(\mathrm{exp})^{+0.47}_{-0.40}(\mathrm{model})\ \mu\mathrm{b},$$

where the experimental uncertainty includes statistical, systematic and normalisation errors added in quadrature, while the model error is due to the uncertainty in the pion flux integral obtained for the different flux parametrisations compatible with the data. Taking a value of $\sigma(\gamma p \to \rho^0 p) = (9.5 \pm 0.5)\ \mu\mathrm{b}$ at the corresponding energy $\langle W \rangle = 24$ GeV, which is an interpolation between fixed target and HERA measurements, one obtains for the ratio $r_{\mathrm{el}} = \sigma_{\mathrm{el}}^{\gamma\pi}/\sigma_{\mathrm{el}}^{\gamma p} = 0.25\pm0.06$. This ratio is significantly smaller than the expected value $r_{\mathrm{el}} = (\frac{b_{\gamma p}}{b_{\gamma \pi}}) \cdot (\sigma_{\mathrm{tot}}^{\gamma\pi}/\sigma_{\mathrm{tot}}^{\gamma p})^2 = 0.57\pm0.03$ which can be deduced by combining the optical theorem, the eikonal approach relating cross sections with elastic slope parameters and the data on pp, π^+p and γp elastic scattering. Such a suppression of the cross section is usually attributed to rescattering, or absorptive corrections, which are essential for leading neutron production. For the exclusive reaction $\gamma p \to \rho^0 n\pi^+$ studied here this would imply an absorption factor of $K_{abs} = 0.44 \pm 0.11$.

Finally, the cross section as a function of the four-momentum transfer squared of the ρ^0 meson, t', is presented in Fig. 4c. It exhibits the very pronounced feature of a strongly changing slope between the low-t' and the high-t' regions, which is characteristic to double peripheral exclusive reactions. In a geometric picture, the large value of slope b_1 suggests that for a significant part of the data ρ^0 mesons are produced at large impact parameter values of order $\langle r^2 \rangle = 2b_1 \cdot (\hbar c)^2 \simeq 2\text{fm}^2 \approx (1.6 R_\text{p})^2$.

6 Summary

Four new measurements are presented in the area of diffractive and exclusive channels, by the H1 and ZEUS collaborations, making use of full HERA data samples. Whenever hard scale is present, perturbative QCD calculations are successful. The data show sensitivity to some QCD models parameters. They can also be used to further constrain DPDFs, especially at high $z_{I\!P}$. Photon-pion elastic cross section is extracted for the first time experimentally in the one-pion-exchange approximation. Strong absorptive effects are confirmed in leading neutron production. Since the nature of those is non-perturbative, new experimental results are essential for the tuning of 'Survival Gap Probability' models.

References

[1] J.C. Collins, *Phys. Rev.* **D57** (1998) 3051. Erratum-ibid. 61, 019902 (2000).
[2] H. Abramowicz et al. [ZEUS Collab.], *Phys. Rev.* **D96** (2017) 032006.
[3] H. Jung, *Comp. Phys. Commun.* **86** (1995) 147.
[4] V. Andreev et al. [H1 Collab.], *Eur. Phys. J.* **C77** (2017) 340.
[5] B.W. Harris and J. Smith, *Phys. Rev.* **D57** (1998) 2806.
[6] A. Aktas et al. [H1 Collab.], *Eur. Phys. J.* **C48** (2006) 715.
[7] H. Abramowicz et al. [ZEUS Collab.], *Nucl. Phys.* **B909** (2016) 934.
[8] C. Adloff et al. [H1 Collab.], *Phys. Lett.* **B421** (1998) 385.
   C. Adloff et al. [H1 Collab.], *Eur. Phys. J.* **C10** (1999) 373.
[9] V. Andreev et al. [H1 Collab.], *Eur. Phys. J.* **C76** (2016) 41.
[10] B. List and A. Mastroberardino, Proc. of the Workshop on Monte Carlo Generators for HERA Physics, eds. A.T. Doyle et al., DESY-PROC-1999-02 (1999) 396.

THE OLYMPUS EXPERIMENT — TWO-PHOTON EXCHANGE IN ELECTRON PROTON SCATTERING

Uwe Schneekloth [a] for the OLYMPUS Collaboration
Deutsches Elektronen-Synchrotron DESY, Hamburg, Germany

Abstract. The OLYMPUS collaboration has recently made a precise measurement of the positron-proton to electron-proton elastic scattering cross section ratio, $R_{2\gamma}$, over a wide range of the virtual photon polarization, $0.456 < \epsilon < 0.978$. This provides a direct measure of hard two-photon exchange in elastic lepton-proton scattering widely thought to explain the discrepancy observed between unpolarized and polarized measurements of the proton form factor ratio, $\mu_p G_E^p/G_M^p$. The OLYMPUS results are small, within 1% on unity, over the range of momentum transfers measured and significantly lower than theoretical calculations that can explain part of the observed discrepancy in terms of two-photon exchange at higher momentum transfers. However, the results are in reasonable agreement with predictions based on phenomenological fits to the available form factor data.
The motivation for measuring $R_{2\gamma}$ will be presented followed by a description of the OLYMPUS experiment. The importance of radiative corrections in the analysis will be shown also. Then we will present the OLYMPUS results and compare with results from two similar experiments and theoretical calculations.

1 Introduction

It has been about 100 years since Ernst Rutherford named the hydrogen nucleus the proton; later discovered to be a fundamental component in all nuclei. Yet many fundamental parameters of the proton are still not completely understood and still excite both theoretical and experimental research. The proton radius [1], the proton spin [2], and how the proton mass arises from the energy of the constituent and current quarks in lattice QCD [3] are all still topical subjects in nuclear physics. The OLYMPUS experiment addressed yet another "proton puzzle" [4,5] concerning the ratio of the charge and magnetic form factors.

Electron scattering has long been a standard technique for studying nucleons and nuclei. The electromagnetic interaction is well understood and the point-like nature of electrons make them ideal for probing electric and magnetic charge distributions. Historically, unpolarized electron-proton scattering has been analysed in terms of one-photon exchange (Born approximation) to determine the electric, G_E^p, and magnetic, G_M^p, form factors for the proton. But recent experiments with polarized electrons, polarized targets, and measurements of the polarization transferred to the proton are in striking disagreement with the unpolarized results for the proton form factor ratio, $\mu_p G_E^p/G_M^p$ (see Fig. 1). The unpolarized results [6–13] obtained using the Rosenbluth technique, are known to be insensitive to the electric form factor, G_E^p, at high momentum transfer while the polarization measurements [14–21] make a direct

---
[a] E-mail: uwe.schneekloth@desy.de

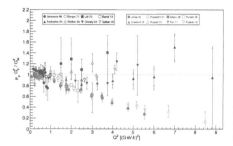

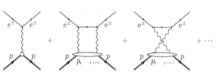

Figure 1: Ratio of proton form factors measured using the unpolarized Rosenbluth technique and using polarization transfer.

Figure 2: Some of the Feynman diagrams for lepton-proton scattering showing one- and two-photon exchange. Diagrams for self-energy, vertex corrections, vacuum polarization, and bremsstrahlung (not shown) must also be included in any calculation.

measurement of the form factor ratio, by measuring the ratio of transverse to longitudinal nuclear polarization.

It has been suggested that two-photon exchange (see Fig. 2) might be able to explain the observed discrepancy in the measured form factor ratio. Radiative corrections must be applied to the measured cross sections to extract the equivalent one-photon exchange value so results from different experiments and theoretical calculations can be compared. These radiative corrections can be significant and are complicated by details of the experimental acceptance, efficiency, and resolution. Two-photon exchange is generally included in the standard radiative corrections but only in the "soft" limit where one of the photons imparts negligible energy to the proton. These calculations generally do not include models for the proton structure. A more complete handling of two-photon exchange contributions might be able to resolve the discrepancy. A proper calculation of "hard" two-photon exchange is more difficult because details of the proton ground state and nucleon resonances for the intermediate state must also be considered.

To determine the contribution of "hard" two-photon exchange, the OLYMPUS experiment proposed to measure the ratio of positron-proton to electron-proton elastic scattering. If two-photon exchange is a significant factor in lepton-proton scattering the ratio will deviate from unity because the interference between one- and two-photon exchange changes sign between electron and positron scattering. Naively, one would expect a small effect, of order $\alpha \approx 1/137$, but that wouldn't explain the striking discrepancy observed.

2 OLYMPUS Experiment

The OLYMPUS experiment [22] ran on the DORIS storage ring at the DESY Laboratory in Hamburg, Germany. DESY undertook significant modifications to the DORIS storage ring to accommodate the OLYMPUS experiment. RF

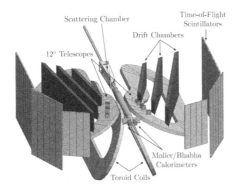

Figure 3: Schematic of OLYMPUS detector. The top four toroid coils are not shown to reveal the horizontal sectors. The drift chambers are shown as three separate chambers in each sector but are actually combined into a single gas volume.

cavities and quadrupoles had to be relocated from the straight section of the storage ring where OLYMPUS was to be located. Services for cooling water and power for the OLYMPUS toroidal magnet had to be installed and the shielding walls extended to make room for the detector. The power supplies for the DORIS ring were also modified so their polarity could be changed quickly when switching between positron and electron running. This capability was crucial for the OLYMPUS experiment, which switched daily between beams of electrons and positrons. A large transport frame was also produced to support the OLYMPUS detector on rails. This allowed the detector to be assembled outside the ring and then rolled into the ring for the experiment.

A hydrogen gas target [23] was installed internal to the storage ring. The target consisted of a thin-walled, elliptical tube 600 mm long without entrance or exit windows. Hydrogen gas was injected into the centre of the tube and allowed to diffuse to either end where series of vacuum pumps were used to maintain the high vacuum required by the storage ring. The nominal target areal density was approximately 3×10^{15} atoms/cm^2. Additionally, the target region required collimators to shield against synchrotron radiation and specially designed transition pieces to minimize wakefield effects.

In 2010, the former BLAST detector [24] from MIT-Bates was disassembled and shipped to DESY where it was reassembled. The detector, shown schematically in Fig. 3, consisted of an eight-sector toroidal magnetic spectrometer with the two horizontal sectors instrumented with large acceptance drift chambers covering polar angles $20° \leq \theta \leq 80°$ and azimuthal angles $-15° \leq \phi \leq 15°$ for 3D particle tracking and walls of time-of-flight scintillator bars for triggering and particle identification. The detector was left-right symmetric and was used as a cross-check during the analysis.

Two new detector systems were built to monitor the luminosity. These were symmetric Møller/Bhabha calorimeters at $\theta = \pm 1.29°$ and two telescopes of three triple GEM (gas electron multiplier) detectors interleaved with three multi-wire proportional chambers mounted at $\theta = \pm 12°$ relative to the beam axis.

The timeline for the OLYMPUS experiment was very tight. OLYMPUS received approval and funding in December, 2009, and faced a fixed deadline of December, 2012, when DORIS was scheduled to be shut down. The detector rolled into the DORIS ring in July, 2011. After a few commissioning tests, it ran for one month in February, 2012, and then for two months at the end of 2012, alternating daily between electrons and positrons at 2.01 GeV with a typical current around 65 mA. In total OLYMPUS collected approximately 4.5 fb^{-1} of data, 25% more than the original proposal.

3 Analysis and Results

The analysis of the OLYMPUS experiment was complicated by an inhomogeneous magnetic field and drift chamber inefficiencies due to the high rate of Møller and Bhabha electrons bent into the innermost drift chambers. It was planned to change the toroid magnet polarity each day to reduce tracking systematics but the background with negative polarity prevented operation at high currents so the current analysis is with positive polarity only. To properly analyse the OLYMPUS data a detailed Monte Carlo simulation of the experiment was written using GEANT4. This allowed the Monte Carlo simulation to account for the differences between electrons and positrons with respect to radiative effects, changing beam position and energy, the spectrometer acceptance, track reconstruction efficiency, luminosity, and elastic event selection. The resulting ratio for the positron-proton to electron-proton cross sections was then determined by calculating:

$$R_{2\gamma} = \frac{\sigma_{e^+p}}{\sigma_{e^-p}} = \frac{N_{e^+p}^{Data}}{N_{e^-p}^{Data}} \times \frac{N_{e^-p}^{MC}}{N_{e^+p}^{MC}}$$

where the N_i are luminosity normalized experimental and Monte Carlo yields.

A custom event generator was developed to convolve the standard radiative corrections with the detector acceptance and resolutions. Since lepton-proton bremsstrahlung interference also changes sign between electrons and positrons, care was taken to calculate contributions as accurately as possible, without resorting to peaking or soft-photon approximations. The calculations for "soft" two-photon exchange employed two standard prescriptions: Mo-Tsai [25] or Maximon-Tjon [26]. The generator could calculate corrections to order α^3, or all orders using exponentiation. The approximate size of the radiative corrections to OLYMPUS is shown in Fig. 4 relative to the Born approximation.

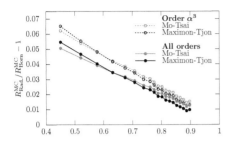
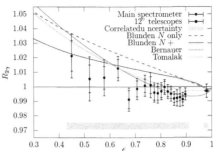

Figure 4: Magnitude of radiative corrections as a function of ϵ for two standard "soft" photon prescriptions for two-photon exchange with exponentiation (all orders) or just to α^3.

Figure 5: OLYMPUS results for $R_{2\gamma}$ as a function of ϵ. Inner error bars are statistical while the outer error bars include uncorrelated systematic uncertainties added in quadrature. The gray band is correlated systematic uncertainty.

The OLYMPUS results [27] are shown in Fig. 5 together with various calculations. The results are below unity at high ϵ but tend to rise with decreasing ϵ. The dispersive calculations of Blunden [28], which can account for part of the discrepancy observed in the form factor ratio at higher Q^2, are systematically above the OLYMPUS results in this energy regime. The phenomenological prediction from Bernauer [29] and the subtractive dispersion calculation from Tomalak [30], using Bernauer's fit, are in reasonable agreement with the OLYMPUS results.

Two other recent experiments, VEPP-3 [31] and CLAS [32], also measured the ratio of positron-proton to electron-proton scattering. However, it is difficult to compare these results directly with OLYMPUS since their measurements were performed at different energies yielding results at different points in the (ϵ, Q^2) plane. To partially account for this, we can compare all the two-photon exchange results by taking the difference with respect to a theoretical calculation (in this case Blunden's N+Δ calculation) evaluated at the correct (ϵ, Q^2) for each data point. This is shown in Fig. 6 plotted versus ϵ. In this view, the results from the three experiments are seen to be in reasonable agreement with each other over the range in ϵ but are systematically below the theoretical calculation. This supports the previous assertion that the theoretical calculation does not reproduce the results in this energy regime. However, the ϵ dependence of both the results and calculations appears to be in agreement.

4 Conclusions

At the momentum transferred range measured by OLYMPUS the effect of "hard" two-photon exchange is small, on the order of 1%. This is good news

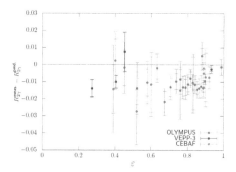

Figure 6: Difference between experimental results and Blunden's N+Δ calculation.

for historical electron scattering measurements made at low energies but does not explain the observed discrepancy in the form factor ratio at higher energies. The rising trend in the ratio with increasing Q^2 (decreasing ϵ) may indicate that two-photon exchange is present and may become significant at higher energies. However, to prove this will require measurements at higher energies that will be difficult due to the rapid decrease in the cross section.

Current theoretical calculations that explain part of the observed discrepancy at higher energies overestimate the effect at the energies measured by the three recent experiments. Possibly higher order radiative corrections are required or nucleon states beyond N+Δ need to be considered.

The discrepancy in the form factor ratio measured using unpolarized and polarized techniques and the possible role played by two-photon exchange continues to be topical within the nuclear physics community. A parallel session at the NSTAR 2017 Workshop [33] was devoted to two-photon exchange. Also, the need for future experiments at higher energy have stimulated discussions at JLab [34] as well as other laboratories. Hopefully, more theoretical and experimental work will bring a better understanding of the proton's structure in the near future.

Acknowledgments

The OLYMPUS experiment was an international collaboration of 55 physicists from 13 institutions plus significant support from engineering and technical staff. We gratefully acknowledge the support from: the Ministry of Education and Science of Armenia, the Deutsche Forschungsgemeinschaft, the European Community-Research Infrastructure Activity, the United Kingdom Science and Technology Facilities Council and the Scottish Universities Physics Alliance, the United States Department of Energy and the National Science Foundation, and the Ministry of Education and Science of the Russian Federation.

References

[1] R. Pohl et al., Ann. Rev. Nucl. Part. Sci. **63**, (2013) 175.
[2] C.A. Aidala *et al.*, Rev. Mod. Phys. (2013), **85**, 655.
[3] S. Durr *et al.*, Science, (2008), **322**, 1224.
[4] A. Afanasev *et al.*, Prog. Part. Nucl. Phys (2017), **95**, 245.
[5] S.K. Blau *Proton structure seen in a new light*, Physics Today (2017), **70**, 14.
[6] T. Janssens *et al.*, Phys. Rev. (1966) **142**, 922.
[7] C. Berger *et al.*, Phys. Lett. (1971), **B35**, 87.
[8] J. Litt *et al.*, Phys. Lett., (1970) **B31**, 40.
[9] W. Bartel *et al.*, Nucl. Phys. (1973), **B58**, 429.
[10] L. Andivahis *et al.*, Phys. Rev. (1994), **D50**, 5491.
[11] R.C. Walker *et al.*, Phys. Rev. (1994) **D49**, 5671.
[12] M.E. Christy *et al.*, Phys. Rev. (2004), **C70**, 015206.
[13] I.A. Qattan *et al.*, Phys. Rev. Lett. (2005) **94**, 142301.
[14] M.K. Jones *et al.*, Phys. Rev. Lett., (2000) **84**, 1398.
[15] T. Pospischil *et al.*, EPJ Web of Conferences, (2001) **A12**, 125.
[16] O. Gayou *et al.*, Phys. Rev. (2001), **C64**, 038202.
[17] V. Punjabi *et al.*, Phys. Rev. (2005) **C71**, 055202, Erratum-ibid.C71: 069902,2005.
[18] C. Crawford *et al.*, Phys. Rev. Lett. (2007), **98**, 052301.
[19] A.J.R. Puckett, *et al.*, Phys. Rev. Lett., (2010) **104** 242301.
[20] G. Ron *et al.*, Phys. Rev. (2011) **C84**, 055204.
[21] A.J.R. Puckett *et al.*, Phys. Rev. (2012) **85**, 045203.
[22] R. Milner *et al.*, Nucl. Instrum. Methods (2014) **A741**, 1.
[23] J.C. Bernauer *et al.*, Nucl. Instrum. Methods, (2014), 20.
[24] D. Hasell *et al.*, Nucl. Instrum. Methods (2009), **A603**, 247.
[25] L.W. Mo and Y.S. Tsai, Rev. Mod. Phys. (1969) **41**, 205.
[26] L.C. Maximon and J.A. Tjon, Phys. Rev. (2000) **C62**, 054320.
[27] OLYMPUS Collaboration, B.S. Henderson *et al.*, Phys. Rev. Lett. **118** (2017) 092501.
[28] P.G. Blunden and W. Melnitchouk Phys. Rev. (2017), **nucl-th**
[29] J.C. Bernauer *et al.*, Phys. Rev. **C90**, (2014)
[30] O. Tomalak and M. Vanderhaeghen, EPJ Web of Conferences (2015) **A51**, 24.
[31] I.A. Rachek, I A *et al.*, Phys. Rev. Lett. (2015) **114**, 062005.
[32] D. Adikaram *et al.*, Phys. Rev. Lett. (2015), **114**, 062003.
[33] http://nstar2017.physics.sc.edu/.
[34] *JPos17*, https://www.jlab.org/conferences/JPos2017/index.html.

A FEW-BODY APPROACH TO LOW-ENERGY ATOMIC COLLISIONS INVOLVING MUONS AND ANTIPROTONS

Renat A. Sultanov [a]

St. Cloud State University, St. Cloud, Minnesota, 56301-4498, USA

Abstract. Low-energy rearrangement scattering process with participation of heavy particles such as an antiproton ($\bar{p}$), a proton (p) and a muon (μ^-), *i.e.* $\bar{p} + (p\mu^-) \rightarrow (\bar{p}p)_\alpha + \mu^-$ is computed with inclusion of the additional strong $\bar{p}p$ interaction.

1 Introduction

Antimatter research is not the newest area of modern physics. Some of the first works in the field ascend to the 1930s, right after Dirac's work and his famous quantum relativistic equation. Dirac's theory also predicted positrons — *i.e.* positive electrons, e$^+$ — and a few years later these particles were experimentally confirmed. It is perhaps from these fundamental works that antimatter physics was born. Then with the discovery of antiprotons ($\bar{p}$'s), antimatter physics obtained all the more developments. An antiproton possesses the same mass, gyromagnetic ratio, and lifetime as p, but has a negative charge. In subsequent years after the $\bar{p}$ particle was discovered, many efforts have been made in order to obtain the lower kinetic energy antiprotons. Therefore, a recent breakthrough in stopping energetic antiprotons ($\bar{p}$'s) down to a few Kelvin temperatures and production of a substantial amount of low temperature antihydrogen atoms ($\bar{H}$'s) made it possible to carry out new precision experiments in order to compare fundamental characteristics of matter and antimatter. For example, it would be interesting to compare the spectrum of H and the energy levels of $\bar{H}$. Such comparison measurements could help, for example, to validate or, perhaps, not validate the fundamental CPT symmetry theorem.

Since $\bar{p}$ is an antibaryon particle with the baryonic number $B = -1$, it would be extremely interesting to discover the strong nuclear interaction between, for example, $\bar{p}$ and a proton, *i.e.* the $\bar{p}p$ interaction in a protonium atom — a bound state of the particles: $Pn = (\bar{p}p)_\alpha$. This two-body system is also called antiprotonic hydrogen. Additionally, it would be very useful and interesting to consider and compare results for the following Coulomb-nuclear atomic systems: $\bar{p}D$ and $\bar{p}T$, where $D \equiv {}^2H^+$ is the deuterium nucleus and $T \equiv {}^3H^+$ is tritium. Perhaps it would be even more useful to obtain the atoms in their ground or close to ground states when the systems become more compact and nuclear interaction becomes even more pronounced inside the atoms.

---

[a] E-mail: rasultanov@stcloudstate.edu; r.sultanov2@yahoo.com

It is possible to prepare such atomic systems with a use of muons, for example, in the three-body reaction:

$$\bar{p} + (p\mu^-) \to (\bar{p}p)_\alpha + \mu^-. \tag{1}$$

where μ^- is a negative muon and α is the final atomic state of Pn. At low energy collisions $\alpha = 2p$, $2s$ or $1s$ state of Pn in the reaction (1). The following two reactions are also of our special interest in the study of the antimatter-matter Coulomb-nuclear interaction: $\bar{p} + (D\mu^-) \to (\bar{p}D)_\beta + \mu^-$ and $\bar{p} + (T\mu^-) \to (\bar{p}T)_\gamma + \mu^-$. In the context of muonic physics, we would like to mention a recent result of the precise experiments on the size of a proton [1] (this is also known as the "proton puzzle" problem). In paper [1], the authors found that the size (form-factor) of p in the muonic hydrogen $H_\mu = (p\mu^-)$ differs within ~4% from the size of p in the normal hydrogen atom H. The main reason of this difference is still not known. In light of this very interesting investigation, we think it would be useful to carry out new corresponding experiments with participation of the muonic antihydrogen $\bar{H}_\mu$ atom, which is a bound state of $\bar{p}$ and a positive muon, i.e. μ^+. A combination of slow antiprotons and muons in future particle physics experiments will, probably, produce even more interesting results. However, as a first step, it is necessary to create $\bar{H}_\mu$. In work [2], the authors obtained the first theoretical results for the $\bar{H}_\mu$ formation cross sections and rates. These data have been obtained in the framework of the following three-body low-energy reaction: $\bar{p} + (\mu^+\mu^-) \to \bar{H}_\mu + \mu^-$. However, there is also another reaction which can be used to create $\bar{H}_\mu$: $\bar{p} + Mu \to \bar{H}_\mu + e^-$, where Mu is a muonium atom (*i.e.* a bound state of a muon μ^+ and an electron e^-) and $\bar{H}_\mu$ is a muonic anti-hydrogen (*i.e.* a bound state of $\bar{p}$ and μ^+). In order to create the low temperature $\bar{H}_\mu$ atoms, this second reaction (with Mu) may be more suitable. In the current work we deal only with the reaction (1). Below we provide a brief description of our method and represent our preliminary results on (1). The effect of the strong $\bar{p}p$ nuclear interaction is taken into account in the framework of a detailed few-body Faddeev-like equation approach.

2 A Few-Body Method and Computational Results

For the case of the Coulomb-nuclear three-body problem in this work, we continue to use our detailed few-body approach, which is based on the reduction of the total three-body wave function Ψ of the ($\bar{p}$, p, μ^-) three-body system onto two Faddeev components, *i.e.* $\Psi = \Psi_1 + \Psi_2$. This equation reflects the fact that at low energy collisions — before the break-up threshold — for the particles $\bar{p}$, p, and μ^-, one has only two asymptotic spatial configurations [2]. For example, $\Psi_1(\vec{\rho}_1, \vec{r}_{23})$ represents the case of the $\bar{p}+(p\mu)$ input channel, and $\Psi_2(\vec{\rho}_2, \vec{r}_{13})$ the case of the output channels, *i.e.* $(\bar{p}p)_\alpha + \mu^-$. The vectors $(\vec{\rho}_k, \vec{r}_{ij})$ are the so called Jacobi coordinates in our three-body system. Now, one can determine the Faddeev components by writing down the following set of two equations [3]:

$$\left(E - \hat{T}_{\rho_1} - \hat{h}_{23}(\vec{r}_{23})\right)\Psi_1(\vec{r}_{23}, \vec{\rho}_1) = \left(V_{23}(\vec{r}_{23}) + V_{12}(\vec{r}_{12})\right)\Psi_2(\vec{r}_{13}, \vec{\rho}_2), \quad (2)$$

$$\left(E - \hat{T}_{\rho_2} - \hat{h}_{13}^{\bar{N}N}(\vec{r}_{13})\right)\Psi_2(\vec{r}_{13}, \vec{\rho}_2) = \left(\tilde{\mathbf{V}}_{13}(\vec{r}_{13}) + V_{12}(\vec{r}_{12})\right)\Psi_1(\vec{r}_{23}, \vec{\rho}_1), \quad (3)$$

where $\tilde{\mathbf{V}}_{13}(\vec{r}_{13}) = V_{13}(\vec{r}_{13}) + v_{13}^{\bar{N}N}(\vec{r}_{13})$, $\hat{h}_{23}(\vec{r}_{23}) = \hat{T}_{\vec{r}_{23}} + V_{23}(\vec{r}_{23})$ and $\hat{h}_{13}^{\bar{N}N}(\vec{r}_{13})$ = $\hat{T}_{\vec{r}_{13}} + V_{13}(\vec{r}_{13}) + v_{13}^{\bar{N}N}(\vec{r}_{13})$ are the two-particle target Hamiltonians, an extra strong p̄p nuclear potential is shown explicitly, and $V_{ij}(\vec{r}_{ij})$ are the Coulomb potentials. To solve Equations (2)–(3), a modified close-coupling approach is used. It leads to an expansion of the system's wave function components Ψ_1 and Ψ_2 into eigenfunctions $\varphi_n^{(1)}(\vec{r}_{23})$ and $\varphi_{n'}^{(2)\bar{N}N}(\vec{r}_{13})$ of the subsystems:

$$\begin{cases} \Psi_1(\vec{r}_{23}, \vec{\rho}_1) \approx \left(\int + \sum\right)_n f_n^{(1)}(\vec{\rho}_1)\varphi_n^{(1)}(\vec{r}_{23}), \\ \Psi_2(\vec{r}_{13}, \vec{\rho}_2) \approx \left(\int + \sum\right)_{n'} f_{n'}^{(2)}(\vec{\rho}_2)\varphi_{n'}^{(2)\bar{N}N}(\vec{r}_{13}). \end{cases} \quad (4)$$

This procedure brings a set of coupled one-dimensional integral-differential equations for the unknown functions $f_n^{(i)}(\vec{\rho}_i)$ after the partial-wave projection [2, 3]. The set of coupled equations can be solved in the framework of different close-coupling approximations, such as 2×1s, 2×(1s-2s), 2×(1s-2s-2p) etc. The sign "2×" indicates that we use two independent sets of expansion functions. The full potential between p̄ and p is more complex, because its second part, $v_{13}^{\bar{N}N}(\vec{r}_{13})$, possesses an asymmetric $\bar{N}N$ nuclear interaction We did not explicitly include the strong interaction in the current calculations. Therefore, in the case of the target Pn eigenfunctions we used the two-body pure Coulomb (atomic) wave functions. Nonetheless, the strong p̄p interaction is approximately taken into account in this work through the eigenstates $\mathcal{E}_{n'}$ which have shifted values from the original Coulomb levels $\varepsilon_{n'}$, that is:

$$\varphi_{n'}^{(2)\bar{N}N}(\vec{r}_{13}) \approx \sum_{l'm'} R_{n'l'}^{(2)\bar{N}N}(r_{13})Y_{l'm'}(\hat{r}_{13}) \approx \sum_{l'm'} R_{n'l'}^{(2)}(r_{13})Y_{l'm'}(\hat{r}_{13}), \quad (5)$$

$$\mathcal{E}_{n'} \approx \varepsilon_{n'} + \Delta E_{n'}^{\bar{N}N} = -\mu_2/2n'^2 + \Delta E_{n'}^{\bar{N}N}. \quad (6)$$

Here: $R_{n'l'}^{(2)}(r_{13})$, $Y_{l'm'}(\hat{r}_{13})$ are the radial, angular p̄p wave functions, μ_2 (n') are the p̄p's reduced mass (quantum number). In this work, as a first step, we apply an approximation with the use of an energy shift in the $\mathcal{E}_{n'}$ - eigenstate of Pn, i.e. in Eq. (6), where $\varepsilon_{n'}$ is the pure Coulomb level and $\Delta E_{n'}^{\bar{N}N}$ is its nuclear shift. In this paper, the Deser formula has been applied [4]:

$$\Delta E_{n'}^{\bar{N}N} = -(4/n')(a_s/B_{Pn}) \cdot \varepsilon_{n'}, \quad (7)$$

Table 1: The total Pn formation cross sections $\sigma_{tr}(\varepsilon_{coll})$ and rates $\Lambda(Pn)$ in the three-body reaction (1), where $\alpha = 1s$ and ε_{coll} is the collision energy (eV). The rate is a product of the formation cross sections and the corresponding center-of-mass velocities $v_{c.m.}$ between $\bar{p}$ and H_μ in the input channel of (1). The results are represented in the framework of the different close-coupling approaches, i.e. $1s$, $1s$-$2s$, and $1s$-$2s$-$2p$. The cross section σ_{tr} is given in cm^2 and $\Lambda(Pn)$ in m.a.u. Results with inclusion of the strong nuclear interaction between $\bar{p}$ and p are presented only in the $1s$-$2s$-$2p$ approximation. For convenience, our rates Λ's have been multiplied by factor of "×5" in this table, as $\Lambda'(Pn)$ in Ref. [5]. The rate with inclusion of the nuclear interaction, i.e. $\Lambda^{\bar{p}p}$, is also multiplied by the same factor. $\Lambda'(Pn) \approx 0.2$ m.a.u [5].

	1s	1s-2s		1s-2s-2p			1s-2s-2p+$\bar{p}$p.
ε_{coll}	σ_{tr}	σ_{tr}	$\Lambda(Pn)$	σ_{tr}	$\Lambda(Pn)$	$\sigma_{tr}^{\bar{p}p}$	$\Lambda^{\bar{p}p}(Pn)$
10^{-4}	1.3E-19	1.9E-19	0.1269	4.95E-19	0.3251	7.55E-19	0.5027
10^{-3}	4.1E-20	6.0E-20	0.1269	1.57E-19	0.3249	2.39E-19	0.5025
0.05	5.8E-21	8.5E-21	0.1269	2.18E-20	0.3193	3.33E-20	0.4950
1.0	1.3E-21	1.9E-21	0.1273	2.89E-21	-	5.22E-21	-
10.0	3.9E-22	6.4E-22	0.1343	-	-	-	-

where a_s is the strong interaction scattering length in the $\bar{p}$+p collision, i.e. without inclusion of the Coulomb interaction between the particles, and B_{Pn} is the Bohr radius of Pn. In the current work the total angular momentum $L = 0$. Further computational and numerical details of this work can be found in the arXiv paper [3]. In the literature one can find other approximate expressions to compute $\Delta E_{n'}^{NN}$, and it would be interesting to apply some of these formulas in conjunction, perhaps, with the relativistic effects in protonium.

In this paragraph, we present our results on numerical computation of the three-body reaction (1). As mentioned earlier, the low energy collisions between the particles may be of particular interest, because at these energies, the antiprotonic hydrogen atom (protonium) in (1) will be formed in its ground state, i.e. $1s$ and in $2s$ and $2p$ states. In this paper, our results are represented only for the transition to the ground state. Table 1 depicts our data for the cross sections $\sigma_{tr}(\varepsilon_{coll})$ and the rates $\Lambda(Pn) = \sigma_{tr}(\varepsilon_{coll}) \cdot v$ of the Pn formation. The table includes the results in the framework of different close coupling approximations: $1s$, $1s$-$2s$, and $1s$-$2s$-$2p$. The six state approximation 2×($1s$-$2s$-$2p$) gave surprisingly good results for the cross sections and rates for the muon transfer three-body reactions [2]. Therefore in the current work, we also picked up this approximation to compare with the results of work [5], in which $\Lambda'(Pn) \approx 0.2$ m.a.u. Our corresponding result is $\Lambda(Pn) \approx 0.32$. The adiabatic approximation is not used in the current work since all particles in the reaction (1) have comparable masses. The inclusion of the strong $\bar{p}p$ nuclear interaction in accordance with Eqs. (5)–(7) increased the rate $\Lambda(Pn)$ by $\sim 50\%$, i.e. $\Lambda^{\bar{p}p}(Pn) \approx 0.5$ m.a.u. It represents a quite significant contribution of the final-state nuclear $\bar{p}p$ interaction to the rate of the reaction (1).

References

[1] R. Pohl et al., Nature **466** (2010) 213.
[2] R. A. Sultanov at el., J. Phys. B: At. Mol. Opt. Phys. **46** (2013) 215204.
[3] R. A. Sultanov at el., arXiv:1606.01325v3 [physics.atom-ph].
[4] S. Deser et al., Phys. Rev. **96** (1954) 774.
[5] A. Igarashi, N. Toshima, Eur. Phys. J. **D46** (2008) 425.

FRACTALITY OF STRANGE PARTICLE PRODUCTION IN PP COLLISIONS AT RHIC

M.V. Tokarev [a]
Joint Institute for Nuclear Research, 141980 Dubna, Russia
I. Zborovský [b]
Nuclear Physics Institute, 25068 Řež, Czech Republic

Abstract. Experimental data on transverse momentum spectra of strange particles produced in pp collisions at $\sqrt{s} = 200$ GeV obtained by the STAR and PHENIX collaborations at RHIC are analyzed in the framework of the z-scaling approach. Properties of data z-presentation are discussed. The dependence of the momentum fractions and recoil mass on the transverse momentum and type of the inclusive particle is studied. The constituent energy loss is estimated. The obtained results can be useful in study of strangeness origin, in searching for new physics with strange probes, and can serve for better understanding of fractality of hadron interactions at small scales.

1 Introduction

One of the fundamental principles governing hadron interactions at high energies is self-similarity. A typical manifestation of this principle is possibility of description of the system in terms of self-similarity variables constructed as suitable combinations of its kinematic and dynamical characteristics. The scaling behavior related to the ideas of self-similarity and fractality of hadron interactions at a constituent level is manifested by the z-scaling [1–3]. Universality of the z-scaling is given by its flavour independence. The strange particles are of special interest for they are traditionally used as important probes when studying the medium created in the collisions of protons or nuclei.

We analyzed [4] recent data [5–7] on inclusive cross sections of strange mesons (K_S^0, K^-, K^{*0} and ϕ) measured by the STAR and PHENIX collaborations using the z-scaling approach. Here we present some conclusions on scaling properties of strangeness production in pp collisions. Results of the analysis are compared with other data on meson and hyperon spectra obtained at RHIC.

2 z-Scaling

In the z-scaling approach [1], the collision of hadrons or nuclei is considered at high energies as an ensemble of the individual interactions of their constituents. Structures of the colliding objects are characterized by parameters δ_1 and δ_2. The interacting constituents carry the fractions x_1 and x_2 of the momenta P_1 and P_2 of the colliding particles. The inclusive particle carries the momentum fraction y_a of the scattered constituent with a fragmentation characterized by

---
[a] e-mail: tokarev@jinr.ru
[b] e-mail: zborovsky@ujf.cas.cz

a parameter ϵ_a. A fragmentation of the recoil constituent is described by ϵ_b and the momentum fraction y_b. Multiple interactions are considered to be similar.

The principle of self-similarity of hadron interactions at a constituent level reflects a property that hadron constituents and their interactions are similar at various levels of resolution and different stages of their evolution. This property is manifested by self-similarity variables. For inclusive reactions, the self-similarity variable z depends on the 4-momenta of the colliding and inclusive particles, fractal dimensions of the interacting objects, and is function of dynamical characteristics of the produced system. It is defined as follows

$$z = z_0 \Omega^{-1}, \quad (1)$$

where $z_0 = \sqrt{s_\perp}/[(dN_{ch}/d\eta|_0)^c m_N]$ and $\Omega = (1-x_1)^{\delta_1}(1-x_2)^{\delta_2}(1-y_a)^{\epsilon_a}(1-y_b)^{\epsilon_b}$. The quantity Ω is a volume in the space of the momentum fractions, $\sqrt{s_\perp}$ is the transverse kinetic energy of the underlying subprocess consumed on production of the inclusive particle and its recoil counterpart, m_N is the nucleon mass, and $dN_{ch}/d\eta|_0$ is multiplicity density of charged particles at pseudorapidity $\eta = 0$. The variable z is expressed via the momentum fractions x_1, x_2, y_a, y_b, fractal dimensions δ and $\epsilon_F \equiv \epsilon_a = \epsilon_b$, multiplicity density, and "heat capacity" c. The momentum fractions are obtained by maximization of Ω under the condition $(x_1 P_1 + x_2 P_2 - p/y_a)^2 = M_X^2$, where $M_X \equiv (x_1 M_1 + x_2 M_2 + m_b/y_b)$ is the recoil mass in the constituent subprocesses.

The scaling function $\psi(z)$ is expressed [1] in terms of the experimentally measured inclusive cross section $E d^3\sigma/dp^3$, the multiplicity density of inclusive particles $dN/d\eta$, and the total inelastic cross section σ_{in} as follows

$$\psi(z) = -\frac{\pi}{(dN/d\eta)\,\sigma_{in}} J^{-1} E \frac{d^3\sigma}{dp^3}. \quad (2)$$

The symbol J stands for Jacobian of the transformation from $\{p_T^2, y\}$ to $\{z, \eta\}$. The scale transformation $z \to \alpha_F z$, $\psi \to \alpha_F^{-1} \psi$ was used for comparison of the shapes of the scaling function for different hadron species.

3 Strange particle production at RHIC

Below we demonstrate (see [4] and references therein) that recent RHIC data [5-7] on inclusive cross sections of strange hadron production obtained by the STAR and PHENIX collaborations confirm properties of the z-scaling [1,2]. In particular, the flavor independence of the z-presentation of inclusive spectra is shown to be valid for various types of strange particles over a wide range of z.

Figure 1(a) shows z-presentation of the transverse momentum spectra of strange mesons [5-8] and baryons [9,10] measured in pp collisions at the energy $\sqrt{s} = 200$ GeV in the central rapidity region at RHIC. The multiplicative factors

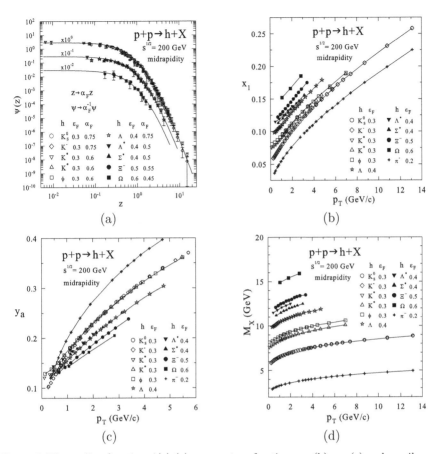

Figure 1: The scaling function $\psi(z)$ (a), momentum fractions x_1 (b), y_a (c) and recoil mass M_X (d) for strange ($K_S^0, K^-, K^{*0}, \phi, \Lambda, \Lambda^*, \Sigma^*, \Xi^-, \Omega$) hadrons produced in pp collisions at $\sqrt{s} = 200$ GeV and $\theta_{cms} = 90^0$ [5-10].

are used to show the z-presentation separately for mesons, single-strange and multi-strange baryons. The $\psi(z)$ for pions ($\alpha_\pi = 1$) is shown by the solid line.

Figure 1(b) demonstrates p_T-dependence of the momentum fraction x_1 for production of different strange hadrons and π^- mesons. For a given strangeness and p_T, the value of x_1 increases with the mass of the strange hadron. Due to the larger energy loss and recoil mass by production of hadrons with strange quark content relative to π^- mesons, the values of x_1 are higher for strange hadrons as compared to those corresponding to pions.

Figure 1(c) illustrates p_T-dependence of the momentum fraction y_a of strange hadrons and π^- mesons produced in pp collisions at $\sqrt{s} = 200$ GeV. The relative energy loss $\Delta E_q/E_q = (1 - y_a)$ depends on the fragmentation dimension ϵ_F. As one can see, the momentum fraction y_a increases with p_T for all particles.

The corresponding energy loss decreases with p_T and depends on the type of the strange hadron. For a given $p_T > 1$ GeV/c, the energy loss is larger for strange baryons than for strange mesons. The growth indicates increasing tendency with larger number of strange valence quarks inside the strange baryon $(\Delta E/E)_\Omega > (\Delta E/E)_{\Xi^-} > (\Delta E/E)_\Lambda \simeq (\Delta E/E)_{\Lambda^*} \simeq (\Delta E/E)_{\Sigma^*}$.

The transverse momentum dependence of the recoil mass M_X for various strange hadrons is illustrated in Fig. 1(d). The values of M_X reveal characteristic growth with p_T for all particles. For a given p_T, the recoil mass M_X increases with the meson masses $M_X(\phi) > M_X(K^{*0}) > M_X(K^0_S) \simeq M_X(K^-) > M_X(\pi^-)$. One can see the similar tendency for Ω, Ξ^- and Λ baryons as well.

The performed analysis shows a hierarchy in the behavior of the parameter ϵ_F for different strange particles (see Fig. 1). The value of ϵ_F is found to be larger for baryons than for mesons. For hyperons, the fractal dimension increases with number of the valence strange quarks, $\epsilon_\Omega > \epsilon_\Xi > \epsilon_\Lambda \simeq \epsilon_{\Lambda^*} \simeq \epsilon_{\Sigma^*}$.

4 Conclusion

The experimental data on p_T-distributions of strange hadrons produced in pp collisions at RHIC have been studied by means of the z-scaling approach. New indication on the self-similarity of strangeness production in pp collisions was obtained. The results demonstrate properties of data z-presentation found in our previous analyses of inclusive spectra of different hadrons measured at the accelerators ISR, SPS, Tevatron and RHIC. The results indicate on flavor independence of the scaling function and support microscopic scenario of hadron production as formation of fractal-like objects in proton-proton collisions.

Acknowledgments

The study has been supported by the RVO61389005 institutional support and by the grant LG 15052 of the Ministry of Education of the Czech Republic.

References

[1] I. Zborovský, M.V. Tokarev, Phys. Rev. **D 75** (2007) 094008.
[2] I. Zborovský, M.V. Tokarev, Int. J. Mod. Phys. **A 24** (2009) 1417.
[3] M.V. Tokarev, I. Zborovský, Phys. Part. Nucl. Lett. **7** (2010) 160.
[4] M.V. Tokarev, I. Zborovský, Int. J. Mod. Phys. **A 32** (2017) 1750029.
[5] A. Adare et al. (PHENIX Collab.), Phys. Rev. **D 83** (2011) 052004.
[6] G. Agakishiev et al. (STAR Collab.), Phys. Rev. Lett. **108** (2012) 072302.
[7] A. Adare et al. (PHENIX Collab.), Phys. Rev. **C 90** (2014) 054905.
[8] J. Adams et al. (STAR Collab.), Phys. Rev. **C 71** (2005) 064902.
[9] B.I. Abelev et al. (STAR Collab.) Phys. Rev. Lett. **97** (2006) 132301.
[10] B.I. Abelev et al. (STAR Collab.), Phys. Rev. **C 75** (2007) 064901.

ANALYSIS OF ANISOTROPIC TRANSVERSE FLOW IN Pb-Pb COLLISIONS AT 40A GEV IN THE NA49 EXPERIMENT

Oleg Golosov[1,a], Ilya Selyuzhenkov[1,2], Evgeny Kashirin[1]

[1] *National Research Nuclear University MEPhI (Moscow Engineering Physics Institute), Kashirskoe highway 31, Moscow, 115409, Russia*
[2] *GSI Helmholtzzentrum fur Schwerionenforschung, Darmstadt, Germany*

Abstract. The results of flow analysis in Pb-Pb collisions at the beam energy of $40A$ GeV recorded with the fixed target experiment NA49 at the CERN SPS are presented. The three-subevent technique is used for the differential measurements of the directed and elliptic flow. Corrections for the detector acceptance anisotropy in the transverse plane are applied using the QnCorrections Framework developed originally for the ALICE experiment at the LHC. The results are compared with those published by the STAR at RHIC and the NA49 at CERN SPS collaborations. In the future, the developed procedure will be used for the analysis of the new Pb-Pb data collected by the NA61/SHINE experiment at the CERN SPS.

Anisotropies in momentum distributions of particles produced in heavy ion collisions are highly sensitive to the properties of the system very early in its evolution. The origin of this phenomenon lies in the initial asymmetries in the geometry of the system. Anisotropic transverse flow is quantified by the Fourier coefficients v_n of a decomposition of the distribution of particle azimuthal angle ϕ relative to that of the reaction plane ψ_{RP}, which is defined by the beam direction and the impact parameter of the colliding nuclei. The aim of this work is to develop a flow reconstruction procedure reproducing previously published results and to use it for the analysis of new data from the NA61/SHINE experiment.

Minimum bias data from Pb+Pb collisions collected by the NA49 experiment at $40A$ GeV (periods 01D and 02C) were used. Event selection yielded 335K events. Event classification was based on the multiplicity of produced particles. Mean energy loss in the tracking detectors was used for particle identification.

Flow analysis was performed using information from the tracking detectors VTPC1, VTPC2, MTPC-L and MTPC-R. Flow coefficients were calculated with the scalar product three subevent technique [2]. Corrections for detector acceptance non-uniformity were introduced using a three-step procedure described in Ref. [1] and implemented in QnCorrections Framework originally developed for the ALICE experiment [3]. The values of the applied corrections varied depending on event class. Errors were calculated using the bootstrap procedure with 100 samples.

Results for $v_1(y)$ and $v_2(p_T)$ for π^- in midcentral collisions (centrality 10-30%) are shown in the Fig. 1. Open points for $v_1(y)$ were reflected antisymmetrically with respect to zero rapidity. In first approximation v_1 appears

---

[a]E-mail: oleg.golosov@gmail.com

to be an odd function of rapidity, though it does not cross the zero exactly. This might represent additional correlations introduced by global momentum conservation. Additional corrections should be made to reduce this effect [4]. The results are consistent with those published by the NA49 [5] (Pb+Pb at $40A$ GeV, π^+ and π^-, random subevent, event plane method) and STAR [6] (Au+Au at $\sqrt{s_{NN}} = 7.7$ GeV, random subevent, event plane method) collaborations.

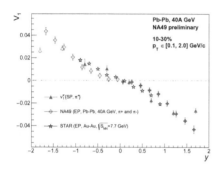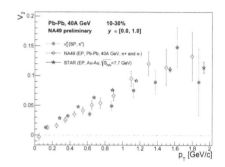

Figure 1: $v_1(y)$ (left) and $v_2(p_T)$ (right) for π^- (centrality 10-30%). Open points for $v_1(y)$ were reflected at midrapidity. Results are compared with those published by the NA49 (Pb+Pb at $40A$ GeV, π^+ and π^-, random subevent, event plane method) and STAR (Au-Au at $\sqrt{s_{NN}} = 7.7$ GeV, random subevent, event plane method) collaborations.

Directed v_1 and elliptic v_2 flow of π^- was measured as a function of transverse momentum and rapidity. The results are qualitatively consistent with those published by the NA49 [5] and STAR [6] collaborations. Further improvement of the method, especially mitigation of effects introduced by global momentum conservation, is under investigation.

Acknowledgments

This work was partially supported by the Ministry of Science and Education of the Russian Federation, grant N 3.3380.2017/4.6, and by the National Research Nuclear University MEPhI in the framework of the Russian Academic Excellence Project (contract No. 02.a03.21.0005, 27.08.2013).

References

[1] I. Selyuzhenkov and S. Voloshin, Phys. Rev. C 77 (2008) 034904.
[2] A. Poskanzer, S. Voloshin, Phys.Rev. C58 (1998) 1671-1678.

[3] V. Gonzalez, J. Onderwaater and I. Selyuzhenkov, GSI Scientific Report 2015 (p. 81); https://github.com/FlowCorrections/FlowVectorCorrections.
[4] N. Borghini *et al*, Phys.Rev. C66 (2002) 014901.
[5] NA49 Collaboration, Phys.Rev. C68 (2003) 034903.
[6] STAR Collaboration, Phys. Rev. Lett. 112 (2014) 162301.

ELLIPTIC FLOW OF π^- MEASURED WITH THE EVENT PLANE METHOD IN Pb-Pb COLLISIONS AT $40A$ GeV

J. Gornaya[a], I. Selyuzhenkov, A. Taranenko

National Research Nuclear University MEPhI (Moscow Engineering Physics Institute), 115522, Moscow, Russia
GSI Helmholtzzentrum für Schwerionenforschung, Germany

Abstract. The measurement of elliptic flow for negatively charged pions in Pb-Pb collisions at beam energy of $40A$ GeV performed with the NA49 experiment at the CERN SPS is presented. Results are compared with measurements at similar energies by the STAR experiment at RHIC.

Hot and dense matter is produced in heavy-ion collisions at relativistic energies. Anisotropic expansion of this matter results in azimuthal asymmetry of particle production relative to the reaction plane, the so-called azimuthal anisotropic flow.

NA49 was one of the fixed target experiments at the CERN SPS accelerator. Four large volume time projection chambers (TPC) were used for measurement and identification of charged particle tracks [1]. The ring ralorimeter (RCAL) was used for transverse energy and elliptic flow measurements.

The analysis used Pb+Pb collision data recorded by NA49 at $40A$ GeV energy from which 120 K events passed the interaction vertex selection cuts. Event classes are based on the TPC multiplicity distribution parameterized with the Modified Wounded Nucleon model also known as MC-Glauber.

Evaluation of the symmetry plane angle is based on flow vectors [2]:

$$Q_2 \cos 2\Phi_2 = Q_{2x} = \sum_{i=1} \omega_i \cos(2\varphi_i), \quad Q_2 \sin 2\Phi_2 = Q_{2y} = \sum_{i=1} \omega_i \sin(2\varphi_i), \quad (1)$$

where the weights ω_i are the transverse momentum of tracks from the TPCs or the energy deposited in the modules of the RCAL and φ_i is the azimuthal angle of positively charged particles or of the calorimeter cells. The Q-vector recentering and flattening procedures [2] were applied for each 5% centrality window.

Two and three subevent methods were used to determine the event plane resolution correction R_2 for elliptic flow measurement from the TPCs and RCAL using the following formulae [2]:

$$M_2\{A, B\} = \langle \cos(2(\Phi_{2,A} - \Phi_{2,B})) \rangle, \quad (2)$$

$$R_2\{A, B\} = \sqrt{2 M_2\{A, B\}}, \quad R_2^C\{A, B\} = \sqrt{\frac{M_2\{B, C\} M_2\{A, C\}}{M_2\{A, B\}}}, \quad (3)$$

---
[a]E-mail: yulyagornaya@mail.ru

where subevents A and B are random subevents from TPC and C is a subevent from RCAL. Elliptic flow coefficients of negatively charged pions measured with the event plane method [2] are shown in Fig. 1 for event classes 10-40% and different detectors. The values of v_2 were calculated as:

$$v_2 = \frac{\langle cos(2(\varphi_\pi - \Phi_2^{corr}))\rangle}{R_2}, \qquad (4)$$

where φ_π is the azimuthal angle of negatively charged pions, Φ_2^{corr} is the event plane angle with applied recentering and flattening corrections and R_2 is the event plane resolution correction factor calculated from Eqs. 2, 3. Consistent

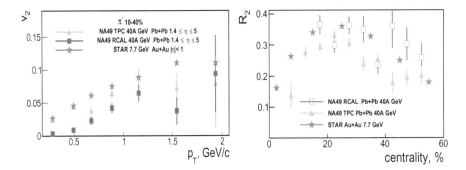

Figure 1: Elliptic flow coefficient v_2 of negatively charged pions (left) and corresponding event plane resolution correction factor R_2 (right).

results were found for v_2 measured by STAR/RHIC and the reanalyzed NA49 data using the TPC event plane. The values of $v_2\{EP\ RCAL\}$ are systematically lower than those of $v_2\{EP\ TPC\}$ and need further investigation.

The presented results will be used as reference for flow measurements from the lead ion energy scan program of the NA61/SHINE experiment at the CERN SPS [4].

Acknowledgments

We thank P. Seyboth for providing information about the RCAL. This work was partially supported by the Ministry of Science and Education of Russia, grant N 3.3380.2017/4.6, and by the NRNU MEPhI in the framework of the Russian Academic Excellence Project (contract No. 02.a03.21.0005, 27.08.2013).

References

[1] S. Afanasev et al. (NA49 Collaboration), NIMA430, 210-244 (1999).
[2] A.M. Poskanzer, S.A. Voloshin, PRC58, 1671-1678 (1998).
[3] L. Adamczyk et al. (STAR Collaboration), PRC93, 014907 (2016).
[4] M.Gazdzicki (NA49 & NA61/SHINE Collaborations), JPG38 124024 (2011).

PRODUCTION OF NUCLEAR FRAGMENTS IN THE ^{12}C + ^{7}Be COLLISIONS AT INTERMEDIATE ENERGIES

B.M. Abramov, P.N. Alekseev, Yu.A. Borodin, S.A. Bulychjov, I.A. Dukhovskoy, A.P. Krutenkova[a], V.V. Kulikov, M.A. Martemianov, M.A. Matsyuk and E.N. Turdakina

NRC "Kurchatov institute - ITEP", Moscow, 117218, Russia

Abstract. Nuclear fragments emitted at 3.5° in ^{12}C fragmentation at 0.3 - 0.95 Gev/nucleon on a Be target have been measured. The spectra obtained are used for testing the predictions of four ion-ion interaction models: INCL++, BC, LAQGSM03.03 and QMD.

The study of fragment emission is important to understand the nature of ion-ion interactions. Different reaction mechanisms contribute to this rather complicated process which can hardly be described in analytical way. For this reason we tested a few Monte-Carlo transport codes against the data of the FRAGM experiment at ITEP TWAC facility [1]–[7].

In the FRAGM experiment, we have measured the fragment yields from the reaction C + ^{9}Be → f + X with a beamline spectrometer set at 3.5° to carbon beam; f stands for all fragments from protons up to isotopes of projectile nucleus. The projectile kinetic energies were T_0= 0.2–3.2 GeV/nucleon. Here we present the data at T_0= 0.95 GeV/nucleon for all possible fragments, see Fig. 1. The analogous data at 0.3 and 0.6 GeV were shown in [7] and in [4], respectively. The fragments were measured at a wide momentum region which include the midrapidity, the fragmentation peak and the cumulative regions. In the last one the fragment momenta per nucleon are much higher than momentum per nucleon of the projectile. This gives a good testing ground for a comparison with the predictions of different ion-ion interaction models. The fragment yields were measured by scanning the beamline spectrometer momentum with a step of 50–100 MeV/c and counting the number of events corresponding to different fragments and normalizing to the monitor. The fragments were well separated on time-of-flight vs dE/dx plots. The cross sections $d^2\sigma/(d\Omega dp)$, where p is the fragment momentum in a laboratory frame, were calculated. They are shown for several fragments in Fig. 2 in comparison with the calculations by four models: INCL++ [8], BC [9], LAQGSM [10] and QMD [11]. Our measurements cover two-to-five orders in the cross section magnitude, depending on the fragment. Model predictions for the cross section at fragmentation peak maxima differ by no more than 2–3 times for light fragments up to ^{4}He being larger for heavier fragments. The largest differences are observed for the QMD model, which predicts smaller width for the fragmentation peaks. The LAQGSM model reproduces the energy dependence of the cross sections at high energy part of the fragmentation peaks, but underestimates the cross

---
[a]E-mail: krutenk@itep.ru

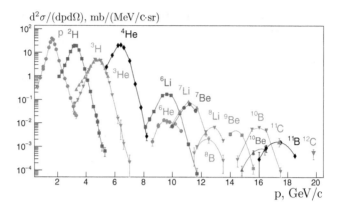

Figure 1: Measured yields of different isotopes in ^{12}C + ^{9}Be interaction at 0.95 GeV/nucleon as a function of fragment laboratory momentum.

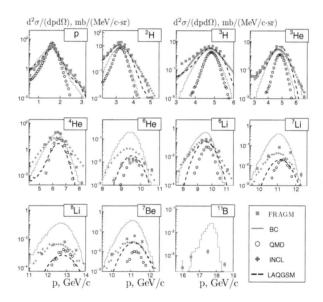

Figure 2: Data/MC comparison for yields isotopes in ^{12}C + ^{9}Be interaction at 0.95 GeV/nucleon as a function of fragment laboratory momentum.

section in the low-energy part. No one of the model describes all the spectra well; the INCL++ looks better than other models. In the framework of thermodynamical picture of nuclear fragmentation, the kinetic energy spectra (T) were parameterized by a sum of two Maxwell-Boltzmann distributions

$Ed^3\sigma/d^3p = E(A_S \exp(-T/T_S) + A_C \exp(-T/T_C))$, where A_S and A_C are normalization factors for low and high energy regions, and the slope parameters T_S and T_C are "temperatures" defined in these regions. The mentioned function provides qualitative description of both the data and the calculations for all models and all fragments. The INCL++ model describes the data for p and d well, but fails to do it for ^{4}He, see Table 1. The experimental results for T_C from [12] obtained at GSI at 1 GeV/nucleon for ^{197}Au + ^{197}Au collisions are also presented.

Table 1: Slopes of kinetic energy spectra T_C in the two exponent approach (see formula in the text) for hydrogen, deuterium and helium isotopes in the rest frame of incoming nucleus.

T_0	950 MeV	950 MeV	1000 MeV
f	data	INCL	Odeh
p	34.2±1.2	32.2±0.3	25.5±1.0
d	21.9±0.2	23.6±0.3	16.0±1.0
^{4}He	15.0±2.0	28.2±0.5	14.0±1.0

In conclusion, fragment yields from the reaction ^{9}Be(^{12}C, f)X (f stands for fragments from p to ^{7}Li) at 0.95 GeV/nucleon were measured and compared to prediction of four models of ion-ion interactions. The INCL++ describes all momentum spectra rather well, both in the region of fragmentation peak and in the cumulative region while all other models underestimate the experimental results in the cumulative (high momentum) region. Kinetic energy spectra in the projectile rest frame can be parameterized as $E(A_S \exp(-T/T_S) + A_C \exp(-T/T_C))$.

Acknowledgments

We thank S.G. Mashnik and K.K. Gudima for the calculations by the LAQGSM model. We are indebted to the personnel of TWAC-ITEP and technical staff of the FRAGM experiment. The work has been supported in part by the RFBR (grant No. 15-02-06308).

References

[1] Abramov B.M. et al., JETP Letters, **97** (2013) 439.
[2] Abramov B.M. et al., Physics of Atomic Nuclei, **78** (2015) 373.
[3] Abramov B.M. et al., Physics of Atomic Nuclei, **79** (2016) 700.
[4] Abramov B.M. et al., Physics of Atomic Nuclei, **79** (2016) 1419.
[5] Kulikov V.V. et al., PoS BaldinISHEPPXXII (2015) 079, arXiv:1502.01662.

[6] B.M.Abramov et al., J. Phys.: Conf. Ser., **381**, 012037 (2012).
[7] B.M.Abramov et al., J. Phys.: Conf. Ser., **798**, 012077 (2017).
[8] Dudouet J. et al., Phys. Rev., **C89** (2014) 054616;
[9] Folger G. et al, Europ. Phys. J., **A21** (2004) 407.
[10] Mashnik S.G. et al., arXiv:0805.0751.
[11] Koi T. et al, AIP Conf. Proc., **896** (2007) 21.
[12] Odeh T. et al., Phys. Rev. Lett., **84** (2000) 4557.

MUON PARTICLE PHYSICS PROGRAM AT J-PARC

Satoshi Mihara [a]
*High Energy Accelerator Research Organization
Institute of Particle and Nuclear Studies, and
SOKENDAI (The Graduate University for Advanced Studies),
305-0801 Tsukuba, Japan*

Abstract. J-PARC provides high-intense proton beam with its unique accelerators Variety of research programs are in progress and in preparation using the proton beam provided here. In this presentation particle physics programs at J-PARC are introduced with a focus on future experiments using muons.

1 Introduction

J-PARC (Japan Proton Accelerator Research Complex) consists of three accelerators; Linac accelerating protons (H^-) up to 400 MeV, RCS (Rapid Cycle Synchrotron) to 3 GeV (electrons are split off when they are injected), and MR (Main Ring) to 30 GeV. The RCS provides pulsed proton beam with 25 Hz acceleration cycle to the Material and Life Science Facility (MLF) and MR with 2.48 – 6.0 seconds acceleration cycle to either of the neutrino facility or nuclear and particle physics experiment facility. There are two types of extraction mode in MR; one is Fast Extraction (FX) for the neutrino program (the T2K experiment) where protons accelerated in 8 acceleration bunches in MR are extracted in one single turn using fast extraction kickers. The other is Slow Extraction (SX) for the other particle and nuclear physics experiments where protons are extracted in a continuous mode and are injected to a primary target to produce secondary particles. Time structure of the proton beam in SX mode, which determines the time structure of the secondary beam in the end, is selected depending on the purpose of the experiment. In case continuous flat time structure is required, which is the case almost all experiments requires, the bunched proton beam structure is flattened after acceleration in MR completes by switching off the acceleration RF voltage. In case pulsed but slowly extracted beam is required, which is the case of the muon experiment as described later, the RF voltage is applied during the extraction period to keep the time structure of the proton beam.

History of beam power upgrade of J-PARC accelerators are summarized in Fig. 1. The RCS beam power reached 500kW in 2015 but has been reduced due to a trouble of the neutron target in which most of the beam is to be stopped. The target recovery work is in progress in parallel to the beam operation as of 2017. The MR beam power increases both in FX and SX modes as detailed tuning of the accelerator proceeds; 470 kW beam power is already achieved in FX mode and 44 kW in SX mode. Continuous efforts both in accelerator and

---

[a] E-mail: satoshi.mihara@kek.jp

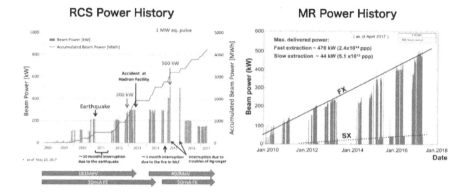

Figure 1: Accelerator power history of RCS providing proton beam to MLF (left) and MR providing proton beam to the neutrino facility in FX mode and to the particle and nuclear experiment facility in SX mode (right).

facilities are being made to reach 750 kW in FX and more than 100 kW in SX modes. Upgrade of MR magnet power supplies are being conducted as of 2017 to this end.

2 Muon Particle Physics Program at J-PARC

J-PARC accelerator is multi-purpose, providing the high-intensity proton beam to various applications. Different kinds of characteristic secondary particles are produced using this proton beam; muons and neutrons are produced in MLF in pulsed mode for material and life science applications and meson beam such as kaons, both of neutral and charged, and pions are produced for applications to particle and nuclear physics experiments. In addition to these existing research programs at J-PARC, we intend to start new projects using the characteristic muon beam in order to investigate fundamental of elementary particles in extreme precision. There are two major activities in preparation at J-PARC; one is the precise measurement of the muon magnetic and electric dipole moments (muon g-2/EDM) and the other is the search for muon-to-electron conversion with a world best sensitivity. The muon g-2/EDM experiment will be carried out using the muon beam provided at MLF and the muon-to-electron (μ-e) conversion search experiment (COMET experiment) will be launched in the particle and nuclear physics experiment facility.

2.1 muon g-2/EDM measurement

It is a well-known experimental fact that the measured value of the muon magnetic dipole moment is significantly deviated from the prediction by the

Standard Model by more than 3σ [1,2]. The theoretical prediction is as precise as the experimental measurement, indicating that there can be effects caused by physics beyond the Standard Model. There are two attempts being made in the world; one is the renewal of the previous experiment using the same experimental set-up at FNAL [4] and the other is a completely new approach to measure the muon magnetic dipole moment at J-PARC.

The J-PARC experiment aims at improving the measurement precision down to the level of 0.1 ppm by using an innovative ultra-cold muon beam being developed at J-PARC. This ultra-cold muon beam can be stored in an extremely uniform magnetic field without using an electric focusing. This enables us also to carry out a measurement of the muon electric dipole moment with the same setup. The ultra-cold muon beam, which is the most important ingredient of the experiment, is generated by accelerating muons at almost rest with a series of linear accelerators. Muons at rest are produced by ionizing muoniums with a high-power laser. The muoniums are generated by stopping the surface muons produced at J-PARC MLF in a specially prepared muon stopping target for this experiment. More details of the experiment is presented in another presentation in this conference [3].

2.2 μ-e conversion search

Lepton flavor violation in the charged lepton sector (cLFV) in the framework of the Standard Model including finite neutrino mass is extremely small to be measured in any experiment. Thus observation of cLFV process indicates the stunning evidence of new physics beyond the Standard Model and the probability of the process happening (branching ratio of the decay or reaction) provides us precious information regarding the energy scale (and the strength of the coupling) of the new physics. There are three muon cLFV processes predicted to exist near the current upper limit by many models of new physics [5] ; $\mu \to e\gamma$, μ-e conversion in muonic atom, and $\mu \to eee$. Searches for $\mu \to e\gamma$ and $\mu \to eee$ are efficiently carried out using DC muon beam as they have multi-particles in the final state. Accidental overlaps of more than one muon decays could be misidentified as the signal event in these searches. On the other hand the μ-e conversion process has only one particle (electron) with a characteristic energy in the final state. Thus the principal source of the experimental search is associated to the proton beam producing muons. The beam has to be pulsed with an extreme precision as explained later.

Among these three muon processes, only the $\mu \to e\gamma$ process has a gamma in the final state; the process is possible only through a loop diagram as shown in Fig. 2 (a). The other two process are possible through both loop (b) and tree (c) diagrams as depicted in Fig. 2 although the branching ratio for the case of (b) would be about 100 times smaller compared to the $\mu \to e\gamma$ process as there is an additional coupling of electron or quark necessary to realize the signature.

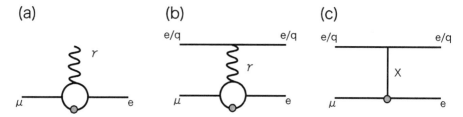

Figure 2: Possible new physics diagrams of muon cLFV processes.

The COMET experiment at J-PARC searches for the μ-e conversion process with a target single-event sensitivity better than 10^{-16} using the high-intensity pulsed proton beam provided at J-PARC [6]. The experiment uses muons produced as decay products of pions that are generated in proton-nucleus interactions at the proton target. The pions are collected efficiently by a solenoid magnet with gradient magnetic field surrounding the proton target and are transported to the experiment area through a chain of solenoid magnets. Most of pions decay to electrons (and neutrinos) while being transported although some of pions reach the experiment setup. The experiment uses aluminum as a muon stopping material to form muonic atoms. If the μ-e conversion occurs in a muonic aluminum, the signal electron should carry a characteristic energy of 108 MeV, which is equivalent to the muon mass minus the binding energy of muons and recoil energy of the residual atom. The lifetime of the muon in the muonic aluminum is known to be 890 nsec. Fully utilizing this lifetime of the muon, we remove the background associated to the primary proton interaction at the target; pions produced at the prompt timing would reach the muon stopping target, providing high-energy (about 100 MeV) photons through π^0 production via the charge exchange process. The photon could be a source of background electron with about 100 MeV energy through a pair creation of electron and position. We simply remove this type of background by masking the data in a few hundred nanoseconds after the prompt pions hit the muon stopping target.

This method to remove the background requires an unique time structure of the proton beam as the muon beam time structure is determined by the proton beam time structure; the muon beam should have a structure of the pulse-to-pulse time width as wide as the muon lifetime in the muonic atom. The background pions arriving at the muon stopping target at the prompt timing are to be removed by masking the data. This means that tiny leakage of protons from a beam pulse would cause pion contamination in between muon beam pulses and potentially mimic the μ-e conversion signal. The ratio of the number of proton leakage to that in a pulse is called a beam extinction factor and should be well smaller than $10^{-9 \sim -10}$ to suppress the pion induced background below the target sensitivity of the experiment. The J-PARC MR is suitable for this purpose. This is because the accelerator is designed with an imaginary transition energy not to induce beam divergence when the beam energy crosses the transition energy during acceleration [7]. This certainly helps to achieve precise beam pulsing for the COMET experiment.

The setup of the experiment is shown in Fig. 3. The production target is located at the center of a solenoidal magnet (Production Solenoid, PS) with a gradient magnetic field. The pions produced backward at the target are captured and transported to a beam-transferring curved solenoid (Transport Solenoid, TS) with a large momentum acceptance. Pions decay to muons (and neutrinos) while being transported and muons are stopped on the muon stopping target located at the end of the transport solenoid. The muon stopping target is composed of a stack of 17 aluminum thin disks so that the signal electron emerging from any one of the disks should not suffer from multiple scattering in the disk.

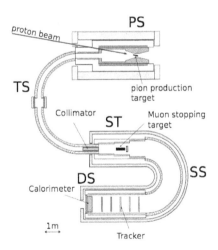

Figure 3: Set up of the COMET experiment.

The COMET experiment takes a staged approach to realize its target sensitivity. In Phase I the setup is constructed until the end of the first 90 degree bend of TS and a spectrometer with a muon stopping target at its center is connected to the end of TS. The spectrometer is composed of a cylindrical drift chamber (CDC) and a trigger hodoscopes located at both ends of the spectrometer magnet. The setup is illustrated in Fig. 4. The single-event sensitivity of COMET Phase I is designed to be better than 10^{-14}, which is two orders magnitude better than the current limit [8]. The muon stopping rate of Phase I is designed to be larger than 10^9 Hz with a proton beam power of 3.2 kW injected to a 16 cm long graphite target. The Phase II will take place after Phase I completes with a proton beam power of 56 kW injecting to a target made of heavy metal such as tungsten to improve the muon beam production rate. The beam background rejection will be improved thanks to the extension of TS (as shown in Fig. 3) and the counting rate of the detector located at the end of another 180 degree bend of a curved solenoid (Spectrometer Solenoid, SS) without losing the signal electron acceptance. The target solenoid (ST) also helps to increase the signal acceptance with its magnetic lens effect realized by the gradient magnetic field. The spectrometer of the Phase II is composed of a series of ultra-low mass straw-tube trackers [9] and LYSO calorimeter [10] to carry out momentum and energy measurements of the electron.

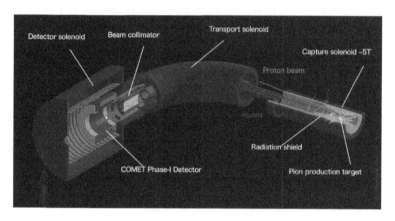

Figure 4: Setup of the COMET experiment Phase I.

The COMET experiment intend to start the engineering run of Phase I in 2019–2020 followed by physics data acquisition for about 150 days. Preparation of the Phase II setup, especially extension of the solenoid magnet system, will be made in parallel to Phase I.

3 Summary

J-PARC is providing high-power proton beam for particle/nuclear physics experiments and material/life science programs. There are many ongoing unique physics programs and many others in preparation. Muon particle physics programs are among them and two major experiments are in preparation; the g-2/EDM measurement using ultra cold muon beam and μ-e conversion search

(COMET experiment) using highly intense pulsed muon beam. The COMET experiment is in preparation to start physics data acquisition with a sensitivity of better than 10^{-14} in Phase I in a few years and better than 10^{-16} in Phase II. It should also be noted that many more results come out from neutrino, kaon and nuclear physics programs in near future as the J-PARC machine power upgrade is realized.

Acknowledgments

The author would like to thank the COMET collaboration.

References

[1] G.W. Bennett et al. (Muon g-2 Collaboration), Phys. Rev. D**73** 072003.
[2] T. Teubner, K. Hagiwara, R. Liao, A.D. Martin, D. Nomura, Nuclear Physics **B225** (2012) 282.
[3] T. Mibe, presentation in this conference.
[4] R. M. Carey et al. (Muon g-2 Collaboration), INSPIRE-1108252.
[5] S. Mihara, J.P. Miller, P. Paradisi, G. Piredda, Annual Review of Nuclear and Particle Science, **63** (2013) 531.
[6] R. Akhmetshin et al. (COMET Collaboration), KEK/J-PARC-PAC 2012-10, (2012).
[7] T. Koseki et al., PTEP, (2012) 02B004.
[8] W. Bertl et al. (SINDRUM-II Collaboration), Eur. Phys. J. C **47** (2006) 337.
[9] H. Nishiguchi et al. Nucl. Instrum and Methods A **845** (2017) 269.
[10] K. Oishi, *"Proceedings of the 2nd International Symposium on Science at J-PARC – Unlocking the Mysteries of Life, Matter and the Universe –* (JPS, Japan, 2015) 025014.

THE MUON g-2 EXPERIMENT AT FERMILAB

Wesley Gohn [a] for the Muon g-2 Collaboration
University of Kentucky, Lexington, KY, USA

Abstract. A new measurement of the anomalous magnetic moment of the muon, $a_\mu \equiv (g-2)/2$, will be performed at the Fermi National Accelerator Laboratory with data taking beginning in 2017. The most recent measurement, performed at Brookhaven National Laboratory (BNL) and completed in 2001, shows a 3.5 standard deviation discrepancy with the standard model value of a_μ. The new measurement will accumulate 21 times the BNL statistics using upgraded magnet, detector, and storage ring systems, enabling a measurement of a_μ to 140 ppb, a factor of 4 improvement in the uncertainty the previous measurement. This improvement in precision, combined with recent improvements in our understanding of the QCD contributions to the muon g-2, could provide a discrepancy from the standard model greater than 7σ if the central value is the same as that measured by the BNL experiment, which would be a clear indication of new physics.

1 Introduction

The anomalous magnetic moment of the muon was last measured by the Brookhaven experiment E821 in 1999–2001, resulting in a 3.5σ discrepancy with the Standard Model of particle physics [1]. The Muon g-2 experiment E989 at Fermilab will improve the precision on a_μ by a factor of four by performing the measurement using 21 times the statistics of E821 [2]. The new experiment will attempt to answer the question, is this discrepancy an indication of new physics beyond the standard model?

The muon has a magnetic dipole moment of $\vec{\mu} = g\frac{q}{2m}\vec{s}$, with $g = 2$ for a point-like particle [3]. The Standard Model predicts effects from QED, electroweak theory, and QCD, such that $g_{SM} = 2_{Dirac} + \mathcal{O}(10^{-3})_{QED} + \mathcal{O}(10^{-9})_{EW} + \mathcal{O}(10^{-7})_{QCD}$. If a discrepancy with the standard model value is found, beyond standard model contributions to g-2 could come from SUSY, dark photons, extra dimensions, or other new physics. Significant theoretical effort has gone into understanding these contributions including [4–8]. The error on the theoretical value is expected to be reduced on the timescale of the new experiment, due to efforts to reduce the hadronic contributions to a_μ from lattice calculations [9,10] and improved measurements of lepton scattering cross-sections. A measurement to this precision at the BNL central value could provide a better than 7σ discrepancy from the standard model, which would provide clear evidence of new physics.

---

[a] E-mail: gohn@pa.uky.edu

2 The Experiment

To perform the experiment, we inject polarized muons into a magnetic storage ring. Muons will precess in the magnetic field, and we measure the frequency ω_a via the timing of muon decays to positrons using 24 electromagnetic calorimeters [11,12]. Measurements of the precession frequency and magnetic field lead to a_μ. The anomalous precession frequency $\omega_a = \omega_s - \omega_c$, where the Cyclotron frequency $\omega_c = \frac{e}{m\gamma}B$ and the spin precession frequency $\omega_s = \frac{e}{m\gamma}B(1+\gamma a_\mu)$.

To reduce the effect of electric fields, the muons are injected at a magic momentum with $\gamma = 29.3$, which cancels the second term in Eq. 1.

$$\vec{\omega}_a = -\frac{Qe}{m}[a_\mu \vec{B} - (a_\mu - (\frac{mc}{p})^2)\frac{\vec{\beta} \times \vec{E}}{c}], \qquad (1)$$

which leaves

$$\vec{\omega}_a = -\frac{Qe}{m}a_\mu \vec{B}. \qquad (2)$$

The proton precession frequency ω_p is measured as a proxy for $\vec{B}$ leading to an expression for the anomalous moment as

$$a_\mu = \frac{\omega_a/\omega_p}{\mu_\mu/\mu_p - \omega_a/\omega_p}. \qquad (3)$$

The Measurement of ω_p is performed using NMR. Fixed NMR probes measure time variations of the field during data taking. A trolley with mounted NMR probes periodically circumnavigates the interior of the ring to perform precision measurements of the field in the muon storage region, performing 6000 magnetic field measurements per trolley run. Probes are calibrated to provide measurement to 35 ppb.

We plan to collect 21 times the BNL statistics, which will reduce our statistical uncertainty by a factor of four. To reduce systematic uncertainty, accelerator facilities will have p_π closer to magic momentum, utilize a longer decay channel, and increase injection efficiency. Systematics on ω_a will be decreased from 180 ppb in E821 to 70 ppb by using an improved laser calibration, a segmented calorimeter, better collimator in the ring, and improved tracker. Systematics on ω_p will be decreased from 170 ppb in E821 to 70 ppb by improving the uniformity and monitoring of the magnetic field, increasing accuracy of position determination of trolly, better temperature stability of the magnet, and providing active feedback to external fields.

3 Current status

Muon g-2 performed a 5 week engineering run during June-July 2017. The first beam was injected into the ring on May 31, 2017. The accelerator performed

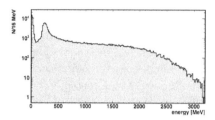

Figure 1: Energy distribution from June 2017 data recorded in the calorimeter. The low energy peaks are from protons and lost muons.

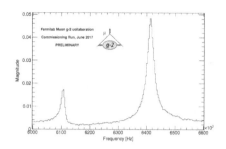

Figure 2: Fourier Transform of data from fiber harps shows the proton cyclotron frequency and the betatron frequency of stored protons.

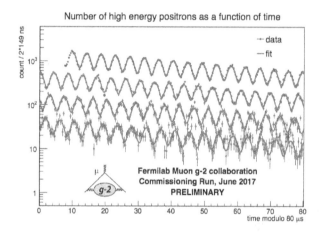

Figure 3: Time distribution of muon decays from June 2017 engineering run.

well during the run [13], though the beam was mostly protons with $\mathcal{O}(1\%)$ muons and had a fill rate of 0.1 Hz (compared nominal rate of 12 Hz expected during production operations). The effort of the engineering run was largely dedicated to optimizing muon injection into the ring by tuning the inflector magnet, electrostatic quadrupoles, and kickers to store the beam. Stored protons, muons, and positrons were detected with all 24 operational calorimeters, one of the three planned tracker stations, and the retractable fiber harp detectors, as shown in Fig. 2. A sufficient number of muon decays were observed during the run to see the muon precession signal, as shown in Fig. 3, though the best precision from this data is expected to yield a result of no more than 50 ppm, which is two orders of magnitude below the E821 statistics.

4 Conclusion

The new Muon g-2 experiment at Fermilab will measure the anomalous magnetic moment of the muon to 4 × the precision of the previous BNL measurement. If the previously measured value holds, this could provide a 7σ discrepancy from the standard model. A successful commissioning run was performed in June of 2017 and the experiment will continue running this November. The 2017 run provided 10^{-5} of the required statistics, which is between the total statistics of the CERN II and CERN III measurements of a_μ. The experiment's goal is for a BNL level (500 ppb) result from the 2018 data and the final 140 ppb measurement from data collected through 2020.

Acknowledgments

This work is supported by Fermi Research Alliance, LLC under Contract No. DE-AC02-07CH11359 with the United States Department of Energy.

References

[1] G.W. Bennett et al. Final Report of the Muon E821 Anomalous Magnetic Moment Measurement at BNL. *Phys. Rev.*, D73:072003, 2006.
[2] J. Grange et al. Muon (g-2) Technical Design Report. 2015.
[3] Paul A.M. Dirac. The Quantum theory of electron. *Proc. Roy. Soc. Lond.*, A117:610–624, 1928.
[4] A. I. Studenikin and I. M. Ternov. Constraints on the Weak Gauge Boson Composite Scale From the Muon G^-2 Factor. *Phys. Lett.*, B234:367–370, 1990.
[5] Kamila Kowalska and Enrico Maria Sessolo. Expectations for the muon g-2 in simplified models with dark matter. *JHEP*, 09:112, 2017.
[6] Michel Davier, Andreas Hoecker, Bogdan Malaescu, and Zhiqing Zhang. Reevaluation of the Hadronic Contributions to the Muon g-2 and to $\alpha(MZ)$. *Eur. Phys. J.*, C71:1515, 2011.
[7] Tatsumi Aoyama et al. Complete Tenth-Order QED Contribution to the Muon g-2. *Phys. Rev. Lett.*, 109:111808, 2012.
[8] C. Gnendiger, D. Stockinger, and H. Stockinger-Kim. The electroweak contributions to $(g-2)_\mu$ after the Higgs boson mass measurement. *Phys. Rev.*, D88:053005, 2013.
[9] Thomas Blum et al. Using infinite volume, continuum QED and lattice QCD for the hadronic light-by-light contribution to the muon anomalous magnetic moment. *Phys. Rev.*, D96(3):034515, 2017.
[10] Thomas Blum et al. Connected and Leading Disconnected Hadronic Light-by-Light Contribution to the Muon Anomalous Magnetic Moment with a Physical Pion Mass. *Phys. Rev. Lett.*, 118(2):022005, 2017.

[11] J. Kaspar et al. Design and performance of SiPM-based readout of PbF$_2$ crystals for high-rate, precision timing applications. *JINST*, 12(01): P01009, 2017.

[12] A.T. . Fienberg et al. Studies of an array of PbF$_2$ Cherenkov crystals with large-area SiPM readout. *Nucl. Instrum. Meth.*, A783:12–21, 2015.

[13] Diktys Stratakis et al. Accelerator performance analysis of the Fermilab Muon Campus. *Phys. Rev. Accel. Beams*, 20(11):111003, 2017.

INTENSE GAMMA RADIATION BY ACCELERATED QUANTUM IONS

Noboru Sasao [a]
Research Institute for Interdisciplinary Science, Okayama University, 700-0082 Okayama, Japan

Abstract. Intense gamma-ray sources using partially stripped ions are a promising alternative to the existing sources based on laser Compton scattering, and are expected to provide a platform for opening up new fields of science. In this report, features of this scheme are discussed and an example of such sources is presented.

1 Introduction

Currently accelerator-based gamma-ray sources use a scheme called Laser Compton Scattering (LCS). In this scheme one accelerates electrons to high-energy and injects a laser beam to collide electrons head-on. They undergo Compton scattering and, in particular, backward scattered photons have energies in the laboratory $4\gamma^2$ times the original laser photon energy, where γ is the Lorentz boost factor of the electrons. With sufficiently large γ those photons are in the range of MeV$\sim$GeV.

It is known for some time a scheme in which electrons are replaced with partially-stipped ions (PSI) may be used to generate intense gamma-rays [1–4]. In this case, laser photons corresponding to a resonance between ground and excited states of PSI is injected, and photons from PSI emitted backwards are observed. Those photons have $4\gamma^2$ times the original laser photon energy as in the case of LCS.

The main advantage of the method is the following. The basic cross section involved in LCS is the Thomson cross section expressed by $\sigma_T = \dfrac{8\pi}{3}r_e^2$, where r_e ($\simeq 2.8 \times 10^{-15}$ [m]) is the classical electron radius. On the other hand, for the case of PSI, the basic cross section is a resonance absorption cross section $\sigma_R = \dfrac{1}{2\pi}\lambda^2$, where λ represents the wavelength corresponding the resonance transition. The ratio of these two cross sections σ_R/σ_T may become 10^8, providing a potential of intense gamma-ray sources. In this report, several features of the PSI scheme are discussed and example of such sources are presented.

2 Possible applications

Before we discuss its details, we first mention about possible applications. They are found to be very useful in (i) nuclear physics, (ii) as a source of tertially beams such as neutrons or positrons, and (iii) for industrial uses such

---

[a] E-mail: sasao@okayama-u.ac.jp

as non-destructive isotope analysis or tomography. Here only a few specific examples, which are related to fundamental science, are presented.

First example is nuclear astrophysics [5], [6]. It is known that as hydrogen burns in a star a core of helium is formed, and it becomes a fuel for further nuclear fusion. The nucleosynthesis of heavier elements begins with the fusion process of triple α ($_2^4$He) particles: $_2^4$He + $_2^4$He $\rightarrow {_4^8}$Be followed by $_2^4$He+$_4^8$Be $\rightarrow {_6^{12}}C$+2γ, producing a stable carbon nuclei. Furthermore $_6^{12}C$ fuses with an additional α into oxygen via $_6^{12}C + {_2^4}$He $\rightarrow {_8^{16}}O + \gamma$. The cross section of these processes are very important because they determine oxygen/carbon ratio and subsequent evolution of stars. Experimentally they have not observed yet, especially in the energy range important for the star evolution (CM energy of $\sim$300 keV) because of extreme small cross sections ($\sim 10^{-17}$ b). On the other hand, theoretical estimates have uncertainties up to $\pm15\%$ (the former process), and $\pm40\%$ (the latter). Using intense gamma-ray sources (e.g. $\sim$ 8 MeV gammas of 10^{17} sec^{-1}), the inverse reactions can be studied. This will allow to reduce uncertainty substantially.

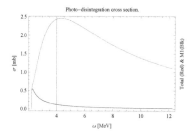 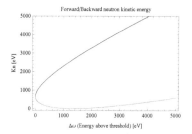

Figure 1: Photo-disintegration total cross section (red) vs incident energy. Near threshold the M1 contribution (black) dominates.

Figure 2: The maximum (black) and minimum (red) of the neutron kinetic energy vs the incident energy from the threshold.

The second example is generation of a cold neutron beam, which is useful for fundamental physics (electric dipole moment etc.) and material science. Here we focus on cold neutron beam produced by a photo-disintegration process of deuterons $\gamma + D \rightarrow p + n$. The threshold of the reaction is about 2.23 MeV, and its total cross section peaks to $\sim$ 2.4 [mb] at about 5 MeV, as shown schematically in Fig.1. Since this process is two-body reaction, energy of produced neutron is unique once scattering angle is chosen. Figure 2 shows the relation between incident energy and the maximum (black) and minimum(red) of the neutron kinetic energy. Since it is an endothermic reaction, the kinetic energy becomes zero at a certain incident energy E_m, giving a potential of cold neutron source. Given $\sim$0.14 mb cross section at E_m, a 60-cm-long liquid deuteron target and gamma flux of 10^{17} s^{-1}, the neutron total flux of 10^{12} s^{-1} is possible. Selecting scattering angle further, a cold neutron (< 25 meV) flux of 10^5 s^{-1} can be obtained.

3 Generation of gamma-rays by partially stripped ions

3.1 Basic features of PSI as a gamma-ray converter

Partially stripped ions act as a converter of photons. Suppose PSI has an excited level with energy $\hbar\omega_e$ and is accelerated to γ. If a laser beam of frequency ω_L is injected in counter propagating direction, it is absorbed by PSI when the resonance condition $\omega_{eg} = 2\gamma\omega_L$ is satisfied. Excited PSI eventually goes back to the ground state by emitting a photon; the energy of those emitted backwards are boosted again by 2γ, resulting in $4\gamma^2\omega_L$. The average rate of the absorption and emission cycle is given by $R = \dfrac{I}{I+I_s}A_{eg}$, where A_{eg} is an Einstein A coefficient of PSI, and I (I_s) denotes incident laser (its saturation) intensity. The saturation intensity is defined by tan intensity at which the absorption rate is equal to A_{eg} [b]. If $I = I_s$, for example, it takes the time period of $1/A_{eg}$ to absorb a laser photon and anther $1/A_{eg}$ to emit a photon, so that the cycle rate is $A_{eg}/2$.

3.2 Selection of partially stripped ions

In order to obtain high gamma-ray flux, ions with a large A_{eg} should be chosen. In addition the resonance condition $\omega_{eg} = 2\gamma\omega_l$ should be satisfied. Here we present an example of hydrogen like ions. It is well known that its energy levels scale as Z^2 with Z being the atomic number, and E1 dipole transition rate scales as Z^4. For example, the transition rate from $1s$ to $2p$ states is $6.27 \times 10^8 Z^4$ sec^{-1}. Heavier elements are advantageous for a large rate.

3.3 Example of PSI gamma-ray sources

In this section, we present an example of high intensity gamma-ray source using an hypothetical accelerator placed at the KEK Belle tunnel. The relevant accelerator parameters are listed in Table 1. In the first example, we plan to use hydrogen-like Ti ions ($Z = 22$). The transition energy of $2p$ state from $1s$ is about 5 keV, and thus a 10 eV light source is needed. Since laser with such frequency is not available at present, we use undulator or free electron laser (FEL) as a light source, whose specification is listed in Table 1. The expected power from this light source is about 100 W in undulator mode and 12kW in the FEL mode. The gamma-rays produced by such combination has an energy of 2.5 MeV and intensity of 8×10^{14} gammas/second and 10^{17} gammas/second, respectively, for the undulator and FEL modes.

---

[b]Using the absorption cross section $\sigma_{abs}(\omega) = \dfrac{2\pi c^2}{\omega^2}A_{eg}g(\omega)$ (g is a line shape function) and incident photon flux per unit frequency $\dfrac{d\Phi(\omega)}{d\omega} = \dfrac{I/(\hbar\omega)}{\Delta\omega}$ ($\Delta\omega$ is a bandwidth of laser), I_s in the ion rest system can be obtained by equating $A_{eg} = \int \sigma_{abs}(\omega)\dfrac{d\Phi(\omega)}{d\omega}d\omega$.

Table 1: Specification of accelerator and light source.

Ion accelerator		Light source	
item	value	item	value
Boost factor γ	250	Electron energy	400 [MeV]
Number of ions per bunch	10^9	Electron current	100 [mA]
Bunch period	100 [nsec]	Undulater period	5 [cm]
Energy spread $\Delta E/E$	0.5×10^{-3}	Number of dipole pairs	1000

3.4 Practical limitations

There are several practical issues which may limit the performacne of such gamma-ray sources. They include
- collision loss by residual gas molecules
- loss by intra-beam scatterings
- two-photon ionization loss
- Stark effect due to magnet fields (bending magnet)

among others. All of them may affect the stability of ions in an accelerator ring. Although there seems no serious problems, each of them should investigated carefully and experimentally.

In conclusion, gamma-ray sources using partially stripped ions are a promising alternative to the existing sources based on LCS, and are expected to provide a platform for opening up new fields of science. Initial studies for such possibilities are now underway in Japan as well as at CERN [7].

Acknowledgments

This work is done with collaboration of members of the QIB group. Special thanks should go to M. Yoshimura and Y.Honda for their valuable discussions. This work was supported by JSPS KAKENHI Grant No. 16H02136.

References

[1] E.G. Bessonov and K.J. Kim, Phys. Rev. Lett. **76** (1996) 431.
[2] E.G. Bessonov, Nucl. Instr. Meth. **B309** (2013) 92.
[3] M.W. Krasny, arXiv:1511.07794v1 [hep-ex] 24 Nov 2015.
[4] M. Yoshimura and N. Sasao, Phys. Rev. **D92** (2015) 073015.
[5] J. Jose and C. Iliadis, Reports on Progress in Physics, **74**, (2011) 9
[6] L. Buchmann L and C. A. Barnes, Nucl. Phys.**A777** (2006) 254
[7] M.W. Krasny *et al.*, Eps-HEP2017, Venice, Italy, 5-12 July 2017, We thank Dr. Frank Zimmermann for informing this activity.

THE INTERNATIONAL LINEAR COLLIDER PROJECT

Daniel Jeans [a]
KEK, Tsukuba, Japan.

Abstract. The International Linear Collider is a proposed electron positron collider, with the potential to operate at centre-of-mass energies from 250 GeV to 1 TeV. We discuss the motivation for such a facility, the design of the accelerator and detectors, and the present status of the project.

1 Introduction

With the discovery of the Higgs boson, the Standard Model (SM) of particle physics is, in some ways, complete. All of the particles and forces within it have been observed. The discovery of the Higgs boson was the ultimate verification of the model.

However, many aspects of our universe are left unexplained by the SM. In particular, dark matter and energy are often used to explain certain aspects of our universe. This dark sector has no explanation within the SM, and leads us to believe that the SM describes only a rather small fraction of the contents of our universe.

The Higgs boson is a unique member of the SM, which sits at its very centre. It has no charge, no spin: the quantum numbers of the vacuum. Does it interact with the dark sector? Is there new physics at higher energy scales which impacts the Higgs sector?

Different models of beyond-the-SM (BSM) physics have different effects on the properties of the Higgs boson. As an example, BSM models can cause the Higgs boson's couplings to be altered from the SM predictions, in ways that reflect the underlying BSM theory. For example, rather different patterns of Higgs boson coupling deviations are induced for supersymmetric and composite Higgs models. For a new physics scale around the TeV scale, deviations are typically at the level of a few percent.

2 Physics at high energy electron-positron collisions

Electron-positron collisions at high energy will allow high precision measurements of the Higgs sector and of top quark properties, and thereby for the indirect effects of BSM physics. It will also allow direct searches for new particles, making use of the clean experimental environment.

The leptonic initial state produces collisions between elementary particles, with a well-defined initial state in terms of particles, energy, (and in some cases also polarisation), and results in a rather clean experimental environment, with

---

[a] E-mail: daniel.jeans@kek.jp

low levels of background. The electro-weak initial state gives "democratic" production of different final states, as opposed to hadron colliders whose collisions are dominated by QCD-induced interactions.

A number of particular energy points are of particular interest for electron-positron collisions. The cross-section for Higgs boson production starts a little over 200 GeV, reaching a (local) maximum at around 250 GeV. At energies below around 450 GeV, Higgs bosons are produced mostly via the so-called "Higgs-strahlung" process $e^+e^- \to ZH$, while at higher energies "WW fusion" $e^+e^- \to H\nu_e\bar{\nu}_e$ dominates. A "Higgs factory" would therefore initially operate around 250 GeV, before proceeding to higher energies, for example 500 GeV, to measure the WW fusion process and the Higgs self-coupling in the process $e^+e^- \to HHZ$.

The threshold of top quark pair production $e^+e^- \to t\bar{t}$ is at around 350 GeV. An energy scan around this energy can give a theoretically well-defined measurement of the top quark mass with excellent experimental precision. The measurement of the top quark Yukawa coupling becomes possible at energies around 500 GeV, via the process $e^+e^- \to t\bar{t}H$. This energy is also optimal for the measurement of other top quark electro-weak couplings.

All energies will also enable high precision tests of the electro-weak sector, providing stringent constraints on models of new physics. Compared to LEP2, the International Linear Collider will provide a three orders of magnitude larger total integrated luminosity, and will add the advantages of polarised beams and higher energy.

As well as the above "guaranteed" measurements, there is also the possibility that new particles not in the SM will be produced. The ILC has several advantages over the LHC, in particular sensitivity to low energy or even invisible decays of new particles, which stem from the low total cross-section of e^+e^- collisions. This enables the use of "trigger-less" detector readout, recording even the lowest energy detector signals, and lead to very low levels of backgrounds due to secondary collisions superimposed on an event's primary collision. Such searches are limited by the collisions' centre-of-mass energy, so higher energies are preferred to maximise the range of new particle masses.

3 The International Linear Collider

The International Linear Collider (ILC) is a proposed energy frontier linear electron-positron collider [1]. The advantage of a linear accelerator is connected to synchotron radiation. In a circular collider, the power loss due to synchotron radiation scales as E_{beam}^4/ρ^2, where E_{beam} is the beam energy, and ρ the radius of the ring. At high beam energies, the power loss becomes prohibitive, even for a ring of very large radius. A linear accelerator (in which the radius $\rho \to \infty$) avoids this limitation, leading to a more-or-less linear scaling of accelerator

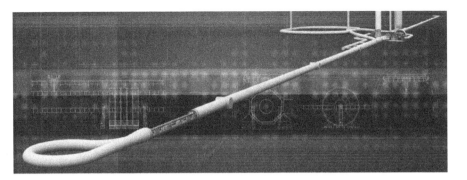

Figure 1: A view of the proposed ILC facility.

cost and power requirements with beam energy. This ensures that the machine which can reasonably be upgraded in energy. A linear accelerator also allows the collision of polarised beams, providing a powerful extra handle with which to investigate the nature of particle interactions. The ILC is designed to provide an 80% (30%) polarised electron (positron) beam.

The ILC will initially operate at a centre-of-mass energy of 250 GeV, and has the potential to be upgraded, probably in several stages, in energy up to 1 TeV. The target luminosity is of order $10^{34} cm^{-2} s^{-1}$ (or around $100 fb^{-1}$ per year). The 250 GeV collider will have a total length of around 20 km, which will be extended to 30 km for operation at 500 GeV. Over a 25 year experimental program, we can expect $2 ab^{-1}$ of data at 250 GeV, $200 fb^{-1}$ at 350 GeV, and $4 ab^{-1}$ at 500 GeV.

The Technical Design Report [2] of the ILC was published in 2013, developed by an international team of physicists and engineers. It is based on 1.3 GHz niobium superconducting accelerating structures, operating at 2 K, and providing an average gradient of 31.5 MeV/m. To maximise the provided luminosity, beams are focused into an extremely small beam spot at the interaction region, with a vertical size of just 6 nm. Since the TDR, the accelerator design has been under strict "change control", so the design is rather stable, well optimised, and mature.

The same niobium superconducting technology on which the ILC relies are now rather widely used in other contexts. The largest-scale example is the European X-FEL light source at DESY, with similar facilities being developed at SLAC (LCLS-II) and Shanghai. The European X-FEL has recently started operation, and houses 2 km of accelerating structures with a nominal beam energy of 17.5 GeV. This corresponds to $\sim 10\%$ of the ILC linacs, demonstrating that the technology is mature and can be produced in large quantities in an industrial setting.

Figure 2: The International Large Detector (ILD), one of two proposed ILC detectors.

4 Detectors

Two detector concepts are being developed for use at the ILC [3]. They are general purpose detectors, aiming to address the whole spectrum of physics measurement possible at the ILC. This includes the identification and measurement of hadrons, charged leptons and photons produced in ILC collisions, over as much as possible of the available solid angle. The leptonic collisions at ILC give rise to relatively low event rates, allowing trigger-less readout, with all detector signals being recorded for off-line analysis. Many physics analyses will make use of W and Z bosons to identify signal events: for example the Z bosons produced together with the Higgs boson in the Higgs-strahlung process, or the W bosons produced in top quark decay. To make the most efficient use of ILC collisions, all decay modes of these bosons should be identified, in particular their dominant hadronic decays.

This consideration is a central consideration in the design of the ILC detectors, which aim to achieve unprecedented jet energy resolution to enable hadronic decays of W and Z bosons to be distinguished. This is done via "particle flow" reconstruction, which individually reconstructs each final state particle, including components of hadronic jets. The measurement of the most precise sub-detector can be used to determine the energy of each particle in the final state: the tracking system for charged particles, and the calorimeters for photons and neutral hadrons. This places ambitious requirements on the calorimeters, which must have extremely high readout granularity (of order $0.1 \to 100\ cm^3$) in order to disentangle the showers induced by nearby particles within jets.

Charged particle tracking is also of utmost importance, in particular to reconstruct leptonic decays of Z bosons produced in the Higgs-strahlung process, which can give model-independent precision measurements of the Higgs boson mass and production cross-section without looking directly at its decay products. Relative resolution on transverse momentum are of order $\sigma_{pT}/p_T \sim 3 \cdot 10^{-5} p_T$. High precision vertex detectors will provide excellent reconstruction of displaced tracks produced in the decay of beauty and charm hadrons, and of τ lepton decay products. Key is the use of low-mass technologies in vertex detector construction to reduce multiple scattering, and finely segmented sensors with very good position resolution. The typical expected impact parameter resolution is $\sigma_{d0} \sim 5\mu m \oplus 10\mu m/(p[GeV]\sin^{3/2}\theta)$.

5 Overview of some expected measurement precisions

The physics capabilities of ILC and of its detectors have been extensively studied, based on full simulation of its proposed detectors, and realistic event reconstruction computer codes. Over an illustrative 25 year ILC program, a wealth of high precision measurements will be possible. Overviews of the prospects for ILC physics measurements can be found in [4–6]. Couplings of the Higgs to Z, W, b, g, τ, c will be measured to at least 1% precision. The photon coupling will be measured to similar precision, when combined with LHC results on the ratio between Z and photon couplings. The total Higgs width can be determined to 2%, and the decay width to invisible final states can be constrained to better than 1%. Couplings to the top quark and muon will be better than 10%. These ILC measurements significantly exceed HL-LHC expectations in most cases, and also require significantly less theoretical model assumptions than those made at LHC. Recent approaches to combining these and other electro-weak measurements in an EFT-inspired approach can be found in [7]. Detailed studies of the Higgs sector's CP properties are also possible, via the Higgs bosons decays to τ leptons, vector bosons, or Higgs production associated with top quarks.

The measurement of top quark electro-weak couplings becomes possible at energies of 350 GeV and above (the optimal energy is around 500 GeV). The measurement of both CP conserving and violating couplings can be achieved with precision up to several orders of magnitude better than what is possible at HL-LHC [8,9]. An energy scan around the $t\bar{t}$ threshold will give access to a top quark mass measurement to better than 100 MeV.

6 Project status

Several linear electron-positron collider projects aiming to provide energies higher than 200 GeV have now been studied for several decades. These efforts were united under the ILC organisation in 2006, with the choice of the superconducting accelerator technology. The technical design was developed under the ILC Global Design Effort, resulting in a Technical Design Report, published in 2013.

Efforts to realise the ILC were given fresh impetus by the discovery of the Higgs boson in 2012, which provided a guaranteed role for lepton colliders

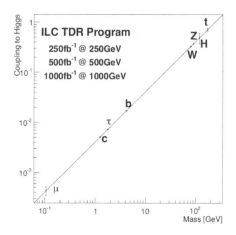

Figure 3: Expected precisions on the measurement of the coupling between the Higgs boson and other particles of the SM. Figure taken from [4].

in precision Higgs measurements. In 2012, the Japanese High Energy Physics community proposed to realise the ILC in Japan, and a preferred site, in the Kitakami mountains in the northern part of the main island, was selected in 2013. 2013 also saw the creation of the Linear Collider Collaboration (LCC) under the auspices of ICFA, the International Committee for Future Accelerators, bringing together the design teams for the ILC and CLIC projects. The LCC has developed the accelerator design, adapting it for the chosen site, and working on realising the political prospects of the project, including investigating possible cost reductions.

Various international community-lead strategy processes (ACFA, AsiaHEP, P5 in the USA, the European Strategy for Particle physics) explicitly supported these efforts to realise the ILC project in Japan. Since 2014, an ILC review committee, set up by MEXT, the relevant Ministry of the Japanese Government, has been studying the case for hosting ILC. Key interim recommendations were to closely monitor the results of LHC experiments, reduce the cost of the project, and develop the international cost sharing model for the facility.

Over the last year, the initial scope of the ILC has reduced from a 500 GeV collider to a 250 GeV Higgs factory, resulting in a significant reduction of the initial project cost [10]. A linear collider is relatively easy to extend in energy, so energy upgrades are considered as possible future extensions, depending on the development of our understand of physics. This Higgs factory project is now under consideration in the MEXT, and a decision is expected within the next six months or so. Strong international participation will be the key to the success of the project.

7 Conclusion

The ILC will make high precision measurement of the Higgs sector, a unique, essential, and as-yet poorly explored aspect of nature. Its properties are rather sensitive to physics beyond the SM, allowing the ILC to make far-reaching statements about the nature of physics beyond the SM. The linear accelerator can be upgraded in energy, giving access to top quark production, enabling wonderful measurements of its electro-weak couplings and mass. Running at 500 GeV will also allow the probing of the Higgs self-coupling and thereby shape of the Higgs potential, and the top-Yukawa coupling.

Efforts to realise the ILC project in Japan are very active at many levels: political, governmental, local, scientific. The large scale of the project will require significant support from other countries. A decision is awaited by a large community of eager future ILC participants from around the globe.

References

[1] T. Behnke et al., *The International Linear Collider Technical Design Report - Volume 1: Executive Summary*, ILC-REPORT-2013-040 (2013). arXiv:1306.6327.

[2] C. Adolphsen et al., *The International Linear Collider Technical Design Report - Volume 3.II: Accelerator Baseline Design*, ILC-REPORT-2013-040 (2013). arXiv:1306.6328.

[3] Ties Behnke et al., *The International Linear Collider Technical Design Report - Volume 4: Detectors*, ILC-REPORT-2013-040 (2013). arXiv:1306.6329.

[4] K. Fujii et al., *Physics Case for the International Linear Collider*, ILC-NOTE-2015-067 (2015). arXiv:1506.05992.

[5] K. Fujii *et al.*, *Physics Case for the 250 GeV Stage of the International Linear Collider*, KEK-PREPRINT-2017-31 (2017). arXiv:1710.07621.

[6] K. Fujii et al., *The Potential of the ILC for Discovering New Particles*, KEK-PREPRINT-2016-60. arXiv:1702.05333.

[7] T. Barklow et al., *Improved Formalism for Precision Higgs Coupling Fits*, DESY-17-120 (2017). arXiv:1708.08912.

[8] M. S. Amjad et al., *A precise characterisation of the top quark electroweak vertices at the ILC*, Eur. Phys. J. C **75**, no. 10, 512 (2015). arXiv:1505.06020.

[9] W. Bernreuther et al, *CP-violating top quark couplings at future linear e^+e^- colliders.* arXiv:1710.06737 (2017).

[10] L. Evans, S. Michizono, *The International Linear Collider Machine Staging Report 2017*, KEK-2017-3 (2017). arXiv:1711.00568.

Astroparticle Physics and Cosmology

THE DETECTION OF GRAVITATIONAL WAVES: 50 YEARS OF EXPERIMENTAL EFFORTS

Eugenio Coccia
Gran Sasso Science Institute, viale F. Crispi 7, I-67100 L'Aquila, Italy
Istituto Nazionale di Fisica Nucleare, Italy.

Abstract. The last years marked the beginning of a new era of observations of the Universe. Gravitational waves were detected from several binary black-hole mergers and also from a spectacular binary neutron star merger by the LIGO-VIrgo Collaboration. In the same period, LISA Pathfinder demonstrated the technology for gravitational-wave observation in space beyond its planned requirements. Many gravitational observations and discoveries are expected in the next years with the Advanced LIGO and Virgo detectors, with strong impact on many astrophysical fields, from the physics governing compact object formation and evolution to the physics of the emission process and to nuclear astrophysics. I summarize here some historical milestones that lead to the first direct detection, and discuss the importance of the so called multimessenger astronomy in which gravitational-wave sources will be observed in all bands of the electromagnetic spectrum with ground and space telescopes.

1 Introduction

Long is the path of knowledge that has led Homo Sapiens to detect for the first time gravitational waves on September 14, 2015 [1]. The waveform agreed with the prediction of general relativity for the emission of gravitational waves during the inspiraling and merging of two black holes of, respectively, 36 and 29 solar masses, distant 1 billion and three hundred million light years from the Earth. A single final black hole of 62 solar masses was formed from the collision. The 3 missing solar masses are equivalent to the energy emitted in the form of Gravitational waves during the event. This was the first direct detection of a gravitational wave, 100 years from Albert Einstein's theoretical prediction and after more than 50 Years of experimental efforts. It was also the first time that the merging of a binary black hole system was observed. These measurements give us for the first time direct access to the space-time properties in a regime of strong gravitational field and high speed (the two black holes at the time of the fusion have a speed of more than half that of light). It has been a historical moment for science and for mankind. We actually are able to perceive the vibrations of space time, which can be compared to the ability to "listen" to the Universe, so far only "seen" with photons. Let's look at some of the steps that have led to this result, and at some future perspectives.

2 50 years of experimental efforts

In 1916, the year after the formulation of the equations of general relativity, Einstein predicted the existence of gravitational waves. He found that in the weak field approximation, where the metric tensor was that of Minkowski's flat space-time with the addition of a small perturbation, its field equations linearized and had simple solutions: transverse waves of spatial deformation traveling at speed of light, generated by time variations of the moment of mass quadrupole of the source [2].

Einstein immediately understood that the gravitational waves would have been of very small amplitude ("a practically vanishing value") and for many years the issue of gravitational waves fell in the forgetting. The steady progress of astronomy, in particular the discovery of compact objects like neutron stars and black holes, and the significant developments in technology, changed the perspectives. It should be considered, however, that up to the 1950s a debate was alive, that today can look surprising: were gravitational waves a real, measurable effect, or were they just a fictitious effect, eliminable with a transformation of the coordinates? Einstein oscillated between these two positions and was often on the second front [3].

A turning point in the history of the search for gravitational waves occurred in 1957 at the "Conference on the Role of Gravitation in Physics" held at Chapel Hill, University of North Carolina [4], see also [5] and references therein. The question on the agenda was the physical reality of gravitational waves and their interaction with a possible detector. Felix Pirani proposed to observe the gravitational effects by measuring the relative acceleration of two bodies in free fall. He clearly connected the equation of geodetic deviation of General Relativity with Newton's second law, identifying some components of the Riemann tensor with the second derivatives of the Newtonian potential, that is, with the tidal field. Hermann Bondi, present at the conference, immediately understands the essence of the message: in the presence of a gravitational perturbation, or rather a gravitational wave, a system of two bodies connected with a spring absorbs energy, a measurable effect, because of their relative acceleration. Or if one of the two bodies is equipped with a source of light and a detector and the other with a mirror, upon the arrival of the wave the time taken by the light to travel back and forth the distance between the two bodies will change, as it will change their distance, and this effect can also be measured. Richard Feynman, also present, blessed this conclusion.

There was another physicist present at that conference, following all the talks. His name was Joseph Weber. He was professor of Electrical engineering at the University of Maryland, and had already a brilliant idea for the realization of what would then became the maser. In 1955, he was in Princeton with a Guggenheim fellowship, working with John Archibald Wheeler. Shortly after the conference at Chapel Hill, Weber and Wheeler wrote a paper where they illustrated how to extract energy from a gravitational wave. In February 1959, Weber went a long way farther away, publishing on Physical Review "Detection and Generation of Gravitational Waves," in which he depicted his program for the realization of gravitational wave detectors [6]. His detector consisted of an aluminum cylinder of a couple of tons of mass, where the vibrations of the fundamental longitudinal mode were monitored. The cylinder played the role of the two test masses of Pirani connected by a spring. Ultimately, it was a giant diapason equipped with piezoelectric ceramic to convert the mechanical vibrations in electrical signals.

Weber realized two detectors, placed them at 1000 km distance and analyzed the data in search for coincident signals. In fact a gravitational wave would put in vibration simultaneously the two oscillators (by 3 thousandths of a second) by allowing distinguish the signal from the possible causes of local noise.

It took great faith to start this research at a time where black holes and neutron stars were objects just barely imagined, but not known in any astrophysical context.

In the late 1960s, Weber found some coincidences in its antennas and believed to have detected gravitational waves coming from from the center of our Galaxy [7]. Years of emotions followed, with the birth of new groups all over the world to replicate the experiment. Years of controversy. Finally, from the mid-1970s, a growing consensus was reached in the newborn community on the incorrectness of the conclusions of Weber. However, in the meantime the seeds planted by the Chapel Hill conference and by Weber with his announcement were flourishing.

Then came Hulse and Taylor with the observation of the loss of energy of the binary pulsar PSR 1913 + 16 to provide the earlier possible evidence, albeit indirect, of the existence of gravitational waves [8]. Experimenters, since then and on four continents, have been engaged in these researches without stopping using sophisticated cryogenic versions of Weber resonant antennas, or by building giant detectors based on laser interferometry.

These researchers are heirs of two great experimental traditions. One is the tradition of the precision mechanical experiments, exemplified by the work of Cavendish, Eotvos, Dicke, Braginski. At the heart of any experiment on gravitational waves there are masses isolated from the external noise, in conditions as similar as possible to the ones of ideal bodies in free fall. The other tradition is the one of precision optical measurements, started with Michelson, and supported by the developers of lasers and by the pioneers of microwave technology.

Several countries have a first row position in the quest for gravitational waves: USA, Italy, Germany, Great Britain, France, Japan, Australia. Some words about Italy, where the activity is as old as 1970, when Guido Pizzella and Edoardo Amaldi in Rome started an activity on resonant antennas destined to become protagonist for three decades at international level, first at CERN with the antenna Explorer, then at the INFN Laboratories in Frascati with Nautilus and at the INFN Laboratories of Legnaro with Auriga (conducted by Massimo Cerdonio). In the 80s Adalberto Giazotto proposed successfully to

INFN the realization of an ambitious project: a great laser interferometer in Italy. From the work of Giazotto and a few pioneers, including Alain Brillet who led the French CNRS to cofinance the project, the Virgo collaboration flourished in Cascina, near Pisa, where an international lab was established: EGO, the European Gravitational Observatory. The USA was the first country where the leading vision of large laser interferometers by MIT and Caltech won the competition with resonant bar detectors. A vision which become a reality at two different sites with the two LIGO detectors, that with Virgo is recognized today as the most advanced GW Observatory. For a significant picture of the activity of the community in mid '90s, where both the past and the future GW experiments were represented, see [9].

3 The first detection

A gravitational wave observatory is based on detectors widely spaced to distinguish signals from local instrumental and environmental noises. This also allows to locate the position of the source from the time of arrival of the signal received by the individual detectors, and also measure the wave polarizations thanks to the measured amplitude of the signals at the different sites. The detectors involved in the discovery are Michelson interferometers, and measure the deformation imposed in space by the passing gravitational wave through the different effect in the lengths of their perpendicular arms. In the LIGO detection, the distance L = 4 km of each arm varied of about $dL = 10^{-18}\,m$. The wave amplitude h (by definition it is a deformation and therefore adimensional) is: $h = dL/L = 10^{-21}$. 50 years of experimental efforts were needed to measure a change of a billionth of a billionth of a meter.

The discovery paper illustrates the results of the analysis of 16 days of observation in coincidence between the two LIGO detectors, from september 12 to October 20, 2015 [1]. That's just a part of the overall data taking period, lasting until January 12, 2016 and subject to further analysis. The signal was baptized as GW150914, from its date of arrival, and has emerged from two types of analysis. One is optimized to detect signals of coalescence of compact objects, using filters fitted with the waveforms predicted by general relativity. The other identifies a broad range of generics transient signals with minimal assumption on the expected waveform. Both analyzes have clearly identified the GW150914 event. The features of GW150914 indicate that its origin is the coalescence of two black holes, that is, the final phase of their mutual orbiting motion, their collision and the formation of the final black hole. In about $0,2\,s$, the signal increases in frequency and amplitude, running in 8 cycles from $35\,Hz$ to $150\,Hz$, frequency to which the amplitude reaches the maximum value. The natural explanation of this evolution is the spiraling of two masses, due to gravitational waves emission. This waveform is called conventionally chirp and its evolution is characterized by the parameter known as chirp mass, depending on the values of the two masses.

To reach a gravitational frequency of 150 Hz, and hence an orbital frequency of 75 Hz (the gravitational signal frequency is twice the orbital frequency, consequence of the quadrupular nature of the gravitational radiation) the two objects must be very close and therefore very compact.

Two equal masses (resulting in the measured chirp mass) would be orbiting at 75 Hz when approaching a distance of 350 km. Two neutron stars, though compact, would be much lighter (typically 1.4 solar mass each one) and on the other hand a binary system consisting of a black hole and a neutron star would have a very large total mass and would collide at much lower frequency. This leaves two black holes as the only pair of compact objects able to reach as distint objects an orbital frequency of 75 Hz. In addition, the waveform amplitude decay after the peak is consistent with the damped oscillation of a rotating black hole that reaches the stationary configuration. I leave aside the detailed results and the description of the analysis, reported in the discovery paper. It is however worth mentioning here that several analyses have been accomplished with the aim of determining whether GW150914 is consistent with a black holes system in general relativity, with affirmative answer. Gravitational waves were not "dispersed" in the observed signal, that is, all the Fourier components of the signal have propagated at the same speed (within the limits of sensitivity of the experiment, of course). This limits the Compton wavelength of the graviton to be greater than 10^{13} km. The data can also be interpreted as a limit on the graviton mass, which must therefore be less than 10^{-22} eV/c^2.

This improves all other previous limits due to measures in the solar system and to the Hulse and Taylor pulsar system. In short, all tests on GW150914 are consistent with the predictions of general relativity [10].

GW150914 demonstrates for the first time the existence of black holes of stellar mass greater than 25 solar masses, more massive than those hitherto identified by the study of X-ray binaries where a compact object accretes matter taken by a star companion (as in the case of the famous Cygnus X-1). And it also establishes, obviously, that a binary system of black holes can form and merge into a time lower than the age of the Universe. None of these statements was obvious.

The study of the astrophysical implications of the existence of such a system, only possible as a result of stellar evolution in environments of low-metallicity, it's only at the beginning.

4 Multimessenger astronomy with gravitational waves

One of the most promising issue of contemporary astrophysics is the investigation of the most powerful and violent events in the Universe, taking advantage of the simultaneous observations of all possible cosmic messengers: photons at all wavelengths, cosmic rays, neutrinos and gravitational waves. The goals are to gain a more complete understanding of cosmic processes through the combination of information from the different probes, and to increase search sensitivity over an analysis using a single messenger.

The first direct detection of GWs has created excitation. Some of the most promising astrophysical sources of GWs are expected to produce broadband electromagnetic (EM) emission and also neutrinos. The presence or absence of any EM or neutrino signature will provide constraints on emission mechanisms, progenitors and energetics of the GW source, as well as its environment. New windows in unexplored domains of the physics of supranuclear density matter and very strong, time-varying gravitational fields can be opened.

Focusing the attention on joint GW and EM observations, one can say, in general, that EM observations are key to localize and characterize the astrophysical source, to probe the physics of its environment and the distribution of magnetic fields, while GWs provide insight into its mass distribution and gravitational fields in the strong regime. Several detectable GW sources, like core-collapse supernovae, binary NS (BNS) or NS - BH (NSBH) mergers, and the early evolution of new born highly magnetized NSs, are expected to be accompanied by EM emission across the spectrum and over time scales ranging from seconds to years.

For transient GW sources, multiwavelength observations are crucial to find an EM counterpart and improve the source localization down to the arcsecond level, leading to the identification of the host galaxy and measurement of the redshift. This will allow not only to determine the EM intrinsic luminosity of the GW source, but also improve upon the measurement of its extrinsic GW properties, among them the GW luminosity distance. This will allow to break degeneracies present in the signal and to obtain for the first time an independent measure of the Hubble constant using general relativity as only calibrator [11] [12]. EM observations will also help in supporting and guiding searches for continuous GW sources, like asymmetric spinning NSs (e.g. pulsars), both isolated and in binary systems, ranging from the radio to the gamma-ray band, and will be crucial to probe the physics of matter at supra-nuclear densities that can not be tested on Earth laboratories and the environment of strongest gravitational fields in the Universe, leading to a breakthrough of paramount importance to both physics, astrophysics and cosmology.

Let us be more specific considering the important example of a compact binary coalescence (CBC). In a CBC event, a tight binary comprised of two neutron stars, two black holes, or a NS and a BH, experiences a runaway orbital decay due to gravitational radiation. In a binary including at least one NS a binary neutron star (BNS) or neutron starblack hole (NSBH) merger we expect EM signatures due to energetic outflows at different timescales and wavelengths. The coincident detection of a short GRB and GW signal would provide the first direct evidence that short GRBs are associated to the merging of two compact objects and will discriminate on their nature (BH or NS). It will further yield a wealth of information on the mechanism powering the GRB. If a relativistic jet forms, the prompt short gamma-ray burst (GRB) lasting on the order of one second or less, will be followed by X-ray, optical, and radio afterglows of hours to days duration. Rapid neutron capture in the sub-relativistic ejecta is hypothesized to produce a kilonova or macronova, an optical and near-infrared signal lasting hours to weeks. Eventually, we may observe a radio blast wave from this sub-relativistic outflow, detectable for months to years. Furthermore, several seconds prior to or tens of minutes after merger, we may see a coherent radio burst lasting milliseconds. As it is evident from these examples, a NS binary can produce EM radiation over a wide range of wavelengths and time scales.

On the other hand, in the case of a stellar-mass BBH, the current consensus is that no significant EM counterpart emission is expected (and the GW150914 observation confirms this view) except for those in highly improbable environments pervaded by large magnetic fields or baryon densities.

In short, the benefit of coupling GW and EM observations will be tremendous, and will bring the study of the GW signals fully into the realm of astrophysics and cosmology. During the Lomonosov Conference, another gift from the sky arrived, to make the above considerations reality. On 2017 August 17 the merger of two compact objects with masses consistent with two neutron stars was discovered through gravitational-wave (GW170817), gamma-ray (GRB170817A), and optical observations. The optical source was associated with the early-type galaxy NGC 4993 at a distance of just 40 Mpc, consistent with the gravitational-wave measurement. This extraordinary event signes the birth of the Multimessenger astronomy.

The new field of multimessenger astronomy includes also neutrino observations. While GWs are produced by the bulk motion of the progenitor, typically carrying information on the dynamics of the source's central region, high-energy neutrinos, require hadron acceleration in, e.g., relativistic outflows from a central engine. Astrophysical processes that produce GWs may also drive relativistic outfows, which can emit high-energy radiation, such as GeV–PeV neutrinos or gamma rays. which can emit high-energy radiation, such as GeV–PeV neutrinos or gamma rays. The search for common sources of GWs and high-energy-neutrino has recently become possible with the construction and upgrade of large-scale observatories and high-energy neutrinos of cosmic origin have been observed, for the first time, by IceCube [13] [14]. Their detection represents a major step towards multimessenger astronomy.

IceCube is also sensitive to low energy (MeV) thermal neutrinos from nearby supernovae, and contributes to the Supernova Early Warning System (SNEWS) network along with several other neutrino detectors in underground laboratories in Kamioka, Gran Sasso and Sudbury. Supernovae have been at the forefront of astronomical research for the better part of a century, and yet no one is sure how they work. Hence there are important scientific motivations for a joint analysis of GWs and low-energy neutrino data to probe the processes powering a supernova explosion.

5 Conclusion

The detection of the first GW signal signes the birth of the GW astronomy.

The LIGO Hanford and Livingston sites are just the first two Advanced detectors nodes [15] of a growing global network of highly sensitive GW facilities, including since August 2017 Advanced Virgo [16], and later KAGRA [17], and in the future LIGOIndia. In the next decade, oservatories based on interferometers of arms as long as 10 km, located underground to reduce seismic and Newtonian noise, and probably cooled to low-temperatures to reduce thermal noise, are planned. The European project named Einstein Telescope, is an example [18].

These Earth based instruments are limited at low frequency, say below 1 Hz, by the Newtonian noise. There are fascinating GW signals (and sources) at lower frequencies that these instruments can't perceive.

The most fascinating GW enterprise in preparation to study the GW spectrum at lower frequency is LISA, the first observatory in space to explore the Gravitational Universe [19]. LISA can be thought of basically as a high precision Michelson interferometer in space with three spacecrafts in heliocentric orbit and arm lengths of 1 million km. Its frequency range goes from 0.1 mHz to 10 mHz. Expected to fly in 2034 as an ESA mission, this observatory will be dominated by the signals (a dream for a GW physicist). Known binary compact systems, still far from the coalescence phase, will calibrate the observatory and signals emitted by supermassive black holes and possibly even the stochastic background of gravitational waves can be detected and studied. The LISA demonstrator of technology, LISA PathFinder, under the leadership of Stefano Vitale, is now flying successfully, surpassing all expectations [20].

To complete the experimental panorama, one more set of running facilities must be mentioned: the pulsar timing array (PTA) projects, where a set of pulsars are analyzed to search for correlated signatures in the pulse arrival times [21]. The most well known application is to use an array of millisecond pulsars to detect and analyse gravitational waves. Such a detection would result from a detailed investigation of the correlation between arrival times of pulses emitted by the millisecond pulsars as a function of the pulsars' angular separations. Millisecond pulsars are used because they appear not to be prone to the starquakes and accretion events which can affect the period of classical pulsars. PTA can be used to study low-frequency gravitational waves, with a frequency of $10^{-9} Hz$ to $10^{-6} Hz$; the expected astrophysical sources of such gravitational waves are supermassive black hole binaries in the centres of merging galaxies, where tens of millions of solar masses are in orbit with a period between months and a few years.

Globally there are three active pulsar timing array projects, which have begun collaborating under the title of the International Pulsar Timing Array project: the Parkes Pulsar Timing Array at the Parkes radio-telescope; the European Pulsar Timing Array(EPTA) using data from the largest radio telescopes in Europe (Lovell Telescope, Westerbork Synthesis Radio Telescope, Effelsberg Telescope, Nancay Radio Telescope, and soon the Sardinia Radio Telescope will be added); the North American Nanohertz Observatory for Gravitational Waves using data collected by the Arecibo and Green Bank radio telescopes [22].

A new golden age is announced for experimental gravitation, also thanks to the theoretical work consisting in source modelling, numerical simulations and interpretation of the observations.

Let me conclude this contribution reporting two dates. I like to see them as two milestones in the history of our understanding of the Universe.

January 7, 1610. By raising his telescope to the sky, Galileo Galilei watched the moons of Jupiter. A small solar system appeared up there, where nothing was expected to disturb the clear serenity of the Jupiter crystal sphere. The big planet dared mimic, in the eyes of the great scientist, the special role that the Sun had just assumed in the Copernican vision. Galileo was the first witness of the universality of gravitation. Nothing would be as before. New eyes, looking increasingly distant and sensitive also to new electromagnetic windows (radio, microwave, infrared, X, gamma) would then be open to observations, and unimagined surprises manifested: powerful radio emissions from galactic centers, stars dense as atomic nuclei and heavier than the Sun in swirling rotation, the echo of the Big Bang, ultra-high energy photons generated by particles accelerators that we can not even dream of here on Earth, stellar explosions capable of fertilizing the cosmos forming planetary systems like our.

September 14, 2015. Two great sensational microphones on Earth recorded for the first time the vibrations of spacetime, the most elusive waves that ever have been imagined. Generated over a billion light-years away, the first cosmic "sounds" detected by humanity indicate the existence, otherwise inperceivable, of pairs of large black holes merging. And it's just the beginning. We had the sight, now we have the also the hearing. We can finally listen to the Universe. This will be less obscure from today and, again, nothing will be as before.

References

[1] B. P. Abbott et al. Observation of Gravitational Waves from a Binary Black Hole Merger. *Phys. Rev. Lett.*, 116(6):061102, 2016.

[2] Albert Einstein. Approximative Integration of the Field Equations of Gravitation. *Sitzungsber. Preuss. Akad. Wiss. Berlin (Math. Phys.)*, 1916:688–696, 1916.
[3] Albert Einstein and N. Rosen. On Gravitational waves. *J. Franklin Inst.*, 223: 43–54, 1937.
[4] Cecile M. De Witt, editor. *Proceedings: Conference on the Role of Gravitation in Physics, Chapel Hill, North Carolina, Jan 18-23, 1957*, 1957.
[5] Peter R. Saulson. Josh Goldberg and the physical reality of gravitational waves. *Gen. Rel. Grav.*, 43:3289–3299, 2011.
[6] J. Weber. Detection and Generation of Gravitational Waves. *Phys. Rev.*, 117:306–313, 1960.
[7] J. Weber. Evidence for discovery of gravitational radiation. *Phys. Rev. Lett.*, 22:1320–1324, 1969.
[8] R. A. Hulse and J. H. Taylor. Discovery of a pulsar in a binary system. *Astrophys. J.*, 195:L51–L53, 1975.
[9] E. Coccia, G. Pizzella, and F. Ronga, editors. *Gravitational wave experiments. Proceedings, 1st Edoardo Amaldi Conference, Frascati, Italy, June 14-17, 1994*, 1995.
[10] B. P. Abbott et al. Tests of general relativity with GW150914. *Phys. Rev. Lett.*, 116(22):221101, 2016.
[11] Bernard F. Schutz. Determining the Hubble Constant from Gravitational Wave Observations. *Nature*, 323:310–311, 1986.
[12] Samaya Nissanke, Daniel E. Holz, Scott A. Hughes, Neal Dalal, and Jonathan L. Sievers. Exploring short gamma-ray bursts as gravitational-wave standard sirens. *Astrophys. J.*, 725:496–514, 2010.
[13] M. G. Aartsen et al. First observation of PeV-energy neutrinos with Ice-Cube. *Phys. Rev. Lett.*, 111:021103, 2013.
[14] M. G. Aartsen et al. Evidence for High-Energy Extraterrestrial Neutrinos at the IceCube Detector. *Science*, 342:1242856, 2013.
[15] J. Aasi et al. Advanced LIGO. *Class. Quant. Grav.*, 32:074001, 2015.
[16] F. Acernese et al. Advanced Virgo: a second-generation interferometric gravitational wave detector. *Class. Quant. Grav.*, 32(2):024001, 2015.
[17] Eiichi Hirose, Dan Bajuk, GariLynn Billingsley, Takaaki Kajita, Bob Kestner, Norikatsu Mio, Masatake Ohashi, Bill Reichman, Hiroaki Yamamoto, and Liyuan Zhang. Sapphire mirror for the KAGRA gravitational wave detector. *Phys. Rev.*, D89(6):062003, 2014.
[18] B. Sathyaprakash et al. Scientific Objectives of Einstein Telescope. *Class. Quant. Grav.*, 29:124013, 2012. [Erratum: Class. Quant. Grav. 30,079501(2013)].
[19] Heather Audley et al. Laser Interferometer Space Antenna. 2017.
[20] M. Armano et al. Charge-induced force-noise on free-falling test masses: results from LISA Pathfinder. *Phys. Rev. Lett.*, 118(17):171101, 2017.

[21] G. Hobbs et al. The international pulsar timing array project: using pulsars as a gravitational wave detector. *Class. Quant. Grav.*, 27:084013, 2010.

[22] Z. Arzoumanian et al. The NANOGrav Nine-year Data Set: Limits on the Isotropic Stochastic Gravitational Wave Background. *Astrophys. J.*, 821(1):13, 2016.

PRIMORDIAL BLACK HOLES AND COSMOLOGICAL PROBLEMS

Alexander Dolgov [a]
Novosibirsk State University, Novosibirsk, 630090, Russia
ITEP, Moscow, 117218 Russia

Abstract. It is argued that the bulk of black holes (BH) in the universe are primordial (PBH). This assertion is strongly supported by the recent astronomical observations, which allow to conclude that supermassive BHs with $M = (10^6 - 10^9) M_\odot$ "work" as seeds for galaxy formation, intermediate mass BHs, $M = (10^3 - 10^4) M_\odot$, do the same job for globular clusters and dwarf galaxies, while black holes of a few solar masses are the constituents of dark matter of the universe. The mechanism of PBH formation, suggested in 1993, which predicted such features of the universe, is described. The model leads to the log-normal mass spectrum of PBHs, which is determined by three constant parameters. With proper adjustment of these parameters the above mentioned features are quantitatively explained. In particular, the calculated density of numerous superheavy BHs in the young universe, $z = 5 - 10$, nicely fits the data. The puzzling properties of the sources of the LIGO-discovered gravitational waves are also naturally explained assuming that these sources are PBHs.

1 Introduction

Recent, and not only recent, astronomical observations revealed many mysterious features of the universe, which were not expected in frameworks of conventional cosmology and astrophysics. All these problems are neatly solved if practically all black holes (BH) in the universe are primordial ones with a wide spread mass spectrum. Primordial black holes by definition are those which were formed in the universe at prestellar epoch, i.e. before stars appeared in sky. The mechanism of their formation was suggested by Zeldovich and Novikov (ZN) [1]. According to them, PBH was formed if the density fluctuation in the early universe was of order unity, $\delta\rho/\rho \sim 1$, at the cosmological horizon scale. In this case the piece of space happened to be inside its gravitational radius, so it decoupled from the overall Hubble expansion and a black hole appeared. In the original version PBH created by such mechanism were rather light (a small fraction of the solar mass) and had narrow (delta-function) mass spectrum. At least such form of the spectrum was mostly assumed in subsequent analysis of observational manifestation of such PBHs.

In 1993 [2], a generalization of ZN mechanism was proposed (see also later work [3]), which could lead to very massive PBH with log-normal mass spectrum:

$$\frac{dN}{dM} = \mu^2 \exp\left[-\gamma \ln^2(M/M_0)\right], \quad (1)$$

---
[a] E-mail: dolgov@fe.infn.it

with only 3 parameters: μ, γ, M_0. The form of the spectrum is practically universal. It is completely determined by the exponential cosmological expansion during inflationary stage.

Fitting the parameters of distribution (1) one can explain the accumulated astronomical data about black holes in contemporary and young universe. In this sense the surprising results of the new precise observations performed in the recent decade are *predicted* in the papers [2, 3].

In this talk I briefly review the following observational data.
Young universe, $z \approx 5 - 10$, overpopulated by:

1. Bright quasi stellar objects (QSO), super-massive BHs.
2. Superluminous young galaxies.
3. Supernovae and gamma-bursters.
4. Very high level of dust.

Contemporary universe:
1. Supermassive BH (SMBH) in every large galaxy.
2. SMBH in small galaxies and in almost *empty* space.
3. Stars older than the Galaxy and even older than the Universe.
4. MACHOs (low luminosity solar mass objects).
5. Problems with the BH mass spectrum in the Galaxy: unexpected maximum at $M \sim 8 M_\odot$.
6. Problems with the sources of the observed gravitational waves (GW).
7. Intermediate mass, $\geq 10^3 M_\odot$, BHs in globular clusters and dwarf galaxies.

More details and references can be found in [4, 5].

2 Young universe

2.1 SMBH

About 40 quasars with $z > 6$ are already known, each quasar containing SBH with $M \sim 10^9 M_\odot$. The maximum redshift value among these quasars reaches $z = 7.085$ i.e. the quasar was formed before the universe reached 0.75 Gyr. Its luminosity and mass are respectively $L = 6.3 \cdot 10^{13} L_\odot$ and $M = 2 \cdot 10^9 M_\odot$ [6]. Other high z quasars have similar properties. The formation of SMBHs, which fuel these quasars through the standard accretion mechanism, demands much more time than the universe age at $z \sim 6$. The unsolvable problem with creation of these SMBHs was multiply deepened with the discovery of a real "monster" of 12 billions solar masses [7], i.e. an order of magnitude more massive, than the mentioned above forty. Even the formation of the contemporary SMBH, which had in their disposal the whole universe age, 14 Gyr, is difficult to explain, see the next section.

After this Conference was over, a new discovery of a SMBH at now new maximum redshift $z \approx 7.5$ and the mass 0.8 billion solar masses was announced [8].

This is not the largest mass in the family of high redshift quasars, but what makes it particularly interesting is that the surrounding matter is neutral, not ionized. This is a very strong argument against formation of this SMBH by the usual accretion process, so its primordial origin remains the only natural possibility.

2.2 Early bright galaxies

Several galaxies have been observed at high redshifts, with natural gravitational lens "telescopes". A few examples are:
1) A galaxy at $z \approx 9.6$ which was created when the universe was younger than 0.5 Gyr [9].
2) A galaxy at $z \approx 11$ [10] which already existed when the universe age was $t_U \sim 0.4$ Gyr. It is particularly impressive that this very young galaxy is three times more luminous in UV than other galaxies at $z = 6 - 8$. This is a striking example of unexpectedly early burn and powerfull creature.
3) Not so young but extremely luminous galaxy was found three years ago. Its luminosity reaches gigantic magnitude, $L = 3 \cdot 10^{14} L_\odot$. The universe age when the galaxy already existed was $t_U \sim 1.3$ Gyr. According to the authors of the discovery: "The new study outlines three reasons why the black holes in the extremely luminous infrared galaxies, could have grown so massive. First, they may have been born big. In other words, the galactic seeds, or embryonic black holes, might be bigger than thought possible." One of the authors, P. Eisenhardt said: "How do you get an elephant? One way is start with a baby elephant." The BH was already billions of $M_\odot$, when our universe was only a tenth of its present age of 13.8 billion years. "Another way to grow this big is to have gone on a sustained binge, consuming food faster than typically thought possible." For the realization of these conditions low spin is necessary!

According to the paper "Monsters in the Dark" [11] density of galaxies at $z \approx 11$ is 10^{-6} Mpc^{-3}, an order of magnitude higher than estimated from the data at lower z. Origin of these galaxies is unclear.

These data strongly support the idea that initially primordial SMBHs appeared and later galaxies were seeded by these PBHs. To the best of my knowledge this idea was first pronounced in Ref. [2] and the recent observations do confirm the early creation of very massive black holes.

2.3 Early miscellanea

The universe at $z = 5 - 10$ was filled with supernovae, gamma-bursters, and was very dusty. To make dust a long succession of events is necessary: first, supernovae exploded to deliver heavy elements into space (metals), then metals cool and form molecules, and lastly molecules make macroscopic pieces of matter. Abundant dust is observed in several early galaxies, e.g. in HFLS3 at

$z = 6.34$ [12] and in A1689-zD1 [13] at $z = 7.55$. The second galaxy is the earliest one where interstellar medium is observed The universe age at this redshift is below 0.5 Gyr.

Catalogue of the observed dusty sources [14] indicates that their number is an order of magnitude larger than predicted by the canonical theory of galaxy evolution.

Hence, prior to or simultaneously with the QSO formation a rapid star formation should take place. These stars should evolve to a large number of supernovae enriching interstellar space by metals through their explosions which later make molecules and dust. (We all are dust from SN explosions, but probably at much later time.)

Observations of high redshift gamma ray bursters (GBR) also indicate a high abundance of supernova at large redshifts. The highest redshift of the observed GBR is 9.4 and there are a few more GBRs with smaller but still high redshifts. The necessary star formation rate for explanation of these early GBRs is at odds with the canonical star formation theory.

3 Mysteries in the sky today and in the nearest past

3.1 Supermassive black holes

Every large galaxy and some smaller ones contain a central supermassive BH with mass typically larger than $10^9 M_\odot$ in giant elliptical and compact lenticular galaxies, and $\sim 10^6 M_\odot$ in spiral galaxies like Milky Way. The origin of these BHs is unclear. The accepted faith is that these BHs are created by the matter accretion to galactic center with an excessive mass density. However, the usual accretion efficiency is insufficient to create them during the Universe life-time, $t_U \approx 14$ Gyr. Even more puzzling is that SMHBs are observed in small galaxies and even in almost empty space, where no material to make a SMBH can be found.

Below several examples are presented demonstrating serious inconsistencies between observation and theoretical picture.

The mass of BH is typically 0.1% of the mass of the stellar bulge of galaxy but some galaxies may have huge BH: e.g. NGC 1277 has the central BH of $1.7 \times 10^{10} M_\odot$, or 60% of its bulge mass [15]. This creates serious problems for the standard scenario of formation of central supermassive BHs by accretion of matter in the central part of a galaxy.

According to Ref. [16] the galaxies, Henize 2-10, NGC 4889, and NGC1277 are examples of SMBHs at least an order of magnitude more massive than their host galaxy suggests. The dynamical effects of such ultramassive central black holes are unclear.

A recent discovery [18] of an ultra-compact dwarf galaxy older than 10 Gyr, enriched with metals, and probably with a massive black hole in its center also seems to be at odds with the standard model.

In the paper entitled "An evolutionary missing link? A modest-mass early-type galaxy hosting an over-sized nuclear black hole" [17], a black hole with the mass $M_{BH} = (3.5 \pm 0.8) \cdot 10^8 M_\odot$, is found inside the host galaxy with mass of the stars $M_{stars} = 2.5^{+2.5}_{-1.2} \cdot 10^{10} M_\odot$, and huge accretion luminosity: $L_{AGN} = (5.3 \pm 0.4) \cdot 10^{45}$ erg/s $\approx 10^{12} L_\odot$, equal to 12% of the Eddington luminosity. The active galactic nuclei (AGN) is more prominent than expected for a host galaxy of this modest size. The data are in tension with the accepted picture in which this galaxy would recently have transformed from a star-forming disc galaxy into an early-type, passively evolving galaxy.

Probably the most impressive in this list is a discovery of "A Nearly Naked Supermassive Black Hole" [19]. According to the paper, a compact symmetric radio source B3 1715+425 is too bright (brightness temperature $\sim 3 \times 10^{10}$ K at observing frequency 7.6 GHz) and too luminous (1.4 GHz luminosity $\sim 10^{25}$ W/Hz) to be powered by anything but a SMBH, but its host galaxy is much smaller.

There are more example of such puzzling galaxies with superheavy black holes but even with the presented ones the inverted picture of galaxy formation looks more plausible, when first a supermassive black hole was formed and later it attracted matter serving as a seed for subsequent galaxy formation.

3.2 MACHOs

MACHO is the name of some invisible or low luminosity objects discovered through gravitational microlensing by Macho [20, 21] and Eros [22] groups respectively in the Galactic halo and in the direction to the center of the Galaxy. Later they were registered in the Andromeda (M31) galaxy [23]. The masses of the registered objects are about one half of the solar mass. The up to date situation with MACHOs is summarized in Ref. [24]:

Macho group: $0.08 < f < 0.50$ (95% CL) for $0.15 M_\odot < M < 0.9 M_\odot$;
EROS: $f < 0.2$, $0.15 M_\odot < M < 0.9 M_\odot$;
EROS2: $f < 0.1$, $10^{-6} M_\odot < M < M_\odot$;
AGAPE: $0.2 < f < 0.9$, for $0.15 M_\odot < M < 0.9 M_\odot$;
EROS-2 and OGLE: $f < 0.1$ for $M \sim 10^{-2} M_\odot$ and $f < 0.2$ for $\sim 0.5 M_\odot$.

Thus, the MACHO density is comparable to the density of the halo dark matter but their nature is unknown. They could be brown dwarfs, dead stars, or primordial black holes. The first two options are in conflict with the accepted theory of stellar evolution, if such invisible stars were created in the conventional way.

The only remaining option is that MACHOs are low mass black holes, but one can hardly imagine that such low mass black holes, abundant in the Galactic halo, were created as a result of stellar collapse of normal stars. So the natural conclusion is that MACHOs are primordial black holes as it is stated in Ref. [3]. The log-normal spectrum of the PBH allows to make much larger contribution to DM from heavier PBH, to which the microlensing method is not sensitive. So ultimately 100% of DM may be made out of PBHs with different masses.

3.3 Properties of the sources of gravitational waves

Direct registration of gravitational waves (GW) by LIGO [25] revealed intriguing properties of the GW sources [26]. The shape of the signal in the interferometer is well described by the assumption that the observed GWs are produced by the binary of coalescing BHs, but:

1. The origin of heavy BHs with masses $\sim 30 M_\odot$ is unclear. Such BHs are believed to be created by massive star collapse, though a convincing theory is still lacking. To form so heavy BHs, the progenitors should have $M > 100 M_\odot$ and a low metal abundance to avoid too much mass loss during the evolution. Such heavy stars might be present in young star-forming galaxies but they are not yet observed in sufficiently high number.

2. In all events, but one, the spins of the coalescing BHs are very small, compatible with zero. It strongly constrains astrophysical BH formation from close binary systems. However, the dynamical formation of double massive low-spin BHs in dense stellar clusters is not excluded, but difficult.

3. Formation of BH binaries from the original stellar binaries has very low probability. Stellar binaries were formed from common interstellar gas clouds and are quite frequent in galaxies. If BH is created through stellar collapse, a small non-sphericity of the collapse results in a huge velocity of the BH and the binary is destroyed. BH formation from PopIII stars and subsequent formation of BH binaries with $(36 + 29) M_\odot$ is analyzed and found to be negligible.

All these problems are solved if the observed sources of GWs are the binaries of primordial black holes (PBH).

3.4 Globular clusters and intermediate mass BHs.

Recently the so called intermediate mass black holes (IMBH) with masses $M \approx 2000 M_\odot$ and $M \sim 20000 M_\odot$ were presumably observed in the centers of globular clusters [27, 28]. These observations nicely fit our conjecture [29] that IMBH play an important role in the formation and evolution of globular clusters. Using the parameters of the mass distribution (1), found in our paper [26], we find that the density of the primordial IMBH is sufficient to seed the formation of all globular clusters observed in galaxies.

In addition to globular clusters, IMBHs are probably also contained in centers of dark stellar clusters [30,31]. These clusters have high mass-luminosity ratio. They may be the remnants of dwarf spheroids with the masses between those of globular clusters and large galaxies. It looks natural that these spheroids were seeded by the primordial IMBH [27].

3.5 Solar mass Black holes in the Milky Way

The mass spectrum of black holes observed in the Galaxy demonstrates some peculiar features, which are difficult to explain in the standard model of BH formation by stellar collapse. In particular, it is found [32] that the masses of black holes in the Galaxy are concentrated in the narrow range $(7.8 \pm 1.2)M_\odot$. This result agrees with another paper where a peak around $8M_\odot$, a paucity of sources with masses below $5M_\odot$, and a sharp drop-off above $10M_\odot$ are observed [33].

On the other hand, such mass spectrum is well described by the log-normal form. This is an argument in favor of primordial origin of the black holes in the Galaxy.

3.6 Old stars in the Milky Way

Recently several groups presented substantially more accurate determinations of stellar ages in the Galaxy. Surprisingly quite a few stars happened to be considerably older than expected.

According to Ref. [34]: employing thorium and uranium abundances in comparison with each other and with several stable elements the age of metal-poor, halo star BD+17° 3248 was estimated as 13.8 ± 4 Gyr. This star is much older than the inner halo of the Galaxy, which has the age equal to 11.4 ± 0.7 Gyr [35].

The age of another star in the galactic halo, HE 1523-0901, was estimated to be about 13.2 Gyr [36]. First time many different chronometers, such as the U/Th, U/Ir, Th/Eu, and Th/Os ratios to measure the star age, have been employed.

And at last a star older than the universe was found [37]. Metal deficient high velocity subgiant in the solar neighborhood HD 140283 has the age 14.46 ± 0.31 Gyr. The central value exceeds the universe age by two standard deviations, if $H = 67.3$, and $t_U = 13.8$ Gyr; while if $H = 74$, and $t_U = 12.5$ Gyr, the star would be older than the universe by more than 10 σ. This is of course impossible, but the star may look older that it is, if initially the star was enriched by heavy elements and evolve to its present state faster than the normal one.

Our model of PBH formation [2,3] leads also to creation of compact primordial stellar-like objects consisting not only from hydrogen and helium but enriched with plenty of heavier elements.

4 Mechanism of massive PBH formation

In this Section the main features of the mechanism [2,3] of massive PBH formation are described. We assume that a slightly modified baryogenesis scenario suggested by Affleck and Dine (AD) [38] is realized. The main ingredient of this AD-scenario is a scalar field χ with non-zero baryonic number B. It is assumed that the potential of χ has the so called flat directions along which the potential does not rise. In the course of the cosmological expansion χ might acquire large expectation value turning practically into a classical field with large B. Later after decay of χ this accumulated baryonic number turned into baryonic number of quarks, leading to a large baryon asymmetry β of the universe. It may be even of order unity, while the observed value of β is about 10^{-9},

We modified the AD mechanism by introduction of general renormalizable coupling of χ to the inflaton field Φ (the first term in the r.h.s. of the equation below), which can be written in the form:

$$U = g|\chi|^2(\Phi - \Phi_1)^2 + \lambda|\chi|^4 \ln\left(\frac{|\chi|^2}{\sigma^2}\right) + \lambda_1\left(\chi^4 + h.c.\right) + (m^2\chi^2 + h.c.). \quad (2)$$

With this interaction the flat direction of the potential U are open only when $\Phi \approx \Phi_1$, which was taken by Φ in the course of inflation before it was over. If the window to flat direction, when $\Phi \approx \Phi_1$ is open only during a short period, cosmologically small but possibly astronomically large bubbles with high β could be created, These bubbles with large β might occupy only a small fraction of the universe volume, while the rest of the universe would have the normal small baryon-to-photon ratio $\beta \approx 6 \cdot 10^{-10}$, created by small χ, which did not succeed to penetrate through the briefly open window to a large value.

After the QCD phase transition, when massless quarks turned into heavy nucleons, the initial isocurvature perturbation created by inhomogeneities in the chemical content turned into (large) density perturbations. This would lead to an early formation of PBH or compact stellar-type objects with high baryonic density. As a result, the bulk of baryons and maybe antibaryons would be contained in compact cosmologically tiny stellar-like objects or PBH. These high-B density bubbles would live in huge by size but not so dense universe with low baryonic background density, which initially was practically homogenenous.

The formation of PBHs or compact stellar type objects took place at very high z after the QCD phase transition at $T \sim 100$ MeV down to $T \sim$ keV.

As a byproduct, the mechanism of Refs. [2,3] may lead, though not necessarily, to abundant compact antimatter objects in the universe and, in particular, in the Galaxy [24,39,40].

5 Conclusion

The problems emerged from the multitude of astronomical observations, some of which are mentioned in this talk, are uniquely and simply resolved if the universe is populated by the primordial massive black holes and stellar-like compact objects with wide mass spectrum. The mechanism which leads to an abundance of such objects in the universe was put forward in 1993 [2,3] and essentially predicted the subsequent surprising discoveries.

All the multitude of the various astronomical data are well explained by the natural baryogenesis model which leads to formation of PBHs and compact stellar-like objects in the early universe after the QCD phase transition, $t \leq 10^{-5}$ sec. These objects are predicted to have log-normal mass spectrum. They can be numerous enough to give significant contribution to the cosmological dark matter or even make all of it.

The model opens the possibility for the inverted picture of the galaxy formation, when firstly supermassive black holes are formed which later accrete matter creating galaxies. The new observations persuasively indicate in this direction. Lighter PBHs with 2000 $M_\odot$ are predicted in sufficient amount to explain the origin of globular clusters, while heavier PBHs, with $M \sim 10^4 M_\odot$ can seed formation of dwarf spheroids. There seem to be strong indications in favor of this scenario.

PBHs formed through such mechanism can explain the peculiar features of the sources of GWs observed by LIGO.

The considered mechanism resolves the numerous mysteries of $z \sim 10$ universe: abundant population of supermassive black holes, early created gamma-bursters and supernovae, early bright galaxies, and evolved chemistry including dust.

Existence of high density invisible "stars" (MACHOs) is explained.

"Older than t_U" stars may exist. The old age is mimicked by the unusual initial chemistry.

The model can possibly lead to the prediction of numerous compact antimatter objects (antistars). The observational data allow for large amount of such objects in the Galaxy. However, their density is model dependent and the prediction is uncertain.

Acknowledgments

This work was supported by RSF Grant No. 16-12-10037.

References

[1] Ya.B. Zeldovich, I.D. Novikov, Sov. Astron. 10, 602 (1967).

[2] A. Dolgov, J. Silk, *Phys. Rev.*, **D47** (1993), 4244.
[3] A.D. Dolgov, M. Kawasaki, N. Kevlishvii, *Nucl. Phys.*, **B807** (2009), 229.
[4] A.D. Dolgov, e-Print: arXiv:1605.06749.
[5] A.D. Dolgov, e-Print: arXiv:1701.05774.
[6] D.J. Mortlock, *et al*, Nature 474 (2011) 616; arXiv:1106.6088.
[7] Xue-Bing Wu *et al*, Nature 518, 512 (2015).
[8] E. Bañados, et al, arXiv:1712.01860v1.
[9] W. Zheng, *et al*, *Nature*, **489** (2012) 406; arXiv:1204.2305.
[10] D. Coe *et al* Astrophys. J. 762 (2013) 32.
[11] D. Waters, et al, Mon. Not. Roy. Astron. Soc. 461 (2016), L51.
[12] Clements, D.L., *et al*, arXiv:1511.03060.
[13] Mattsson T., arXiv:1505.04758.
[14] Asboth V., et al, arXiv:1601.02665.
[15] Bosch van den R. C. E., et al., *Nature*, **491** (2012), 729.
[16] F. Khan, K. Holley-Bockelmann, P. Berczik arXiv:1405.6425.
[17] J. Th. van Loon, A.E. Sansom, arXiv:1508.00698v1.
[18] J. Strader, *et al* Ap. J. Lett. 775, L6 (2013).
[19] J.J. Condon, et al arXiv:1606.04067.
[20] Alcock A., *et al*, *Astrophys. J.*, **542** (2000), 281.
[21] Bennett D.P., *Astrophys. J.* **633** (2005), 906.
[22] Tisserand P., et al, *A&A*, **469** (2007), 387.
[23] Riffeser A., Seitz S., Bender R., *Astrophys. J.*, **684** (2009), 1093.
[24] S.I. Blinnikov, A.D. Dolgov, K.A. Postnov, *Phys. Rev.*, **D92** (2015), 023516.
[25] Abbott B.P., et al., *Phys. Rev. Lett.*, **116** (2016), 061102.
[26] S. Blinnikov, *et al*, *JCAP*, **1611** (2016), 036.
[27] Kiziltan B., Baumgardt H., Loeb A., *Nature*, **542** (2017,) 203
[28] Perera B.B.P., et al., arXiv:1705.01612.
[29] A. Dolgov, K. Postnov, *JCAP*, **1704** (2017), 036.
[30] M. A. Taylor M.A., et al., *Astrophys. J.*, **805** (2015), 65, [1503.04198].
[31] M. S. Bovill, M.S., et al., *Astrophys. J.*, **832** (2016), 88; [1608.06957].
[32] F. Ozel F., *et al.*, arXiv:1006.2834.
[33] L. Kreidberg, *et al*, arXiv:1205.1805.
[34] J.J. Cowan, *et al* Ap. J. 572 (2002) 861.
[35] J. Kalirai, Nature 486 (2012) 90, arXiv:1205.6802.
[36] A. Frebe, *et al* Astrophys. J. 660 (2007) L117.
[37] H. E. Bond, et al, Astrophys. J. Lett. 765, L12 (2013), arXiv:1302.3180
[38] Affleck I., Dine M., *Nucl. Phys.* **B 249** (1985), 361.
[39] C. Bambi, A. D. Dolgov, Nucl. Phys. B 784, 132 (2007).
[40] A. D. Dolgov, S. I. Blinnikov, Phys. Rev. D 89, 021301(R) (2014).

SOLAR MODULATION OF GALACTIC COSMIC RAYS: PHYSICS CHALLENGES FOR AMS-02

Nicola Tomassetti [a]

Università degli Studi di Perugia & INFN-Perugia, I-06100 Perugia, Italy

Abstract. The Alpha Magnetic Spectrometer (AMS) is a new generation high-energy physics experiment installed on the International Space Station in May 2011 and operating continuously since then. Using an unprecedently large collection of particles and antiparticles detected in space, AMS is performing precision measurements of cosmic ray energy spectra and composition. In this paper, we discuss the physics of solar modulation in Galactic cosmic rays that can be investigated with AMS my means of dedicated measurements on the time-dependence of cosmic-ray proton, helium, electron and positron fluxes.

1 Introduction

The Alpha Magnetic Spectrometer (AMS) is a state-of-the-art particle physics experiment operating on the International Space Station (ISS) since May 2011. In the first 6 years of missions, AMS has detected over 100 billion cosmic ray (CR) particles. Recently, it has released new results on proton, antiproton, lepton, and nuclei energy spectra at unexplored energies and with an unmatched level of accuracy [1, 2, 4, 3, 5]. A collection of these data is available on the ASI/SSDC-CR database [6]. The long duration of mission, planned to last for the whole ISS lifetime, will cover a complete solar cycle from the ascending phase of cycle 24, through its maximum, and the descending phase into the next solar minimum. This makes AMS an excellent multichannel CR monitor of solar activity [7, 8]. Precision measurements of the CR time evolution, in connection with the changing solar activity, may give us strong insight on the so-called *solar modulation* effect.

Solar modulation is a time-, space-, energy-, and particle-dependent phenomenon that arises from basic transport processes of CRs in the heliosphere [9]. Along with its connection with solar and CR physics, understanding CR modulation addresses a prerequisite for modeling space weather effects, which is an increasing concern for space missions and air travelers. The study of these effects has been limited for long time by the scarcity of long-term CR data on different species, and by the poor knowledge of the local interstellar spectra (LIS). The very first LIS measurements on the CR fluxes have been recently provided by Voyager-1 probe in the interstellar space. A continuous stream of time-resolved and multichannel CR data is being provided by the AMS on the ISS. In light of these milestones, new objectives of low-energy physics investigation can be devised: (i) to advance solar modulation observations of

---

[a] E-mail: nicola.tomassetti@cern.ch

CR particles and antiparticles, and (ii) to develop improved and measurement-validated models of CR transport in the heliosphere.

2 Physics challenges in cosmic-ray modulation

We now discuss some important physics topics that can be investigated with dedicated analysis of the AMS data on CR modulation. The first one deals with puzzling anomalies detected in the energy spectra of CR proton and helium nuclei. These spectra are found to harden at rigidity $R = pc/Ze \gtrsim 200$ GV, while their p/He ratio as function of rigidity falls off steadily as p/He $\propto R^{-0.08}$ [3]. The decrease of the p/He ratio is usually interpreted in terms of particle-dependent acceleration, but this interpretation is in contrast with the *universal* nature of diffusive-shock-acceleration mechanisms, *i.e.*, the conception that CRs are injected in the Galaxy by composition-blind rigidity-dependent accelerators. A model based on two-source components has been proposed in [10] and further investigated in [11]. In this model, the p/He anomaly is explained by a flux transition between two source components (*L*-type and *G*-type) having by different injection spectra and composition. The universality of particle acceleration is not violated in this model, since each class of source is assumed to provide elemental-independent acceleration spectra. A possible signature of this scenario is a progressive flattening of the p/He ratio at multi-TeV energies,

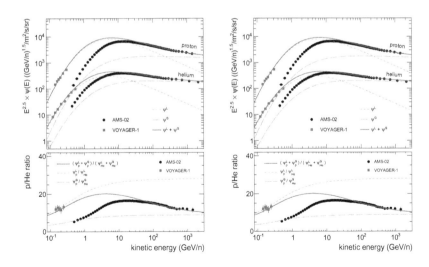

Figure 1: Proton and helium energy spectra (top) and p/He ratio (bottom) from [12]. The calculations (solid lines) are shown in comparisons with the AMS and Voyager-1 data [4,3,13]. The dashed lines are the single source components. Calculations are shown for LIS (left) and modulated (right) fluxes. Solar modulation uncertainties are shown as shaded bands.

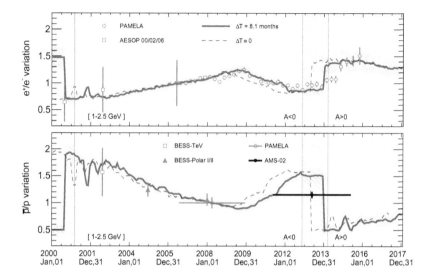

Figure 2: Time profile of the ratios e^+/e^- (top) and $\bar{p}/p$ (bottom) at $E = 1 - 2.5\,\text{GeV}$ [14]. Model predictions and their corresponding uncertainties are shown in comparisons with the data [15, 1, 16]. The shaded bars indicate the magnetic reversals of the Sun's polarity.

i.e., where the G-type source dominates the CR flux. Such a signature is hinted at by the recent data, but the situation is unclear. Interestingly, a complementary observational test can be done in the sub-GeV energy regime, *i.e.*, where the L-type source is expected to dominate the flux. However, in this energy window, the CR fluxes are significantly affected by the time-dependent effect of solar modulation, hence a careful modeling of the CR transport in the heliosphere is essential. In Fig. 1, from [10], it is shown that the $\bar{p}$-Hedata reported by AMSand Voyager-1 are in good agreement with the model predictions, once the modulation effect is accounted, therefore supporting the universality of CR acceleration mechanism. However, this study was limited by simple model approach and by the lack of time-dependent measurements on CR in the period of investigation. These measurements are being carried out by AMS over its first 6 years of mission. This can provide a conclusive observational test for such a scenario of CR origin and propagation.

Another important topic is the recent observation of a eight-month *time lag* in solar modulation of CR [14]. This effect reveals important properties on the dynamics of the formation and changing conditions of the heliospheric plasma. While the analysis of [14] is based on CR proton, the key observational tests for the model are given for the evolution of CR antimatter/matter ratios, shown in Fig. 2. Crucial tests can be performed by AMS via monthly-resolved measurements of these ratios, or even better, by measurements of individual particle fluxes for p, $\bar{p}$, e^+, and e^- under both polarity conditions and across the magnetic reversal. This demonstrates that time-dependent measurements on CR antimatter can provide precious information on the physics of the heliosphere.

Understanding charge-sign dependence of CR modulation is also essential to search for dark matter signatures in CR antimatter fluxes. Other problems related to uncertainties in solar modulation are studied in recent works [7, 17, 18]. Without dedicated data on the time-dependence of the CR flux, solar modulation models show degeneracies with the CR propagation parameters and fall short of predicting the level of astrophysical antimatter background.

3 Conclusions

We have briefly discussed some physics challenges, in modern astrophysics, that will benefits from a dedicated *multichannel investigation of solar modulation effects in Galactic CRs*: the origin of the anomalies in the spectra of CR nuclei, the dynamics of the changing conditions of the heliospheric plasma, the parameters constraints in astrophysical models of CR propagations, the origin of CR antimatter and its connection with the dark matter puzzle. In order to develop reliable and data-driven models of CR modulation, however, the availability of time-resolved measurements over the period of interest is crucial. In this respect, monthly-resolved data from AMS will be very precious.

Acknowledgments

The European Commission is acknowledged for support under the H2020-MSCA-IF-2015 action, grant N. 707543- MAtISSE, *Multichannel Investigation of Solar Modulation Effects in Galactic Cosmic Rays*.

References

[1] M. Aguilar et al., Phys. Rev. Lett. 117 (2016) 091103.
[2] M. Aguilar et al., Phys. Rev. Lett. 117 (2016) 231102.
[3] M. Aguilar et al., Phys. Rev. Lett. 115 (2015) 211101.
[4] M. Aguilar et al., Phys. Rev. Lett. 114 (2015) 171103.
[5] M. Aguilar et al., Phys. Rev. Lett. 110 (2013) 141102.
[6] ASI/SSDC CR database: https://tools.asdc.asi.it/CosmicRays.
[7] V. Bindi et al., Adv. Space Res. 60 (2017) 865-878.
[8] S. Della Torre, Proc. XXV ECRS (2016) [arXiv:1612.08441].
[9] M. S. Potgieter, Living Rev. Sol. Phys. 10 (2013) 3.
[10] N. Tomassetti, ApJ Lett. 815 (2015) L1.
[11] Y. Zhang et al., ApJ Lett. 844 (2017) L3.
[12] N. Tomassetti, Adv. Space Res. 60 (2017) 815-825.

[13] A. C. Cummings et al., ApJ 831 (2016) 18.
[14] N. Tomassetti et al., ApJ Lett. 849 (2017) L32.
[15] O. Adriani et al., Phys. Rev. Lett 116 (2016) 241105.
[16] K. Abe et al., Phys. Rev. Lett. 108 (2012) 051102.
[17] N. Tomassetti, Astrophys. Space Sci. 342 (2012) 131-136.
[18] N. Tomassetti, Phys. Rev. D 96 (2017) 103005.

CRITICAL INDICES AND LIMITS ON SPACE-TIME DIMENSIONS FROM COSMIC RAYS

Y. Srivastava[a], A. Widom, J. Swain
Physics Department, Northeastern University, Boston, MASS USA

Abstract. Two major results [increase with energy of total hadronic cross-sections & power law spectrum in energy of produced nuclei] -for which high energy cosmic ray data have been precursors- shall be reviewed. First result is that the highest energy cross-section data from cosmic rays are consistent with the Froissart bound and with extrapolations from LHC and lower accelerator energies. We show that this puts very stringent constraints on the allowed dimension (D) of space-time, viz. $D = 4$. The second result follows from the cosmic particle energy spectrum showing isotropy & stable critical indices. Their explanation in terms of thermal field theory where particles evaporate from a hot compact source as in Landau-Fermi liquid theory is quite satisfactory for both bosonic and fermionic critical indices, in particular for the very precise data for (e^-e^+) spectrum. A quantum phase transition is found that explains the onset and the development of the knee in the cosmic spectrum. Estimates of energy losses show that for electrons and positrons of energy below 1 TeV, they are minuscule and thus the predicted critical indices remain robust and of fundamental importance. Recently, the DAMPE collaboration has confirmed a break in the (e^-e^+) spectrum near 1 TeV. We conclude with work in progress regarding the expected particle composition at far asymptotic energies (most likely helium).

1 Introduction

There are two central results in this presentation:

1. The dimensionality $D = 4$ of space-time is phenomenologically robust [see, Sec.(2)] until $\sqrt{s} \leq 100\ TeV$ [1,2].

2. Observed power law fall off of cosmic ray energy spectrum [*Critical Indices*] can be theoretically understood in terms of thermal field theory [3–7]. We shall be brief as some of the results have been presented at IHEP in Chicago [8] and at the 17th Lomonosov in Moscow [9]. Given the paucity of space, we shall concentrate on (i) the radiation losses, for which we have new compact, manifestly relativistic, expressions [see, Sec.(3)] and (ii) about the composition of very high energy cosmic rays [see, Sec.(4)].

2 Is $D = 4$?

For $D = 3 + 1$, there are rigorous proofs that all hadronic cross-sections are [Froissart-Martin] bounded by $\sigma_{tot}(s) \leq (\frac{\pi}{m_\pi^2})[ln(s/s_o)]^2$. Highest energy cross-section data from cosmic rays are consistent with the Froissart bound and with extrapolations from LHC and lower accelerator energies.

---
[a]E-mail: yogendra.srivastava@gmail.com

Readers should be reminded that the above asymptotic behavior is crucially dependent on space-time dimension D [11]. For instance, from D-dimensional string thermodynamics, one finds that both for fermionic and bosonic strings, the total cross-section increases as a power law in s [1], violating the Froissart bound and available accelerator and cosmic ray data [2].

3 Radiation losses and Energy spectrum of cosmic rays

For fermions, we had predicted [3, 4] that the critical power law index would be $\alpha_{Fermi}(theory) = 3.151$. Recently, AMS Collaboration has measured the energy distribution of combined electrons & positrons: $\alpha(experiment) = 3.17 \pm 0.08 \pm 0.08$ for $e^{\mp}$ in the energy interval 30 $GeV < E < 1$ TeV (for a critical discussion see [6]).

We have investigated conditions under which energy losses due to EM radiation would not change the observed energy power law spectrum from the critical indices at the source [7]. Our analysis is of particular relevance for the AMS electron/positron data as energy losses for these are (potentially) the largest. We find that for $E < 1$ TeV, the measured index remains the same and reflects the index at the source. Hence, AMS data are quite robust and of fundamental importance.

An exact, new classical covariant result for the damping force $\mathcal{F}^\mu$ has been obtained in [7]:

$$\mathcal{F}^\mu = \sigma_o \left[T^{\mu\nu}(\frac{v_\nu}{c}) + \{(\frac{v_\alpha}{c})T^{\alpha\beta}(\frac{v_\beta}{c})\}(\frac{v^\mu}{c})\right], \tag{1}$$

in terms of σ_o the Thomson cross-section and $T^{\mu\nu}$ the EM energy-momentum tensor.

As an electron moves a distance l from the source, the Lorentz factor at the source $\gamma(0)$ changes to $\gamma(l)$

$$\gamma(l) \approx \frac{\gamma(0)}{1 + [\gamma(0)(\frac{l}{L})]}; \quad L \approx 2.2 \times 10^{31} cm. \approx 7.17 \times 10^9 \; Kpc. \tag{2}$$

Thus, energy loss through bremsstrahlung from cosmic background radiation can be neglected for distances $l << (L/\gamma)$. As the size of our galaxy is $l_G \sim 30$ Kpc, the above energy loss can be neglected for $\gamma \leq 10^7$ ($E \leq 1$ TeV).
The estimate of magnetic bremsstrahlung can be made through considerations of the average magnetic density in our galaxy, $u_{mag} \sim [eV/cm^3]$. For electrons and positrons that travel galactic distances, synchrotron radiation is negligible for $\gamma \leq 10^6$. The energies for which the leptons will reflect the energy distribution at the source should be $\gamma \leq 10^6$. Hence, for distance to sources of galactic length, the leptons should have energies less than about a TeV for their measured power law indices to that at the source. For larger energies, their energy

power law index measured by detectors in our Solar system would need to be corrected for radiation losses. Such a predicted break at 0.9 TeV has been verified by DAMPE [10].

4 Composition of UHE cosmics: Helium or Iron?

We apply entropy computations due to Landau for the Landau-Fermi liquid drop model of a heavy nucleus [3–6, 8, 9]. The energy distribution of the decay products are evaporating nuclei from the bulk liquid drops excited by a heavy nuclear collision. We proposed as the sources of cosmic rays, the evaporating stellar winds from (say) neutron star surfaces. The structure of neutron stars consists of a big nuclear droplet facing a very dilute gas, i.e. the "vacuum". Neutron stars differ from being simply very large nuclei in that most of their binding is gravitational rather than nuclear, But, the droplet model of the nucleus should still offer a good description of nuclear matter near the surface where it can evaporate. A gist of our mechanism follows:

1. The evaporation is from the effective fields in the form of (i) scalar nuclei $^4He \equiv \alpha$ and (ii) vector deuterons $^2H \equiv d$. What evaporates from the neutron star via scalar and vector fields are then energetic protons and alpha particles along with other nuclei to a much lesser extent.

2. Deuterons are only weakly bound and would be expected to photo-dissociate on ubiquitous background photons rapidly to produce protons and neutrons which in turn would produce protons when they decay.

3. The energy distribution comes directly from the entropy of evaporation of scalar (spin zero) and vector (spin 1) combinations of baryons.

4. There is a break at an energy $E_c \sim 1 \ PeV$ [The Knee]. For energies of nuclei lower than the knee, the index $\alpha = 2.7$(Boson value) and for energies of nuclei higher than the knee, the index $\alpha = 3.1$ (Fermion value).

5. There exists a cosmic quantum phase transition at the crossing energy. It is evidently a quantum phase transition since the order parameter involves the difference between Bose and Fermi statistical phases. In virtue of the experimental continuity of the energy distribution and thereby the entropy, the phase transition is higher than first order.

6. The boson phase arises from evaporating nuclei described by the collective Boson models of nuclear physics. There exist pairing correlations in odd-odd nuclei made up of deuterons. Correlations between two spin one deuterons lead to spin zero alpha particles and so forth all within the pairing condensate. Photo-disintegration of boson odd-odd nuclei into nucleon and alphas is important for the phase transition. The condensate resides near the surface of evaporating high baryon number $A >> 1$ nuclei. For example, a neutron

star itself is merely a nucleus of extremely high baryon number ($A \gg 1$) with superfluidity (and superconductivity) in the neighborhood of the nuclear surface.

7. The "partons" from the pairing condensate are evidently the nucleons. For energies above the knee, the cosmic rays must be composed mainly of protons and alpha particles.

8. For cosmic energies far beyond the knee, we expect a dramatic change in the composition of the nuclei so much so that at asymptotic energies we expect to observe essentially only [protons + Alphas]. Data (favouring a preponderance of alpha particles at UHE) were presented at this conference by I. Maris(Auger) and Y. Zhezher(TA) Collaborations.

Acknowledgments

YS would like to thank G. Ambrosi and B. Bertucci for useful discussions.

References

[1] J. Swain, A. Widom, Y. Srivastava, *Asymptotic High Energy Total Cross Sections and Theories with Extra Dimensions*, arXiv: hep-ph 1104.2553v4.
[2] G. Pancheri and Y. Srivastava, *Introduction to the physics of total cross-section at LHC*, Eur. Phys. J., **C77** (2017) 150.
[3] A. Widom, Y. Srivastava, J. Swain, *Entropy and the Cosmic Ray Particle Energy Distribution Power Law Exponent*: arXiv:hep-ph 1410.1766 v1.
[4] A. Widom, Y. Srivastava, J. Swain, *Concerning the Nature of the Cosmic Ray Power Law Exponents*: arXiv: hep-ph1410.6498.
[5] A. Widom, Y. Srivastava, J. Swain, *A Quantum Phase Transition in the Cosmic Ray Energy Distribution*: arXiv:1501.07809v2[hep-ph](2015).
[6] A. Widom, Y. Srivastava, J. Swain, *The Theoretical Power Law Exponent for Electron and Positron Cosmic Rays, A Comment on the Recent Letter of the AMS Collaboration*: arXiv: hep-ph 1501.07810v2.
[7] A. Widom, J. Swain, Y. Srivastava, *Bremsstrahlung Energy Losses for Cosmic Ray Electrons and Positrons*; arXiv:hep-ph 1509.08365.
[8] J. Swain, A. Widom, Y. Srivastava, *The Origin of the Broken Power Law Spectrum for Cosmic Rays*, PoS ICHEP2016 (2016) 082, Presented at the 38th International Conference on High Energy Physics 3-10 August 2016, Chicago, USA.
[9] Y. Srivastava, A. Widom, J. Swain, *Nature of the cosmic ray power law exponents*, Proceedings, 17th Lomonosov Conference on Elementary Particle Physics, Moscow, Russia, August 2015, A. Studenikin Editor, [World Scientific 2017]. Conference: C15-08-20.

[10] G. Ambrosi *et al*, DAMPE Collaboration, arXiv 1711.10981v1[astro-phHE] 29 Nov 2017.
[11] For gauge theories (QED, QCD etc.), $D = 4$ is special because the gauge coupling constant $(dim[g(D)] \sim M^{(4-D)/2})$ is *a-dimensional and scale invariant* only for $D = 4$. Planck was able to do photon thermodynamics using scale invariance of $D = 4$ alone: $U(T,V) = aT^4$, where $a = (\pi k_B^2)^2/[15(\hbar c)^3]$.

DIRECT MEASUREMENTS OF COSMIC RAYS

Piero Spillantini [a]
*Istituto Nazionale di Astrofisica, via fosso del cavaliere 100, 00133 Rome, Italy;
NRNU MEPhI, Kashirskaya chosse 31, 115409 Moscow, Russia*

Abstract. Direct measurements of Cosmic Rays were achieved only three decades after their discovery, when plastic balloons, airplanes and rockets could bring instruments at a so high altitude (not less than 20 km) where most of cosmic rays yet didn't interact with the atmosphere. The content of heavy nuclei was observed in 1948 and primary electrons much later, in 1959. Observations went on by high altitude borne instruments in the late 40's and satellite borne from the late 50's. A long shortage of satellite launches and of good quality balloons followed the Challenger disaster. A series of successful balloon launches in the 90's preceded the recovery of satellite borne experiments from 1998 (NINA, AMS-1, PAMELA and AMS-2 experiments). In last year's other satellite experiments (CALET and DAMPE) begun taking data of the electronic component in the TeV region, the NUCLEON satellite borne experiment begun measuring fluxes of proton and nuclei toward knee energy, and EUSO-TUS prototype exploring the possibility of measuring the flux at extreme energy. Further satellite or space station borne experiments are in preparation to improve the observation of the electronic component (Gamma-400 on board of Russian satellite) and of proton and nuclei in the PeV region (HERD on board of the Chinese Space station). A LDB flight of the GAPS instrument is in program in Antarctica in 2020 for measuring antideuteron flux. Proposals for extending the measurement of the energy spectra of positrons beyond the TeV and of antiprotons beyond the PeV are under study.

1 Cosmic radiation: exploratory phase

At altitudes lower than 20 km primary particles are drowned by secondary particles they produce in the atmosphere. Many of the early results were confusing and came slowly. The exploratory phase lasted for nearly fifty years. It has been a long and passionate scientific adventure, conducting from the study of natural radioactivity to the elementary particle physics.

For direct measurements of cosmic radiation, instruments had to be operated at more than 20 km, an altitude not attainable with manned balloons and airplanes. In 1935, a manned balloon flight brought a 30 kg payload up to 23 km: it first registered neutrons in cosmic rays [1] and remained for years a record altitude for man-carrying flights. Huge and light plastic balloons had to be developed together with the systems for controlling their trajectory, the transmission to ground of the collected information and possibly the recovery of the payload. Also the parallel development of detectors was of capital importance. The electrometers, developed and refined in the 20's, allowed to measure the rate of production of ions in closed volumes, and therefore measure the total flux of the ionizing radiation. Individual particles could be identified and measured only when nuclear emulsions, cloud chambers, and Geiger-Muller counters

---

[a] E-mail: spillantini@fi.infn.it

could be developed and operated. Particularly important was the possibility of triggering cloud chambers in a strong magnetic field by counter telescopes. Latitude [2] and east-west [3] effects could be studied and the particle nature of the cosmic radiation recognized. Using cloud chambers in strong magnetic field several investigators studied secondary interactions and led Anderson to the accidental discovery of the positron [4].

2 History and results of the early direct measurements

The development of high altitude plastic balloons, aircrafts and V-2 rockets greatly expanded the experimental opportunities in the late 40's and early 50's, especially for identifying primary particles and the secondary particles created in the interactions within the atmosphere. Heavy primary nuclei could be identified in 1948 [5], primary electrons in two independent experiments in 1959 [6] and the ratio of electrons on positrons measured three years later [7].

The radiation belts wrapping Earth were discovered by van Allen with the first USA satellite Explorer1 in 1958. The knee of the energetic spectrum was discovered in the Proton series of Russian satellites of Grigorov in 1966–1968. The higher energy spectra of separate group of elements were obtained by the Chicago group, with the large instrument 'Chicago egg' (CNR), on board of the Space-lab in 1985. The measurement of the spectra of separate elements went on by several long duration balloon (LDB) flights of JACEE (in Antarctica) and RUNJOB instruments (from Kamchatka to Europe), and in last year's ATIC and CREAM (in Antarctica).

Antiprotons in cosmic rays were first measured with balloon borne magnetic spectrometers in 1979 by Golden in USA and Bogomolov in Russia. These measurements continued with several balloon borne magnetic spectrometers (WIZARD, BESS, PBAR, HEAT, IMAX) waiting for the occasion to continue the measurements in space. Light elements and isotopes spectra were measured with these same instruments and by the Cosmic Ray Isotope Spectrometer (CRIS) instrument on board of the ACE explorer (see below) starting from 1997.

Most of first direct detection experiments were conducted by small teams or single researchers; a transition to experiments conducted by teams of several institutions in collaboration began in the 90's. The reason goes back to the development of the access to space. At the beginning of 80's, taking advantage of the availability of the fleet of Space Shuttles, NASA implemented the strategy for space astronomy and astrophysics, recommended by the USA National Academy of Sciences [8] for the continuous and simultaneous observation of the Universe, planning four Space Great Observatories CGRO, AXAF, HUBBLE, SIRTF and the Long Base Interferometer LBI on ground, for covering the whole e.m. spectrum. The observatories were complemented by the simultaneous observation of the charged component in its whole energy range, from a few tens MeV up to the ultrahigh energies: it was foreseen the Advanced Composition Explorer ACE [9] for measuring the energy spectra of rare elements and radioactive isotopes, and two CR facilities on board of the planned Freedom Space Station (FSS), ancestress of the present International Space Station (ISS): (a) the Particle-Antiparticle Superconducting Magnet Facility, named ASTROMAG [10] where could be operated CR experiments; (b) a large area passive detector (HNC)) [11] for collecting the heavy nuclei component; besides a facility for collecting cosmic dust. The ACE explorer and the two CR facilities ATROMAG and HNC were the main points of the global CR program issue by NASA [12]. The difficulties of the NASA space program at the end of 80's and beginning of 90's greatly affected the realization of the Great Observatories, whose launches lasted from the 1990 of HST to 2003 of SIRFT, and ACE explorer, first proposed in 1983, had to wait several years to be finally launched in 1997.

Because of the Challenger disaster in 1986 the FSS program was delayed and finally stopped in 1991. The two facilities HNC and ASTROMAG planned on board of the FSS were cancelled. The teams gathered during the preparation of the experiments had to pursue their objectives by adapting their programs to balloon borne instruments.

However it is important to remark that the study of the ASTROMAG facility and the elaboration of its experimental program collected in one large international working group scientists from USA, Europe and Japan, allowing individuals and small teams to converge in large collaborations joining several experimental techniques and many researchers, in a style similar to that operating in the Elementary Particle field at accelerators.

After the cancellation of the FSS program, the thematic of the first experiment to be operated in the ASTROMAG facility, named WIZARD [13], mainly dedicated to the measurement of the antiparticle components of cosmic rays and hunting for antinuclei, evolved through eight balloon borne experiments from 1989 to 1998. From 1993, the Wizard collaboration included also three Russian institutions and constructed with them the Russian Italian Mission RIM program which realized the satellite borne experiment NINA dedicated to the study of the high energy tail of solar events, developed the GILDA project (progenitor of AGILE [14]) for identifying the sources of high energy gamma's, conducted a series of life science experiments on board of the Russian MIR Space Station, lasting with the satellite borne experiment PAMELA [15].

Subsequent experiments foreseen to be operated in ASTROMAG facility were: LISA [16] for the systematic measurement of the energy spectra of elements and isotopes beyond 1 GeV/nucleon; MAGIC-SCHINAT [17] for measuring the elemental composition in the knee region, ASTROGAM [18] for extending up to several hundred GeV the observation of the energy spectrum of the electromagnetic component and identifying its sources.

Independently from the ASTROMAG program the study of the antiproton component at low energies and the hunting for antinuclei was also the thematic pursued by the Japanese-USA collaboration BESS [19] gathered around an alternative design (a light and transparent to particles superconducting solenoid) of the magnetic spectrometer of the ASTROMAG facility.

The thematic of experiments LISA, MAGIC-SCHINAT and ASTROGAM planned for the ASTROMAG facility and that for the HNC facility underwent the following evolution:

(a) The measurement of the elemental and isotopic energy spectra was afforded by ballooning with the IMAX superconducting spectrometer. The superconducting spectrometer ISOMAX [20] realized with an activity lasted several years for improving the quality and the energy range of IMAX was lost by accident and never reconstructed.

(b) the extension in energy of the proton and ion fluxes programmed with MAGIC-SCHINAT continued to be pursued with the above mentioned long duration balloon (LDB) flights of the JACEE and RUNJOB instruments and more recently by LDB flights in Antarctica of the ATIC and CREAM [21] instruments;

(c) the observation and measurement of the gamma ray sources of the ASTROGAM program has been recovered only two decades later by the missions AGILE [14] launched in 2006 and FERMI-GLAST [22] launched in 2008;

(d) for what concerns the measurements of the flux of the high Z elements up to actinides foreseen for the HNC facility, several balloon flights of the TIGER [23] and super-TIGER [24] instruments could measure the fluxes of trans-iron elements up to Zirconium, but for what concerns the heavier nuclei up to actinides the Heavy Nuclei Explorer (HNX) [25] with its ENTICE [26] and ECCO [27] instruments couldn't be flown because of the termination of the Space Shuttle program.

3 PAMELA and AMS-2 observatories

The development of the PAMELA concept, the last step of the RIM program, kept several years. It had the ambition to be conceived as CR observatory at 1 AU from the Sun, devoted to antiparticle study but extending its measurements to all elementary particles and light nuclei down to very low energies (50 MeV for electrons and 80 MeV for protons) and up to energies (10 TeV/n) well beyond the 1.2 TV MDR of its spectrometer, recovering a large part of the cosmic ray observations foreseen for ASTROMAG.

In the limited mass (450 kg), power (345 W) and volume ($1.3 \times 0.7 \times 0.7$ m^3) imposed by the satellite and launcher, this was obtained by several instrumental choices (for detail see Ref. [15] and section 2 of Ref. [28]).

The 'observatory' concept of the instrument could also make use of the long time duration of the mission, that was in continuous operation for 10 years, and the long period of low solar activity between the 23rd and 24th solar cycles.

Two important remarks are mandatory: NINA and PAMELA were first CR experiments in space after the Challenger disaster of 1986 that halted for several years all possibilities of launching experiments to orbit; furthermore they have been first space experiments in collaboration with Russian institutions after the Cold War era.

In the 90's a worldwide collaboration of physicists coming from the field of elementary particle physics at accelerators gathered for hunting for antinuclei (and antiparticles in CR) by using a large volume permanent magnet. After a test of a silicon tracker in the field of this magnet in 1998 on board of a Space Shuttle flight (AMS-1) [29] it was constructed around this spectrometer a large and complex telescope, equipped with several modern ancillary detectors [30], installed in 2011 on board of the International Space Station ISS. The quality and number of the ancillary detectors and the good performance of the magnetic spectrometer make AMS-2 an observatory as PAMELA has been in the decade 2006-2016, capable of a higher statistics in all the observed channels, due to the much larger acceptance, and the forecast of several further years of data collection.

Until now AMS-2 results confirmed the two most important discoveries of PAMELA, the increase with energy of the positron fraction [31], see Fig. 1, and the two slopes of the energy spectra of proton and helium [32], see Fig. 2, and extended them to higher energy and other ions.

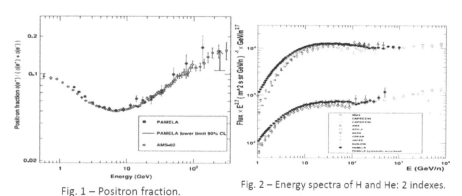

Fig. 1 – Positron fraction.

Fig. 2 – Energy spectra of H and He: 2 indexes.

The positron fraction [33] was the most unexpected result. Already after the first publication it originated a rich discussion. The discrepancy with the expected behavior based on the current models of the diffusion of CR and of the structure of the Galaxy is very significant. At high energy these models, also if deeply modified, cannot reproduce its sudden increase. A more complex scenario must be invoked based on the possible contribution of nearby pulsars or on the indirect signal of dark matter annihilation. But for this the absence of a signal in the antiproton/proton ratio [34] [35] requires ad-hoc extreme hypotheses.

Important (and unexpected) results are obtained in the measurement of the energy spectrum of the two most abundant CR components, the proton and helium, mostly of primary origin. The high stability of the instrument and high statistics collected in many years allowed to put in evidence a change of slope in the spectra occurring at about 230GV/c [32], confirmed with much higher statistics and extended to other ions by AMS-2 [36]. This result is very important because challenges the 'paradigm' of the interpretation of the diffusion of primary CR in the Galaxy and the models of the Galaxy itself. By these measurements PAMELA and AMS-2 are heralding the era of GCR precision measurements, allowing to check different models of their diffusion and of the structure of the Galaxy.

AMS-2 also extended in energy and enriched in statistics the PAMELA measurements in other channels.

After the H and He the most abundant component are electrons. They lose energy in several processes, possible structures in the shape of their energy spectrum should be a consequence of these losses and possible new sources. The PAMELA results [37], confirmed by AMS-2 [38], are particularly significant as they cover about 10 years of data acquisition and span the largest energy range (three decade) covered in single CR experiments. The sources discussed as possible explanations for the excess of positrons should also explain the hardening of energy spectrum of electrons above 70 GeV [39].

Important for the study of the GCR diffusion in the Galaxy is the precise measurement of energy spectra of isotopes of H and He. They are of secondary origin, tracers of the CR transport in the Galaxy, crucial for analyzing antiproton and positron spectra. The PAMELA [40] and AMS-2 [41] results are the most precise so far.

Also in the light nuclei energy spectra [42], and in particular in the B/C ratio the PAMELA and AMS-2 results are unrivaled for precision on a wide energy range [43].

Unrivaled precision results were also obtained by PAMELA for the modulation with the solar activity of protons, helium [44], electrons and positrons [45], all confirmed by AMS. In particular PAMELA resulted to be an unique instrument for its low energy threshold and the wide interval of measured energies for Solar Energetic Particle (SEP) observations, filling the energy gap between the measurements in space (up to a few hundred MeV/n) and the ground-based observation (above 1 GeV/n), a key energy range for understanding the particle acceleration processes. The most significant results concern the two events occurred in December 13th and 14th 2006 [46].

Furthermore for PAMELA the altitude (350–610 km) of the quasi polar orbit, as well the high acquisition rate allowed by electronics, allowed to perform a very detailed study of cosmic radiation in different regions of the magnetosphere, where CR transport and interaction models as well trapped particle radiation models can be tested with unprecedented statistical significance and in energy ranges where no experimental data are available. Antiprotons created in the decay of antineutrons are stably trapped in the radiation belts. Geomagnetically trapped antiprotons were observed in the SAA region [47].

4 Other new results

In last year's two new experiments have been launched to space for measuring the energy spectrum of the electromagnetic component of the cosmic radiation up to several TeV energy and identifying possible contributions of its sources: CALET [48] launched in August 2015 and installed on the Japanese sector of ISS, and DAMPE [49] launched in December 2015 on board of the Wukong Chinese satellite. The results of the two experiments for the electron energy spectrum based on the first collected data are reported in Fig. 3 and Fig. 4.

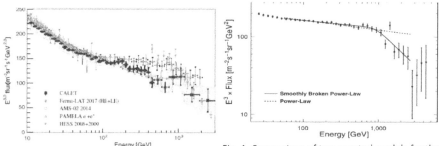

Fig. 3 Electron + positron flux of CALET experiment.

Fig. 4 Comparison of two spectral models for the DAMPE electron + positron spectrum.

A specific comment must be dedicated to the long standing effort for measuring the energy spectra of proton and ions beyond the 'so-called' knee (3 × 1016 eV) where the index of the global CR flux changes. The ground experiments still give inconclusive results. LDB experiments (RUNJOB, JACEE, ATIC, CREAM) are slowly and patiently collecting information in the region up to 1015 eV, but the needed 1016 eV frontier is still beyond their possibilities. A noticeable step forward is expected by the space version of the CREAM instrument to be installed next year on board of the ISS. From space the answer to this problem could be straightforward, but the steep energy spectrum of the flux requires an instrument accepting particles on a few $m^2 sr$ for several years. The large acceptance instruments proposed in USA (ACCESS in a two versions) and in Russia (INCA and PROTON5) have not been realized.

A noticeable step toward this goal will be the new Russian instrument NUCLEON [50] launched in Dec 2014 on board of the Resurs-P satellite. The results of the analysis of proton and helium data collected in last two years are reported in Fig. 5.

It must be finally reminded that in ground EAS experiments have been observed showers generated by primaries of very high energy, up to 1020 eV. At so high energies the extremely faint flux doesn't allow any kind of direct detection, but it comes to help the possible registration of the copious fluorescence light emitted by the shower, allowing to follow its whole development in atmosphere.

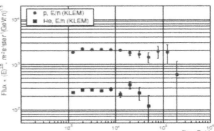

Fig. 5 Spectra of protons and helium nuclei measured in the NUCLEON experiment.

This light can supply a global image fostering (at least on a statistical basis) the identification and measurement of the energy of the primary. At the end of 90's three projects based on the detection from space of florescence light emitted by the shower were developed, OWL [51] in USA, EUSO [52] in Europe and KLYPVE [53] in Russia. Nowadays the proposing teams converged in a unique program of test of the instrumentation and flight of prototypes. The prototype instrument TUS [54] launched in April 2016 on board of the Russian Lomonosov satellite should register a few possible events of extreme energy, and the analysis of collected data is in progress. A further prototype MINI-EUSO will be installed in 2018 in the Russian sector of the ISS. Next steps of the program should be the mission K-EUSO [55] also on board of ISS, and later the two satellite constellation of the project POEMMA [56] now under consideration by NASA.

5 Next and possible future experiments

Besides the 'direct detection' of CR extreme energy for understanding their nature, a few important thematic are still open to be understood in next future:

(1) energy spectra of proton and ions in the 'knee' energy region (up to several PeV/nucleon);

(2) the energy trend of the ratios antiproton/proton and positron/electron (for the possible indication of the dark matter contribution);

(3) as well the fluxes of very high Z ions up to actinides in particular for understanding the mechanisms of neutronic accretion and the rate of production in their sources.

For the first thematic the most important experimental requirement is to maximize the acceptance and duration of experiments in space. The first attempt will be the installation next year on ISS of an adapted version of the CREAM instrument flown until now in Antarctica. For this thematic it must also be mentioned the development of large and deep calorimeters suitable for identifying and measuring on nearly the whole solid angle energy and direction of the incoming particle (as for example on 5 faces of a cubic calorimeter [57]). This is the kind of instrument proposed for the experiment HERD [58] planned to be installed in few years on the Chinese Space Station.

Another experiment in preparation for the next future is the Russian telescope GAMMA-400 [59] characterized by an unprecedented angular precision for resolving high energy gamma sources in dense regions, e.g. the center of the Galaxy.

In program for next future is also the LDB flight in Antarctica in 2022 of the GAPS [60] experiment dedicated to dark matter identification by measuring the antideuteron flux.

More difficult it appears to afford the second thematic of measuring up to high energy the antiproton and positron spectra, because the instrument, besides maintaining a very large acceptance, must be surrounded by a strong magnetic field (such as the proposed magnetic bubble [61]) for identifying the sign of the charge of the incoming particle up to very high energies, beyond PeV/n for ions and TeV for electrons and positrons. A preliminary design proposed a few years ago, consisting on a large acceptance calorimeter surrounded by a toroidal superconducting magnet [62], is nowadays in phase of design and evaluation for a project of experiment summited to ESA.

A quasi-4? acceptance calorimeter of suitable dimensions could also contribute to the third thematic, the measurement of fluxes of very high charge CR's up to actinides, provided that it could be complement by a suitably precise measurement of the charge of the incoming particle.

References

[1] L.H. Rumbaugh and G.L. Locher, Phys. Rev. 44 (1936) 855.
[2] J.Clay, Proc. R. Acad. Amsterdam 31 (1938) 1091; A.H. Compton, Phys. Rev. 41 (1932) 111-113(L)
[3] T.H. Johnson, Phys. Rev. 43 (1933) 834-835(L); L. Alvarez and A.H. Compton, Phys. Rev. 43 (1933)835.
[4] C.D. Anderson, Phys. Rev. 43 (1933) 491-494.
[5] P. Freier, E.J. Lofgren, E.P. Ney et al., Phys. Rev. 74 (1948) 213-217.
[6] J. Earl, Phys. Rev. Letters 6 (1961) 125-128; P. Meyer and R. Vogt, Phys. Rev. Letters 6 (1961) 193-196
[7] J.A. DeShong Jr., R.M. Hidelbrand and P. Meyer, Phys. Rev. Letters 12 (1964)3-6.
[8] National Research Council of the USA NAS, USA NAS report, 1979.
[9] E.C. Stone et al., Space Science Reviews, 86(1998)285.

[10] M.A. Green et al., IEEE Trans. in Magnetics, MAG-23 (1987) 1240.
[11] P.B. Price, The NASA CR program for the 1990s and beyond, New York, 1990, AIP, pp 63-66.
[12] NASA CR Program Working Group, "Cosmic Ray Program for the 1980's", NASA , Aug 1982.
[13] R.L.Golden, "WiZard, a proposal to measure CR including.", NMSU, November 1988.
[14] A. Morselli et al., Nuclear Physics B85(2000)22.
[15] O. Adriani et al., 24th ICRC (Rome) 3(1995)591; P. Picozza et al., Astrop. Physics 27(2007)296.
[16] J.F. Ormes et al., NASA GSFC, Nov. 1988.
[17] T.A. Parnell et al., NASA MSFC, Nov. 1988.
[18] J. Adams and D. Eichler, Astrophys. J. 317 (1987) 551.
[19] A. Yamamoto et al., Adv. Space Res., 14 (1994) 75.
[20] J.W. Mitchell et al., 26th ICRC (Salt Lake City) 3 (1999) 113; M. Hof et al., NIM. A454 (2000) 180.
[21] E.S. Seo et al., 26th ICRC 3(1999)207; O. Ganel et al., 27th ICRC (2001)2163.
[22] A.A. Abdo et al., Astropys. J. 697 (2009) 1071.
[23] B.F. Rauch et al., Ap J 501 (2009) 911.
[24] R.P. Murphy et al., Astrophysical Journal 831:140 (2016) 7pp.
[25] W.R. Binns et al., 27th Int. Cosmic Ray Conf., 4 (2001) 2181, Hamburg.
[26] M.H. Israel et al., 27th Int. Cosmic Ray Conf. (2001) 2231, Hamburg.
[27] A.J. Westphal et al., 26th Int. Cosmic Ray Conf., 4 (1999) 160, Salt Lake City.
[28] A. Galper and P. Spillantini, Adv. Space Res. 53 (2017) in press.
[29] J. Alcaraz et al., Physics Reports 334:331 (2002) 74pp.
[30] R. Battiston et al., 29th ICRC (Pune) 10 (2005) 151
[31] O. Adriani et al., NATURE 458 (2009) 607-609.
[32] O. Adriani et al., Science 332 (2011) 69-72.
[33] L. Accardo et al., Phys. Rev. Lett. 113 (2014) 121101.
[34] O. Adriani et al., Phys. Rev. Lett. 105 (2010) 121101.
[35] M. Anguilar et al., Phys. Rev. Lett. 117 (2016) 091013.
[36] L. Accardo et al., Phys. Rev. Lett. 114 (2015) 171103.
[37] O. Adriani et al., Phys. Rev. Lett. 106 (2011) 201101
[38] L. Accardo et al., Phys. Rev. Lett. 113 (2014) 121102.
[39] P. Picozza and L. Marcelli, Astropart. Phys. 53 (2014) 160-165.
[40] V. Formato et al., Nucl. Instr. Meth. A 742 (2014) 273-275.
[41] M. Anguilar et al., Phys. Rev. Lett. 117 (2016) 231106.
[42] M. Boezio et al., Astropart. Phys. 39 (2012)95-108.
[43] O. Adriani et al., Astrophys. J. 791 (2014) 93.
[44] O. Adriani et al., Astrophys. J. 765 (2013) 91.

[45] V.V. Mikhailov et al., Bull. Russ. Acad. Sci. Phys. 77 (2013) 1309-1311.
[46] O. Adriani et al., Astrophys. J. 742 (2011) 102.
[47] O. Adriani et al., Astrophys. J. 737 (2011) L29.
[48] O. Adriani et al., Phys. Rev. Lett. 119 (2017) 181101.
[49] Dampe collaboration, Nature 552 (2017) 63-66.
[50] D. Podorozhnyi et al., 27th ICRC (2001) 2188; E. Atkin et al., APJ Web of Conf. 105 (2015) 01002.
[51] J.F. Krizmanic et al., AIP Conf. Proc. (1998) 433.
[52] L. Scarsi et al., "EUSO proposal for ESA F2/F3 missions" ESA/MSM-GU/2000.462/RDA, Dec. 2000.
[53] B.A. Khrenov et al., Int. Worksh. on UHECR from Space and Earth, Metepec, Aug 2000, AIP 2001.
[54] V.V. Alexandrov et al., 27th ICRC (2001) 831; G.K. Garipov et al., Bull. RAS 66 (2002) 1817.
[55] P.A. Klimov et al., 35th ICRC 2017 (Busan) proceedings in press.
[56] A. Olinto et a., 35th ICRC 2017 (Busan) proceedings in press.
[57] N. Mori et al., Nucl. Instr. Meth. A732 (2013) 311-315; O. Adriani et al., Astrophys. Physics 96 (2017) 11
[58] S.N. Zhang et al., Proc. SPIE Int. Soc. Opt. Eng. 9144 (2014) 91440x.
[59] A.M. Galper et al., Adv. Space Res. 51 (2013) 297-300.
[60] T. Aramaki et al., Physics Reports 618 (2016) 1.
[61] P. Spillantini, Adv. Space Res. 55 (2014) 428-433.
[62] P. Papini and P. Spillantini, ECRS (Kiel) Journal of Physics: conferences series 632 (2015) 012030.

FAST RADIO BURSTS: A NEW MAJOR PUZZLE IN ASTROPHYSICS

Maxim Pshirkov [a], Sergei Popov [b], and Konstantin Postnov [c]
Sternberg Astronomical Institute, M.V. Lomonosov Moscow State University, 119234 Moscow, Russia

Abstract. We briefly review main observational properties of fast radio bursts (FRBs) and discuss two most popular hypothesis for the explanation of these puzzling intense millisecond radio flares. FRBs are likely to happen on extragalactic distances, and their rate on the sky is about a few thousand per day. Two leading scenarios describing these events include hyperflares of magnetars and supergiant pulses of young energetic radio pulsars.

1 Introduction

Fast radio bursts (FRBs) are millisecond radio flares demonstrating large peak fluxes ranging from a few hundredths of Jy to ~ 100 Jy. All known FRBs are characterized by large dispersion measures (DM) $\sim 200 - 2600$ pc cm^{-3}, which cannot be explained by the interstellar medium in our Galaxy. The first FRB010724 was reported in 2007 [1]. However, active FRB research began only after 2013 when four other similar events have been discovered [2]. The high DM suggests extragalactic distances ($\sim$ Gpc), so the bright peak flux and short millisecond duration imply a very high brightness temperature of the observed radio emission (about $10^{35} - 10^{36}$ K) and hence a non-thermal radiation mechanism.

The occurrence rate of FRBs estimated from observations is very high: several thousand events a day over the sky, much exceeding that of many other known transients (e.g., gamma-ray bursts or coalescence of compact objects). For extragalactic FRBs this rate corresponds to ~ 100 events a day from the local 1 Gpc^3 volume. However, due to a small field of view of the existing radio telescopes, only a small fraction of the bursts is detected.

Presently, about 30 FRBs have been recorded (see the on-line catalog on http://frbcat.org, see [3]). Most of them ($> 70\%$) was discovered by the 64-m Parkes radio telescope in Australia. One source (FRB121102) is found to produce repeating bursts without any detected periodicity. More than few hundred events are already detected from this source. For this object, the host galaxy was identified [4]. For other sources no repetitions of activity are identified, despite intensive searches. All available data are consistent with cosmological origin of these events.

---

[a] E-mail: pshirkov@gmail.com
[b] E-mail: polar@sai.msu.ru
[c] E-mail: kpostnov@gmail.com

The nature of FRBs remains unknown. About twenty different hypothesis have been proposed (and with variations their number reaches a few dozens!). Some of them already can be considered as being "refuted" (for example, because of too low rate or due to undetection of any transient counterparts of FRBs), some look rather exotic (like models related to cosmic strings or charged black holes). Most conservative approaches link FRBs to some kind of activity of neutron stars. Presently, the two leading approaches relate these millisecond bursts of radio emission to the activity of magnetars or to very strong pulses of energetic radio pulsars. Below we briefly describe both models and then present our conclusions.

2 Hyperflares of magnetars and FRBs

Already in 2007, immediately after the first FRB was reported, the paper [5] suggested that such short radio bursts can be related to hyperflares of magnetars — neutron stars with superstrong magnetic fields (see a review of magnetars in [6]). This scenario seems very plausible, as statistical and energetic considerations allow one to explain basic properties of FRBs with the transformation of just a tiny fraction ($\sim 10^{-5}$) of the magnetar hyperflare energy ($\sim 10^{44} - 10^{46}$ erg) into a short radio burst. Also the time scale of two types of transients match as strong magnetar bursts have very sharp rising fronts.

In addition to statistical and energy properties of FRBs, the magnetar model can easily explain absence of counterparts in other wavelengths. If a magnetar flare happens at ~ 1 Gpc distance, then it is impossible to detected the burst at high energy range. Thus, the FRB might not be accompanied by a gamma-ray burst or any kind of afterglow directly related to the emission of the magnetar flare.

A working model of the radiation mechanism for FRBs from magnetar flares was elaborated by Yuri Lyubarsky [7]. After that, several modifications of this scenario were proposed, see for example [8].

In this model, the coherent radio emission is generated by the synchrotron maser mechanism operating at a relativistic shock. The forward shock is produced when a strong electromagnetic pulse from the magnetar hyperflare propagates from the twisted magnetar's magnetosphere outwards and meets a density discontinuity (for example, the boundary between magnetar's pulsar-wind nebula and the ambient medium). Note that pulsar-wind nebulae have been indeed found around some Galactic magnetars [9]. On other hand, not necessarily every magnetar is a source of FRB, as not all of them are situated in proper surroundings, from the point of view of the model [7].

An important prediction of this scenario is a strong TeV burst arising simultaneously with the radio pulse due to synchrotron radiation of electrons at the forward relativistic shock. Observations by ground-based gamma-ray telescopes (such as VERITAS and H.E.S.S.) can verify this prediction.

3 Supergiant pulses from radio pulsars

The idea that FRBs can be analogues of giant pulses of energetic radio pulsars was proposed in 2015 [10]. Indeed, scaling of the most powerful "shots" observed in the Crab pulsar to a neutron star with millisecond spin period and magnetic field above $\sim 10^{13}$ G results in bursts potentially similar to FRBs (we underline here that the precise distance scale for FRBs is still unknown). Giant pulses of the Crab pulsar are known to be very short (below one millisecond), so similar events can fit timing properties of FRBs.

This scenario was later discussed and developed in many papers. For example, in paper [11] the authors investigated evolution of the DM due to a young supernova remnant around the pulsar. This can explain large total DM even if distances are significantly below Gpc scale. In this model, any detected repeating burst will display a DM quickly decreasing a time scale of several years as young supernova remnants evolve on this scale. This prediction can be tested soon, if consequent bursts from the same sources are detected (it is expected that in this model each source could produce new pulses every few days, see [11] for details). Also this model naturally explains the lack of FRB detections at low frequencies — the free-free absorption in the dense medium of the remnant precludes radiation at frequencies lower than $700 - 800$ MHz from leaving the shell.

The model of supergiant bursts leads to an important prediction: as the FRB sources are "local" in some sense, $D < 100 - 200$ Mpc, one would expect some degree of correlation with the positions of galaxies in this volume. Large localization errors and small number of detected bursts does not allow us to robustly check this prediction at the moment, but it is expected that observation of $\mathcal{O}(100)$ sources will be sufficient for this purpose. Also, these young neutron stars might be strong X-ray sources due to huge rotational energy losses, as it is well-known in the case of Galactic energetic radio pulsar [12]. This prediction will also be checked with better localization of the FRBs. The increasing statistics of FRBs and rapid localization within small error boxes can be done soon with new radio astronomical facilities within few years at most.

4 Conclusion

After 10 years of exploration, fast radio bursts remain a major puzzle. Despite new important discoveries (real-time detections with intensive follow-up at different wavelengths, detection of circular and linear polarization, the detection of just one repeating source and identification of its host galaxy — see a recent review in [13]), it is unclear whether all these sources represent a single population, what is their exact distance scale, and hence the typical luminosity.

Most likely, the engine of these enigmatic bursts is related to neutron stars. Either the magnetic (in the magnetar scenario) or rotational (in the case of supergiant pulses) energy of a neutron star is transformed into a strong radio flare with enormously high brightness temperature up to $\sim 10^{36}$ K. General properties of FRBs can be reproduced in any of these models. Both models can also explain the repeating FRB 121102 making it just a peculiar source (may be at a specific evolutionary stage, for example, a very young neutron star) of generally the same nature. Interestingly, properties of the host galaxy of the repeating source are consistent with both models. The galaxy demonstrates high star formation rate, so the appearance of a young neutron star (a magnetar, or a very energetic pulsar) seems to be quite natural [4].

However, some important questions remain unanswered. For example, why up to now FRB pulses have not been detected at frequencies below ~ 600 MHz? What mainly contributes to the observed DM: the intergalactic medium, or immediate surroundings of the source (for example, a dense supernova remnant shell or may be interstellar matter in a star formation region)? Hopefully, in the near future, the increasing FRB statistics which will be obtained with new observational facilities (FAST, Apertif, UTMOST, CHIME, etc.) can help to solve this puzzle.

Acknowledgments

This work has been supported by the Russian Science Foundation grant 14-12-00146. S.B.P. is grateful to M. Lyutikov for numerous discussions on FRBs.

References

[1] D.R. Lorimer et al., Science **318** (2007) 777.
[2] D. Thornton et al., Science **341** (2013) 53.
[3] E. Petroff et al., Publ. Astron. Soc. Austr. **33** (2016) e045.
[4] S.P. Tendulkar et al., Astrophys. J. **834** (2016) L7.
[5] S.B. Popov, K.A. Postnov, arXiv:0710.2006 (2007).
[6] R. Turolla, S. Zane, A.L. Watts, Rep. Prog. Phys., **78** (2015) id. 116901.
[7] Y. Lyubarsky, Mon. Not. Royal Astron. Soc. **442** (2014) L9.
[8] K. Murase, K. Kashitama, P. Meszaros, Mon. Not. Royal Astron. Soc. **461** (2016) 1498.
[9] J. Vink, A. Bamba, ApJ Lett. **707** (2009) L148.
[10] J.M. Cordes, I. Wasserman, Mon. Not. Royal Astron. Soc. **457** (2016) 232.
[11] M. Lyutikov, L. Burzawa, S.B. Popov, Mon. Not. Royal Astron. Soc. **462** (2016) 941.
[12] S.B. Popov, M.S. Pshirkov, Mon. Not. Royal Astron. Soc. **462** (2016) L16.
[13] A. Rane, D. Lorimer, J. Astroph. Astron. **38** (2017) 55.

A CANDIDATE FOR AN UV COMPLETION: QUADRATIC GRAVITY IN FIRST ORDER FORMALISM

Enrique Alvarez [a] , Jesus Anero [b] , Sergio Gonzalez-Martin [c] and Raquel Santos-Garcia [d]
Departamento de Física Teórica and Instituto de Física Teórica (IFT-UAM/CSIC), Universidad Autónoma de Madrid, Cantoblanco, 28049, Madrid, Spain

Abstract. We consider the most general action for gravity which is quadratic in curvature. In this case first order and second order formalisms are not equivalent. This framework is a good candidate for a unitary and renormalizable theory of the gravitational field; in particular, there are no propagators falling down faster than $\frac{1}{p^2}$. The UV regime is in a conformal invariant phase; only when Weyl invariance is broken through the coupling to matter can an Einstein-Hilbert term (and its corresponding Planck mass scale) be generated.

1 Introduction

It is well-known that general relativity is not renormalizable (cf. [2] and references therein for a general review). However, quadratic (in curvature) theories are renormalizable, albeit not unitary [3] -at least in the standard second order formalism- although they have been widely studied over the years. When considering the Palatini version of the Einstein-Hilbert Lagrangian the connection and the metric are treated as independent variables and the Levi-Civita connection appears only when the equations of motion are used.

It is however the case that when more general quadratic in curvature metric-affine actions are considered in first order formalism the deterministic relationship between the affine connection and the Levi-Civita one is lost, even on shell [5]. That is, the equations of motion do not force the connection to be the Levi-Civita one.

This is quite interesting because it looks as if we could have all the goods of quadratic Lagrangians [3] (mainly renormalizability) without conflicting with Källen-Lehmann's spectral theorem. This justifies our claim of this theory being a candidate for an ultraviolet (UV) completion of quantum gravity. A preliminary exploration of these ideas has been done in [4], to which we refer for a more detailed discussion.

---

[a] E-mail: enrique.alvarez@uam.es
[b] E-mail: jesusanero@gmail.com
[c] E-mail: sergio.gonzalez.martin@uam.es
[d] E-mail: ir.raquel.santos.garcia@gmail.com

2 The action principle

The action of the theory reads

$$S = \int d^n x \sqrt{|g|} \mathcal{L}\left(g_{\mu\nu}, A^{\tau}_{\gamma\epsilon}\right) = \int d^n x \sqrt{|g|} \sum_{I=1}^{I=16} g_I R^{\mu}{}_{\nu\rho\sigma}(D_I)^{\nu\nu'\rho\rho'\sigma\sigma'}_{\mu\mu'} R^{\mu'}{}_{\nu'\rho'\sigma'} \quad (1)$$

where the Riemann tensor is defined in terms of an arbitrary (although torsion-free) connection, Γ, which differs from the Levi-Civita value by a three-index field, A:

$$\Gamma^{\mu}_{\rho\lambda} \equiv \left\{{}^{\mu}_{\rho\lambda}\right\} + A^{\mu}_{\rho\lambda} \quad (2)$$

The D_I are straightforward tensors built out of the metric tensor, $g_{\alpha\beta}$. This theory is conformally invariant in $n = 4$ dimensions, because under $g_{\mu\nu} \to \Omega^2 g_{\mu\nu}$ (and Γ inert fields)

$$\mathcal{L} \to \Omega^{-4} \mathcal{L} \quad (3)$$

It is always possible to choose the background connection as the metric one (that is, the background A field vanishes), in such a way that only quantuum fluctuations in the A field need to be considered

3 Dynamical generation of the Einstein-Hilbert term

The theory so far considered is always in the *conformal phase*; it is Weyl invariant. This is the symmetry that prevents the appearance of a cosmological constant on the theory and ensures that all counterterms must be inside our list of quadratic operators. This ultraviolet regime of the theory is then a candidate for an UV completion of quantum gravity.

This symmetry is not to be found at low energies, however; which means that it must be broken at some scale, which we will relate to Planck's. Once this happens, both a cosmological constant and an Einstein-Hilbert term in the lagrangian are not forbidden anymore. Several scenarios for this breaking can be proposed (cf. for example, [1]); may be the simplest possibility is through interaction with a minimally coupled scalar sector

$$L_s \equiv \sqrt{|g|} \left(\frac{1}{2} g^{\mu\nu} \partial_\mu \phi \partial_\nu \phi - V(\phi)\right) \quad (4)$$

Quantum corrections will include a term

$$\Delta L = \frac{C}{n-4} R \phi^2 \quad (5)$$

Were the scalar field to get a nonvanishing vacuum expectation value $\langle \phi \rangle = v$ the counterterm implies an Einstein-Hilbert term

$$L_{EH} = M^2 \sqrt{|g|} R \quad (6)$$

Once generated, this term dominates the infrared (IR) phase of the theory.
To conclude, when considering quadratic in the Riemann tensor gravity theories in the first order formalism, quartic propagators never appear. The ensuing theory naively appears to be both renormalizable and unitary.

Acknowledgments

E.A. is indebted to Alexander Studenikin for the kind invitation to the Lomonosov conference. Two of us (E.A and S.G-M) are grateful to the LBNL and UC Berkeley for hospitality in the initial stages of this project. S. G-M. is also grateful to the University of Southampton (U.K.) for their kind hospitality in the final stages of this work. We are grateful to Bert Janssen, C.P. Martín, Tim R. Morris and Jos Vermaseren for illuminating discussions. Comments by Stanley Deser are always greatly appreciated. This work has received funding from the European Unions Horizon 2020 research and innovation programme under the Marie Sklodowska-Curie grants agreement No 674896 and No 690575. We also have been partially supported by FPA2012-31880 and FPA2016-78645-P(Spain), COST actions MP1405 (Quantum Structure of Spacetime) and COST MP1210 (The string theory Universe). The authors acknowledge the support of the Spanish MINECO *Centro de Excelencia Severo Ochoa* Programme under grant SEV-2012-0249, as well as by the Spanish Research Agency (Agencia Estatal de Investigacion) through the grant IFT Centro de Excelencia Severo Ochoa SEV-2016-0597.

References

[1] A. Salvio and A. Strumia, "Agravity," JHEP **1406** (2014) 080.
[2] E. Alvarez, "Quantum Gravity: An Introduction To Some Recent Results," Rev. Mod. Phys. **61** (1989) 561. E. Alvarez, A. F. Faedo and J. J. Lopez-Villarejo, "Ultraviolet behavior of transverse gravity," JHEP **0810** (2008) 023 [arXiv:0807.1293 [hep-th]].
[3] K. S. Stelle, "Renormalization of Higher Derivative Quantum Gravity," Phys. Rev. D **16**, 953 (1977). "Classical Gravity with Higher Derivatives," Gen. Rel. Grav. 9, 353 (1978). J. Julve and M. Tonin, "Quantum Gravity with Higher Derivative Terms," Nuovo Cim. B **46** (1978) 137.
[4] E. Alvarez, J. Anero and S. Gonzalez-Martin, "Quadratic gravity in first order formalism," arXiv:1703.07993 [hep-th].
E. Alvarez and S. Gonzalez-Martin, "Weyl Gravity Revisited," JCAP **1702** (2017) no.02, 011.
[5] M. Borunda, B. Janssen and M. Bastero-Gil, "Palatini versus metric formulation in higher curvature gravity," JCAP **0811** (2008) 008.

CALOCUBE: A NOVEL APPROACH FOR A HOMOGENEOUS CALORIMETER FOR HIGH-ENERGY COSMIC RAYS DETECTION IN SPACE

Gabriele Bigongiari [a] on behalf of the CaloCube collaboration
INFN Pisa, Largo Bruno Pontecorvo 3, I-56127 Pisa, Italy

Abstract. Unambiguous measurements of the energy spectra and of the composition of cosmic rays up to the PeV region could provide some of the long sought answers to the pending questions on their origin, acceleration mechanism, and composition. The requirements for a space instrument to be lightweight and compact are two of the most severe limiting factors. That contrasts with the large geometrical acceptances needed to collect enough statistics in a reasonable time. To overcome this limitation an innovative cubic calorimeter is presented, which addresses these issues while limiting the mass and volume of the detector. The large acceptance needed is obtained by maximizing the number of entrance windows, while thanks to its homogeneity and high segmentation this new detector allows to achieve an excellent energy resolution, an enhanced separation power between hadrons and electrons. A prototype of the proposed detector, instrumented with CsI(Tl) cubic crystals, has already taken data with ion beams at CERN.

1 Introduction

The direct measurement of cosmic ray (CR) spectrum in the PeV region is one of the instrumental challenge for the future CR experiments. Indirect measurements on ground show, around this energy region, a sudden steeping in the inclusive spectrum of particles and a progressively heavier composition, a feature known as the CR knee. So a precise knowledge of particle spectra and composition in this spectral region would allow to address key items in the field of high-energy CR physics. The direct CR detection can permit unambiguous elemental identification and a more precise energy measurement, but suffers from low exposure due to the steepness of the CR spectrum. This limitation prevented the past experiments to go beyond 100 TeV/n for the nuclei and 1 TeV for electron + positron spectra. Direct measurements of cosmic ray proton and nuclei spectra up to 1 PeV/n and electron spectrum above 1 TeV require an acceptance of few m^{2}str, an energy resolution better than 40% for nuclei and 2% for electrons, a good charge identification and a high electron proton rejection power (at least 10^5).

2 A novel calorimeter

To achieve the above discussed performances, the major constraint comes from the limitation in weight for the detectors (few tons), which severely affects both the geometrical factor and the energy resolution. The R&D project CaloCube

---

[a] E-mail: gabriele.bigongiari@pi.infn.it

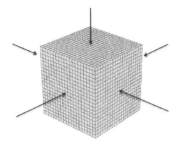 

Figure 1: the CaloCube design.   Figure 2: the partially completed prototype used for the tests at CERN.

aims to optimize the design of a space-borne calorimeter to extend the range of direct CR measurements up to the PeV region in order to measure the knee of the lightest components [1]. The proposed solution consists of a segmented calorimeter made of a large number of cubic scintillating crystals, readout by photodiodes (PDs), arranged to form a cube (see Fig. 1). The cubic geometry and the homogeneity provides the possibility to collect particles from either the top or the lateral faces, thus allowing to maximize the geometrical acceptance for a fixed mass budget. The active material provides good energy resolution, while the high granularity allows shower imaging and provides criteria for both leakage correction and h/e separation [2].

3 The Monte Carlo simulations

A FLUKA-based model of the calorimeter has been developed, in order to evaluate the expected performances and to optimize the design. A comparative study of different scintillating crystals has been done, among CsI(Tl), BaF 2, YAP(Yb), BGO and LYSO(Ce). For the hadron detection, the best choice is dictated by the balance between size (density of the absorber) and shower containment (interaction length), which determine energy resolution. The geometric parameters have been defined by assuming about 2 tons of active material in total; the size of the single cube has been fixed to one Moliere radius and the gap among adjacent elements has been rescaled to obtain the same active volume fraction ($\sim$78%). The signal induced in the PDs by the scintillation light has been evaluated by accounting for the light yield of the scintillators, the light collection efficiency on one face, the size and the quantum efficiency of the PD at the emission peak. Direct ionization on the PD has been also considered. Isotropic fluxes of protons hitting one face of the calorimeter have been generated and the effective geometrical factor evaluated. All the five geometries satisfy the basic requirements, by providing an effective geometrical factor of at least 2.5 m^{2}sr with an energy resolution better than 40%.

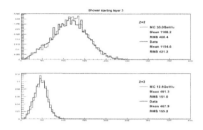
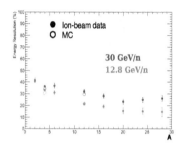

Figure 3: distribution of the energy deposit (in MIP units) of 30 GeV/n (upper panel) and 13 GeV/n (bottom panel) He ions.

Figure 4: energy resolution as a function of the ion mass number and of the beam energy, for showers having the same containment.

4 The prototype

To test the CaloCube concept, a prototype with 135 CsI(Tl) cubic crystals of 3.6 cm size, arranged in 15 planes of 3 × 3 cubes each, with a gap of 0.4 cm between them, has been constructed (see Fig. 2). This prototype results to have a lateral shower containment of about 1.5 Moliere radius and a total depth of 1.35 interaction lengths, corresponding to 28.4 radiation lengths, and has been tested at CERN with different particle beams . Signals are readout by means of polyimide flexible printed circuit boards and routed to the front-end board, placed on the side of the calorimeter. The front-end electronics is based on a high dynamic-range, low-noise ASIC, developed by members of the CaloCube collaboration. The chosen PD is a large-area ($\sim$100 mm^2) sensor that, coupled to CsI(Tl) crystals and readout electronics, allows to clearly detect minimum-ionizing protons with a signal-to-noise ratio of about 15. One of the most challenging requirements is the very large dynamic range (10^7) ranging from 20 MeV for minimum ionizing protons to 10% of the energy of a PeV proton. This will be accomplished by using also a second small area PD ($\sim$1 mm^2).

5 Tests with Ion and Electron beams

The first version of the prototype was tested in 2013 and 2015 with ion beams extracted from the H8 line of CERN SPS. The beam contained A/Z = 2 fragments produced by a primary (Pb/Ar) beam colliding with a target (Be/Poly). The experimental set-up included a Si tracking system in front of the calorimeter to provide tracking information and Z tagging. The single-crystal performances were studied, by selecting non-interacting ions. The responses were equalized by normalizing to the energy deposit of non- interacting He nuclei, the most abundant fragments. The showers developing inside the calorimeter were classified on the basis of the starting point, which can determine the shower containment. Figure 4 shows the energy resolution for different ions at the same shower containment while Fig. 5 shows the response of the calorimeter as a function of projectile energy for showers initiated before the fifth layer [2] [3]. A Fluka-based model of the prototype has been developed and its predicted response is shown in Fig. 3 and Fig. 4 (open circles) in comparison with real data. A fine tuning of the Monte Carlo was necessary in order to reproduce the beam-test data. In particular, an additional spread of 4.5% on the single-crystal responses and an optical cross-talk of 14% were introduced. During the beam test at CERN in the summer 2015, the prototype was initially exposed to μ beams to equalize the response of all the cubes that compose the calorimeter. Then an estimate of the energy resolution has been determined exposing the calorimeter to electron beams of different energy and determining the total deposited energy. A preliminary result is shown in Fig. 6, referring to a beam of 50 GeV electrons. The corresponding resolution is at level of 1.5%, in good agreement with the expectation.

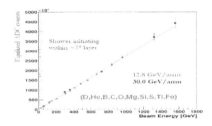
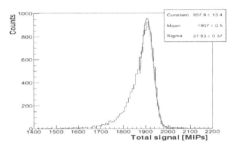

Figure 5: linearity of response as a function of projectile energy, with the nuclei identified by a separate silicon tracker.

Figure 6: measured distribution of total energy (expressed in MIP units) released with a 50 GeV electrons beam, fit with the expected distribution (red curve).

6 Recent developments

A new prototype has been constructed with a completely redesigned mechanics characterized by 18 layers, each equipped with a matrix of 5 × 5 crystals. The total depth of this new prototype is 1.6 interaction lengths, corresponding to 35 radiation lengths. In this prototype the light signal of each crystal is read out by two photodiodes, with different sensitive area, in order to cover the full expected dynamic range. This prototype was tested in 2016 with electron and hadron beams at the H4 line of CERN SPS. The analysis of data collected during the test is currently under way. Further beam tests are planned in the August and November 2017 at the CERN SPS.

Acknowledgments

This work was supported by the Istituto Nazionale di Fisica Nucleare (INFN), through the CaloCube project, and by the H2020 project AIDA-2020, GA no. 654168. The authors thank CERN for the allocation of beam time at the North Area test facility.

References

[1] Bongi M. et al., *J. Phys.: Conf. Ser.*,587 (2015) 012029.
[2] D'Alessandro R. et al., *Nucl. Instrum. Meth.*, A824 (2016), pp. 609-613.
[3] Vannuccini E. et al., *Nucl. Instrum. Meth.*, 845 (2017), pp. 421-424.

CONFORMAL INVARIANCE AND PHENOMENOLOGY OF COSMOLOGICAL PARTICLE PRODUCTION

Victor Berezin[1],[a] Vyacheslav Dokuchaev[1,2],[b] and Yury Eroshenko[1],[c]
[1] *Institute for Nuclear Research, RAS, 117312 Moscow, Russia*
[2] *National Research Nuclear University MEPhI, 115409, Moscow, Russia*

Abstract. Starting with the idea to describe phenomenologically the particle creation in the strong gravitational fields, we introduced explicitly the particle number nonconservation (= creation law) into the action integral with the corresponding Lagrange multiplier. Following the fundamental result by Ya. B. Zel'dovich and A. A. Starobinsky (1977) we then postulated that the rate of particle creation is proportional to the square of the Weyl tensor. Concerning the conformal invariance, yet another question arises: how the scalar field could know about the surgery made on the metric tensor (if such an invariance is the fundamental law of Nature and not just the mathematical exercise)? The only way is that the scalar field is itself the part of the metric, namely, the conformal factor. We showed, that such an identification results in the natural appearance of the quartic self-interaction term in the scalar field Lagrangian, which is needed to make particle massive. And it is just quartic, because our space-time is four-dimensional.

The quantum field theory predicts the phenomenon of particle creation from the vacuum fluctuations in the presence of the external fields. The modern cosmology based on the Friedmann's model of the expanding Universe provides us with the strong and rapidly varying gravitational field in the vicinity of the Big Bang singularity. In 1970's there was a a big activity in studying the particle creation by the quantized scalar field on the background scalar field on the background metric of the homogeneous and slightly anisotropic cosmological models [1–12]. Here we are interested in two main results of these investigations. First, it is the necessity of inclusion the additional terms into the gravitational part of the total action integral. Namely, the terms which are quadratic in Riemann curvature tensor and its counterparts, Ricci tensor and curvature scalar. And second, the role of particle production, as was shown by Ya. B. Zel'dovich and A. A. Starobinsky [13], is proportional to the square of the Weyl tensor, the latter being the completely traceless part of Riemann tensor. The appearance of the quadratic terms was foreseen by A. D. Sakharov [14] in 1967 (50 years ago!). However the attempts to solve the self-consistent problem, with the account for the back reaction of the (averaged) energy-momentum tensor of the quantized scalar fields on the space-time metric encounter serious obstacles. The matter is that in order to solve the quantum part of the problem, one needs to impose the appropriate boundary conditions, but this would be done only after solving the gravitational equations, for which the averaged quantum fields serve as the source. That is why we have to use some phenomenological approach. Our goal

---
[a] E-mail: berezin@inr.ac.ru
[b] E-mail: dokuchaev@inr.ac.ru
[c] E-mail: eroshenko@inr.ac.ru

is not only to take into account the energy-momentum tensor of the already created particles, but also the reaction of the space-time structure on the very process of the matter creation. There is a hope that this may be helpful in solving the singularity problem by violating the energy momentum-condition.

We are using the hydrodynamical description of the particle content elaborated by J. R. Ray [15]. The details can be found also in [16] and [17]. The convenience of the chosen approach is that the particle number conservation law infers the hydrodynamical action integral explicitly as the constraint with the corresponding Lagrange multiplier λ_1. And we simply replace it with the particle number creation law, namely

$$\int \lambda_1 (nu^\sigma)_{;\sigma} \sqrt{-g}\, dx \quad \to \quad \int \lambda_1 [(nu^\sigma)_{;\sigma} - \beta C^2] \sqrt{-g}\, dx, \qquad (1)$$

where we already inserted the square of the Weyl tensor, C^2, exploring the already mentioned fundamental result [13] for the cosmological particle creation. It is quite interesting to note that the Lagrange multiplier λ_1 is defined, actually, up to the additive constant. Indeed, let us replace $\lambda_1 \to \lambda_1 + \gamma_0$, $\gamma_0 = const$. Then, $\gamma_0(nu^\sigma)_{;\sigma}\sqrt{-g} = \gamma_0(nu^\sigma)_{,\sigma}$ is the full derivative which does not change the equation of motion, while the term $-\gamma_0 \beta C^2 = \alpha_0 C^2$ is just the Lagrangian density of the Weyl conformal gravity. And this is in spirit of the Sakharov's idea that the gravitational field is not fundamental, but simply the tensions of the quantum vacuum fluctuations of all the matter fields.

And what about the scalar field, χ, the source of the particle creation? We will choose for it the simplest possible Lagrangian which gives the linear equations of motion (without any self-interactions),

$$\mathcal{L}_{\text{scalar}} = \frac{1}{2}\chi_\sigma \chi^\sigma - \frac{1}{2}m^2 \chi^2, \quad \chi_\sigma = \chi_{,\sigma}, \quad \chi^\sigma = g^{\sigma\lambda}\chi_\lambda. \qquad (2)$$

Note that it is not that scalar field which is to be quantized, but some its residual part, because the already created particles are just the quanta of the genuine scalar field.

The conformal gravity was invented by H. Weyl in 1918 [18]. Then it was recognized that it allows only massless particle to exist, and on this ground the theory was rejected. But nowadays, such an unpleasant feature can be "corrected" by the Brout-Englert-Higgs mechanism for the spontaneous symmetry breaking [19]. Besides, the vacuum space-time with very high symmetry is a good candidate for the creation of the universe from "nothing" [20]. The idea that the initial state of the universe should be conformally invariant is advocated also by R. Penrose [21, 22]. We will consider the conformal invariance as the fundamental postulate.

By the conformal transformation we will understand the space-time depending scaling of the metric tensor $g_{\mu\nu}$:

$$ds^2 = g_{\mu\nu}(x)dx^\mu dx^\nu = \Omega^2(x)\hat{g}_{\mu\nu}(x)dx^\mu dx^\nu = \Omega^2(x)d\hat{s}^2. \qquad (3)$$

The, local, conformal invariance means $\delta S_{\text{tot}}/\delta\Omega = 0$. Therefore, we can consider the conformal factor Ω as an additional (to $g_{\mu\nu}$) dynamical variable.

It is known for a long time that, if the conformal transformation of the metric tensor (written above) is supplemented by the transformation $\hat{\chi} = \Omega\chi$ for the massless scalar field, then, by adding the term $\chi^2 R/12$, to the Lagrangian (2) with $m^2 = 0$, one makes the latter conformal invariant (of course, up to the full derivative which is usually neglected). In this way the Einstein-Hilbert Lagrangian (modified by the "dilaton" field) appeared in the total action integral. Remember, that we started with the pure matter (hydrodynamical) term and demanded the particle creation law, from which the Weyl tensor square was extracted (again the Sakharov's idea). But, even for massless scalar field (i.e., when $m^2 = 0$) one serious problem is remained and should be somehow solved. It is the problem of the common sign in front of the curvature scalar R and the kinetic term $\chi^\sigma \chi_\sigma$. Indeed, if one chooses the "correct" sign, i.e., $-\chi^2 R/12$, in order to get the attractive Einstein gravity, then the kinetic term for the scalar field appeared to be wrong, and vice versa. Our choice is the "correct" sign for the gravity. This requires some explanation. First, we do not care about the "correct" sign for the kinetic term, because our scalar filed χ is not the genuine (i.e., fundamental) one: some part of it we have already "used" as the created particles. Second, the "wrong" sign is, in a sense, good since it allows even number of particles to be created.

One more thing. We are now working in the framework of the Riemann geometry, This means that the only fundamental geometric quantity is the metric tensor $g_{\mu\nu}$, the connections $\Gamma^\lambda_{\mu\nu}$ being constructed exclusively from $g_{\mu\nu}$ and its derivatives (as well as the Riemann curvature tensor $R^\mu_{\nu\lambda\sigma}$ and its contractions, $R_{\mu\sigma}$ and R). All the matter fields are considered as the gravitating sources and something external. Then, how the field χ could "know" that it must be conformally transformed following the transformation of the metric tensor? Only, if it is the part of it. For the specific conformal factor $\Omega = \phi$ one can make $\hat{\chi} = \phi/l$, where l is some factor with the dimension of length. Then, the scalar part of the totla action takes the form (with the massive term restored)

$$S_{\text{scalar}} = -\frac{1}{l^2}\int \left(\frac{1}{2}\phi^\mu\phi_\mu + \frac{R}{12}\phi^2 - \frac{\Lambda}{6}\phi^4\right)\sqrt{-g}\,dx, \qquad (4)$$

where we introduced the cosmological term $\Lambda = 3m^2$. We see that there appears the self-interaction term ϕ^4, badly needed in order to switch on the Brout-Englert-Higgs mechanism! And the power 4 in this term is only in the case of the four-dimensional space-time!

We would like to pay tribute to P. I. Fomin whose paper "Gravitational instability of vacuum and the cosmological problem", published in 1973 [24], made him the precursor not only of the idea of the quantum birth of the Universe (from "nothing"), but also of the "emergent" universe scenarios [25–28]. One of us (V.B) expresses gratitude to G. Bisnovatyi-Kogan for reminding about this [29]. Authors acknowledge A. Smirnov and A. Starodubtsev for helpful discussions. This research supported in part by the RFBR under grant No. 15-02-05038-a.

References

[1] L. Parker, *Phys. Rev.* **183** (1969) 1057.
[2] A. A. Grib, S. G. Mamaev, *Sov. J. Nucl. Phys.* **10** (1970) 722.
[3] Ya. B. Zeldovich, *JETP Lett.* **9** (1970) 307.
[4] Ya. B. Zeldovich, L. P. Pitaevsky, *Comm. Math. Phys.* **23** (1971) 185.
[5] Ya. B. Zeldovich, A. A. Starobinskii, *Sov. Phys. JETP* **34** (1972) 1159.
[6] L. Parker, S. A. Fulling, *Phys. Rev. D* **7** (1973) 2357.
[7] S. A. Fulling, *Phys. Rev. D* **7** (1973) 2850.
[8] B. L. Hu, S. A. Fulling, L. Parker, *Phys. Rev. D* **8** (1973) 2377.
[9] L. Parker, S. A. Fulling, *Phys. Rev. D* **9** (1974) 341.
[10] S. A. Fulling, L. Parker, B. L. Hu, *Phys. Rev. D* **10** (1974) 3905.
[11] S. A. Fulling, L. Parker, *Ann. Phys.* **87** (1974) 176.
[12] V. N. Lukash, A. A. Starobinskii, *Sov. Phys. JETP* **39** (1974) 742.
[13] Ya. B. Zeldovich, A. A. Starobinskii, *JETP Lett.* **26** (1977) 252.
[14] A. D. Sakharov, Doklady Akad. Nauk SSSR, **70** (1967) (in Russian); [English translation in Sov. phys. Doklady, **12** (1968) 1040].
[15] J. R. Ray, J. Math. Phys. **13** (1972) 1451.
[16] V. A. Berezin, Int. J. Mod. Phys. A. **2** (1987) 1591.
[17] V. A. Berezin, V. I. Dokuchaev, Yu. N. Eroshenko, JCAP, **01** (2016) 019.
[18] H. Weyl, Reine Infinitesimalgeometrie, Math. Zeit. 2 (1918) 384.
[19] G. tHooft, arXiv:1410.6675 [gr-qc].
[20] A. Vilenkin, Phys. Lett. B **117** (1982) 25.
[21] R. Penrose, *Cycles of time: An extraordinary new view of the Universe* (The Random House, London, 2010).
[22] R. Penrose, Found. Phys. **44** (2014) 557.
[23] V. A. Berezin, V. I. Dokuchaev, Yu. N. Eroshenko, Intern. J. Mod. Phys. A **31** (2016) 1641004.
[24] P. I. Fomin, Preprint ITP-73-137P (1973), English translation in Problems of Atomic Science and Technology No. 3 (85), Series Nuclear Phys. Investigations (60), (2013) 6.
[25] R. Brout, F. Englert, E. Gunzig, Ann. Phys. **115** (1978) 78.
[26] R. Brout, F. Englert, J. -M. Frere, E. Gunzig, P. Nardone, C. Truffin, Ph. Spindel, Nucl. Phys. B **170** (1980) 228.
[27] I. Prigogine, J Gehenian, E. Gunzig, P. Nardone, Proc. Nat. Acad. Sci. USA, **85** (2088) 7420.
[28] I. Prigogine, J Gehenian, E. Gunzig, P. Nardone, Gen. Rel. Grav. **21** (2089) 8.
[29] G. Bisnovatyi-Kogan, private communication.

PROBLEMS OF SPONTANEOUS AND GRAVITATIONAL BARYOGENESIS

Elena Arbuzova [a] and Alexander Dolgov [b]

[a,b] *Novosibirsk State University, Pirogova ul., 2, 630090, Novosibirsk, Russia*
[a] *Dubna State University, Universitetskaya ul., 19, 141983, Dubna, Russia*
[b] *ITEP, Bol. Cheremushkinsaya ul., 25, 117259, Moscow, Russia*

Abstract. Spontaneous and closely related to it gravitational baryogenesis are critically analyzed. It is shown that the coupling of the curvature scalar to baryonic current, which induces nonzero baryonic asymmetry of the universe, simultaneously leads to higher order gravitational equations, which have exponentially unstable solutions. It is shown that this instability endangers the standard cosmology.

Different scenarios of baryogenesis are based, as a rule, on three well known Sakharov principles [1]: a) non-conservation of baryonic number; b) breaking of symmetry between particles and antiparticles; c) deviation from thermal equilibrium. For details see e.g. reviews [2]. However, none of these conditions is obligatory. Only one from three Sakharov principles, namely, non-conservation of baryons, is necessary for spontaneous baryogenesis (SBG) in its classical version [3], but this mechanism does not demand an explicit C and CP violation. It can proceed in thermal equilibrium and it is usually most efficient in thermal equilibrium.

The term "spontaneous" is related to spontaneous breaking of a global $U(1)$-symmetry, which ensures the conservation of the total baryonic number in the unbroken phase. When the symmetry is broken, the baryonic current becomes non-conserved and the Lagrangian density acquires the term

$$\mathcal{L}_{SBG} = (\partial_\mu \theta) J_B^\mu, \quad (1)$$

θ is the (pseudo)Goldstone field and J_B^μ is the baryonic current of matter fields.

For a spatially homogeneous field $\theta = \theta(t)$ the Lagrangian (1) is reduced to $\mathcal{L}_{SB} = \dot{\theta} n_B$, where $n_B \equiv J_B^0$ is the baryonic number density of matter, so it is tempting to identify $(-\dot{\theta})$ with the baryonic chemical potential, μ_B, of the corresponding system. As is argued in Refs. [4, 5], such identification is questionable and depends upon the representation chosen for the fermionic fields, but still the scenario is operative and presents a beautiful possibility to create an excess of particles over antiparticles in the universe.

Subsequently the idea of gravitational baryogenesis (GBG) was put forward [6], where the scenario of SBG was modified by the introduction of the coupling of the baryonic current to the derivative of the curvature scalar R:

$$\mathcal{L}_{GBG} = \frac{f}{m_0^2} (\partial_\mu R) J_B^\mu, \quad (2)$$

---
[a] E-mail: arbuzova@uni-dubna.ru
[b] E-mail: dolgov@fe.infn.it

where m_0 is a constant parameter with dimension of mass and f is dimensionless coupling constant which is introduced for the arbitrariness of the sign.

GBG scenarios possess the same interesting and nice features of SBG, namely generation of cosmological asymmetry in thermal equilibrium without necessity of explicit C or CP violation in particle physics. However, an introduction of the derivative of the curvature scalar into the Lagrangian of the theory results in high order gravitational equations which are strongly unstable. The effects of this instability may drastically distort not only the usual cosmological history, but also the standard Newtonian gravitational dynamics. We discovered such instability for scalar baryons [7] and found similar effect for the more usual spin one-half baryons (quarks) [8].

Let us start from the model where baryonic number is carried by scalar field ϕ with potential $U(\phi, \phi^*)$. The action of the scalar model has the form:

$$A = \int d^4x \sqrt{-g} \left[\frac{m_{Pl}^2}{16\pi} R + \frac{1}{m_0^2} (\partial_\mu R) J^\mu - g^{\mu\nu} \partial_\mu \phi \, \partial_\nu \phi^* + U(\phi, \phi^*) \right] - A_m, \quad (3)$$

where $m_{Pl} = 1.22 \cdot 10^{19}$ GeV is the Planck mass, A_m is the matter action, $J^\mu = g^{\mu\nu} J_\nu$, and $g^{\mu\nu}$ is the metric tensor of the background space-time. We assume that initially the metric has the usual GR form and study the emergence of the corrections due to the instability described below.

In the homogeneous case the equation for the curvature scalar in the FRW metric takes the form:

$$\frac{m_{Pl}^2}{16\pi} R + \frac{q^2}{6m_0^4} \left(R + 3\partial_t^2 + 9H\partial_t\right) \left[\left(\ddot{R} + 3H\dot{R}\right) T^2\right] + \frac{1}{m_0^2} \dot{R} \langle J^0 \rangle = -\frac{T^{(tot)}}{2}, \quad (4)$$

where $\langle J^0 \rangle$ is the thermal average value of the baryonic number density of ϕ, q is the baryonic number of ϕ, $H = \dot{a}/a$ is the Hubble parameter, and $T^{(tot)}$ is the trace of the energy-momentum tensor of matter including contribution from the ϕ-field. In the homogeneous and isotropic cosmological plasma

$$T^{(tot)} = \rho - 3P, \quad (5)$$

where ρ and P are respectively the energy density and the pressure of the plasma. For relativistic plasma $\rho = \pi^2 g_* T^4/30$ with T and g_* being the plasma temperature and the number of particle species in the plasma. The Hubble parameter is expressed through ρ as $H^2 = 8\pi\rho/(3m_{Pl}^2) \sim T^4/m_{Pl}^2$.

Keeping only the linear in R terms and neglecting higher powers of R, such as R^2 or HR, we obtain the linear fourth order differential equation:

$$\frac{d^4 R}{dt^4} + \mu^4 R = -\frac{1}{2} T^{(tot)}, \text{ where } \mu^4 = \frac{m_{Pl}^2 m_0^4}{8\pi q^2 T^2}. \quad (6)$$

The homogeneous part of this equation has exponential solutions $R \sim \exp(\lambda t)$ with $\lambda = |\mu| \exp(i\pi/4 + i\pi n/2)$, where $n = 0, 1, 2, 3$. There are two solutions with positive real parts of λ. This indicates that the curvature scalar is exponentially unstable with respect to small perturbations, so R should rise exponentially fast with time and quickly oscillate around this rising function.

Now we need to check if the characteristic rate of the perturbation explosion is indeed much larger than the rate of the universe expansion, that is:

$$(\operatorname{Re}\lambda)^4 > H^4 = \left(\frac{8\pi\rho}{3m_{Pl}^2}\right)^2 = \frac{16\pi^6 g_*^2}{2025} \frac{T^8}{m_{Pl}^4}, \quad (7)$$

where $\rho = \pi^2 g_* T^4/30$ is the energy density of the primeval plasma at temperature T and $g_* \sim 10 - 100$ is the number of relativistic degrees of freedom in the plasma. This condition is fulfilled if

$$\frac{2025}{2^9 \pi^7 q^2 g_*^2} \frac{m_{Pl}^6 m_0^4}{T^{10}} > 1, \quad (8)$$

or, roughly speaking, if $T \leq m_{Pl}^{3/5} m_0^{2/5}$. At these temperatures the instability is quickly developed and the standard cosmology would be destroyed.

Let us now generalize results, obtained for scalar baryons, to realistic fermions. We start from the action in the form

$$A = \int d^4x \sqrt{-g} \left[\frac{m_{Pl}^2}{16\pi} R - \mathcal{L}_m\right],$$

$$\mathcal{L}_m = \frac{i}{2}(\bar{Q}\gamma^\mu \nabla_\mu Q - \nabla_\mu \bar{Q}\gamma^\mu Q) - m_Q \bar{Q}Q$$

$$+ \frac{i}{2}(\bar{L}\gamma^\mu \nabla_\mu L - \nabla_\mu \bar{L}\gamma^\mu L) - m_L \bar{L}L$$

$$+ \frac{g}{m_X^2}\left[(\bar{Q}Q^c)(\bar{Q}L) + (\bar{Q}^c Q)(\bar{L}Q)\right] + \frac{f}{m_0^2}(\partial_\mu R)J^\mu + \mathcal{L}_{other}, \quad (9)$$

where Q is the quark (or quark-like) field with non-zero baryonic number, L is another fermionic field (lepton), ∇_μ is the covariant derivative of Dirac fermion in tetrad formalism, $J^\mu = \bar{Q}\gamma^\mu Q$ is the quark current with γ^μ being the curved space gamma-matrices, $\mathcal{L}_{other}$ describes all other forms of matter. The four-fermion interaction between quarks and leptons is introduced to ensure the necessary non-conservation of the baryon number with m_X being a constant parameter with dimension of mass and g being a dimensionless coupling constant. In grand unified theories m_X may be of the order of $10^{14} - 10^{15}$ GeV.

Varying the action (9) over metric, $g^{\mu\nu}$, and taking trace with respect to μ and ν, we obtain the following equation of motion for the curvature scalar:

$$-\frac{m_{Pl}^2}{8\pi} R = m_Q \bar{Q}Q + m_L \bar{L}L + \frac{2g}{m_X^2}\left[(\bar{Q}Q^c)(\bar{Q}L) + (\bar{Q}^c Q)(\bar{L}Q)\right]$$

$$- \frac{2f}{m_0^2}(R + 3D^2)D_\alpha J^\alpha + T_{other}, \quad (10)$$

where T_{other} is the trace of the energy momentum tensor of all other fields. At relativistic stage, when masses are negligible, we can take $T_{other} = 0$. The average expectation value of the interaction term proportional to g is also small, so the contribution of all matter fields may be neglected.

We used the kinetic equation, which leads to an explicit dependence on R of the current divergence, $D_\alpha J^\alpha$, if the current is not conserved. As a result we obtain 4th order differential equation for R:

$$\frac{d^4 R}{dt^4} = \lambda^4 R, \text{ where } \lambda^4 = \frac{5 m_{Pl}^2 m_0^4}{36\pi g_s B_q f^2 T^2}. \quad (11)$$

Here g_s and B_q are respectively the number of the spin states and the baryonic number of quarks. Deriving this equation we neglected the Hubble parameter factor in comparison with time derivatives of R.

Evidently Eq. (11) has extremely unstable solution with instability time by far shorter than the cosmological time. This instability would lead to an explosive rise of R, which may possibly be terminated by the nonlinear terms proportional to the product of H to lower derivatives of R. Correspondingly one may expect stabilization when $HR \sim \dot{R}$, i.e. $H \sim \lambda$. Since $\dot{H} + 2H^2 = -R/6$, H would also exponentially rise together with R, $H \sim \exp(\lambda t)$ and $\lambda H \sim R$. Thus stabilization may take place at $R \sim \lambda^2 \sim m_{Pl} m_0^2/T$. This result should be compared with the normal General Relativity value $R_{GR} \sim T_{matter}/m_{Pl}^2$, where T_{matter} is the trace of the energy-momentum tensor of matter.

Acknowledgments

This work was supported by the RSF Grant N 16-12-10037.

References

[1] A.D. Sakharov, Pis'ma ZhETF **5** (1967) 32.
[2] A.D.Dolgov, Phys. Repts **222** (1992) No. 6; A.D. Dolgov, Surveys in High Energy Physics, **13** (1998) 83; V.A. Rubakov, M.E. Shaposhnikov, Usp. Fiz. Nauk, **166** (1996) 493.
[3] A. Cohen, D. Kaplan, Phys. Lett. **B 199**, (1987) 251; A. Cohen, D. Kaplan, Nucl. Phys. **B 308** (1988) 913; A. G. Cohen, D.B. Kaplan, A.E. Nelson, Phys. Lett. **B 263** (1991) 86; A. G. Cohen, D.B. Kaplan, A.E. Nelson, Phys. Lett. **B 336** (1994) 41.
[4] A.D. Dolgov, K. Freese, Phys. Rev. **D 51** (1995) 2693-2702; A.D. Dolgov, K. Freese, R. Rangarajan, M. Srednicki, Phys. Rev. **D 56** (1997) 6155.

[5] E.V. Arbuzova, A.D. Dolgov, V.A. Novikov, Phys. Rev. **D 94** (2016) 123501.
[6] H. Davoudiasl, R. Kitano, G. D. Kribs, H. Murayama, P. J. Steinhardt, Phys. Rev. Lett. **93** (2004) 201301.
[7] E.V. Arbuzova, A. D. Dolgov, Phys. Lett. **B 769** (2017) 171.
[8] E.V. Arbuzova, A. D. Dolgov, JCAP **1706** (2017) 001.

DIRECTIONAL DETECTION OF DARK MATTER WITH A NUCLEAR EMULSION BASED DETECTOR

Andrey Alexandrov[a] on behalf of the NEWSdm collaboration
INFN sezione di Napoli, I-80126 Napoli, Italy

Abstract. The directional detection of Dark Matter requires very sensitive experiment combined with highly performing technology. The NEWSdm experiment, based on nuclear emulsions, is proposed to measure the direction of WIMP-induced nuclear recoils and it is expected to produce a prototype in 2017. We discuss the discovery potential of a directional experiment based on the use of a solid target made by newly developed nuclear emulsions and read-out systems reaching sub-micrometric resolution.

1 Introduction

A variety of experiments have been developed over the past decades, aiming at detecting Weakly Interactive Massive Particles (WIMPs) via their scattering in a detector medium [1]. Doubling their sensitivity roughly every 18 months, ultimately, direct detection experiments will start to see signals from coherent scattering of solar, atmospheric and diffuse supernova neutrinos. Hence, direct DM searches below the so-called "neutrino floor" will eventually require background subtraction or directional capability in WIMP direct detection detectors to separate neutrinos from the dark matter signals. Therefore, the directionality becomes particularly valuable when the residual background energy spectrum mimics the one from a possible WIMP signal [2]. Directional detection is a next-generation strategy that offers a unique opportunity to conclusively identify WIMP events and aims to reconstruct both the energy and the track direction of a recoiling nucleus following a WIMP scattering.

2 Experimental concept

The NEWSdm experiment (Nuclear Emulsions for WIMP Search directional measurement) [3] proposes an innovative approach for a high sensitivity experiment to directly detect WIMPs. The detector is based on innovative nuclear emulsions with an extremely high spatial resolution allowing to see the direction of the recoiled nuclei. Nuclear emulsions act both as target and as tracking detector with nanometric resolution to reconstruct the direction of the recoiled nucleus. Nuclear emulsions are unique detectors for the measurement of WIMP direction: given the extremely higher sensitivity and density compared to gas detectors, they are the ideal candidates to stand out in the landscape of the direct dark matter searches with the directionality approach.

---

[a]E-mail: andrey.alexandrov@na.infn.it

Nuclear emulsions are made of small crystals of silver halide (usually bromide, AgBr) immersed in an organic gelatine. The energy loss of ionizing particles crossing the films induces along its path atomic-scale perturbations that, after a chemical treatment, produce a sequence of visible grains in the emulsion. The so-called Nano Imaging Trackers (NIT) and Ultra-Nano Imaging Trackers (U-NIT) [4], have grains of 44 and 24 nm diameter, respectively. The granularity corresponds to the average distance between the crystals. These distances are 71 nm for NIT and 40 nm for U-NIT. Therefore, tracks of these ranges are detectable, assuming a detection efficiency of 100%. These ranges are equivalent to 25 keV for NIT and 13 keV for U-NIT in the carbon recoil case. Both emulsion types make the reconstruction of trajectories with path lengths shorter than 100 nm possible, if analyzed by means of microscopes with enough resolution.

Fast automatic optical microscopes are used to read out nuclear emulsions [8]. After the exposure, the search for signal candidates requires the scanning of the whole emulsion volume. The analysis of NIT emulsions is performed with a two-step approach: a fast scanning, based on techniques developed for the OPERA experiment [5–7] for the signal preselection: an ellipse is fitted to the optically reconstructed track, and candidate events are selected by applying a cut on the ratio of the lengths of the major and minor axes [9]. After elliptical shape recognition with optical microscopy candidate events are confirmed by super-resolution microscopy techniques.

Directional detection in solids is an extremely challenging task due to low energy of recoils produced in WIMP interactions. The expected recoil lengths in emulsions do not exceed 1 μm and the major part of recoils is shorter than 200 nm, which is the resolution limit of optical microscopy. Detailed simulations of the realistic NEWSdm detector [10] show that the reduction of the detection threshold by the factor of 2 reduces the required exposure by one order of magnitude, the factor of 10. And, therefore, overcoming the readout resolution limit is crucial for the overall sensitivity.

NIT emulsions with controllable grain size and shape open a possibility to take advantage of the Localized Surface Plasmon Resonance (LSPR) optical phenomenon, generated by an interaction of an incident light wave and conductive nanoparticles (NPs), silver grains [11,12]. Properties of light scattered by NPs strongly depend on the composition, size, geometry, dielectric environment and NPs separation distance. Moreover, non-spherical shape of NPs leads to a strong polarization effect that could be used to improve the optical resolution of a microscope and overcome the diffraction limit bringing it to super-resolution. The brightness of grain images (clusters) changes with the rotation of the polarization angle, with the phase being quite random and independent even for grains belonging to the same track. Analysis of the cluster brightness change allows distinguishing tracks composed of several close grains from single grains as well as isolating individual grains inside the track with a 10 nm accuracy [8].

The final sensitivity of low-energy rare event searches is strongly limited by the background induced by radioactivity. Two main categories have to be taken into account: the environmental or external background and the intrinsic one. The main background sources α, β, γ-rays and and neutron induced recoils, while NIT are essentially not sensitive to minimum ionizing particles (MIP). α originating from U and Th isotopes having energies of the order of MeV, can be identified by measuring the track length. Moreover, γ-rays and β can be rejected by properly regulating the emulsion response, in terms of number of sensitized crystals per unit path length (i.e. the sensitivity), through a chemical treatment of the emulsion itself, therefore they are considered as reducible background. However, the contribution from neutrons is usually considered as irreducible, since they induce nuclear recoils as WIMPs do. The measured neutron-induced background due to the intrinsic radioactive contamination allows the design of an emulsion detector with an exposure of about 10 kg×year [13]. A careful selection of the emulsion components and a better control of their production could further increase the radiopurity, thus extending the detector exposure.

The directional detection has the unique capability of distinguishing the WIMP signal from the background by exploiting the feature of the signal, expected to be peaked in the direction of motion of the Sun. On the contrary, there is no reason for the neutron-induced nuclear recoil spectrum to present the same feature. In particular, the neutron background is expected to be isotropically distributed. Neutrons in underground laboratories originate either from cosmic muon interactions, or from environmental radioactivity, or from spontaneous fissions and (α,n) reactions. The first two sources can be reduced by an appropriate shielding; the latter one, coming from the intrinsic radioactivity of the target materials, is associated with an isotropic distribution in the laboratory frame.

The 90% C.L. upper limit in case of null observation is shown in Fig. 1 for an exposure of 1 kg.year of NIT emulsions, with a minimum detectable track length ranging from 200 nm down to 50 nm and in the hypothesis of zero background [3]. Even not including the directionality discrimination of the signal and assuming to reach a negligible background level, such an experiment would cover a large part of the parameter space indicated by the DAMA/LIBRA results with a small (1 kg) detector mass, using a powerful and complementary approach.

3 Conclusions and outlook

Both innovative NIT emulsions and the fast super-resolution miroscopy are key ingredients for next-generation directional DM searches. Study of new phenomena and technologies is now ongoing: an R&D on sintetic polymer emulsions is started to eliminate the intrinsic radioactivity; NIT and U-NIT emulsions are being tested under low temperature conditions to reduce thermal excitation; the optical readout is being extended to provide both the 3D and wavelength

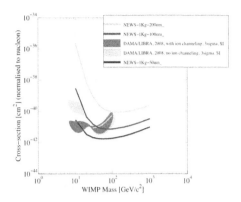

Figure 1: The 90% C.L. upper limits for a NIT detector with an exposure of 1 kg X year, a threshold ranging from 200 nm down to 50 nm, in the zero background hypothesis. The directionality information is not included [3].

information; novel techniques of computational microscopy are under testing to boost the readout; a dedicated deep learning analysis method is being developed to provide a gain in sensitivity, background discrimination and head-tail sense recognition.

As a result of the continuous improvements in the field of nuclear emulsion technologies and super-resolution readout, relevant progresses in the reduction of the track length threshold are foreseen in the near future, thus allowing the NEWSdm experiment to start building a few kg mass detector as a demonstrator of a technology scalable to multi-ton masses.

References

[1] T. M. Undagoitia et al., J. Phys. G: Nucl. Part. Phys. **43** (2016) 013001.
[2] J. Billard et al., Phys. Rev. **D 91** (2015) 023513.
[3] A. Aleksandrov et al., LNGS-LOI **48/15** (2015), arXiv:1604.04199.
[4] T. Asada et al., PTEP **6** (2017) 063H01.
[5] A. Alexandrov et al, JINST **10** (2015) P11006.
[6] A. Alexandrov et al, JINST **11** (2016) P06002.
[7] A. Alexandrov et al, Nature Scientific Reports **7** (2017) 7310.
[8] A. Alexandrov et al, NIM **A824** (2016) 600.
[9] J. B. R. Battat et al, Physics Reports **662** (2016) 1.
[10] N. Agafonova et al, arXiv:1705.00613.
[11] J. J. Mock et al, The Journal of Chemical Physics **116** (2002) 6755.
[12] V. Myroshnychenko et al., Chem. Soc. Rev. **37** (2008) 1792-1805.
[13] A. Alexandrov et al, Astroparticle Physics **80** (2016) 16.

DAMA/LIBRA RESULTS AND PERSPECTIVES

R. Bernabei[a], P. Belli, R. Cerulli, S. d'Angelo[b],
A. Di Marco, V. Merlo, F. Montecchia[c]
*Dip. di Fisica, Univ. "Tor Vergata" and
INFN sez. Roma "Tor Vergata", I-00133 Rome, Italy*
A. d'Angelo, A. Incicchitti, F. Cappella
*Dip. di Fisica, Univ. "La Sapienza" and
INFN sez. Roma, I-00185 Rome, Italy*
V. Caracciolo
*INFN, Laboratori Nazionali del Gran Sasso,
I-67100 Assergi (AQ), Italy*
C.J. Dai, H.L. He, H.H. Kuang, X.H. Ma, X.D. Sheng, R.G. Wang, Z.P. Ye[d]
*IHEP, Key Laboratory of Particle Astrophysics,
100049 Beijing, China*

Abstract. The DAMA/LIBRA experiment ($\sim$ 250 kg of highly radio-pure NaI(Tl) sensitive mass) is in data taking in the Gran Sasso National Laboratory (LNGS). During the first phase of this experiment (DAMA/LIBRA-phase1) and the former DAMA/NaI experiment ($\sim$ 100 kg of highly radio-pure NaI(Tl)) data have been collected over 14 independent annual cycles, (total exposure 1.33 ton × yr) exploiting the model-independent Dark Matter (DM) annual modulation signature. A peculiar annual modulation effect has been observed at 9.3 σ C.L. satisfying all the several specific requirements of the exploited DM signature and, thus, supporting the presence of DM particles in the galactic halo. No systematic or side reaction able to mimic the observed DM annual modulation has been found or suggested by anyone. Recent analyses on possible diurnal effects, on the Earth shadowing effect and on corollary analyses in terms of Mirror DM will be mentioned. At present DAMA/LIBRA is running in its phase2 with increased sensitivity.

1 Introduction

The DAMA project develops and uses low background scintillators. In particular, the second generation DAMA/LIBRA set-up [1–21], as the former DAMA/NaI (see for example Ref. [8, 22–29] and references therein), is further investigating DM particles in the galactic halo by exploiting the DM model-independent annual modulation signature (originally suggested in the mid 80's [30]) with increased sensitivity. The detailed description of the DAMA/LIBRA set-up during the phase1 has been discussed in details in Ref. [1–4, 8, 17–21].

The signature exploited by DAMA/LIBRA (the model independent DM annual modulation) is a consequence of the Earth's revolution around the Sun

---
[a]E-mail: rita.bernabei@roma2.infn.it
[b]Deceased
[c]Also Dip. di Ingegneria Civile e Ingegneria Informatica, Università di Roma "Tor Vergata", I-00133 Rome, Italy
[d]Also University of JingGangshan, Jiangxi, China

which is moving in the galaxy; in fact, the Earth should be crossed by a larger flux of DM particles around $\simeq 2$ June (when the projection of the Earth orbital velocity on the Sun velocity with respect to the Galaxy is maximum) and by a smaller one around $\simeq 2$ December (when the two velocities are opposite). This DM annual modulation signature is very effective since the effect induced by DM particles must simultaneously satisfy many requirements: the rate must contain a component modulated according to a cosine function (1) with one year period (2) and a phase peaked roughly $\simeq 2$ June (3); this modulation must only be found in a well-defined low energy range, where DM particle induced events can be present (4); it must apply only to those events in which just one detector of many actually "fires" (*single-hit* events), since the DM particle multi-interaction probability is negligible (5); the modulation amplitude in the region of maximal sensitivity must be $\simeq 7\%$ for usually adopted halo distributions (6), but it can be larger (even up to $\simeq 30\%$) in case of some possible scenarios such as e.g. those in Ref. [31, 32]. Thus, this signature is model-independent, very discriminating and, in addition, it allows the test of a large range of DM candidates, cross sections and halo densities. This DM signature might be mimicked only by systematic effects or side reactions able to account for the whole observed modulation amplitude and to simultaneously satisfy all the requirements given above. No one is available [1–4, 7, 8, 12–16, 19, 21–23, 33].

2 The Results of DAMA/LIBRA–phase1 and DAMA/NaI

The total exposure of DAMA/LIBRA–phase1 is 1.04 ton × yr over seven annual cycles; when also including that of the first generation DAMA/NaI experiment, it is 1.33 ton × yr, corresponding to 14 independent annual cycles [2–4, 8].

Many different independent data analyses have been performed obtaining always consistent results. In particular, for simplicity we just report in Fig. 1 the time behaviour of the experimental residual rates of the *single-hit* scintillation events in DAMA/LIBRA–phase1 in the (2–6) keV energy interval. The χ^2 test excludes the hypothesis of absence of modulation in the data: $\chi^2/\text{d.o.f.} = 154/87$ for the (2–6) keV energy interval (P-value = 2.2×10^{-3}). When fitting the *single-hit* residual rate of DAMA/LIBRA–phase1 and DAMA/NaI, with the function: $A\cos\omega(t-t_0)$, considering a period $T = \frac{2\pi}{\omega} = 1$ yr and a phase $t_0 = 152.5$ day (June 2nd) as expected for the DM annual modulation signature, the following modulation amplitude is obtained: $A = (0.0110 \pm 0.0012)$ cpd/kg/keV corresponding to 9.2 σ C.L.. When the period, and the phase are kept free in the fitting procedure, the modulation amplitude is (0.0112 ± 0.0012) cpd/kg/keV (9.3 σ C.L.), the period $T = (0.998 \pm 0.002)$ year and the phase $t_0 = (144 \pm 7)$ day, values well in agreement with expectations for a DM annual modulation signal. In particular, the phase is consistent with about June 2nd and is fully consistent with the value independently determined by Maximum

Likelihood analysis [4][e]. The run test and the χ^2 test on the data have shown that the modulation amplitudes singularly calculated for each annual cycle of DAMA/NaI and DAMA/LIBRA–phase1 are normally fluctuating around their best fit values [2-4, 8].

A power spectrum analysis of the *single-hit* residuals of DAMA/LIBRA–phase1 and DAMA/NaI [8] has been performed, obtaining a clear principal mode in the (2–6) keV energy interval at a frequency of 2.737×10^{-3} d^{-1}, corresponding to a period of $\simeq 1$ year, while only aliasing peaks are present in the energy region above 6 keV.

Absence of any significant background modulation in the energy spectrum has been verified in energy regions not of interest for DM [4]; it is worth noting that the obtained results account for whatever kind of background and, in addition, no background process able to mimic the DM annual modulation signature (that is able to simultaneously satisfy all the peculiarities of the signature and to account for the whole measured modulation amplitude) is available (see also discussions e.g. in Refs. [1–4, 7, 8, 14, 15]).

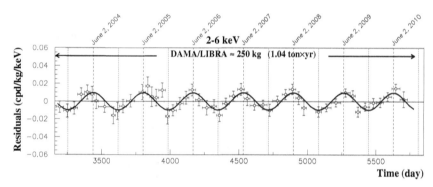

Figure 1: Experimental residual rate of the *single-hit* scintillation events measured by DAMA/LIBRA–phase1 in the (2–6) keV energy interval as a function of the time. The superimposed curve is the cosinusoidal function behaviour $A \cos \omega (t - t_0)$ with a period $T = \frac{2\pi}{\omega} = 1$ yr, a phase $t_0 = 152.5$ day (June 2nd) and modulation amplitude, A, equal to the central values obtained by best fit on the data points of the entire DAMA/LIBRA–phase1. The dashed vertical lines correspond to the maximum expected for the DM signal (June 2nd), while the dotted vertical lines correspond to the minimum.

A further relevant investigation in the DAMA/LIBRA–phase1 data has been performed by applying the same hardware and software procedures, used to acquire and to analyse the *single-hit* residual rate, to the *multiple-hit* one. In fact, since the probability that a DM particle interacts in more than one detector is negligible, a DM signal can be present just in the *single-hit* residual rate.

---

[e]For completeness, we recall that a slight energy dependence of the phase could be expected e.g. in case of possible contributions of non-thermalized DM components to the galactic halo, such as e.g. the SagDEG stream [34–36] and the caustics [37].

Thus, the comparison of the results of the *single-hit* events with those of the *multiple-hit* ones corresponds practically to compare between them the cases of DM particles beam-on and beam-off. This procedure also allows an additional test of the background behaviour in the same energy interval where the positive effect is observed. In particular, the residual rates of the *single-hit* events measured in the (2–6) keV energy interval over the DAMA/LIBRA–phase1 annual cycles, as collected in a single cycle, are reported in Ref. [4] together with the residual rates of the *multiple-hit* events in the same energy interval. A clear modulation satisfying all the peculiarities of the DM annual modulation signature is present in the *single-hit* events, while the fitted modulation amplitude for the *multiple-hit* residual rate is well compatible with zero: $-(0.0005 \pm 0.0004)$ cpd/kg/keV in the same energy region (2–6) keV. Thus, again evidence of annual modulation with the peculiar features required by the DM annual modulation signature is present in the *single-hit* residuals (events class to which the DM particle induced events belong), while it is absent in the *multiple-hit* residuals (event class to which only background events belong). Similar results were also obtained for the last two annual cycles of the DAMA/NaI experiment [23]. Since the same identical hardware and the same identical software procedures have been used to analyse the two classes of events, the obtained result offers an additional strong support for the presence of a DM particle component in the galactic halo.

By performing a maximum-likelihood analysis of the *single-hit* scintillation events, the modulation amplitude, S_m, as a function of the energy has been derived by considering $T = 1$ yr and $t_0 = 152.5$ day. These results have also reiterated the presence of a positive signal in the (2–6) keV energy interval, while S_m values compatible with zero are present above [4]. Moreover, the observed annual modulation effect is well distributed in all the 25 detectors, the annual cycles and the energy bins at 95% C.L. [2–4, 8], Further analyses have been performed; all of them confirm the evidence for the presence of an annual modulation in the data satisfying all the requirements of a DM signal.

Sometimes naive statements were put forwards as the fact that in nature several phenomena may show some kind of periodicity. The point is whether they might mimic the annual modulation signature in DAMA/LIBRA (and former DAMA/NaI), i.e. whether they might be not only quantitatively able to account for the observed modulation amplitude but also able to contemporaneously satisfy all the requirements of the DM annual modulation signature. The same is also for side reactions. This has already been deeply investigated in Ref. [1–4] and references therein. Additional arguments can be found in Ref. [7, 8, 14, 15]. Cautious upper limits on possible contributions to the DAMA/LIBRA measured modulation amplitude are summarized in Ref. [2–4]. It is worth noting that they do not quantitatively account for the measured modulation amplitudes, and also are not able to simultaneously satisfy all the

Table 1: Summary of the contributions to the total neutron flux at LNGS; the value, $\Phi^{(n)}_{0,k}$, the relative modulation amplitude, η_k, and the phase, t_k, of each component is reported. It is also reported the counting rate, $R_{0,k}$, in DAMA/LIBRA for *single-hit* events, in the $(2-6)$ keV energy region induced by neutrons, muons and solar neutrinos, detailed for each component. The modulation amplitudes, A_k, are reported as well, while the last column shows the relative contribution to the annual modulation amplitude observed by DAMA/LIBRA, $S^{exp}_m \simeq 0.0112$ cpd/kg/keV [4]. For details see Ref. [15] and references therein.

	Source	$\Phi^{(n)}_{0,k}$ (neutrons cm^{-2} s^{-1})	η_k	t_k	$R_{0,k}$ (cpd/kg/keV)	$A_k = R_{0,k}\eta_k$ (cpd/kg/keV)	A_k/S^{exp}_m
SLOW neutrons	thermal n ($10^{-2}-10^{-1}$ eV)	1.08×10^{-6}	$\simeq 0$ however $\ll 0.1$	–	$< 8 \times 10^{-6}$	$\ll 8 \times 10^{-7}$	$\ll 7 \times 10^{-5}$
	epithermal n (eV-keV)	2×10^{-6}	$\simeq 0$ however $\ll 0.1$	–	$< 3 \times 10^{-3}$	$\ll 3 \times 10^{-4}$	$\ll 0.03$
FAST neutrons	fission, $(\alpha, n) \to n$ (1-10 MeV)	$\simeq 0.9 \times 10^{-7}$	$\simeq 0$ however $\ll 0.1$	–	$< 6 \times 10^{-4}$	$\ll 6 \times 10^{-5}$	$\ll 5 \times 10^{-3}$
	$\mu \to $ n from rock (> 10 MeV)	$\simeq 3 \times 10^{-9}$	0.0129	end of June	$\ll 7 \times 10^{-4}$	$\ll 9 \times 10^{-6}$	$\ll 8 \times 10^{-4}$
	$\mu \to $ n from Pb shield (> 10 MeV)	$\simeq 6 \times 10^{-9}$	0.0129	end of June	$\ll 1.4 \times 10^{-3}$	$\ll 2 \times 10^{-5}$	$\ll 1.6 \times 10^{-3}$
	$\nu \to $ n (few MeV)	$\simeq 3 \times 10^{-10}$	0.03342*	Jan. 4th*	$\ll 7 \times 10^{-5}$	$\ll 2 \times 10^{-6}$	$\ll 2 \times 10^{-4}$
	direct μ	$\Phi^{(\mu)}_0 \simeq 20\ \mu$ m^{-2}d^{-1}	0.0129	end of June	$\simeq 10^{-7}$	$\simeq 10^{-9}$	$\simeq 10^{-7}$
	direct ν	$\Phi^{(\nu)}_0 \simeq 6 \times 10^{10}\ \nu$ cm^{-2}s^{-1}	0.03342*	Jan. 4th*	$\simeq 10^{-5}$	3×10^{-7}	3×10^{-5}

* The annual modulation of solar neutrino is due to the different Sun-Earth distance along the year; so the relative modulation amplitude is twice the eccentricity of the Earth orbit and the phase is given by the perihelion.

many requirements of the signature. Similar analyses have also been done for the DAMA/NaI data [22, 23].

In particular, we recall here that in Refs. [7, 15] a quantitative evaluation why the neutrons, the muons and the solar neutrinos cannot give any significant contribution to the DAMA annual modulation results and cannot mimic this signature is outlined. Table 1 summarizes the safety upper limits on the contributions to the observed modulation amplitude due to the total neutron flux at LNGS, either from (α, n) reactions, from fissions and from muons' and solar-neutrinos' interactions in the rocks and in the lead around the experimental set-up; the direct contributions of muons and solar neutrinos are reported there too.

In conclusion, DAMA/LIBRA has confirmed the presence of an annual modulation satisfying all the requirements of the DM annual modulation signature, as previously pointed out by DAMA/NaI; in particular, the evidence for the presence of DM particles in the galactic halo is cumulatively supported at 9.2 σ C.L.

The obtained model independent evidence is compatible with a wide set of scenarios regarding the nature of the DM candidate and related astrophysical, nuclear and particle Physics. For examples some given scenarios and parameters are discussed in Refs. [22–29, 34] and in Appendix A of Ref. [2]. Many other interpretations of the annual modulation results are available in literature; others are open.

It is worth noting that no other experiment exists, whose result can be directly compared in a model-independent way with those by DAMA/NaI and DAMA/LIBRA. Moreover, concerning those activities claiming for some model dependent exclusion under some set of largely arbitrary assumptions (see for example discussions in [2, 22, 23, 44, 45]), some important critical points exist in some of their experimental aspects (energy threshold, energy scale, multiple selection procedures, disuniformity of the detectors response, absence of suitable periodical calibrations in the same running conditions and in the claimed low energy region, stabilities, etc.).

Finally, as regards the indirect detection searches, let us note that no direct model-independent comparison can be performed between the results obtained in direct and indirect activities, since it does not exist a biunivocal correspondence between the observables in the two kinds of experiments. Anyhow, if possible excesses in the positron to electron flux ratio and in the γ rays flux with respect to an assumed simulation of the assumed background contribution, which is expected from standard sources, might be interpreted in terms of Dark Matter (but huge and still unjustified boost factor and new interaction types are required), this would also be not in conflict with the effect observed by DAMA experiments.

For completeness, other rare processes [9–11] have also been searched for by DAMA/LIBRA.

2.1 Diurnal modulation

The low energy *single-hit* data collected by DAMA/LIBRA–phase1 have been analysed in terms of a DM second order model-independent effect due to the Earth diurnal rotation around its axis [14]. Also this daily modulation of the rate on the sidereal time — expected when taking into account the contribution of the Earth rotation — presents some specific peculiarities. The interest in this latter signature is also based on the fact that the ratio R_{dy} of this DM diurnal modulation amplitude over the DM annual modulation amplitude is a model independent constant at a given latitude. Considering the LNGS latitude: $R_{dy} = S_d/S_m \simeq 0.016$. Thus, taking into account the DM annual modulation effect pointed out by DAMA/LIBRA–phase1 for single-hit events in the low energy region, the expected value of the diurnal modulation amplitude for the (2-6) keV energy interval is $\simeq 1.5 \times 10^{-4}$ cpd/kg/keV. No diurnal variation with a significance of 95% C.L. is found at the reached level of sensitivity, as reported in [14]. Considering the (2-6) keV energy interval the obtained upper limit on the DM diurnal modulation amplitude is 1.2×10^{-3} cpd/kg/keV (90% C.L.) [14]; thus, the effect of DM diurnal modulation, expected because of the Earth diurnal rotation on the basis of the DAMA DM annual modulation results, was out of the DAMA/LIBRA-phase1 sensitivity [14]. DAMA/LIBRA-phase2, presently running, with a lower software energy threshold [6] can also offer the possibility to increase sensitivity to such an effect.

2.2 Daily effect on the sidereal time due to the shadow of the earth

The low energy *single-hit* scintillation data of DAMA/LIBRA–phase1 have been analysed in terms of Earth Shadow Effect, a model-dependent effect that is expected for those DM candidates inducing just nuclear recoils and having a relative high cross-section (σ_n) with ordinary matter [20]. In fact a diurnal variation of the low energy *single-hit* rate could be expected for these specific candidates, because of the different thickness of the shield due to the Earth during the sidereal day. The induced effect should be a daily variation of their velocity distribution, and therefore of the signal rate measured deep underground. However, this effect is very small and would be appreciable only in case of high cross-section candidates. By the fact, this analysis decouples ξ (fraction of galactic dark halo in form of the considered candidate) from σ_n. Considering the measured DM annual modulation effect and the absence at the present level of sensitivity of diurnal effects, the analysis selects allowed regions in the three-dimensional space: ξ, σ_n and DM particle mass in some model scenarios; for details see Ref. [20].

2.3 Mirror dark matter

The DM model-independent annual modulation effect observed by the DAMA experiments has also been investigated in terms of a mirror-type dark matter candidates in some scenarios [46, 47]. In the framework of asymmetric mirror matter, the DM originates from hidden (or shadow) gauge sectors which have particles and interaction content similar to that of ordinary particles. It is assumed that the mirror parity is spontaneously broken and the electroweak symmetry breaking scale v' in the mirror sector is much larger than that in the Standard Model, $v = 174$ GeV. In this case, the mirror world becomes a heavier and deformed copy of our world, with mirror particle masses scaled in different ways with respect to the masses of the ordinary particles. Then, in this scenario dark matter would exist in the form of mirror hydrogen composed of mirror proton and electron, with mass of about 5 GeV which is a rather interesting mass range for dark matter particles. The data analysis in the Mirror DM model framework allows the determination of the $\sqrt{f}\epsilon$ parameter (where f is the fraction of Mirror DM in the Galaxy in form of mirror atoms and ϵ is the coupling constant). In the analysis several uncertainties on the astrophysical, particle physics and nuclear physics models have been taken into account in the calculation. The obtained values of the $\sqrt{f}\epsilon$ parameter in the case of mirror hydrogen atom ranges between 7.7×10^{-10} to 1.1×10^{-7} and they are well compatible with cosmological bounds [46]. In addition, releasing the assumption $M_{A'} \simeq 5m_p$, allowed regions for the $\sqrt{f}\epsilon$ parameter as function of $M_{A'}$, mirror hydrogen mass, obtained by marginalizing all the models for each considered scenario, have been obtained [46]; they also are well compatible with cosmological bounds.

In the framework of symmetric mirror DM model considered in Ref. [47], 3 different scenarios have been considered depending on: i) the adopted quenching factors; ii) either inclusion or not of the channeling effect; iii) either inclusion or not of the Migdal effect. For each scenario a different halo compositions reported have been considered, with halo temperature in the range 10^4–10^8 K and with halo velocity from -400 to $+300$ km/s. The results achieved in [47] for the symmetric mirror DM considered demonstrate that many configurations and halo models favoured by the annual modulation effect observed by DAMA corresponds to $\sqrt{f}\epsilon$ values well compatible with cosmological bounds. It is worth noting that our analysis predict in most halo models an increase of the DM Mirror signal below 2 keV. These behaviours can be tested with the present DAMA/LIBRA–phase2 that now is running.

3 DAMA/LIBRA–phase2 and Perspectives

An important upgrade was done at end of 2010 replacing all the PMTs with new ones having higher Quantum Efficiency (QE); details on the developments and on the reached performances in the operative conditions are reported in Ref. [6]. They have allowed us to lower the software energy threshold of the experiment to 1 keV and to improve also other features as e.g. the energy resolution [6].

Since the fulfilment of this upgrade and after some optimization periods, DAMA/LIBRA–phase2 is continuously running in order e.g.: (1) to increase the experimental sensitivity thanks to the lower software energy threshold; (2) to improve the corollary investigation on the nature of the DM particle and related astrophysical, nuclear and particle physics arguments; (3) to investigate other signal features and second order effects; (4) to investigate other rare processes as also the former DAMA/NaI apparatus did.

Future improvements to further increase the sensitivity of the set-up (possible DAMA/LIBRA-phase3) can be considered by using new metallic high QE and ultra-low background PMTs to be directly coupled to the NaI(Tl) crystals. In this way a further large improvement in the light collection and a further lowering of the software energy threshold would be obtained. The developments with Hamamatsu Co. are successful and 4 PMT prototypes are at hand. New protocols, and also alternative ideas, are ongoing that regards the possibility to dismount the light guide of the detectors which they act also as optical guides.

Finally, the possibility of a pioneering experiment with anisotropic $ZnWO_4$ detectors to further investigate, with the directionality approach, those DM candidates that scatter off target nuclei is in progress [48]; as e.g. the use of the present detectors untouched but with the metallic PMTs and new ultra low background crystal scintillators (e.g. $ZnWO_4$) placed among the DAMA/LIBRA detectors in order to reach a high sensitivity and contemporarily to perform directionality measurements is on going.

References

[1] R. Bernabei et al., Nucl. Instr. and Meth. A **592**, 297 (2008).
[2] R. Bernabei et al., Eur. Phys. J. C **56**, 333 (2008).
[3] R. Bernabei et al., Eur. Phys. J. C **67**, 39 (2010).
[4] R. Bernabei et al., Eur. Phys. J. C **73**, 2648 (2013).
[5] P. Belli et al., Phys. Rev. D **84**, 055014 (2011).
[6] R. Bernabei et al., J. of Instr. **7**, P03009 (2012).
[7] R. Bernabei et al., Eur. Phys. J. C **72**, 2064 (2012).
[8] R. Bernabei et al., Int. J. of Mod. Phys. A **28**, 1330022 (2013).
[9] R. Bernabei et al., Eur. Phys. J. C **62**, 327 (2009).
[10] R. Bernabei et al., Eur. Phys. J. C **72**, 1920 (2012).
[11] R. Bernabei et al., Eur. Phys. J. A **49**, 64 (2013).
[12] R. Bernabei et al., Adv. High Energy Phys. 605659 (2014).
[13] R. Bernabei et al., Phys. Part. Nucl. **46**, 138-146 (2015).
[14] R. Bernabei et al., Eur. Phys. J. C **74**, 2827 (2014).
[15] R. Bernabei et al., Eur. Phys. J. C **74**, 3196 (2014).
[16] R. Bernabei et al., Int. J. Mod. Phys. A **31**, 1642006 (2016).
[17] R. Bernabei et al., Int. J. Mod. Phys. A **31**, 1642005 (2016).
[18] R. Bernabei et al., Int. J. Mod. Phys. A **31**, 1642004 (2016).
[19] R. Bernabei et al., Int. J. Mod. Phys. A **30**, 1545006 (2015).
[20] R. Bernabei et al., Eur. Phys. J. C **75**, 239 (2015).
[21] R. Bernabei et al., Phys. Part. Nucl. **46**, 138-146 (2015).
[22] R. Bernabei el al., La Rivista del Nuovo Cimento **26** n.1, 1 (2003).
[23] R. Bernabei et al., Int. J. Mod. Phys. D **13**, 2127 (2004).
[24] R. Bernabei et al., Phys. Lett. B **389**, 757 (1996); Phys. Lett. B **424**, 195 (1998); Phys. Lett. B **450**, 448 (1999); Phys. Rev. D **61**, 023512 (2000); Phys. Lett. B **480**, 23 (2000); Phys. Lett. B **509**, 197 (2001); Eur. Phys. J. C **23**, 61 (2002); Phys. Rev. D **66**, 043503 (2002).
[25] R. Bernabei et al., Int. J. Mod. Phys. A **21**, 1445 (2006).
[26] R. Bernabei et al., Int. J. Mod. Phys. A **22**, 3155 (2007).
[27] R. Bernabei et al., Eur. Phys. J. C **53**, 205 (2008).
[28] R. Bernabei et al., Phys. Rev. D **77**, 023506 (2008).
[29] R. Bernabei et al., Mod. Phys. Lett. A **23**, 2125 (2008).

[30] K.A. Drukier et al., Phys. Rev. D **33**, 3495 (1986); K. Freese et al., Phys. Rev. D **37**, 3388 (1988).
[31] D. Smith and N. Weiner, Phys. Rev. D **64**, 043502 (2001); D. Tucker-Smith and N. Weiner, Phys. Rev. D **72**, 063509 (2005); D. P. Finkbeiner et al, Phys. Rev. D **80**, 115008 (2009).
[32] K. Freese et al., Phys. Rev. D **71**, 043516 (2005); K. Freese et al., Phys. Rev. Lett. **92**, 11301 (2004).
[33] R. Bernabei et al., Eur. Phys. J. C **18**, 283 (2000).
[34] R. Bernabei et al., Eur. Phys. J. C **47**, 263 (2006).
[35] K. Freese et al., Phys. Rev. D **71**, 043516 (2005); New Astr. Rev. **49**, 193 (2005); astro-ph/0310334; astro-ph/0309279.
[36] G. Gelmini, P. Gondolo, Phys. Rev. D **64**, 023504 (2001).
[37] F.S. Ling, P. Sikivie and S. Wick, Phys. Rev. D **70**, 123503 (2004).
[38] M. Ambrosio et al., *Astropart. Phys.* **7**, 109 (1997).
[39] LVD Collab. (M. Selvi et al.), to appear in *Proceed. of The 31st Int. Cosmic Ray Conference (ICRC 2009)* (Lodz, Poland, 2009).
[40] Borexino Collab., D. D'Angelo talk at the *Int. Conf. Beyond the Desert 2010* (Cape Town, South Africa, 2010).
[41] G. Bellini, talk at the Scientific Committee II of INFN, sept. 2010, available on the web site of the Committee: *http://www.infn.it/*.
[42] Borexino Collab., to appear in *Proceed. of Int. Cosmic Ray Conference (ICRC 2011)* (Beijing, China, 2011).
[43] D. Nygren, *arXiv:1102.0815*.
[44] R. Bernabei et al., "Liquid Noble gases for Dark Matter searches: a synoptic survey", Exorma Ed., Roma, ISBN 978-88-95688-12-1, 2009, pp. 1–53 [arXiv:0806.0011v2].
[45] J.I. Collar and D.N. McKinsey, [arXiv:1005.0838]; [arXiv:1005.3723]; J.I. Collar, [arXiv:1006.2031]; [arXiv:1010.5187]; [arXiv:1103.3481]; [arXiv:1106.0653]; [arXiv:1106.3559].
[46] A. Addazi et al., Eur. Phys. J. C **75**, 400 (2015).
[47] R. Cerulli et al., Eur. Phys. J. C **77**, 83 (2017).
[48] R. Bernabei et al., Eur. Phys. J. C **73**, 2276 (2013).

GLOBAL GEOMETRY OF THE VAIDYA SPACE-TIME

Victor Berezin[1,a], Vyacheslav Dokuchaev[1,2,b] and Yury Eroshenko[1,c]
[1]*Institute for Nuclear Research, RAS, 117312 Moscow, Russia*
[2]*National Research Nuclear University MEPhI, 115409, Moscow, Russia*

Abstract. We find the exact analytical expressions for the classical Vaidya metric with the linear mass function in the special diagonal coordinates. Using these diagonal coordinates, we elaborate the maximum analytic extension of the Vaidya metric and construct the corresponding Carter-Penrose diagrams. The derived global geometry seemingly is valid also for a more general behavior of the black hole mass in the Vaidya metric.

The classical spherically symmetric Vaidya metric [1–3] describes the nonstationary spacetime produced by a radial radiation flow of null particles [4]:

$$ds^2 = \left[1 - \frac{2m(z)}{r}\right] dz^2 + 2dzdr - r^2(d\theta^2 + \sin^2\theta d\varphi^2). \tag{1}$$

Here $m(z)$ is an arbitrary mass function depending (in the case of accretion) on coordinate $z = -v$, where v is the advanced light coordinate or (in the case of emission of radiation) on coordinate $z = u$, where u is the retarded light coordinate. For $m(z) = m_0 = const$, metric (1) describes a Schwarzschild black hole of mass $m = m_0 = const$. It is used units with the velocity of light $c = 1$ and the gravitational constant $G = 1$ (except the Figs.). There is also the representation of the Vaidya solution of the Einstein equations in the double-null (v, u)-coordinates [5].

Bu using the crucial ansatz: $m(z) = -\alpha z + m_0$, $dm/dz = -\alpha = const$, we find (see [6,7] for details) the transformation of the Vaidya metric to the diagonal coordinates (η, y, θ, ϕ):

$$ds^2 = f_0(\eta, y)d\eta^2 - \frac{dy^2}{f_1(\eta, y)} - r^2\left(d\theta^2 + \sin^2\theta d\phi^2\right), \tag{2}$$

where

$$f_0 = -\frac{(y^2 - y + 4\alpha)}{1 - y}\frac{C_{,\eta}^2}{\alpha^2}\Phi^2, \quad f_1 = -\frac{(1-y)^3(y^2 - y + 4\alpha)}{(2C\Phi)^2}, \tag{3}$$

$$\Phi(y) = \exp\left[-2\alpha \int \frac{dy}{(1-y)(y^2 - y + 4\alpha)}\right], \tag{4}$$

$$m = C(\eta)\Phi(y), \quad r = \frac{2C(\eta)}{1-y}\Phi(y), \quad C(\eta) = \alpha\eta + C_0, \quad C_0 = const. \tag{5}$$

---
[a]E-mail: berezin@inr.ac.ru
[b]E-mail: dokuchaev@inr.ac.ru
[c]E-mail: eroshenko@inr.ac.ru

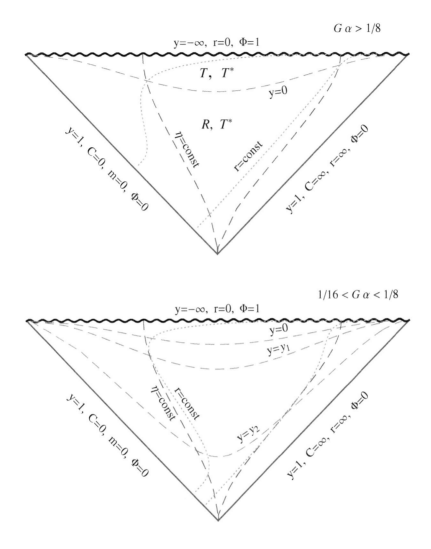

Figure 1: Carter-Penrose diagrams for the global geometry of the Vaidya metric in the cases of "Superpowerful accretion" at $G\alpha > 1/8$ and "powerful" accretion at $1/16 < G\alpha < 1/8$.

The specific form of the function Φ in (4) depends on the value of α. "Powerful" accretion at $\alpha > 1/16$:

$$\Phi = \frac{\sqrt{1-y}}{(y^2 - y + 4\alpha)^{1/4}} \exp\left[-\frac{1}{2\sqrt{16\alpha - 1}}\left(\arctan\frac{2y-1}{\sqrt{16\alpha - 1}} + \frac{\pi}{2}\right)\right]. \quad (6)$$

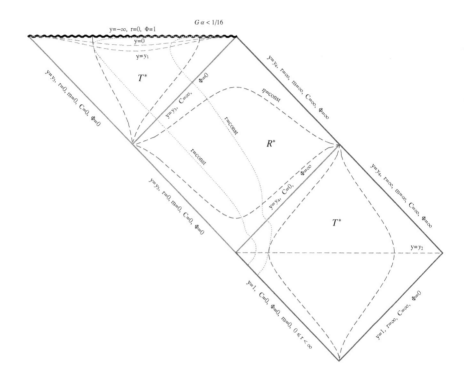

Figure 2: Carter-Penrose diagram for the global geometry of the Vaidya metric in the physically most reasonable case of "weak" accretion $G\alpha < 1/16$.

Transient case at $\alpha = 1/16$:

$$\Phi = \sqrt{\left|\frac{y-1}{y-(1/2)}\right|} \exp\left\{\frac{1}{4[y-(1/2)]}\right\}. \tag{7}$$

"Weak" accretion at $\alpha < 1/16$:

$$\Phi = \sqrt{1-y}\,|y-y_3|^{y_3/[2(y_4-y_3)]}\,|y-y_4|^{-y_4/[2(y_4-y_3)]}, \tag{8}$$

where

$$y_3 = (1-\sqrt{1-16\alpha})/2, \quad y_4 = (1+\sqrt{1-16\alpha})/2. \tag{9}$$

To construct the conformal Carter-Penrose diagrams, describing the global geometry of the Vaidya metric, it is useful to define the invariant

$$Y = \gamma^{ik} y_{,i} y_{,k} = -f_1 = \frac{1}{4m^2}(1-y)^3(y^2-y+4\alpha), \tag{10}$$

where γ^{ik} — 2-dim part of the Vaidya metric (without angle variables). The parts of spacetime, where $Y < 0$, correspond to R^*-regions, where η is the time coordinate, and y is a space coordinate. In other words, the metric signature of in the R^*-regions is $(\eta, y, \theta, \phi) \Rightarrow (+, -, -, -)$. Respectively, the parts of spacetime, where $Y > 0$, correspond to T^*-regions, where η is the space coordinate, y is a time coordinate and metric signature $(\eta, y, \theta, \phi) \Rightarrow (-, +, -, -)$.

In the intermediate step we made the reduction to the simple form of the described metric in the coordinates $\log C(\eta)$ and $\log \Phi(y)$:

$$ds^2 = \frac{C^2}{\alpha^2} \frac{\Phi^2}{(1-y)} \left\{ (y^2 - 2y + 4\alpha) \left[(d\log\Phi)^2 - (d\log C)^2 \right] \right\} - r^2 d\Omega^2 \qquad (11)$$

Finally, the Carter-Penrose diagrams are constructed in the coordinates $\log C(\eta)$ and $\log \Phi(y)$ under the standard arctan-type transformation

$$t = \arctan\left[\log C + \log \Phi(y)\right] - \arctan\left[\log C - \log \Phi(y)\right], \qquad (12)$$
$$x = \arctan\left[\log C + \log \Phi(y)\right] + \arctan\left[\log C - \log \Phi(y)\right]. \qquad (13)$$

Some of the resulting Carter-Penrose diagrams for Vaidya metric are shown in Figs. 1 and 2 (see [6, 7] for more details). At these diagrams the corresponding level lines for the temporal and space coordinates are defined, as well as the lines $r = const$. The diagrams contain quite different space-time regions, separated by horizons and marginal lines of coordinate systems. The space-time on the constructed diagrams geodesically incomplete because we imposed the physically reasonable condition of the non-negativity of the running black hole mass, $m > 0$. Our consideration was limited only by the linear growth of the black hole mass $m(z)$. Nevertheless, the derived global geometry seemingly is valid also for a more general behavior of the black hole mass in the Vaidya metric.

References

[1] P. C. Vaidya, Current Sci. **12** (1943) 183.
[2] P. Vaidya, Proc. Indian Acad. Sci. **A 33** (1951) 264.
[3] P. Vaidya, Nature **171** (1953) 260.
[4] R. W. Lindquist, R. A. Schwartz, C. W. Misner, Phys. Rev. **137** (1965) 1364.
[5] B. Waugh, K. Lake, Phys. Rev. **D 34** (1986) 2978.
[6] V. A. Berezin, V. I. Dokuchaev, Yu. N. Eroshenko, Class. Quantum Grav. **33** (2016) 145003.
[7] V. A. Berezin, V. I. Dokuchaev, Yu. N. Eroshenko, JETP, **124** (2017) 446.

ENTERING THE COSMIC RAY PRECISION ERA

Pasquale Dario Serpico [a]
LAPTh, Université Savoie Mont Blanc & CNRS,
9 Chemin de Bellevue - BP 110, 74941 Annecy Cedex, France

Abstract. I briefly review some current trends in theoretical and phenomenological studies of Galactic cosmic rays (GeV-TeV energy range), in the light of the greatly improved precision of cosmic ray measurements in recent years, which have also revealed some novel features requiring an explanation. The importance of a more refined modeling, to achieve a better assessment of theoretical uncertainties associated to the models, and to test key predictions against the wealth of data available is emphasized.

1 Introduction

The evolution of cosmic ray (CR) astrophysics has been relatively slow, if compared with other branches of astronomy and astrophysics. This is not surprising, giving the lack of positional information and the complicated propagation that makes the source identification and the interstellar transport characterization such difficult and indirect problems. About a decade ago, when I started being seriously interested in this field, it is fair to say that a few main questions in CR physics and the consensual answers to them had crystallized into a "standard framework":

- How is CR acceleration taking place?

  Primarily via "diffusive shock acceleration".

- In what type of objects?

  Predominantly (Galactic) supernova remnants.

- Where are they located? When did the events happen?

  Randomly in the Galaxy, well approximated by a continuum injection term, with a size much smaller than typical source-Earth distance.

- How do CRs get to us, after leaving their acceleration sites?

  Diffusing into an externally assigned, roughly scale-invariant turbulent magnetized interstellar medium (ISM).

Obviously, this does not mean that alternative scenarios had not been occasionally considered. And, certainly, a number of the above-listed topics has been developed into a remarkable detail. For instance, the study of non-linear effects

---

[a] E-mail: serpico@lapth.cnrs.fr

and their impact on some expectations of diffusive shock acceleration models is now a mature sub-field of theoretical research of its own. But it is fair to say that no stringent test of either the standard or more exotic models had been possible *via charged CR measurements* till recently, when a wealth of 21st century experiments has significantly improved the precision of the observations, while extending their dynamical range. Take the following list of statements:

- We only have access to cosmic ray fluxes "modulated" by heliosphere.
- Primary cosmic ray fluxes have power-law spectra.
- Primary spectra have universal (species independent) spectral indices.
- The positron flux is dominated by secondaries.
- Propagation parameters (as opposed to assumptions on the source and model framework) are the dominating uncertainty in theory predictions.

The first three "observational" facts have been disproven by the much more precise measurements available or, for the first item, by the *unique exploit* of the Voyager Interstellar mission[b]. The last two statements, usually taken from granted in most of the decade-old phenomenological studies, have not only been shaken by new data, but appear nowadays very doubtful.

Also in the light of the importance of some of these issues for other astroparticle physics applications (notably, indirect dark matter searches), a new scrutiny of our simplest scenarios is ongoing. Ideally, theorists would like to match theoretical uncertainties with experimental ones, refining the level of predictions and improving our understanding of these high-energy phenomena.

2 Spectral breaks

In order to illustrate this theoretical trend, I will describe a specific example: the impact of the fact that primary cosmic ray fluxes in the GeV to TeV energy range, in particular protons and He nuclei, *do not* manifest a simple power-law spectrum. The evidence in favour of spectral shapes closer to broken power-laws has been accumulating over almost a decade, from the first indications in balloon-borne experiments, such as ATIC [1] and especially CREAM [2], through the first measurements in a single (space-based) experiment, PAMELA [3], till the recent high-precision determinations by AMS-02 on board the International Space Station [4,5].

What is wrong with these observations? To assess that, take the simplest expectation (which, nonetheless, matched data till recently): For stationary, homogeneous and isotropic diffusive propagation problems, and observations

---

[b]https://voyager.jpl.nasa.gov/mission/interstellar-mission/

taken at a single location (i.e. the Earth) the diffusion operator ruling the flux Φ can be effectively replaced by an "diffusive confinement" time $\tau_{\text{diff}}(E)$:

$$\frac{\partial \Phi}{\partial t} - K\nabla^2\Phi = Q \Rightarrow \frac{\Phi}{\tau_{\text{diff}}} = Q \text{ (at steady state)}. \quad (1)$$

If both the source term Q and the diffusion coefficient K (with $\tau_{\text{diff}} \propto K^{-1}$) are power-laws in rigidity R (momentum over charge ratio), as customarily believed/theorized, then a puzzle arises. This schematic exercise naturally suggests (classes of) solutions, where one drops one or several of the following assumptions (examples of actual physical motivations for that below, in italics):

- K homogeneity (and possibly isotropy).
  Ex.: Multi-phase character of the Galactic ISM.
- Power-law behaviour in K.
  Ex.: Multiple sources/mechanisms for the MHD turbulence in the ISM.
- Power-law behaviour in Q.
  Ex.: Multiple classes of sources or spectral feature of a single source class.
- Homogeneity in Q.
  Ex.: Prominent local, discrete sources.

For the sake of brevity, I will briefly concentrate on the latter option to illustrate some recent theoretical efforts within the above-mentioned strategy, while addressing the reader to [6] for a broader overview of the alternatives.

3 Local sources

A number of publications have studied the possibility that the CR spectral breaks come from the emergence of a "local" source contribution over a diffuse/unresolved contribution representative of a Galactic average. Usually, but not always, the local contribution is considered to dominate at high-energy. The emphasis has often been on: i) finding a viable fit; ii) some "qualitative" assessment on the goodness of the model. For instance, one typically needs fast diffusion and low supernova rate explosions for these scenarios to work, which has often been argued to be in tension with other observations.

One may however ask the more general question: How likely is such a hypothesis in itself, given "Galactic variance", i.e. the spatial discreteness (and impulsive time-dependence) of the sources? Conventional models, in fact, replace the actual sources with a continuum "source jelly", with a smoothly varying injection rate per unit volume and time. This corresponds to a "coarse-grained" ensemble average of the actual physical model, and by construction the average theoretical expectation matches the prediction obtained in such a simplified model. But, assuming that a discrepancy between observations and data is found, how safely can we attribute it to a failure of the model? Couldn't it be due to a relatively large statistical "fluctuation" with respect to the average prediction, which is in fact compatible with the model in a more realistic calculation?

In [7], we have outlined the first elements of such a theory. The task is made non-trivial by the fact that the theoretical probability distribution for the flux is of "fat-tail" type, with infinite variance: The Central Limit Theorem does not apply and the familiar Gaussian statistics toolbox cannot be used. We have argued that a generalized Central Limit Theorem holds, and that the flux probability distribution functions are remarkably well approximated by "stable laws", characterized by analytically computable parameters. We tested these conclusions with extensive numerical simulations. As a result, we arrived at two interesting conclusions:

- For currently viable *homogeneous and isotropic* diffusion models, the observed breaks only emerge rarely, certainly in less than 1% of the realizations (actually, a more realistic *upper limit* is around 0.1%). Hence, such explanations appear to require a high-degree of fine-tuning.

- These "irreducible theoretical errors" are not negligible anymore, given the precision of the data: by taking the experimental error $\sigma_{\rm exp}$ reported by AMS-02 on the proton flux measurement [4], we estimate for instance that a 3 $\sigma_{\rm exp}$ deviation from the average expectation is obtained around 50 GeV in about 5% of the theoretical realizations. Till recently, on the contrary, the precision was sufficiently low to make these effects negligible (still true for instance for PAMELA data), justifying a posteriori that such fine effects were neglected in phenomenological analyses.

This study represents more of a beginning than an end to the story. Extending the theory to account for energy bin-to-bin correlations or to include anisotropy observables is certainly something to wish for, to bypass the need for extensive MC in order to compare theoretical predictions with data.

4 Testing break models

Of course, besides refining theory models and the uncertainty assessments, one is also interested in discriminating among competing models for the new features that have been uncovered by the data. As already stressed in [6] regarding break models, the most promising channels to probe the different classes of explanations for the spectral breaks are the so-called "secondary" nuclei. These are relatively fragile nuclei such as Li, Be, B, easily destroyed in stellar processes and thus present but in traces in stellar astrophysical environments. It turns out that they are comparatively abundant in CRs. This fact is interpreted as the result of spallation of "primary" nuclei accelerated at sources (e.g. supernova remnants) onto interstellar material during the CR diffusive propagation to the Earth. While CR are sensitive to both acceleration and propagation effects, the ratio of secondary/primary species is used to constrain propagation parameters, being largely insensitive to injection spectra. Since the Lorentz factor of a nucleus is approximately conserved in a spallation event, with a bit of oversimplification, for ratios plotted in terms of energy/nucleon (or, to some extent, rigidity) one expects:

- For a "source" origin for the break, no feature in the secondaries/primaries ratios.

- For a "propagation" origin for the break, the same break should be seen in secondaries/primaries (i.e. secondary spectra should show a more pronounced break than primary ones.)

- For "local source" models, no deterministic prediction can be made, but the ratio should more likely show a *softening* than a hardening, since secondaries are contributed to (often in a dominant way) by the "unbroken" average Galactic source spectrum.

In [8], we have recently performed a first test *a priori* in this sense on the AMS-02 B/C data published last year [9], comparing a baseline model with a power-law function $K(R)$ vs. a case with a break in $K(R)$, whose parameters are fixed by the proton and Helium data, so that the fit to B/C data (above 15 GV) has the same number of free parameters in the two cases [c].

In all cases tested, a significant preference for the scenario with a break in $K(R)$ is obtained ($\Delta\chi^2 > 10$). This result is robust with respect to: i) subleading transport phenomena (at high rigidity) such as convection or reacceleration, treated as nuisance parameters; ii) different treatments of AMS-02 systematic errors; iii) assumed spallation cross-sections (from existing libraries) as well as a physically motivated logarithmic growth with energy "inherited" from hadronic elastic and total cross-section; iv) the expected amount of "grammage at the source" (so-called secondaries at the source).

The next step is obviously to include other nuclei in the analysis and check for self-consistency, with the encouraging perspective of the forthcoming AMS-02 publications of Li, Be, B, C, O, N data. Also, it will be interesting to see if this "relative preference" persists in more inclusive fits (e.g. from sub-GeV energies in B/C) and in more extended transport models.

---

[c]Note that the two cases are not "epistemologically" identical: the first leaves the *p*-He data unexplained (source effect?) the second attributes the breaks to propagation.

A confirmation of these emerging indications, while important, would be by no way conclusive. Even a pure diffusive origin of the feature may have in fact different interpretations (see for instance [10] and [11] for a couple of models of how this could be realized). Further probes, notably multi-messenger ones, will remain a crucial ingredient for diagnostics even in the future.

5 Conclusion

The observational improvements in cosmic ray (CR) astrophysics have shown the first cracks in the simplest models for CR production and propagation. Many ideas have been proposed for their origin, and still more will likely appear. In general, however, we face a double theoretical challenge: To provide a more refined modeling to account for new facts *and* to keep theoretical errors under control, or at least assess them. Without the former, the models become less and less interesting, but without the latter, the newly attained experimental precision becomes worthless.

I focused on the case of spectral breaks, which can be "naturally" explained within broad classes of models. In fact, my personal opinion is that finding a model that fits the data is not the hardest task, notably if including a lot of additional free parameters. Much more challenging is to find models that provide relatively likely explanations in a statistical sense, or that *predict* (as opposed to *postdict*) features that one can test.

We recently provided [7] a first estimate of the irreducible ("Galactic variance") theoretical error due to space-time discreteness of the CR sources, whose exact location and times are obviously unknown. Alone, this effect cannot explain the spectral breaks firmly observed at least in p and He, but it introduces an uncertainty comparable or even larger than the AMS-02 statistical ones, and should be taken into account.

We also presented a first test on the newly published AMS-02 B/C data, to investigate if they prefer a propagation origin for the breaks, finding intriguing hints in that sense [8]. More precision and CR species plus an extended energy range will help refining such studies, but one certainly needs further theoretical and phenomenological progress as well. For instance, a multimessenger perspective (since some fine details could be "accidental", better to explain approximately all channels than precisely one!) or accounting for "non-local" observables as well (like CR anisotropies or diffuse gamma-rays) can break model degeneracies and bring us closer to a global understanding of the Galactic CR phenomenon. Such a progress would also prove beneficial (if not essential) to sharpen the CR channel potential for astroparticle applications, such as indirect dark matter searches in CR antimatter fluxes.

Acknowledgments

I would like to thank all my collaborators on the topics covered in this manuscript, as well as the organizers of the 18th Lomonosov Conference on Elementary Particle Physics—and in particular Alexander I. Studenikin—for their kind invitation, the smooth organization, and the warm atmosphere.

References

[1] A. D. Panov et al., "Energy Spectra of Abundant Nuclei of Primary Cosmic Rays from the Data of ATIC-2 Experiment: Final Results," Bull. Russ. Acad. Sci. Phys. **73**, 564 (2009) [arXiv:1101.3246].
[2] Y. S. Yoon et al., "Cosmic-Ray Proton and Helium Spectra from the First CREAM Flight," Astrophys. J. **728**, 122 (2011) [arXiv:1102.2575].
[3] O. Adriani et al. [PAMELA Collaboration], Science **332**, 69 (2011) [arXiv:1103.4055].
[4] M. Aguilar et al. [AMS Collaboration], "Precision Measurement of the Proton Flux in Primary Cosmic Rays from Rigidity 1 GV to 1.8 TV with the Alpha Magnetic Spectrometer on the International Space Station," Phys. Rev. Lett. **114**, 171103 (2015).
[5] M. Aguilar et al. [AMS Collaboration], "Precision Measurement of the Helium Flux in Primary Cosmic Rays of Rigidities 1.9 GV to 3 TV with the Alpha Magnetic Spectrometer on the International Space Station," Phys. Rev. Lett. **115**, 211101 (2015).
[6] P. D. Serpico, "Possible physics scenarios behind cosmic-ray "anomalies"," PoS ICRC **2015**, 009 (2016) [arXiv:1509.04233].
[7] Y. Genolini, P. Salati, P. D. Serpico and R. Taillet, "Stable laws and cosmic ray physics," Astron. Astrophys. **600**, A68 (2017) [arXiv:1610.02010].
[8] Y. Genolini et al., "Indications for a high-rigidity break in the cosmic-ray diffusion coefficient," Phys. Rev. Lett. , in press [arXiv:1706.09812].
[9] M. Aguilar et al. [AMS Collaboration], "Precision Measurement of the Boron to Carbon Flux Ratio in Cosmic Rays from 1.9 GV to 2.6 TV with the Alpha Magnetic Spectrometer on the International Space Station," Phys. Rev. Lett. **117**, 231102 (2016).
[10] N. Tomassetti, "Origin of the Cosmic-Ray Spectral Hardening," Astrophys. J. **752**, L13 (2012) [arXiv:1204.4492].
[11] P. Blasi, E. Amato and P. D. Serpico, "Spectral breaks as a signature of cosmic ray induced turbulence in the Galaxy," Phys. Rev. Lett. **109**, 061101 (2012) [arXiv:1207.3706].

CURRENT STATUS OF WARM INFLATION

Raghavan Rangarajan [a]
Physical Research Laboratory, Ahmedabad 380009, India

Abstract. Warm inflation is an inflationary scenario in which a thermal bath coexists with the inflaton during inflation. This is unlike standard cold inflation in which the Universe is effectively devoid of particles during inflation. The thermal bath in warm inflation is maintained by the dissipation of the inflaton's energy through its couplings to other fields. Many models of warm inflation have been proposed and their predictions have been compared with cosmological data. Certain models of inflation that are disallowed in the context of cold inflation by the data are allowed in the warm inflationary scenario, and vice versa.

1 Introduction

In this brief article we shall provide a review of warm inflation and its current status. We shall first discuss what is warm inflation and how it is different from the standard cold inflation. We shall then discuss how to construct a warm inflation model. Finally we shall consider the compatibility of various warm inflation models with the cosmic microwave background data.

In the warm inflation scenario the Universe inflates as in cold inflation. However one considers the decay of the inflaton during inflation. Hence the number density of particles does not go to 0 during inflation. If the dissipation is fast enough so as to maintain a thermal bath with $T > H$ then one has a warm inflation scenario [1, 2]. In some warm inflation models there is no need of a separate reheating era.

There are several models of inflation — over 70 single field inflation models are listed in Encyclopedia Inflationaris [3]. So why should one consider a new scenario like warm inflation? Firstly, it is natural to consider the effects of the inflaton couplings not just during reheating but also during inflation. (Whether or not one will get a sufficiently hot thermal bath is a different matter, as we shall see.). Furthermore, for some warm inflation models, the eta problem, namely, the presence of large quantum corrections to the inflaton potential that ruins its flatness, is resolved. Also, some potentials that are excluded by cosmic microwave background (CMB) data in the cold inflation scenario are allowed in the warm inflation scenario (though the converse is also true).

2 How is Warm Inflation Different from Cold Inflation?

It may be noted that warm inflation constitutes a different paradigm of inflation. The presence of a thermal bath differentiates it from the cold inflation scenario. While studying inflation one considers the homogeneous background

---
[a] E-mail: raghavan@prl.res.in

field $\phi(t)$ and its spatial perturbations $\delta\phi(\mathbf{x},t)$, or their Fourier transform, $\delta\phi_k(t)$. Both the background field and the perturbations are affected by the presence of the thermal bath.

We first consider the homogeneous inflaton field $\phi(t)$. The equation of motion of this background field is given by

$$\ddot{\phi} + (3H + \Gamma)\dot{\phi} + \frac{dV}{d\phi} = 0, \qquad (1)$$

where Γ is a dissipation coefficient due to inflaton couplings to other fields, which is not considered in cold inflation during the inflaton slow roll phase. When $\Gamma > H$ it helps to slow down the inflaton. The slow roll parameters for warm inflation are given by

$$\epsilon = \frac{M_{Pl}^2}{16\pi}\left(\frac{V_\phi}{V}\right)^2, \quad \eta = \frac{M_{Pl}^2}{8\pi}\frac{V_{\phi\phi}}{V}, \quad \beta = \frac{M_{Pl}^2}{8\pi}\left(\frac{\Gamma_\phi V_\phi}{\Gamma V}\right) \qquad (2)$$

where $M_{Pl} = 1.2 \times 10^{19}$ GeV is the Planck mass. The slow roll conditions needed for the inflationary phase are $\epsilon \ll 1+Q$, $|\eta| \ll 1+Q$, $|\beta| \ll 1+Q$, where $Q = \Gamma/(3H)$. The presence of Q on the right hand side of these inequalities, which is obviously absent in cold inflation, implies that the slow roll conditions can be satisfied even if the slow roll parameters are large, if $Q \gg 1$, as it happens in some models of warm inflation. In these models of warm inflation, the eta problem is solved. Finally in some models of warm inflation, before the slow roll conditions break down the inflaton energy density becomes smaller than the radiation energy density. In that case inflation ends but then there is no separate reheating phase because one has an automatic transition to the radiation dominated era (though the inflaton will eventually oscillate and decay).

The thermal bath affects the inflaton perturbations and thereby the primordial curvature perturbations. The curvature perturbations affect the CMB anisotropy and the large scale structure that we observe today. The equation of motion for the inflation perturbations in the presence of the thermal bath is given by [4–6]

$$\delta\ddot{\phi}_k + (3H + \Gamma)\delta\dot{\phi}_k + \left(\frac{k^2}{a^2} + V_{\phi\phi}\right)\delta\phi_k = \sqrt{2\Gamma T}\, a^{-3/2}\, \xi_k, \qquad (3)$$

where ξ_k represents thermal noise. The above is a form of the Langevin equation with the fluctuation term on the r.h.s. related to the dissipation term on the l.h.s. The primordial curvature power spectrum is proportional to $|\delta\phi_k|^2$ (in the spatially flat gauge), where $\delta\phi_k$ is evaluated when the physical wavelength of the perturbation ($\lambda_{phys} = 2\pi a(t)/k$) becomes large enough that $\delta\phi_k$ becomes constant, or freezes out [7].

We are concerned only with the perturbations that correspond to cosmologically relevant length scales today, from 10^{-3} Mpc to 14000 Mpc [8]. This corresponds to about 16 e-foldings of inflation, starting from about 60 e-foldings of inflation before the end of inflation (for GUT scale inflation). The observed CMB anisotropy reflects perturbations on scales of 10 Mpc and larger. When $Q \ll 1$ ($Q \gg 1$) it is referred to as weak (strong) dissipative warm inflation. The inflaton perturbations for cold inflation, weak dissipative warm inflation and strong dissipative warm inflation are given by $\delta\phi_k \sim H, \sqrt{HT}, \sqrt{T(H\Gamma)^{\frac{1}{2}}}$ for Cold Inflation, Weak Dissipative Warm Inflation and Strong Dissipative Warm Inflation respectively [5, 9].

The primordial curvature power spectrum (or scalar power spectrum) is given in Ref. [6] (based on Refs. [1, 4, 5, 7, 10]) as

$$P_{\mathcal{R}}(k) = \left(\frac{H_k}{\dot{\phi}_k}\right)^2 \left(\frac{H_k}{2\pi}\right)^2 \left[1 + 2n_k + \left(\frac{T_k}{H_k}\right) \frac{2\sqrt{3}\pi Q_k}{\sqrt{3 + 4\pi Q_k}}\right], \quad (4)$$

where the subscript k indicates that the variable is evaluated at the time of horizon crossing of the k mode perturbation $\delta\phi_k$, and n_k represents the distribution of inflaton particles in the thermal bath. In the literature, one considers either $n_k = 0$ or the Bose-Einstein distribution, $n(k) = [\exp\{k/(aT)\} - 1]^{-1}$. For the latter case, $1 + 2n_k = \coth[H_k/(2T_k)]$, using $k/a_k = H_k$. In addition to the explicit temperature dependence in the square bracket above, the prefactor (whose form is the same as that for cold inflation) will reflect the influence of dissipation. Note that $[H_k/(2\pi)]^2$ times the first term in the square bracket reflects the standard quantum contribution, as in cold inflation, its product with $\coth[H_k/(2T_k)]$ reflects the weak dissipative warm inflation result for $T \gg H$, and the product with the last term indicates the strong dissipative warm inflation result, as in the expressions for $\delta\phi_k$ above.

Inflation gives rise to both scalar and tensor perturbations of the metric. Gravitational waves are weakly coupled to the thermal bath and so the tensor power spectrum has the same form as in cold inflation, namely, $P_T(k) = 16H_k^2/(\pi M_{Pl}^2)$. The tensor-to-scalar ratio r is defined, as usual, as $r = P_T(k_P)/P_{\mathcal{R}}(k_P)$, where k_P refers to the pivot scale, a fiducial scale for which there is greater observational accuracy for any particular experiment.

3 Constructing a Warm Inflation Model

The dissipation coefficient Γ reflects the transfer of energy from the inflaton field to the thermal bath. If one couples the radiation, i.e. light fields with mass $m < T$, directly to the inflaton then one gets a $\Gamma\dot{\phi}$ term in the equation of motion for ϕ but one also gets large thermal corrections to the inflaton potential and too few e-foldings of inflation [11, 12]. There are two approaches to avoiding this. In the first approach one couples the inflaton only to heavy fields through a superpotential of the form [13]

$$W = f(\Phi) + \frac{g}{2}\Phi X^2 + \frac{h}{2}XY^2 ,\qquad(5)$$

where ϕ is associated with the scalar component of the Φ superfield, and the X fields are heavy, i.e $m_X > T$, while the Y fields are light ($m_Y < T$). The inflaton field can decay to Y particles either through virtual X when $T \ll m_X$, or through decay to real X which then decay to Y when $T < m_X$ and $h\sqrt{N_Y}$ is small [14]. The form of the dissipation coefficient is $\Gamma = C_\phi T^3/\phi^2$. The heavy X ensure that the thermal corrections are small and supersymmetry ensures that the vacuum corrections are small. However, viable models of warm inflation need 10^6 or 10^4 X fields to satisfy warm inflation requirements, particularly $T > H$ during inflation [15, 16]. Such a large number of fields are obtained by considering brane-antibrane models of inflation where the X fields correspond to strings stretched between brane and antibrane stacks [17], or extra-dimensional scenarios with a tower of Kaluza-Klein modes [18].

In the second approach to constructing a warm inflation model, one makes the inflaton field a pseudo-Nambu-Goldstone boson. This has been realised in warm natural inflation models [19], and the warm little inflation model (which is similar to the little Higgs model) [20]. In these models it is sufficient for the inflaton to couple to a few fields. In Ref. [19], there is one additional pseudo-Nambu Goldstone boson besides the inflaton and another light field, and $\Gamma \sim \dot\phi^2 T$ and $Q \gg 1$. In Ref. [20] the inflaton field is coupled to two light fields, and $\Gamma = C_T T$ and both weak and strong dissipative warm inflation scenarios are considered.

4 Comparing with Data

Various models of warm inflation, as identified by the inflaton potential and the form of the dissipation coefficient, have been studied and compared with the cosmological data. $\Gamma = C_\phi T^3/\phi^2$ and $\Gamma = C_T T$ are the usual forms of the dissipation coefficient considered in the literature. In general, $\Gamma = C_\phi T^c \phi^{2a}/M_X^{2b}$ with $c + 2a - 2b = 1$ [5].

Limits from cosmological data are often written in terms of allowed values of the spectral index n_s defined as $n_s - 1 = d\ln P_\mathcal{R}/d\ln k|_{k_P}$ and the tensor-to-scalar ratio r. In Fig. 1 one sees the region in the $r-n_s$ plane allowed by WMAP in teal [21, 22]. Also plotted are the $r - n_s$ values obtained for warm inflation models with monomial potentials ($V \sim \phi^n$) as separate curves in the figure. $n = 2, 4$ and 6 are considered, for two values of the number of e-foldings of inflation from when the perturbation associated with the pivot scale crosses the horizon till the end of inflation, i.e. N_e equal to 60 and 40. We can focus on the $N_e = 60$

curves for warm inflation. Along each curve, the different points correspond to different values of $Q(k_P)$ with the values increasing as one goes down the curve. The cold inflation curves (CI) are also shown. For cold inflation models the different points correspond to values of N_e varying from 50 to 60 (from left to right). We notice that quadratic warm inflation has too large a value of n_s and so is disallowed, while it is consistent with the data for cold inflation. On the other hand, quartic and sextic cold inflation are ruled out by the data while they are allowed in warm inflation for appropriate values of $Q(k_P)$.

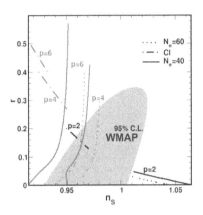

Figure 1: Allowed region by WMAP in $r - n_s$ plane is shown in teal. Also plotted are the $r - n_s$ values obtained for quadratic, quartic and sextic warm inflation models for N_e=40 and 60 (solid and dotted lines), and cold inflation models (dot-dash) for varying N_e.

In Ref. [23] the authors perform a Markov Chain Monte Carlo analysis of the parameters of warm inflation using the publicly available CosmoMC programme [24] and the Planck data. They perform this analysis for quartic, sextic, hilltop, Higgs and plateau sextic warm inflation models for $\Gamma \propto T^3$ and T and find parameters compatible with the Planck data for $Q(k_P)$ between 10^{-3} and 1.4 and r between 10^{-9} and 0.036 for different models. Another CosmoMC analysis of quartic warm inflation using Planck data obtains the joint probability distribution for the inflaton self-coupling λ and $Q(k_P)$ for $N_e = 50$ and 60 [25], as shown in Fig. 2. From the marginalised distributions of the parameters of the model the preferred range of values for λ for $N_e = 50$ is 1.5×10^{-14} to 1.9×10^{-14} with a mean value of 1.6×10^{-14}, and the preferred range of values for $N_e = 60$ is 9.2×10^{-15} to 1.1×10^{-14} with a mean value of 1.0×10^{-14}. The preferred range of values for Q_P is 9.5×10^{-4} to 1.4×10^{-2} with a mean value of 3.7×10^{-3} for $N_e = 50$, and the preferred range of values for $N_e = 60$ is 1.6×10^{-3} to 1.2×10^{-2} with a mean value of 4.4×10^{-3}.

Warm natural inflation models too have been compared with the cosmological data. In the model studied in Ref. [26] it is found that warm natural inflation

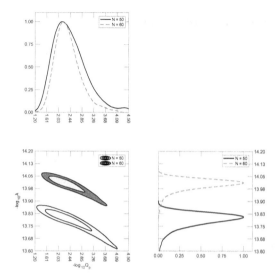

Figure 2: The joint probability distribution for the inflaton self-coupling λ and $Q(k_P)$ for $N_e = 50$ and 60, as discussed in the text.

is viable for the scale of symmetry breaking (that creates the pseudo-Nambu-Goldstone boson) between the GUT scale and the Planck scale, while Ref. [19] finds that the symmetry breaking scale in their model should be the GUT scale. In both models what is significant is that symmetry breaking scales well below the Planck scale are allowed. Planck scale symmetry breaking was one of the less attractive features of cold natural inflation.

Ref. [27] shows that hybrid inflation, which involves the interplay of two fields during inflation, is consistent with the data for warm inflation, in contrast with cold inflation. The viability of brane inflation, G(alileon) inflation and non-canonical inflation has also been studied in Refs. [28–30].

5 Conclusion

In summary, warm inflation is a viable paradigm of inflation. Various warm inflation models are compatible with cosmological data. Models such as monomial quartic and sextic warm inflation and hybrid warm inflation are allowed by the data while the corresponding models in the cold inflationary scenario are disallowed by the data. On the contrary, the quadratic inflationary model is disallowed in warm inflation while consistent with the data for cold inflation. In the case of natural warm inflation the relevant energy scale can be brought down from the Planck scale, as in cold inflation, to the GUT scale. While the requirement of a large number of fields coupled to the inflaton to satisfy the conditions for warm inflation is unattractive this issue has been resolved in warm inflation models where the inflaton is a pseudo-Nambu-Goldstone boson.

References

[1] A. Berera and L.-Z. Fang, Phys. Rev. Lett. **74**, 1912 (1995).
[2] A. Berera, Phys. Rev. Lett. **75**, 3218 (1995).
[3] J. Martin, C. Ringeval and V. Vennin, Phys. Dark Univ. **5-6**, 75 (2014).
[4] L. M. Hall, I. G. Moss and A. Berera, Phys. Rev. D **69**, 083525 (2004).
[5] R. O. Ramos and L. A. da Silva, JCAP **1303**, 032 (2013).
[6] S. Bartrum, M. Bastero-Gil, A. Berera, R. Cerezo, R. O. Ramos and J. G. Rosa, Phys. Lett. B **732**, 116 (2014).
[7] A. Berera, Nucl. Phys. B **585**, 666 (2000).
[8] D. H. Lyth and A. R. Liddle, Cambridge, UK: Cambridge Univ. Press (2009) [Sec. 7.1].
[9] A. Berera, Contemp. Phys. **47**, 33 (2006).
[10] I. G. Moss, Phys. Lett. B **154**, 120 (1985).
[11] A. Berera, M. Gleiser and R. O. Ramos, Phys. Rev. D **58**, 123508 (1998).
[12] J. Yokoyama and A. D. Linde, Phys. Rev. D **60**, 083509 (1999).
[13] I. G. Moss and C. Xiong, hep-ph/0603266.
[14] M. Bastero-Gil, A. Berera, R. O. Ramos and J. G. Rosa, JCAP **1301**, 016 (2013).
[15] A. Berera, M. Gleiser and R. O. Ramos, Phys. Rev. Lett. **83**, 264 (1999).
[16] M. Bastero-Gil, A. Berera and N. Kronberg, JCAP **1512**, no. 12, 046 (2015).
[17] M. Bastero-Gil, A. Berera and J. G. Rosa, Phys. Rev. D **84**, 103503 (2011).
[18] T. Matsuda, Phys. Rev. D **87**, 026001 (2013).
[19] H. Mishra, S. Mohanty and A. Nautiyal, Phys. Lett. B **710**, 245 (2012).
[20] M. Bastero-Gil, A. Berera, R. O. Ramos and J. G. Rosa, Phys. Rev. Lett. **117**, 151301 (2016).
[21] M. Bastero-Gil and A. Berera, Int. J. Mod. Phys. A **24**, 2207 (2009).
[22] M. Bastero-Gil, Talk at "Cosmological perturbations post-Planck", 4-7 June 2013, Helsinki.
[23] M. Benetti and R. O. Ramos, Phys. Rev. D **95**, no. 2, 023517 (2017).
[24] A. Lewis and S. Bridle, Phys. Rev. D **66** (2002) 103511.
[25] R. Arya, A. Dasgupta, G. Goswami, J. Prasad and R. Rangarajan, arXiv:1710.11109 [astro-ph.CO].
[26] L. Visinelli, JCAP **1109**, 013 (2011).
[27] M. Bastero-Gil, A. Berera, T. P. Metcalf and J. G. Rosa, JCAP **1403**, 023 (2014).
[28] M. A. Cid, S. del Campo and R. Herrera, JCAP **0710**, 005 (2007).
[29] S. del Campo and R. Herrera, Phys. Lett. B **653**, 122 (2007).
[30] R. Herrera, JCAP **1705**, no. 05, 029 (2017).

KINEMATICS OF SPIRALS AS A PORTAL TO THE NATURE OF DARK MATTER

Paolo Salucci [a]

SISSA, Via Bonomea 265, Trieste, Italy

Abstract. The gravitational field of Spiral galaxies is well traced by their rotation curves. Only recently it has become of extreme interest that the latter form a family ruled by two parameters of the *luminous* component: the disk length-scale R_D and the magnitude M_I. This evidence is so strong and consequential that it must be taken as the starting point for the investigation on the issue of the dark matter in galaxies. The emerging fact is that structural quantities deeply rooted in the luminous components, like the disk lenghtscales are found to tightly correlate with structural quantities deeply rooted in the dark component, like the DM halo core radii. These unexpected evidences may strongly call for a shift of paradigm with respect to the current Cold collisionless Dark Matter one.

1 Introduction

Recently, a number of discoveries seems to have weakened our certainties about the dark matter, the elusive substance believed to be a particle, that constitutes about 25% of the mass energy of the Universe, plays a crucial role in the formation of structures, and, since late 70's, is thought to surround the luminous disks of Spirals ([8]), interacting with the ordinary matter only by gravitation. In disk systems, the circular velocity balances the radial variations of the total gravitational potential Φ: $V^2 = r\, d\Phi/dr$, while the Poisson Equation relates the galaxy gravitational potential $\Phi = \sum_i \Phi_i$ and the density of the mass components $\nabla^2 \Phi_i = 4\pi G \rho_i$, where ρ_i stands for dark halo, stellar disk, stellar bulge, HI disk surface/volume densities (more specifically, $\rho_h(r), \rho_{bu}(r), \mu_d(R), \mu_{HI}(R)$) with R, z the cylindrical coordinates.

The main component of the luminous matter in spirals is the well-known Freeman stellar disk of exponential surface density: [3]

$$\mu_D(r) = \frac{M_D}{2\pi R_D} e^{-r/R_D} \quad (1)$$

with R_D the disk length-scale and M_D the disk mass).[b] We have within r.m.s. of 0.1 dex ([11]): $\log\left(\frac{R_D}{kpc}\right) = 0.633 + 0.379 \log\left(\frac{M_D}{10^{11} M_\odot}\right) + 0.069 \left(\log \frac{M_D}{10^{11} M_\odot}\right)^2$, that links the stellar disk masses M_D with their length-scales R_D. The HI disk has a surface density profile like Eq. (1), with a length scale $R_{HI} = 3\, R_D$ [c]. However, since the HI component is negligible inside R_{opt}, it not will be further considered.

---

[a] E-mail: salucci@sissa.it
[b] It is useful to define $R_{opt} \equiv 3.2 R_D$ as the optical size of the stellar disk, as the radius encompassing 83 % of the total galaxy luminosity.
[c] The M_{HI} vs M_D relationship is in [2]

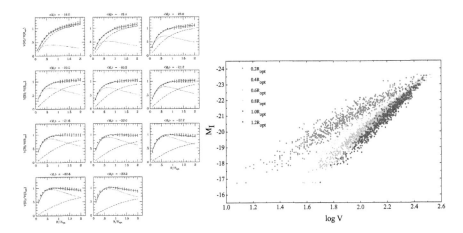

Figure 1: (*left*) The URC (*black lines*) and the coadded RCs (*points*).(*right*) The Radial TF.

2 The Universal Rotation Curve from the coadded rotation curves of 967 Spirals

We can represent the rotation curves of Spirals by means of the Universal Rotation Curve (URC) pioneered in [6, 8] and set in [7, 10]. By adopting the normalized radial coordinate $x \equiv r/R_{opt}$ the RCs of Spirals are well described by a Universal profile, function of x and of λ, where λ is one among M_I, the I magnitude, M_D, the disk mass and M_{vir}, the halo virial mass ([10]).

An universal magnitude-dependent profile is evident in the 11 *coadded* rotation curves $V_{coadd}(x, M_I)$ (see Fig. 1 and [7]) built from the individual RCs of a sample of 967 Spirals. I-band surface photometry measurements [5] provide these objects with their stellar disk length scales R_D. The URC is built in a three steps way **1)** The I magnitude range of Spirals is divided in 11 successive bins centred at M_I as listed in Table 1 of [7]. **2)** The RC of each galaxy is assigned to its corresponding I magnitude bin, normalized by its $V(R_{opt})$ value and then expressed in terms of its normalized radial coordinate x. **3)** The double-normalized RCs $V(x)/V_{opt}$ curves are coadded in 11 magnitude bins and in 20 radial bins of length 0.1, and then averaged to get: $V_{coadd}(x, M_I))/V_{coadd}(1, M_I)$, see Fig. 1. The 11 values of $V_{coadd}(1, M_I)$ are given in Table 1 of [7].

The Universal Rotation Curve $V_{URC}(x, M_I)$ is the (halo+disk) *physical* velocity model that very well fits the above $V_{coadd}(x, M_I)$. At any x and for any M_I: $V_{URC}^2 = V_{URCd}^2 + V_{URCh}^2$. From Eq. (1) the disk velocity component is:

$$V_{URCd}^2(x) = \frac{GM_D}{2\,R_D}(3.2x)^2 (I_0 K_0 - I_1 K_1) \qquad (2)$$

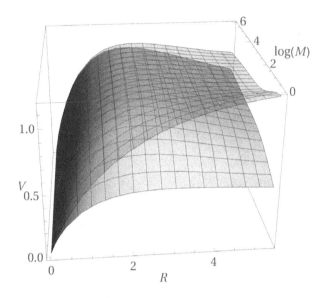

Figure 2: The URC obtained from the coadded RCs *blue*. V is in units of V_{opt}, R in units of R_D and the virial halo mass M in units of $10^{11} M_\odot$. Also shown the baryonic component (*orange*) Auxiliary to Eq. (5) we have: $log\ (M_D/M_\odot) = -0.52 M_I - 0.45$ and $log\ (M_{vir}/M_\odot) = 25.53 - 10/3\ Log\ M_D + 1/5\ (log\ M_D)^2$ that connects M_I, M_D, M_{vir}.

with the Bessel functions evaluated at 1.6 x. For the DM halo we assume the Burkert density profile [9] :

$$\rho_{URCh}(r) = \frac{\rho_0 r_0^3}{(r+r_0)(r^2+r_0^2)} \quad (3)$$

where r_0 the core radius and ρ_0 the central halo density are its free parameters. The velocity profile reads[d]:

$$V_{URCh}^2(r) = 6.4 \frac{\rho_0 r_0^3}{r} \left(\ln\left(1+\frac{r}{r_0}\right) - \arctan\frac{r}{r_0} \right) + \frac{1}{2} \ln\left(1+\frac{r^2}{r_0^2}\right). \quad (4)$$

In detail, $V_{URC}(x, \rho_0, r_0, M_D) = (V_{URCd}(x; M_D)^2 + V_{URCh}(x; \rho_0; r_0)^2)^{1/2}$ fits very well the coadded RC, see Fig. 1. The free parameters result all related to each other [10] (masses in $M_\odot$, densities in g/cm^3, distances in kpc):

$$log\frac{\rho_0}{\text{g cm}^{-3}} = -23.5 - 0.96 \left(\frac{M_D}{10^{11} M_\odot}\right)^{0.31}$$

---

[d]In this framework, the galaxy virial radius R_{vir} is defined as the solution of the equation: $G^{-1}\ V_{URCh}^2(R_{vir})R_{vir} = 100 \times 4/3\pi\ 100\ R_{vir}^3$ where $\rho_c = 1.8\ 10^{-29} g/cm^3$ is the critical density of the Universe. The virial mass is then: $M_{vir} = 10^{12} M_\odot (R_{vir}/260\ kpc)^3$.

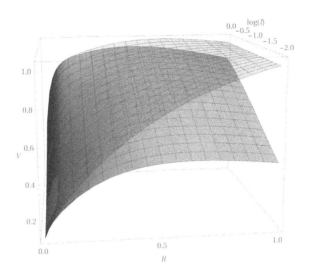

Figure 3: The URC from the Radial Tully Fisher (*blue*). V is expressed in units of V_{opt}, R in units of R_{opt}. Also shown the baryonic component (*orange*). To compare it with the URC in Fig. 2, one uses: $\rho_0 = 1/(4\pi G)(V_{opt}/R_{opt})^2 \frac{1+\alpha}{\alpha}$ and $r_0 = 0.8 \, \alpha \, R_{opt}$.

$$log\left(\frac{r_0}{\text{kpc}}\right) \simeq 0.66 + 0.58 \log\left(\frac{M_{vir}}{10^{11} M_\odot}\right) \quad (5)$$

3 The Universal Rotation Curve from the Radial Tully Fisher relationship in three large samples of Spirals

Yegorova and Salucci (2007), by analysing three samples of 794, 86 and 91 spirals of various Hubble types, discovered what was named as the Radial Tully-Fisher (RTF) relation (see [12] and Fig. 1). This is an ensemble of tight relationships between the galaxy absolute magnitude M_I and $\log V_n$, the logarithm of circular velocity measured, in each object, at the same fixed fraction of the optical radius R_n/R_{opt} ($R_{opt} \equiv 3.2 R_D$). In detail, from [12] we have:

$$V_n \equiv V(R_n), \quad R_n = (n/10) \, R_{opt}, \quad 1 \leq n \leq 10, \quad M_I = a_n \log V_n + b_n \quad (6)$$

where n indicates the position of the center of the radial velocity bin and a_n, b_n the values of the parameters of the nth fit. All relationships of Eq. (10) result statistically relevant (see Eq. (1) and [12]) and, remarkably, in all three samples, a_n shows the same strong radial variation with radius: [12]

$$a_n = -2.3 - n + 0.04 \, n^2 \quad (7)$$

with a small statistical uncertainty. Equation (2) implies that, in spirals, light does not trace the distribution of the gravitating matter which would require: $a_n \simeq -7.5$. Eqs. (6)–(7), instead, imply the existence of a Universal velocity model, function of x and M_I. The URC we build from of Eq. (6) has 3 components (disk, bulge, halo). At any x we have: $V^2_{URC} = V^2_{URCd} + V^2_{URCh} + V^2_{URCbu}$. In the following, without any loss of generality, we assume ([7, 12]): $R_D = R_1 l^{0.5}$ and $M_D = M_1 l^{1.3}$. From Eq. (8) we have:[e] $V^2_{URCd}(x,l) = V_1^2 l^{0.8} f_d(x)$ with $f_d(x) = \frac{1}{2}(3.2x)^2(I_0(1.6x)K_0(1.6x) - I_1(1.6x)K_1(1.6x))$ and: $V_d^2(1,1) = 0.347\, V_1^2$.

For the bulge we set that, at R_{opt}, V^2_{URCbu} is $c_{bu}/(3.2 \cdot 0.347)\, l^{0.5}$ times the disk contribution, c_{bu} is a free parameter and the exponent 0.5 is suggested by the bulge-to disk vs total luminosity relation in spirals. Since all kinematical data refer to radii outside the bulge half light radius, we can consider this component as a point mass. Then: $V^2_{RTFbu}(x,l) = c_{bu}\, V^2_{RFTd}(1,l)\, l^{0.5} x^{-1} f$

For the halo velocity contribution we adopt: $V^2_{URCh}(x) = V_1^2\, f_h(x;\alpha) A(l)$, where α is the halo velocity core radius in units of R_{opt}, and: $f_h(x;\alpha) = (\frac{x^2}{x^2+\alpha^2})(1+\alpha^2)$. This is the simplest profile that describes the DM halo density distribution inside R_{opt}. Notice that for values of $\alpha \simeq 1.2$ and $\alpha < 1/3$, f_h reproduces the Burkert and the NFW velocity profiles. We assume a power law between halo mass and luminosity: $M_h(1,l) \propto l^{k_h}$, where k_h is a free parameter. We normalize the latter by setting that, at $x = 1$, $V^2_{URCh}(1,l)$ is $c_h/(3.2\ 0.347)l^{(k_h-0.5)}$ times the disk contribution $V^2_{URCd}(1,l)$. The quantity $(0.9\, c_h)^{1/2}$ is then the halo-to-disk fraction at $(l,x) = (1,1)$ and $A(l) = c_h l^{k_h-0.5}$. Then, the URC, aimed to best fit Eq. (7), is written as:

$$V^2_{URC}(x,l;\alpha,c_{bu},c_h,k_h) = V_1^2(c_{bu} l^{1.3}/x + l^{0.8} f_d(x) + A(l) f_h(x;\alpha)) \quad (8)$$

The best fit values of the parameters α, c_{bu}, c_h and k_h, that specify completely the URC are ([12]): $k_h = 0.79 \pm 0.04$, $c_{bu} = 0.13 \pm 0.03$, $c_h = 0.13 \pm 0.06$ and $\alpha = 1^{+1}_{-0.5}$. They imply that, inside R_{opt}, less luminous galaxies have a larger fraction of dark matter: $M_D/M_h(1,l) \propto l^{0.5}$ and point towards *cored* DM halos.

4 Discussion and Conclusions

We realize that, once expressed in terms of the same variables, the URCs of Fig. 2 and Fig. 3, built in very different ways, almost coincide. The only difference is in the very inner regions of highest luminosity objects for which the latter URC has an additional contribution from a stellar bulge. In the 3D space,

---

[e] $l \equiv 10^{(M_I - M_I^{max})/2.5}$ with $M_I^{max} = -23.5$. The constants $R_1 = 5.6$ kpc, $M_1 = 2 \times 10^{11} M_\odot$ and $V_1^2 \equiv G\, M_1/R_1 = 500^2 (Km/s)^2$ play no role in the evaluation of the a_n's.

[f] No result changes by adopting a more realistic Sersic profile.

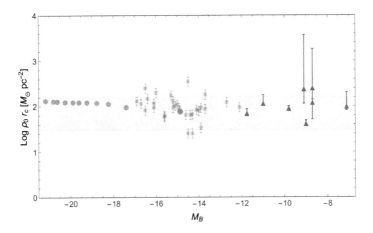

Figure 4: The halo central surface density $\rho_0 r_c$ for different samples of Spirals.

whose axes are V, x, M_I: $i)$ at any magnitude M_I, the coadded RC $V_{coadd}(x, M_I)$ and $ii)$ at any fixed normalized radius x, the Radial Tully Fisher relationship: $M_I = a_x + b_x log(V(x))$, are the normal sections of a surface $V(x, M_I)$, whose analytical representation, in terms of dark and luminous velocity components, is what we call the Universal Rotation Curve.

This result, obtained by means of two independent lines of investigation of the kinematics of thousands of disk systems, has very important consequences:

• The non-dimensional radial coordinate x gauges and links together all spirals: but this occurs independently on whether, at a radius x, $V(x)$ is dominated by the dark or by the luminous matter component.

• All spirals show a dark halo component with a constant density inner core of size r_0 that surrounds a stellar thin disk of lenghtscale R_D. If this would not be enough, the two lenghtscales result closely correlated see [1].

• The central halo density ρ_0 and the core radius r_0 emerge very close correlated, so that their product is constant in galaxies of very different luminosity.

• The URC frames the Mc Gaugh et al relationship, emerging in spirals between the total acceleration $g(r) = V^2(r)/r$ at a radius r from the center and its baryonic component: $g_b(r) = V_b^2(r)/r$, into the cored dark halo scenario. In fact, we can derive the quantities $log\ g_{URC}$ and $log\ g_{URCb}$ from the URC, entirely rooted in such a scenario, and realize that they lie in the region marked by the McGaugh relationship see Fig. 6.

These concordant observational evidences are very difficult, if not impossible, to have them explained in the simple Dark Matter framework in vogue so far. On the other hand, they are likely beacons and maybe portals to the new Physics that seems to lurk behind the phenomenon called "Dark Matter"

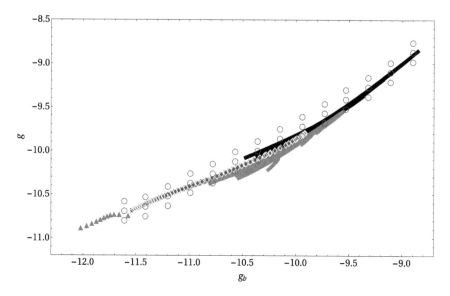

Figure 5: The McGaugh et al. 2016 relationship with its 1-σ uncertainty (*blue circles*) and that obtained by the URC (lines). Accelerations are in units of $Log\ m/s^2$.

Acknowledgments

PS thanks the organizers of the 18 Lomonosov Conference where this work has taken place.

References

[1] Donato, F., Gentile, G., Salucci, P., 2004, MNRAS, 353L, 17.
[2] Evoli, C., Salucci, P., Lapi, A., & Danese, L. 2011, ApJ, 743, 45.
[3] Freeman, K. C., 1970, ApJ 160, 811.
[4] McGaugh, S., Lelli, F. et al, Phys. Rev. Lett. 2016, 117, 201101.
[5] Persic, M & Salucci, P. 1995, ApjS, 99, 501.
[6] Persic, M & Salucci, P. 1991, ApJ, 368, 60.
[7] Persic, M., Salucci, P., Stel, F., 1996, MNRAS 281, 27.
[8] Rubin, V. C., Ford, W. K., Jr., Thonnard, N., et al 1982, ApJ, 261, 439.
[9] Salucci, P., Burkert, A.2000, ApJ, 537,9.
[10] Salucci, P.; Lapi, A.; Tonini, C et al. 2007, MNRAS, 378, 41.
[11] Tonini, C., Lapi, A., Shankar, F., & Salucci, P. 2006, ApJ, 638, L13.
[12] Yegorova, I, Salucci., P. 2007, MNRAS.377, 507.

SEARCH FOR DARK SECTOR PHYSICS IN MISSING ENERGY EVENTS IN THE NA64 EXPERIMENT

Mikhail Kirsanov [a]
Instutute for Nuclear research, Moscow, Russia

Abstract. A search is performed for a new sub-GeV vector boson (A') mediated production of Dark Matter (χ) in the fixed-target experiment, NA64, at the CERN SPS. The A', called dark photon, could be generated in the reaction $e^- Z \to e^- Z A'$ of 100 GeV electrons dumped against an active target which is followed by the prompt invisible decay $A' \to \chi \bar{\chi}$. The experimental signature of this process would be an event with an isolated electron and large missing energy in the detector. From the analysis of the data sample collected in 2016 corresponding to 4.3×10^{10} electrons on target no evidence of such a process has been found. New stringent constraints on the A' mixing strength with photons, $10^{-5} \lesssim \epsilon \lesssim 10^{-2}$, for the A' mass range $m_{A'} \lesssim 1$ GeV are derived.

The results are obtained by using tree level, exact calculations of the A' production cross-sections, which turn out to be significantly smaller compared to the one obtained in the Weizsäcker-Williams approximation for the mass region $m_{A'} \gtrsim 0.1$ GeV.

1 Introduction

Since the first evidence for the existence of Dark Matter (DM) in 1930's, understanding its origin has been one of the major focuses in particle physics. However, despite a wide variety of theoretical and experimental efforts DM still is a great puzzle. Though stringent constraints obtained on DM coupling to standard model (SM) particles in non-accelerator and collider experiments ruled out many DM models, the mass values of DM candidates still have big uncertainty and are currently spread over fifteen orders of magnitude from μeV cosmic axions to TeV scale SUSY WIMP miracles.

The dark sector is generally assumed to be decoupled from the SM and communicates to the visible sector presumably only by gravity. In recent years, various phenomenological models attempting to unify the known interactions into a single gauge scheme, such as the Grand Unified Theories (GUT), motivate the existence of dark sector of particles and fields, which are singlets with respect to the SM gauge group. The coupling of the dark sector to the visible one could occur through fields whose masses are naturally close to the visible scale [1–4]. An exciting possibility is that in addition to gravity, a new force between the two sectors transmitted by a new vector boson, A', called dark photon, might exist. The A' could have a mass $m_{A'} \lesssim 1$ GeV, and couple to the SM through kinetic mixing with the ordinary photon, described by the term $\frac{\epsilon}{2} F'_{\mu\nu} F^{\mu\nu}$ and parameterized by the mixing strength ϵ. The Lagrangian of the

---

[a] E-mail: Mikhail.Kirsanov@cern.ch

SM is extended by the dark sector in the following way:

$$\mathcal{L} = \mathcal{L}_{SM} - \frac{1}{4}F'_{\mu\nu}F'^{\mu\nu} + \frac{\epsilon}{2}F'_{\mu\nu}F^{\mu\nu} + \frac{m_{A'}^2}{2}A'_{\mu}A'^{\mu}$$
$$+i\bar{\chi}\gamma^{\mu}\partial_{\mu}\chi - m_{\chi}\bar{\chi}\chi - e_D\bar{\chi}\gamma^{\mu}A'_{\mu}\chi, \qquad (1)$$

where the massive vector field A'_{μ} is associated to the spontaneously broken $U_D(1)$ gauge group, $F'_{\mu\nu} = \partial_{\mu}A'_{\nu} - \partial_{\nu}A'_{\mu}$, e_D is the coupling constant of the $U(1)_D$ gauge interactions, and $m_{A'}$, m_{χ} are the masses of the dark photons and DM particles, respectively. Here, we consider as an example the Dirac spinor fields χ which are treated as Dark Matter fermions coupled to A'_{μ} by the dark portal coupling constant e_D. The mixing term of (1) results in the interaction:

$$\mathcal{L}_{int} = \epsilon e A'_{\mu}J^{\mu}_{em} \qquad (2)$$

of dark photons with the electromagnetic current J^{μ}_{em} with a strength ϵe, where e is the electromagnetic coupling and $\epsilon \ll 1$ [5–7]. Such small values of ϵ could naturally be obtained in GUT from loop effects of particles charged under both the dark and SM $U(1)$ interactions with a typical 1-loop value $\epsilon = ee_D/16\pi^2 \simeq 10^{-2} - 10^{-4}$ [7], while 2-loops contribution results in the range $10^{-3} - 10^{-5}$. An additional hint for the existence of the A' is suggested by the 3.6 σ deviation from the SM prediction of the muon anomalous magnetic moment $g_{\mu} - 2$ [8], which can be explained by a sub-GeV A' with the coupling $\epsilon \simeq 10^{-3}$ [9–11], as well as by hints on astrophysical signals of DM [3]. This has motivated a worldwide experimental and theoretical effort towards dark forces and other portals between the visible and dark sectors, see Refs. [4,12–15] for a review.

In presence of light dark states χ, in particular DM with the masses $m_{\chi} < m_{A'}/2$, the A' would predominantly decay invisibly into those particles provided that coupling $g_D > \epsilon e$. The decay rate of $A' \to \bar{\chi}\chi$ in this case is given by

$$\Gamma(A' \to \bar{\chi}\chi) = \frac{\alpha_D}{3}m_{A'}\left(1 + \frac{2m_{\chi}^2}{m_{A'}^2}\right)\sqrt{1 - \frac{4m_{\chi}^2}{m_{A'}^2}}. \qquad (3)$$

In the following we assume that the Dark Matter invisible decay mode is dominant, i.e. $\Gamma(A' \to \bar{\chi}\chi)/\Gamma_{tot} \simeq 1$, and that the A' leptonic decay channel is suppressed, $\Gamma(A' \to \bar{\chi}\chi) \gg \Gamma(A' \to e^-e^+)$. If such A' exists, many crucial questions about its mass scale, coupling constants, decay modes, etc. arise. One possible way to answer these questions, is to search for the invisible A' in accelerator experiments.

The approach considered in this work and proposed in Refs. [35,36], is based on the detection of the large missing energy, carried away by the energetic A' produced in the interactions of high-energy electrons in the active beam dump target, see also [17]. The advantage of this type of experiments is that their sensitivity is proportional to the mixing strength squared, ϵ^2, associated with the A' production in the primary reaction and its subsequent prompt invisible decay, while in the former case it is proportional to $\epsilon^4 \alpha_D$, with ϵ^2 associated with the A' production in the beam dump and $\epsilon^2 \alpha_D$ coming from the χ particle interactions in the detector.

In this work we report new results on the search for the A' and light DM in the fixed-target experiment NA64 at the CERN SPS. The experimental signature of events from the $A' \to invisible$ decays is clean and they can be selected with small background due to the excellent capability of NA64 for the precise identification and measurements of the initial electron state.

2 Method of search and the A' production

As follows from the Lagrangian (1), any source of photons will produce all kinematically possible massive A' states according to the appropriate mixing strength. If the coupling strength α_D and A' masses are as discussed above, the A' will decay predominantly invisibly.

The method of the search for the $A' \to invisible$ decay is as follows [35,36]. If the A' exists it could be produced via the kinetic mixing with bremsstrahlung photons in the reaction of high-energy electrons absorbed in an active beam dump (target) followed by the prompt $A' \to invisible$ decay into DM particles in a hermetic detector:

$$e^- Z \to e^- Z A'; \; A' \to \chi \overline{\chi}, \quad (4)$$

see Fig. 1. A fraction f of the primary beam energy $E_{A'} = f E_0$ is carried away by χ particles, which penetrate the target and detector without interactions resulting in zero-energy deposition. The remaining part of the beam energy $E_e = (1-f) E_0$ is deposited in the target by the scattered electron. The occurrence of the A' production via the reaction (4) would appear as an excess of events with a signature of a single isolated electromagnetic (e-m) shower in the dump with energy E_e accompanied by a missing energy $E_{miss} = E_{A'} = E_0 - E_e$ above those expected from backgrounds. Here we assume that in order to give a missing energy signature the χs have to traverse the detector without decaying visibly. No other assumptions are made on the nature of the $A' \to invisible$ decay.

In previous work [30, 37], the differential cross-section A'-production from reaction (1) was calculated with the Weizsäcker-Williams (WW) approximation, see [38, 39]. The cross-sections were implemented a Geant4 [40, 41] based simulations, and the total number $n_{A'}$ of the produced A' per single electron on target (EOT), depends in particular on ϵ, $m_{A'}$, E_0 and was calculated as

$$n_{A'}(\epsilon, \; m_{A'}, \; E_0) = \frac{\rho N_A}{A_{Pb}} \sum_i n(E_0, E_e, s) \sigma_{WW}^{A'}(E_e) \Delta s_i \quad (5)$$

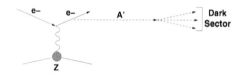

Figure 1: Diagram contributing to the A' production in the reaction $e^-Z \to e^-ZA', A' \to$ dark sector. The produced A' decays invisibly into dark sector particles.

where ρ is density of Pb target, N_A is the Avogadro's number, A_{Pb} is the Pb atomic mass, $n(E_0, E_e, s)$ is the number of $e^{\pm}$ with the energy E_e in the e-m shower at the depth s (in radiation lengths) within the target of total thickness T, and $\sigma(E_e)$ is the cross section of the A' production in the kinematically allowed region up to $E_{A'} \simeq E_e$ by an electron with the energy E_e in the elementary reaction (4). The energy distribution $\frac{\Delta n_{A'}}{\Delta E_{A'}}$ of the A's was calculated by taking into account that the differential cross-section $\frac{d\sigma(E_e, E_{A'})}{dE_{A'}}$ is sharply peaked at $E_{A'}/E_e \simeq 1$ [38]

The numerical summation in Eq. (5) was performed with the detailed simulation of e-m showers done by Geant4 over the missing energy spectrum in the target.

It has been recently pointed out, that for a certain kinematic region of the parameters $m_{A'}, E_{A'}$, the A' yield derived in the WW framework could differ significantly from the one calculated exactly at tree level [43, 44]. Therefore, it is instructive to perform an accurate calculation of the A' cross-section based on precise phase space integration over the final state particles in the reaction $e^-Z \to e^-ZA'$.

In order to derive more accurately the A' yield, first we have performed calculations of the total cross-section of (4) without WW approximation [45]. Then, to implement the exact total cross-section formula into Geant4 [41] simulation of the A' production in the target, we introduce in Eq. (5) a correction k-factor defined by the following ratio

$$k(m_{A'}, E_0, Z, A) = \frac{\sigma^{A'}_{\text{WW}}}{\sigma^{A'}_{\text{EXACT}}}. \qquad (6)$$

Here, the cross-section $\sigma^{A'}_{\text{EXACT}}$ takes into account the exact phase space integration over the final states of the particles and represents the overall uncertainties in the cross-section calculated in simplified WW approach. The A' yield was calculated by using (5) with the replacement

$$\sigma^{A'}_{\text{WW}} \to k(m_{A'}, E_0, Z, A)^{-1} \sigma^{A'}_{\text{WW}} \qquad (7)$$

The details of the calculations will be presented in Ref. [45]. We refer to this

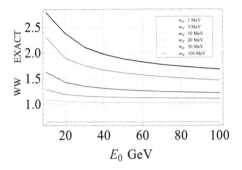

Figure 2: The k-factor for the A' production in the reaction $e^- Z \to e^- Z A'$ as a function of the electron energy E_0 for different values of the A' masses.

method as k-factor approach throughout the paper. For heavy target nuclei k depends rather weakly on Z and A, i.e. for tungsten and lead the deviation is about 0.5%. In Fig. 2, k values are shown as a function of the electron beam energy E_0 for various $m_{A'}$. One can see from Fig. 2 that the exact cross-section of A' production is smaller by a factor 1.7 than the corresponding cross-section in WW approximation for $m_{A'} = 100$ MeV and $E_0 = 100$ GeV. On the other hand, σ_{EXACT} exceeds σ_{WW} for the masses $m_{A'}$ below 5 MeV, such that one can improve the limits on the mixing strength for this mass region.

Once the A' flux (5) was defined, the next step was to simulate the A' emission spectrum from the target. The decay electrons and positrons were tracked through the dump medium including bremsstrahlung photons, their conversion and multiple scattering in the target. The A' reconstruction efficiency in the target was computed and convoluted with the target details and detector geometrical acceptance (see Sec. 3) based on the NA64 Monte Carlo (MC) simulation package used in the our previous searches [30] and corrected for data themselves.

3 H4 beam and NA64 detector

The experiment employs the optimized 100 GeV electron beam from the H4 beam line at the North Area (NA) of the CERN SPS described in details in Ref. [46]. The H4 provided an essentially pure e^- beam for fixed-target experiments. The beam was designed to transport the electrons with the maximal intensity up to $\simeq 10^7$ per SPS spill of 4.8 s in the momentum range between 50 and 150 GeV/c that could be produced by the primary proton beam of 400 GeV/c with the intensity up to a few 10^{12} protons on a beryllium target. The main contribution to the e^- yield from the target was the production of π^0 followed by a process $\pi^0 \to \gamma\gamma \to e^+e^-$. The short-lived π^0 decays inside the target,

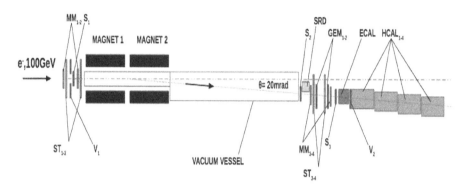

Figure 3: Schematic illustration of the setup to search for $A' \to invisible$ decays of the bremsstrahlung A's produced in the reaction $eZ \to eZA'$ of 100 GeV e⁻ incident on the active ECAL target.

and the electrons were produced through the conversion of the decay photons in a separate converter [47]. Protons and charged secondaries that did not interact in the convertor are separated from the neutrals by deflecting them in a magnetic field to a thick absorber. The electrons produced in the converter are transported to the NA64 detector inside an evacuated beam-line tuned to an adjustable beam momentum. The hadron contamination in the electron beam was $\pi/e^- \lesssim 10^{-2}$. The beam has the transverse size at the detector position of the order of a few cm² and a halo with intensity $\lesssim$ a few %.

The signal event recognition in NA64 must rely on the detection of the incoming and outgoing electron only, since the decay product for the $A' \to invisible$ decay are undetectable. The NA64 detector, which is located at about 500 m from the proton target, is schematically shown in Fig. 3. The setup utilised the beam defining scintillator (Sc) counters S1-S3 and veto V1, and the spectrometer consisting of two successive dipole magnets with the integral magnetic field of $\simeq 7$ T·m and a low-material-budget tracker. The magnets also served as an effective filter rejecting the low energy electrons present in the beam. To improve the high energy electrons selection and suppress background from a possible admixture of low energy electrons, a tagging system utilizing the synchrotron radiation (SR) from high energy electrons in the magnetic field was used, as schematically shown in Fig. 3. The basic idea was that, since the amount SR energy per revolution emitted by a particle with the mass m and energy E_0 is $<E_{SR}> \propto E_0^3/m^4$, the low energy electrons and hadrons in the beam could be effectively rejected by using the cut on the energy deposited in the SR detector (SRD) [35, 49]. A 15 m long vacuum vessel was installed between the magnets and the ECAL to minimize absorption of the SR photons detected immediately at the downstream end of the vessel with a SRD, which was an array of PbSc sandwich calorimeter of a fine longitudinal segmentation. Compared to the previous measurements [30], the SRD was also segmented transversely in three SRD counters, each of 60 × 80mm^2 in lateral size assembled from 80 – 100 μm Pb and 1 mm Sc plates with wave length shifting (WLS) fiber read-out. This allowed to additionally suppress background from hadrons, that could knock off electrons from the output vacuum window of the vessel producing a fake e^- SRD tag, by about two orders of magnitude [49]. The detector was also equipped with an active target, which was a hodoscopic electromagnetic calorimeter (ECAL) for the measurement of the electron energy deposition, E_{ECAL}, with the accuracy $\delta E_{ECAL}/E_{ECAL} \simeq 0.1/\sqrt{E_{ECAL}[\text{GeV}]}$ as well as the X, Y coordinates of the incoming electrons by using the transverse e-m shower profile. The ECAL was a matrix of 6 × 6 Shashlik-type counters assembled from Pb and Sc plates with WLS fiber read-out. Each module was $\simeq$ 40 radiation lengths (X_0) and had an initial part of $\simeq$ 4 X_0 used as a preshower (PS). By requiring the presence of the in-time SR signals in all three SRD counters, and using information on the longitudinal and lateral shower development in the ECAL, the initial level of the hadron contamination in the beam $\pi/e^- \lesssim 10^{-2}$ was further suppressed by more than 4 orders of magnitudes, while keeping the electron ID efficiency at the level $\gtrsim 95\%$ [49]. A high-efficiency veto counter V_2, and a massive, hermetic hadronic calorimeter (HCAL) of $\simeq$ 30 nuclear interaction lengths (λ_{int}) were positioned just after the ECAL. The HCAL which was an assembly of four modules HCAL1-HCAL4, served as an efficient veto to detect muons or hadronic secondaries produced in the e^-A interactions in the ECAL target. Each module was a sandwich of 48 alternating layers of iron and scintillator (Sc) with a thickness of 25 mm and 4 mm, respectively, with a total length of $\simeq 7\lambda_{int}$, and with a lateral size 60 × 60 cm^2. Each Sc layer consisted of 3×3 plates with WLS fiber readout allowing to assemble the whole HCAL module as a matrix of 3 × 3 cells, each of 20 × 20 cm^2. The single electron events were collected with the hardware trigger

$$Tr(A') = \Pi S_i \cdot V_1 \cdot PS(> E_{PS}^{th}) \cdot \overline{ECAL}(< E_{ECAL}^{th}) \qquad (8)$$

designed to accept events with in-time hits in beam-defining counters S_i and clusters in the PS and ECAL with the energy thresholds $E_{PS}^{th} \simeq 0.3$ GeV and $E_{ECAL}^{th} \lesssim 80$ GeV, respectively.

4 Data analysis and selection criteria

The search for the $A' \to invisible$ decay described in this paper uses the full data sample collected during July and October runs in 2016 corresponding to $n_{EOT} = 4.3 \times 10^{10}$ EOT. The results reported here are obtained using three sets of data in which $n_{EOT} = 2.3 \times 10^{10}$, 1.1×10^{10} and 0.9×10^{10} EOT were collected with the beam intensities $\simeq (1.4-2) \times 10^6$, $\simeq (3-3.5) \times 10^6$ and $\simeq (4.5-5) \times 10^6$ e^- per spill, respectively. Data of these three runs (hereafter called respectively the run I, II, and III) were processed with selection criteria similar to the one used in our previous paper [30] and finally combined as described in Sec. 8. Compared to the analysis of Ref. [30], a number of improvements in the event reconstruction, e.g., adding the pileup algorithm, were made in order to increase the reconstruction efficiency.

The strategy of the analysis was to identify $A' \to invisible$ candidates by precise reconstruction of the initial e^- state and an isolated low energy e-m shower in the ECAL that are accompanied by no other activity in the V_2 and HCAL detectors. The measured rate of such events was then supposed to be compared to that expected from known sources.

The spectra of A's produced in the ECAL target by primary electrons were calculated using the approach reported in Ref. [37]. A detailed Geant4 based MC simulation [41] was used to study the detector performance and acceptance losses, to simulate background sources, and to select cuts and estimate the reconstruction efficiency.

The candidate events were preselected with the criteria chosen to maximize the acceptance for simulated signal events and to minimize the numbers of events expected from background sources discussed in Sec. 7. The following selection criteria were applied:

- There must be one and only one incoming particle track having a small angle with respect to the beam axis. This cut rejects low momentum electrons as they were typically correlated with a large-angle incoming tracks originating presumably from the upstream e^- interactions. The reconstructed momentum of the particle was required to be $P_e = 100 \pm 2$ GeV.

- The track should be identified as an electron with the SRD detector. The energy deposited in each of the three SRD modules should be within the SR range emitted by e^-s and in time with the trigger. This was the key cut identifying the pure initial e^- state, with the pion suppression factor $< 10^{-5}$ and electron efficiency $> 95\%$ [49].

- The lateral and longitudinal shower shape in the ECAL should be consistent with the one expected for the signal shower [37,53]. It is also used to distinguish hadrons from electrons providing an additional hadron rejection factor of $\simeq 10$ [54].

- There should be no activity in the veto counter V_2.

In total $\simeq 7 \times 10^4$ events passed these criteria from the combined 2016 data sample. The final selection is a cut-based and uses the cuts on the ECAL

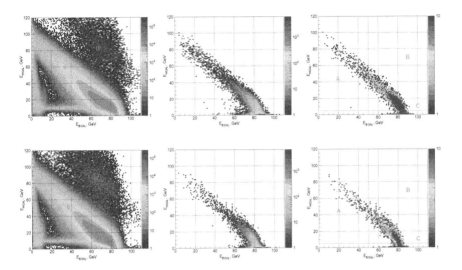

Figure 4: Event distribution in the $(E_{ECAL}; E_{HCAL})$ plane from the runs II(top row) and III (bottom row) data. The left panels show the measured distribution of events at the earlier phase of the analysis. Plots in the middle show the same distribution after applying all selection criteria, but the cut against upstream interactions. The right plots present the final event distributions after all cuts applied. The dashed area is the signal box region which is open. The side bands A and C are the one used for the background estimate inside the signal box. For illustration purposes the size of the signal box along E_{HCAL}-axis is increased by a factor five.

missing energy $E_{miss} = E_{beam} - E_{ECAL}$ and on the energy deposition in the HCAL.

The cut on E_{miss} was optimized as described in Sec. 8.

In Fig. 4 the left panels show an example of the distributions of events from the reaction $e^- Z \to anything$ in the $(E_{ECAL}; E_{HCAL})$ plane measured in the runs II(top) and III(bottom) with moderate selection criteria requiring only the presence of the SRD tag identifying the beam electrons. Here, E_{HCAL} is the sum of the energy deposited in the HCAL1 and HCAL2. The distributions of events from the run I with low intensity are similar to the one shown in Ref. [30]. Events from the areas I in Fig. 4 originate from the QED dimuon production, dominated by the the muon pair photoproduction by a hard bremsstrahlung photon conversion on a target nucleus:

$$e^- Z \to e^- Z \gamma; \gamma \to \mu^+ \mu^- \qquad (9)$$

with some contribution from $\gamma\gamma \to \mu^+\mu^-$ fusion process. The $\mu^+\mu^-$ pairs were characterised by the HCAL energy deposition of $\simeq 10$ GeV. This rare process whose fraction of events with $E_{ECAL} \lesssim 60$ GeV was $\simeq 10^{-5}$/EOT served as a benchmark allowing to verify the detector performance and as a reference for the background prediction. The regions II shows the events from the SM hadron electroproduction in the ECAL which satisfy the energy conservation $E_{ECAL} + E_{HCAL} \simeq 100$ GeV within the detector energy resolution. The leak of these events to the signal region mainly due to the HCAL energy resolution was found to be negligible. The fraction of events from the region III was due to pileup of e^- and beam hadrons. It was beam rate dependent with a typical value from about a few % up to $\simeq 20\%$.

5 Dimuon events from the reaction $e^-Z \to e^-Z\mu^+\mu^-$

To evaluate the performance of the setup, a cross-check between a clean sample of $\gtrsim 10^4$ observed and MC simulated $\mu^+\mu^-$ events was made. The process (9) was used as a benchmark allowing to verify the reliability of the MC simulation and to estimate the corrections to the signal reconstruction efficiency and possible additional uncertainties in the A' yield calculations. Let us first briefly review the description of the gamma conversion into a muon-antimuon pair implemented in Geant4. The dimuon production was also used as a reference for the prediction of background, see Sec. 7.

The maximal difference between the number of observed and MC predicted $\mu^+\mu^-$ events with $E_{ECAL} \lesssim 60$ GeV is $\simeq 20\%$. Part of the difference is explained mainly by additional inefficiency of dimuon reconstruction due to pileup effect at higher beam intensity (see, Sec. 6). The observed discrepancy at lower beam rate of the order of $\simeq 10\%$ can be interpreted as due to the inaccuracy of the dimuon yield calculation for heavy nuclei and, thus can be conservatively accounted for as systematic uncertainty in the A' yield $n_{A'}$.

6 Signal efficiency

Several signal detection efficiencies contribute to the value of $\epsilon_{tot}(m_{A'})$ in the NA64 detector:

$$\epsilon_{tot}(m_{A'}) = \epsilon_e \cdot \epsilon_{A'} \cdot \epsilon_{ECAL} \cdot \epsilon_V \cdot \epsilon_{HCAL} \qquad (10)$$

where ϵ_e, $\epsilon_{A'}$, ϵ_{ECAL}, ϵ_V and ϵ_{HCAL} are the efficiency factors for the incoming e^- detection, the A' acceptance in the signal box range, and the efficiencies for the signal to pass the ECAL, V_2, and HCAL selection criteria, respectively. These factors were determined from the sample obtained with MC simulations and from the data samples of e^- and dimuon events. The flux and spectra of the A's produced in the ECAL target by primary electrons were calculated using the approach reported in Ref. [37] taking into account the development of the signal e-m shower from reaction (4) in the ECAL target (see, Sec. V).

7 Background

The search for the $A' \to invisible$ decays requires particular attention to backgrounds, because every process with a single track and an e-m cluster in the ECAL can potentially fake the signal. In this Section we consider all background sources, which were also partly studied in Refs. [35, 36].

There are several backgrounds resulting in the signature similar to signal which can be classified as being due to detector-, physical- and beam-related sources. The selection cuts to reject these backgrounds have been chosen such that they do not affect the shape of the true E_{miss} spectrum.

The estimation of background levels and the calculation of signal acceptance were both based on the MC simulation, as well as direct measurement with the beam. Because of the small A' coupling strength value, performing a complete detector simulation in order to investigate these backgrounds down to the level of a single event sensitivity $\lesssim 10^{-11}$ would require a very large amount of computing time. Consequently, we have estimated with MC simulations all known backgrounds to the extent that it is possible. Events from particle interactions or decays in the beam line, pileup activity created from them, hadron punchthrough from the target and the HCAL were included in the simulation of background events. Small event-number backgrounds such as the decays of the beam μ, π, K or μs from the reaction of dimuon production were simulated with the full statistics of the data. Large event-number processes, e.g. from e^- interactions in the target or beam line, punchthrough of secondary hadrons were also studied, although simulated samples with statistics comparable to the data were not feasible.

The main detector background sources are related to

- *Instrumental effects*. The leak of energy throughout the possible holes, cracks, etc.

- *Detector hermeticity*.

- *Large transverse fluctuations*.

The beam backgrounds can be subdivided into two categories: upstream interactions and particle decays.

- *Upstream interactions*.

- *Particle decays*.

The remaining physical backgrounds were

- *Dimuon, τ, charm decays*.

- Finally, the electroproduction of a neutrino pair $eZ \to eZ\nu\bar{\nu}$ resulting in the invisible final state accompanied by energy deposition in the ECAL1 from the recoil electron can occur. An estimate showed that the ratio of the cross sections for this reaction to the bremsstrahlung cross section is well below 10^{-13} [35].

The final number of background events estimated from the combined MC and data events is $n_b = 0.12 \pm 0.04$ events for 4.3×10^{10} EOT. The estimated uncertainty of about 30% was due mostly to the uncertainty in background level from upstream beam interactions. It also includes the uncertainties in the amount of passive material for e^- interactions, in the cross sections of the hadron charge-exchange reactions on lead (30%), and systematic errors related to the extrapolation procedure. The total systematic uncertainty was calculated by adding all errors in quadrature.

8 Results and calculation of limits

In the final statistical analysis the three runs I-III were analysed simultaneously using the multi-bin limit setting technique. The corresponding code is based on the RooStats package [57]. First of all, the above obtained background estimates, efficiencies, and their corrections and uncertainties were used to optimize more accurately the main cut defining the signal box by comparing sensitivities, defined as an average expected limit calculated using the profile likelihood method, with uncertainties used as nuisance parameters. Log-normal distribution was assumed for the nuisance parameters [58]. The most important inputs for this optimization were the expected values from the background extrapolation into the signal box for the data samples of the runs I, II, III. The uncertainties for background prediction were estimated by varying the extrapolation functions, as previously discussed. The optimization confirmed the preliminary choice of the cut on the signal region $E_{EC} < 50$ GeV, also previously used in Ref. [30].

After determining and optimizing all the selection criteria and estimating background levels, we examined the events in the signal box and found no candidates, as shown in Fig. 4. We proceeded then with the calculation of the upper limits on the A' production. The combined 90% confidence level (C.L.) upper limits for the corresponding mixing strength ϵ were determined from the 90% C.L. upper limit for the expected number of signal events, $N_{A'}^{90\%}$ by using the modified frequentist approach for confidence levels (C.L.), taking the profile likelihood as a test statistic in the asymptotic approximation [59–61]. The total number of expected signal events in the signal box was the sum of

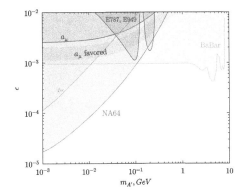

Figure 5: The NA64 90% C.L. exclusion region in the $(m_{A'}, \epsilon)$ plane. Constraints from the BaBar [31], E787 and E949 experiments [25, 26], as well as the muon α_μ favored area are also shown. Here, $\alpha_\mu = \frac{g_\mu - 2}{2}$. For more limits obtained from indirect searches and planned measurements see e.g. Ref. [13, 14].

expected events from the three runs:

$$N_{A'} = \sum_{i=1}^{3} N_{A'}^i = \sum_{i=1}^{3} n_{EOT}^i \epsilon_{tot}^i n_{A'}^i(\epsilon, m_{A'}, \Delta E_e) \quad (11)$$

where ϵ_{tot}^i is the signal efficiency in the run i given by Eq. (10), and the $n_{A'}^i(\epsilon, m_{A'}, \Delta E_{A'})$ value is the signal yield per EOT generated by a single 100 GeV electron in the ECAL target in the energy range ΔE_e. Each i-th entry in this sum was calculated by simulating the signal events for corresponding beam running conditions and processing them through the reconstruction program with the same selection criteria and efficiency corrections as for the data sample from run I. The expected backgrounds and systematic errors estimated for the run were also used in the limits calculation. The combined 90% C.L. exclusion limits on the mixing strength as a function of the A' mass can be seen in Fig. 5. The derived bounds are the best for the mass range $0.001 \lesssim m_{A'} \lesssim 0.1$ GeV obtained from direct searches of $A' \to invisible$ decays [15].

9 Conclusion

From the analysis of the full 2016 data sample, we found no evidence for the existence of dark photon with the mass in the range $\lesssim 1$ GeV which mixes with the ordinary photon and decays dominantly invisibly into light DM particles $A' \to \chi\overline{\chi}$. The limits were obtained by taking into account corrections to the A' production cross section calculated without Weizsäcker-Williams approximation.

For the mass range $m_{A'} \lesssim 0.1$ GeV the most stringent upper limit on the mixing strength, ϵ was obtained.

Acknowledgements

We gratefully acknowledge the support of the CERN management and staff and the technical staffs of the participating institutions for their vital contributions. This work was supported by the HISKP, University of Bonn (Germany), JINR (Dubna), MON, RAS and RF program "Nauka" (Contract No. 0.1764.GZB.2017) (Russia), ETH Zurich and SNSF Grant No. 169133 (Switzerland), and grants FONDECYT 1140471, 1150792, and 3170852, Ring ACT1406 and Basal FB0821 CONICYT (Chile). Part of the work on MC simulations was supported by the RSF grant 14-22-00161. We thank Gerhard Mallot for useful suggestions, Maxim Pospelov for useful communications and help provided for Sec. IX, and Didier Cotte, Michael Jeckel, Vladimir Karjavin, Christophe Menezes Pires, and Victor Savrin for their help. We thank COMPASS DAQ group and the Institute for Hadronic Structure and Fundamental Symmetries of TU Munich for the technical support.

References

[1] P. Fayet, "Effects of the Spin 1 Partner of the Goldstino (Gravitino) on Neutral Current Phenomenology," Phys. Lett. **95B**, 285 (1980).
[2] M. Pospelov, A. Ritz and M. B. Voloshin, "Secluded WIMP Dark Matter," Phys. Lett. B **662**, 53 (2008).
[3] N. Arkani-Hamed, D. P. Finkbeiner, T. R. Slatyer and N. Weiner, "A Theory of Dark Matter," Phys. Rev. D **79**, 015014 (2009).
[4] J. Jaeckel and A. Ringwald, "The Low-Energy Frontier of Particle Physics," Ann. Rev. Nucl. Part. Sci. **60**, 405 (2010).
[5] L. B. Okun, "Limits Of Electrodynamics: Paraphotons?," Sov. Phys. JETP **56**, 502 (1982) [Zh. Eksp. Teor. Fiz. **83** 892 (1982)].
[6] P. Galison and A. Manohar, "Two Z's or not two Z's ?", Phys. Lett. B **136**, 279 (1984).
[7] B. Holdom, "Two U(1)'s and Epsilon Charge Shifts," Phys. Lett. B **166**, 196 (1986).
[8] G. W. Bennett et al. [Muon g-2 Collaboration], "Final Report of the Muon E821 Anomalous Magnetic Moment Measurement at BNL," Phys. Rev. D **73**, 072003 (2006).
[9] S. N. Gninenko and N. V. Krasnikov, "The Muon anomalous magnetic moment and a new light gauge boson," Phys. Lett. B **513**, 119 (2001).
[10] P. Fayet, "U-boson production in e+ e- annihilations, psi and Upsilon decays, and Light Dark Matter," Phys. Rev. D **75**, 115017 (2007).

[11] M. Pospelov, "Secluded U(1) below the weak scale," Phys. Rev. D **80**, 095002 (2009).
[12] R. Essig et al., "Working Group Report: New Light Weakly Coupled Particles", arXiv:1311.0029 [hep-ph].
[13] J. Alexander et al., "Dark Sectors 2016 Workshop: Community Report," arXiv:1608.08632 [hep-ph].
[14] M. Battaglieri et al., "US Cosmic Visions: New Ideas in Dark Matter 2017: Community Report," arXiv:1707.04591 [hep-ph].
[15] C. Patrignani et al. [Particle Data Group], "Review of Particle Physics," Chin. Phys. C **40**, 100001 (2016).
[16] P. deNiverville, M. Pospelov and A. Ritz, "Observing a light Dark Matter beam with neutrino experiments," Phys. Rev. D **84**, 075020 (2011).
[17] E. Izaguirre, G. Krnjaic, P. Schuster and N. Toro, "Testing GeV-Scale Dark Matter with Fixed-Target Missing Momentum Experiments," Phys. Rev. D **91**, 094026 (2015).
[18] E. Izaguirre, G. Krnjaic, P. Schuster and N. Toro, "Analyzing the Discovery Potential for Light Dark Matter," Phys. Rev. Lett. **115**, 251301 (2015).
[19] E. Izaguirre, Y. Kahn, G. Krnjaic and M. Moschella, "Testing Light Dark Matter Coannihilation With Fixed-Target Experiments," arXiv:1703.06881 [hep-ph].
bibitemkuflic1 S. M. Choi, Y. Hochberg, E. Kuflik, H. M. Lee, Y. Mambrini, H. Murayama and M. Pierre, "Vector SIMP dark matter," arXiv:1707.01434 [hep-ph].
[20] E. Kuflik, M. Perelstein, N. R. L. Lorier and Y. D. Tsai, "Phenomenology of ELDER Dark Matter," JHEP **1708**, 078 (2017).
[21] Y. Hochberg, E. Kuflik and H. Murayama, "Dark Spectroscopy," arXiv:1706.05008 [hep-ph].
[22] H. S. Lee, "Muon $g - 2$ anomaly and dark leptonic gauge boson," Phys. Rev. D **90**, 091702 (2014).
[23] E. Izaguirre, G. Krnjaic, P. Schuster and N. Toro, "New Electron Beam-Dump Experiments to Search for MeV to few-GeV Dark Matter," Phys. Rev. D **88**, 114015 (2013).
[24] M. D. Diamond and P. Schuster, "Searching for Light Dark Matter with the SLAC Millicharge Experiment," Phys. Rev. Lett. **111**, 221803 (2013).
[25] H. Davoudiasl, H. S. Lee and W. J. Marciano, "Muon $g2$, rare kaon decays, and parity violation from dark bosons," Phys. Rev. D **89**, 095006 (2014).
[26] R. Essig, J. Mardon, M. Papucci, T. Volansky and Y. M. Zhong, "Constraining Light Dark Matter with Low-Energy e^+e^- Colliders," JHEP **1311**, 167 (2013).
[27] B. Aubert et al. [BaBar Collaboration], "Search for Invisible Decays of a Light Scalar in Radiative Transitions $v_{3S} \to \gamma$ A0," arXiv:0808.0017 [hep-ex].

[28] B. Batell, M. Pospelov and A. Ritz, "Exploring Portals to a Hidden Sector Through Fixed Targets," Phys. Rev. D **80**, 095024 (2009).
[29] B. Batell, R. Essig and Z. Surujon, "Strong Constraints on Sub-GeV Dark Sectors from SLAC Beam Dump E137," Phys. Rev. Lett. **113**, 171802 (2014).
[30] D. Banerjee et al. [NA64 Collaboration], "Search for invisible decays of sub-GeV dark photons in missing-energy events at the CERN SPS," Phys. Rev. Lett. **118**, 011802 (2017).
[31] J. P. Lees et al. [BaBar Collaboration], "Search for invisible decays of a dark photon produced in e+e- collisions at BaBar," arXiv:1702.03327 [hep-ex].
[32] R. Dharmapalan et al. [MiniBooNE Collaboration], "Low Mass WIMP Searches with a Neutrino Experiment: A Proposal for Further MiniBooNE Running," arXiv:1211.2258 [hep-ex].
[33] S.N. Gninenko, "Stringent limits on the $\pi^0- > \gamma X, X- > e^+e^-$ decay from neutrino experiments and constraints on new light gauge bosons" Phys. Rev. D **85**, 055027 (2012).
[34] S.N. Gninenko, "Constraints on sub-GeV hidden sector gauge bosons from a search for heavy neutrino decays" Phys. Lett. B **713**, 244 (2012).
[35] S. N. Gninenko, "Search for MeV dark photons in a light-shining-through-walls experiment at CERN," Phys. Rev. D **89**, 075008 (2014).
[36] S. Andreas et al., "Proposal for an Experiment to Search for Light Dark Matter at the SPS," arXiv:1312.3309 [hep-ex].
[37] S. N. Gninenko, N. V. Krasnikov, M. M. Kirsanov and D. V. Kirpichnikov, "Missing energy signature from invisible decays of dark photons at the CERN SPS," Phys. Rev. D **94**, 095025 (2016).
[38] J. D. Bjorken, R. Essig, P. Schuster and N. Toro, "New Fixed-Target Experiments to Search for Dark Gauge Forces," Phys. Rev. D **80**, 075018 (2009).
[39] Y. S. Tsai, "Axion Bremsstrahlung By An Electron Beam," Phys. Rev. D **34**, 1326 (1986).
[40] S. Agostinelli et al. [GEANT4 Collaboration], "GEANT4: A Simulation toolkit," Nucl. Instrum. Meth. A **506**, 250 (2003).
[41] J. Allison et al., "Geant4 developments and applications," IEEE Trans. Nucl. Sci. **53**, 270 (2006).
[42] Y. S. Tsai, "Pair Production and Bremsstrahlung of Charged Leptons," Rev. Mod. Phys. **46**, 815 (1974) Erratum: [Rev. Mod. Phys. **49**, 521 (1977)].
[43] Y. S. Liu, D. McKeen and G. A. Miller, "Validity of the Weizscker-Williams approximation and the analysis of beam dump experiments: Production of a new scalar boson," Phys. Rev. D **95**, 036010 (2017)

[44] Y. S. Liu and G. A. Miller, "Validity of the Weizscker-Williams approximation and the analysis of beam dump experiments: Production of an axion, a dark photon, or a new axial-vector boson," Phys. Rev. D **96**, 016004 (2017).
[45] D. Kirpichnikov et al., Paper in preparation.
[46] See, for example, http://sba.web.cern.ch/sba/
[47] H.W. Atherton, P. Coet, N. Doble, D.E. Plane, CERN/SPS 85-43 (1985).
[48] D. Banerjee, P. Crivelli and A. Rubbia, "Beam Purity for Light Dark Matter Search in Beam Dump Experiments," Adv. High Energy Phys. **2015**, 105730 (2015).
[49] E. Depero et al., "High purity 100 GeV electron identification with synchrotron radiation," Nucl. Instrum. Meth. A **866**, 196 (2017).
[50] A. B. Mann et al., "The universal sampling ADC readout system of the COMPASS experiment," 2009 IEEE Nuclear Science Symposium Conference Record (NSS/MIC), Orlando, FL (2009)2225.
[51] S. Huber, J. Friedrich, B. Ketzer, I. Konorov, M. Kramer, A. Mann, T. Nagel and S. Paul, "A digital trigger for the electromagnetic calorimeter at the COMPASS experiment," IEEE Trans. Nucl. Sci. **58**, 1719 (2011).
[52] I.Konorov et al., "Overview and future developments of the FPGA-based DAQ of COMPASS," JINST **11**, C02025 (2016).
[53] G. A. Akopdzhanov et al., "Determination of Photon Coordinates in Hodoscope Cherenkov Spectrometer," Nucl. Instrum. Meth. **140**, 441 (1977).
[54] V. A. Davydov, A. V. Inyakin, V. A. Kachanov, R. N. Krasnokutsky, Y. V. Mikhailov, Y. D. Prokoshkin and R. S. Shuvalov, "Particle Identification in Hodoscope Cherenkov Spectrometer," Nucl. Instrum. Meth. **145**, 267 (1977).
[55] H. Burkhardt, S. R. Kelner and R. P. Kokoulin, "Monte Carlo generator for muon pair production," CERN-SL-2002-016-AP, CLIC-NOTE-511.
[56] S. R. Kelner, R. P. Kokoulin and A. A. Petrukhin, "About cross-section for high-energy muon bremsstrahlung," FPRINT-95-36.
[57] I. Antcheva et al., "ROOT: A C++ framework for petabyte data storage, statistical analysis and visualization," Comput. Phys. Commun. **180**, 2499 (2009).
[58] E. Gross, "LHC statistics for pedestrians," CERN-2008-001.
[59] T. Junk, "Confidence Level Computation for Combining Searches with Small Statistics", Nucl. Instrum. Meth. A **434**, 435 (1999).
[60] G. Cowan, K. Cranmer, E. Gross, O. Vitells, "Asymptotic formulae for likelihood-based tests of new physics", Eur. Phys. J. C **71**, 1554 (2011).
[61] A. L. Read, "Presentation of search results: The CL(s) technique," J. Phys. G **28**, 2693 (2002).

RECENT RESULTS FROM THE DAMPE EXPERIMENT

Giovanni Marsella [a] on behalf of DAMPE collaboration
Dipartimento di Matematica e Fisica "E. De Giorgi", Università del Salento and INFN, Lecce. Italy

1 Introduction

The DArk Matter Particle Explorer (DAMPE [1]), was successfully launched into a sun-synchronous orbit at the altitude of 500 km on 2015 December 17th from the Jiuquan launch base. DAMPE offers a new opportunity for advancing our knowledge of cosmic rays, dark matter, and gamma-ray astronomy. In this paper the first scientific results are outlined and discussed. DAMPE is able to detect electrons/positrons, gamma rays, protons, helium nuclei and other heavy ions in a wide energy range with much improved energy resolution and large acceptance (see Table 1 for summary of the instrument parameters) than precedent detectors. The main scientific objectives addressed by DAMPE include: (1) understanding the mechanisms of particle acceleration operating in astrophysical sources, and the propagation of cosmic rays in the the Milky Way; (2) probing the nature of dark matter; and (3) studying the gamma-ray emission from Galactic and extragalactic sources.

1.1 The detector

The DAMPE experiment consists of a Plastic Scintillator strip Detector (PSD), a Silicon-Tungsten tracKer-converter (STK), a BGO imaging calorimeter and a Neutron Detector (NUD). The PSD provides charged-particle background rejection for gamma rays (anti-coincidence detector) and measures the charge of incident particles; the STK measures the charges and the trajectories of charged particles, and allows to reconstruct the directions of incident photons converting into e^+e^- pairs; the hodoscopic BGO calorimeter, with a total depth of about 32 radiation lengths, allows to measure the energy of incident particles with high resolution and to provide efficient electron/hadron identification.

Finally, the NUD provides an independent measurement and further improvement of the electron/hadron identification.

2 The First Results

In 2014 and 2015, the Engineering Qualification Model (EQM) of DAMPE has been extensively tested on different particle beams. This test was fundamental in order to calibrate the charge and energy reconstruction of the various sub-detectors.

---

[a] E-mail: giovanni.marsella@le.infn.it

Table 1: Summary of the design parameters and expected performance of DAMPE instrument.

Parameter	Value
Energy range of γ-rays/electrons	5 GeV – 10 TeV
Energy resolution of γ-rays/electrons	1.5% at 800 GeV
Energy range of protons/heavy nuclei	50 GeV – 100 TeV
Energy resolution of protons	40% at 800 GeV
Effective area at normal incidence (γ-rays)	1100 cm^2 at 100 GeV
Geometric factor for electrons	0.3 m^2 sr above 30 GeV
Photon angular resolution	0.2 at 100 GeV
Field of View (FoV)	1.0 sr

2.1 Charge reconstruction

The measurement of the energy spectra of cosmic-ray nuclei (Z = 1 – 26) in the energy range from 5 GeV to 100 TeV is a major goal of DAMPE. The charge of cosmic rays can be measured by both the PSD and the STK. A charged particle crossing a PSD strip loses energy mainly by ionization, with the energy deposition being proportional to Z^2 and to the path length. From each signal an energy deposition value is calculated, correcting for the path length and the position of the track along the strip to account for light attenuation. The STK, with its 12 layers of silicon strip detectors, can also be used to measure the charge of incident particles. The energy deposition for a cluster can be deduced from the impact point and incidence angle. The charge number can be estimated by combining all those measurements. The PSD and STK are combined to further improve the measurement of Z. In this way a spectrum of charged particle has been obtained between z = 1 and z = 28 with a charge resolution of 0.13e for H and 0.32e for Fe [2].

2.2 Energy reconstruction

The first step of the energy reconstruction algorithm is the conversion of the ADC counts into energy based on the calibration constants, once pedestals have been removed, and choosing the signals from the proper readout dynodes The total deposited energy is then calculated by summing up the energies of all BGO crystal elements. On orbit, cosmic-ray proton MIP events are selected to calibrate the energy response of ADC for each BGO crystal. The resulting ADC distribution of each individual BGO crystal is fitted with a Landau function convolved with a Gaussian distribution. The most probable value (MPV) corresponds to the MPV in energy units taken from the simulation. Thanks to the multi-dynode readout design, the BGO calorimeter enables a measurement of the energy of electrons or γ-rays up to 10 TeV without saturation. The energy deposited in the BGO calorimeter underestimates the true energy of incident particles. Electrons and photons can lose a significant fraction of their energy in the dead materials of calorimeter. The true energy of electrons and photons is evaluated by properly modeling the transversal and longitudinal development of electromagnetic showers in the calorimeter. The details of these procedures can be found in Ref. [3]. The energy measurements for cosmic-ray protons and nuclei are much more complicated than that for electrons or gamma rays, as hadronic showers generally are not fully contained in the BGO. Moreover hadronic showers include an electromagnetic and a hadronic component with large event-by-event fluctuations, which brings relatively large uncertainties in the energy deposition. An unfolding algorithm based on the Bayes theorem has been implemented to estimate the primary energy spectra of cosmic-ray nuclei. DAMPE can measure hadronic cosmic rays up to an energy of 100 TeV without significant saturation.

2.3 Electrons and γ-rays

The measurement of the total spectrum of cosmic ray electrons+positrons is a major goal of DAMPE. Therefore, besides the track and energy reconstruction, a high identification and discrimination power of protons from electron/positrons is required. The basic approach for electron/proton identification is an image based pattern recognition method. MC simulation and beam test data showed that electrons and protons can be indeed well separated. In the GeV–TeV energy range, the proton rejection power reaches a level of 10^5, while keeping at least a 90% electron identification efficiency. In addition, the High Energy trigger has been optimized to suppress the proton events by a factor of 3. Finally, the NUD can be used to further increase the rejection power by a factor of 2.5 at TeV energies. An electron+positron spectrum has been obtained. The contamination of the proton background is estimated to be less than 3% in the energy range of 50 GeV–1 TeV. More details on the systematic uncertainties can be found in published in [4]. Using the PSD as a veto also photons have been selected, allowing the construction of a first γ-rays sky map where the principal sources are evident [5]. Also a first variable γ source has been observed.

2.4 Protons and nuclei

Here are presented the first energy spectra of light components. Figure 1a shows the preliminary proton flux as a function of kinetic energy measured with DAMPE, compared with previous measurements [6]. The statistical errors are plotted as error bars while the systematic errors are showed with a band. Our results show good agreement with the AMS-02 and PAMELA data for

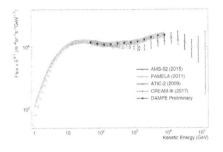 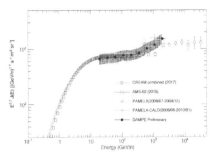

Figure 1: On the left the preliminary flux of CR protons measured by DAMPE. On the right Helium flux as a function of primary energy. Both fluxes are compared with previous results by other experiments.

$E_k \leq 200 GeV$. The spectral hardening at $E_k \geq 200 GeV$ can be observed from our measurements, confirming the previous results. In Fig. 1b, the Helium spectrum is shown [7]. Although the Helium flux measurement reported here is very preliminary, this result by DAMPE is in a good agreement with the previous experiments and with an indication of a hardening of the Helium flux.

3 Conclusions

As shown in this paper, in little more than 1 year, DAMPE reached important results. In the next future, with the increasing of the data collection and improving of the analysis procedures, DAMPE is expected to provide important measurements of the Cosmic Rays spectrum up to 100 TeV particle energy, and of the γ-ray emission from Galactic and extragalactic sources. This will contribute to a better understanding of the origin and propagation mechanism of high energy Cosmic Rays and the possibility of probing the nature of dark matter.

References

[1] J. Chang et al., "The DArk Matter Particle Explorer mission", Astrop. Physics **95** (2017) 6–24.
[2] Y. Zhang et al., PSD performance and charge reconstruction with DAMPE, POS(ICRC2017)168.
[3] C. Yue et al., Nucl. Instrum. Methods A, **856** (2017) 11.
[4] G. Ambrosi et al., Nature, **552**, (2017) 63–66
[5] S. Lei, Gamma-ray Astronomy with DAMPE, POS(ICRC2017)616.
[6] C. Yue et al., Studies of Cosmic-Ray Proton Flux with the DAMPE Experiment, POS(ICRC2017)1076.
[7] V. Gallo et al., Studies on Helium flux with DAMPE, POS(ICRC2017)169.

SCHWARZSCHILD GEODESICS AND THE STRONG FORCE

Dimitrios Grigoriou[1,a] and Constantinos G. Vayenas[1,2,b]
[1] *LCEP, 1 Caratheodory St., University of Patras, Patras GR 26504, Greece*
[2] *Division of Natural Sciences, Academy of Athens, 28 Panepistimiou Ave., GR-10679 Athens, Greece*

Abstract. It has been found recently that the gravitational attraction between ultrarelativistic neutrinos leads to the formation of rotational gravitationally confined composite particles with the properties of hadrons. Here we compare the special relativistic and general relativistic treatments of such composite particles using in the latter case the Schwarzschild geodesics. Both treatments are found to give the same result for the relativistic gravitational force acting on the rotating particles.

1 Introduction

Gravitational interactions between neutrinos were first discussed by Wheeler [1]. It was found recently that gravitational forces between ultrarelativistc neutrinos with energies above 150 MeV are very strong [2] and lead to the formation of rotating neutrino composite structures with the masses of hadrons [2,3] or, when e^+e^- also participate to the rotating neutrons ring, with the masses of bosons [4–6]. Here we compare the SR and GR treatment of a three rotating neutrino model (RNM) structure for a neutron.

The SR treatment of the RNM has been presented already elsewhere [2,3]. In brief, the special relativity equation of motion for a circular orbit [8] is used in conjunction with Newton's gravitational law to obtain

$$F = \gamma m_o \frac{v^2}{r} = \frac{Gm_g^2}{\sqrt{3}r^2}, \quad (1)$$

where m_g is the gravitational mass of each light particle, which according to the equivalence principle equals its inertial mass, m_i. The latter is given by $m_i = \gamma^3 m_o$ where $\gamma (= (1 - v^2/c^2)^{-1/2})$ is the Lorentz-factor [2,7,8].

Upon combining Eq. (1) with the de Broglie wavelength equation, $\lambdabar = \hbar/\gamma m_o v$, and assuming $r = \lambdabar$ one obtains the following results for the mass of the composite particle, m_c, for the gravitational mass $\gamma^3 m_o$ and for γ^6.

$$m_c = 3\gamma m_o = 3^{13/12} m_o^{2/3} m_{Pl}^{1/3} \; ; \; \gamma^3 m_o = 3^{1/4} m_{Pl} \; ; \; \gamma^6 = 3^{1/2} \left(\frac{m_{Pl}}{m_o}\right)^2 \quad (2)$$

where $m_{Pl}(= (\hbar c/G)^{1/2})$ is the Planck mass. Thus, for $m_o = 0.0437$ eV/c^2 [2,3,9], which is within the current experimental limits of the highest neutrino mass eigenstate [9], the first equation in (2) gives m_c=939.5 MeV/c^2 which

---
[a] E-mail: cgvayenas@upatras.gr
[b] E-mail: dgrigoriou@chemeng.upatras.gr

differs less than 0.2% from the proton and neutron masses [10]. It also follows that

$$F = \frac{Gm_o^2\gamma^6}{\sqrt{3}r^2} = \frac{Gm_{Pl}^2}{r^2} = \frac{\hbar c}{r^2} \ ; \ \frac{F}{F_N} = \gamma^6 = 3^{1/2}\left(\frac{m_{Pl}}{m_o}\right)^2 \approx 1.35 \cdot 10^{59}, \quad (3)$$

where F_N is the common ($\gamma \approx 1$) Newtonian force at the same distance.

2 GR treatment

In order to apply the Schwarzschild geodesics equation of GR [11] to the RNM, which comprises three equidistant particles around a common center, we adjust the RNM model to the standard geometry of the Schwarzschild metric, which involves a test particle of mass m^* rotating around a central mass M.

This is done via an intermediate two antidiametric particle model in which each rotating particle with mass m^* has the same rotational radius and gravitational potential as the particles of the RNM model. This implies [12] $m^* = 2^{1/2}3^{-1/4}m_o$. We then consider the one-dimensional Schwarzschild effective potential, $V_s(r)$, with $M >> m^*$ and we set the angular momemntum L equal to $\hbar$.

$$\frac{V_s(r)}{(m^*c^2/2)} = -\frac{r_s}{r} + \frac{a^2}{r^2} - \frac{a^2 r_s}{r^3}, \quad (4)$$

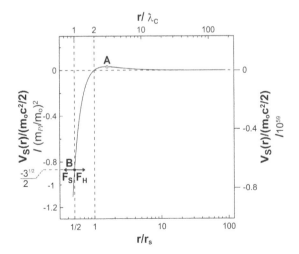

Figure 1: Plot of Eq. (4). At point B the gravitational force, F_s, is counterbalanced by an opposing force, F_H, due to the Heisenberg uncertainty principle.

where $r_s(=2GM/c^2)$ is the Schwarzschild radius of the central mass M and $a(=\hbar/m^*c)$ is the Compton wavelength of the rotating mass m^*. Thus

$$F_s(r) = dV_s(r)/dr = (m^*c^2/2)\left[\frac{r_s}{r^2} - \frac{2a^2}{r^3} + \frac{3a^2 r_s}{r^4}\right]. \quad (5)$$

The first term in the bracket of (5) corresponds to normal gravitational attraction F_N according to Newton's gravitational law, thus one obtains

$$F_s(r)/F_N(r) = 1 + \frac{a^2}{r_s^2}\left[3\left(\frac{r_s}{r}\right)^2 - 2\left(\frac{r_s}{r}\right)\right]. \quad (6)$$

Accounting for $a^2 = 3^{1/2}\hbar^2/2m_o^2c^2$ it follows that at $r = r_s/2$, which is the minimum r value for which $v \leq c$, it is

$$F_s/F_N = 1 + 3^{1/2}\frac{m_{Pl}^4}{M^2 m_o^2} \quad ; \quad F_s/F_N \approx 3^{1/2}\left(\frac{m_{Pl}}{m_o}\right)^2 \approx 1.35 \cdot 10^{59} \quad (7)$$

where the second equality holds for $M = m_{Pl}$. Remarkably, this is the same with the SR equation (3).

It is worth noting that the equality $M = m_{Pl}$ also implies that $r_s/2 = \hbar/Mc$ as shown in Fig. 1, i.e. half the black hole event horizon limit, $r_s/2$, coincides with the Compton wavelength of the confined central mass. This is consistent with the generalized uncertainty principle (GUP) [12, 13]. In summary both SR and GR give the same result regarding the ratio of the strong force to the common ($\gamma = 1$) Newtonian gravitational force and confirm that the strong force is a relativistic gravitational force. It is worth noting that GR and the γ^3 law of SR also give the same results for the Mercury perihelion precession [14].

References

[1] R.B. Dieter, J.A. Wheeler, Rev. Mod. Phys. **29(3)** (1957) 465.
[2] C.G. Vayenas, S. Souentie, *Gravity, special relativity and the strong force: A Bohr-Einstein-de-Broglie model for the formation of hadrons* (Springer, New York, 2012).
[3] C.G. Vayenas, S. Souentie & A. Fokas, Physica A **405** (2014) 360.
[4] C.G. Vayenas, A.S. Fokas, D. Grigoriou, Physica A **450** (2016) 37.
[5] A.S. Fokas, C.G. Vayenas, Physica A **464** (2016) 231.
[6] A.S. Fokas, C.G. Vayenas, D. Grigoriou, Physica A, https://doi.org/10.1016/j.physa.2017.11.003 (2017).
[7] A. Einstein, Ann. der Physik. **17** (1905) 891.
[8] A.P. French, *Special relativity* (W.W. Norton and Co., New York, 1968).
[9] R.N. Mohapatra et al, Rep Prog Phys **70** (2007) 1757.

[10] D. Griffiths, *Introduction to Elementary Particles*. (Wiley-VCH Verlag, Weinheim, 2008).
[11] R.M. Wald, *General relativity*, (The University of Chicago Press, Chicago, 1984)
[12] C.G. Vayenas, D. Grigoriou, J. Phys.: Conf. Ser. **574** (2015) 012059.
[13] S. Das & E.C. Vagenas, Can. J. Phys. **87** (2009) 233.
[14] A.S. Fokas, C.G. Vayenas, D. Grigoriou, arXiv:1509.03326v2 (2016).

CURRENT STATE OF DYNAMICAL DARK ENERGY VERSUS OBSERVATIONS

Joan Solà [a]

Departament de Física Quàntica i Astrofísica (FQA),
and Institute of Cosmos Sciences (ICCUB)
Universitat de Barcelona, Av. Diagonal 647, E-08028 Barcelona, Catalonia, Spain

Abstract. After 100 years of the cosmological term, Λ, in Einstein's field equations, and upon strenuous theoretical and observational efforts, we still ignore if Λ is a fundamental constant or a mildly dynamical variable. Recent analyses show that the $\Lambda =$ const. hypothesis, despite being the canonical assumption, could well be an unfavored one. Occam's razor is not always necessarily true. The combined fit to the modern cosmological data on the SNIa+BAO+$H(z)$+LSS+BBN+CMB observables shows that the performance of the "concordance" ΛCDM model (with $\Lambda =$ const.) is not optimal, in fact it lags behind the outcome of dynamical dark energy (DE) models. A simple XCDM parameterization as well as typical quintessence models perform better; but the running vacuum model (RVM) seems to mark the highest peak supporting dynamical DE, at a level no lesser than $\sim 3.5\sigma$ c.l. In what follows I will pinpoint the most crucial data supporting the signal and I will explain why it has escaped detection until recently.

1 Introduction

On middle February 1917, the famous seminal paper in which Einstein introduced the cosmological term Λ was published [1]. Fourteen years later, the idea of Λ as a fundamental piece of these equations was abandoned by Einstein himself [2]; and only one year later the Einstein-de Sitter model, modernly called the cold dark matter model (CDM), was proposed with no further use of the Λ term for the description of the cosmological evolution [3]. The situation with Λ did not stop here and in the works of Lemaître [4] Λ was associated to the concept of vacuum energy density through the expression $\rho_\Lambda = \Lambda/(8\pi G)$, in which G is Newton's constant. The problem (not addressed by Lemaître) is to understand the origin of the vacuum energy in fundamental physics, namely in the quantum theory and in general in quantum field theory (QFT). Here is where the cosmological constant (CC) problem first pops up, see [5,6] and references therein. The CC problem was first formulated in preliminary form by Zeldovich 50 years ago [7]. Rather than attempting to solve this problem here, we adopt a practical point of view and consider the possibility to test whether the vacuum energy (and in general de dark energy) is dynamical or not in the light of the current observations.

---

[a] E-mail: sola@fqa.ub.edu

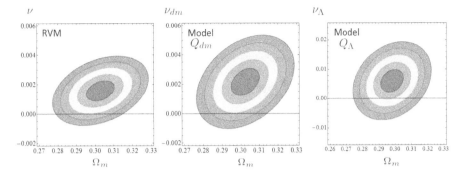

Figure 1: Likelihood contours for the DVMs in the (Ω_m, ν_i) plane up to 5σ c.l. after marginalizing over the rest of the fitting parameters. The ΛCDM ($\nu_i = 0$) appears clearly disfavored in front of the DVMs. In particular, the RVM and the Q_{dm} are favored by more than $\sim 3\sigma$. Subsequent marginalization over Ω_m increases the confidence level of the RVM up to 3.8σ [8].

2 Dynamical vacuum models

The Friedmann and acceleration equations when the DE evolves with the cosmic expansion, $\rho_D = \rho_D(t)$, read formally similar to the standard case:

$$3H^2 = 8\pi G(\rho_m + \rho_D(t)), \qquad 2\dot{H} + 3H^2 = -8\pi G(w_m \rho_m + w_D \rho_D(t)), \quad (1)$$

where w_m and w_D are the equation of state parameters (EoS) of the matter fluid and of the DE, respectively. It is well known that $w_m = 1/3, 0$ for relativistic and nonrelativistic matter, respectively We explore DE models with dynamical vacuum evolution, hence $\rho_D(t)$ is actually $\rho_\Lambda(t)$ with $w_D = -1$. The total matter density ρ_m can be split into the contribution from baryons and cold dark matter (DM), namely $\rho_m = \rho_b + \rho_{dm}$. We assume that the dark matter (DM) density is the only one that interacts with vacuum, whereas radiation and baryons are self-conserved, so that their energy densities evolve in the standard way: $\rho_r(a) = \rho_r^0 a^{-4}$ and $\rho_b(a) = \rho_b^0 a^{-3}$. Being the DM component the one that exchanges energy with the vacuum, the local conservation law reads as follows:

$$\dot{\rho}_{dm} + 3H\rho_{dm} = Q, \qquad \dot{\rho}_\Lambda = -Q. \quad (2)$$

The source function Q depends on the assumptions made on the dynamical vacuum model. Let us compare the running vacuum model (RVM) [6,8] with two alternative dynamical vacuum models (DVMs) with different forms of the interaction sources. Let us list the three DVMs under comparison, which we may denote RVM, Q_{dm} and Q_Λ according to the structure of the interaction source, or also for convenience just I, II and III:

$$\text{Model I (RVM)}: \quad Q = \nu H(3\rho_m + 4\rho_r) \quad (3)$$

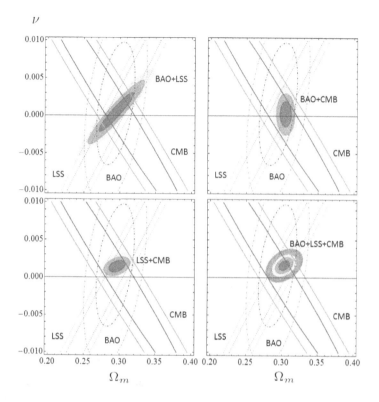

Figure 2: Contour lines for the RVM by considering the effect of only the BAO, LSS and CMB in all the possible combinations. It is only when such a triad of observables is combined that we can see a clear sign that a dynamical vacuum effect at a confidence level of $\sim 4\sigma$.

$$\text{Model II } (Q_{dm}): \quad Q_{dm} = 3\nu_{dm} H \rho_{dm} \qquad (4)$$
$$\text{Model III } (Q_\Lambda): \quad Q_\Lambda = 3\nu_\Lambda H \rho_\Lambda. \qquad (5)$$

Each model has a characteristic (dimensionless) parameter $\nu_i = \nu, \nu_{dm}, \nu_\Lambda$ as a part of the interaction source, which must be fitted to the observational data. Remarkably, in the RVM case, the source function Q in (3) is not just put by hand since it is a calculable expression from the given form of its energy density

$$\rho_\Lambda(H) = \frac{3}{8\pi G} \left(c_0 + \nu H^2 \right). \qquad (6)$$

The fact that $\rho_\Lambda(H)$ is a function of the Hubble rate, not just of the cosmic time t, is specific for the RVM. Such form can be motivated in the context of QFT in curved spacetime, see [6,9] and references therein. Let us provide the energy densities of the DM and vacuum energy for the RVM as a function of the scale factor:

$$\rho_{dm} = \rho_{dm}^0 a^{-3(1-\nu)} + \rho_b^0 \left(a^{-3(1-\nu)} - a^{-3}\right) - \frac{4\nu\rho_r^0}{1+3\nu}\left(a^{-4} - a^{-3(1-\nu)}\right) \quad (7)$$

and

$$\begin{aligned}\rho_\Lambda &= \rho_{\Lambda 0} + \frac{\nu \rho_m^0}{1-\nu}\left(a^{-3(1-\nu)} - 1\right) \\ &+ \frac{\nu}{1-\nu}\rho_r^0\left(\frac{1-\nu}{1+3\nu}a^{-4} + \frac{4\nu}{1+3\nu}a^{-3(1-\nu)} - 1\right).\end{aligned} \quad (8)$$

For the corresponding expressions for Models II and III, see [8] and references therein. As can be easily checked, for $\nu \to 0$ we recover the corresponding results for the ΛCDM, as it should. The Hubble function can be immediately obtained from these formulas after inserting them in Friedmann's equation, together with the conservation laws for baryons and radiation, $\rho_r(a) = \rho_r^0 a^{-4}$ and $\rho_b(a) = \rho_b^0 a^{-3}$.

A crucial issue is to understand the role played by the LSS data: $f(z)\sigma_8(z)$. There seems to be no way at present to describe correctly both the CMB and the LSS data within the ΛCDM. This is of course at the root of the so-called σ_8-tension, one of the important phenomenological problems of the ΛCDM [10]. In the case of the DVMs it is necessary to use more general perturbations equations, which we shall not address here [8,11]. They are needed to correctly fit the structure formation data in the presence of vacuum dynamics.

The recent analysis of [12] provides a possible efficient solution to the σ_8-tension within the RVM. The analysis is based on computing the effect of the dynamical vacuum (6) on the matter transfer function. With such approach it is possible to apply an analytic treatment and therefore a better understanding of the origin of the results. At the same time it provides further support for dynamical vacuum energy and singles out the RVM once more as a most gifted option within the class of the dynamical vacuum models I, II and III.

The numerical fit results for the DVMs can be found in [8]. The contour lines are given in Fig. 1. See also [13] for alternative fits using data sets in which the weak-lensing effects from the CMB are taken into account, the differences turn out to be small. From Fig. 2, we note that the LSS+CMB pair is crucial. Failing to combine LSS and CMB usually misses the signal, which can be made crisper with the addition of BAO, as seen in that figure. In most previous analyses in the literature this feature has not been taken into account or an insufficient set of LSS data has been used [14]. In fact, one cannot expect to fit the overall data better than the ΛCDM if the σ_8-tension is not previously relaxed. This is indeed the touchstone of dynamical DE.

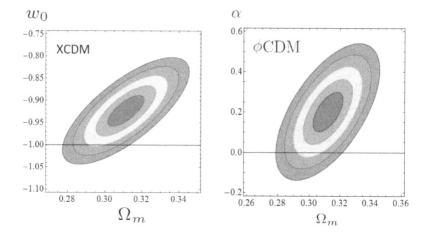

Figure 3: As in Fig. 1, but for models XCDM [15] and ϕCDM with Peebles & Ratra potential [16]. Although they are rather different (for the first $w = w_0$ =const. and for the second w is not constant and computable from the potential), dynamical DE is favored in both cases up to 3.3σ.

3 XCDM and ϕCDM

The simplest parametrization for dynamical DE is the XCDM [15]. In it matter and DE are self-conserved (i.e., they are not interacting), thus the DE density as a function of the scale factor is simply given by $\rho_X(a) = \rho_{X0}\, a^{-3(1+w_0)}$, with $\rho_{X0} = \rho_{\Lambda 0}$ (the current CC density value), where w_0 is the (constant) EoS parameter of the generic DE entity X in this parametrization. A step further is represented by the ϕCDM models (quintessence and the like). As a representative potential we take the original Peebles & Ratra (PR) form [16]:

$$V(\phi) = \frac{1}{2}\kappa M_P^2 \phi^{-\alpha},\qquad(9)$$

in which κ and α are dimensionless parameters and $M_P = 1/\sqrt{G}$ is the Planck mass. The fit results for these models can be found in [8,17]. In Fig. 3, we can see the corresponding contour lines. They both favor dynamical DE since $w_0 > -1$ and $\alpha > 0$ at 3.3σ c.l. after marginalizing over Ω_m.

4 Conclusions

We have reviewed the status of the dynamical DE models, in particular the dynamical vacuum models (DVMs) in their ability to compete with the ΛCDM model (namely the standard or concordance model of cosmology) to fit the overall SNIa+BAO+$H(z)$+LSS+BBN+CMB cosmological observations. We find that the the current cosmological data disfavors the ΛCDM, and hence the

Λ =const. hypothesis, in a very significant way. The model that best fits the data is the running vacuum model: RVM. These results confirm the first hints of dynamical DE first reported by Solà et al. [18,19] and Zhao et al. [20].

Acknowledgments

I would like to thank the chairman of the conference, A. Studenikin for the invitation, and all the organizers for the smooth organization of the 18th Lomonosov conference in Moscow. I am thankful to Javier de Cruz Pérez and Adrià Gómez-Valent for discussions and scientific collaboration. Support by MINECO, Consolider, SGR (Generalitat de Catalunya) and MDM (ICCUB) is also acknowledged.

References

[1] A. Einstein, Sitzungsber. Königl. Preuss. Akad. Wiss. phys.-math. Klasse VI (1917) 142 (submitted on February 8th 1917).
[2] A. Einstein, Sitzungsber. Königl. Preuss. Akad. Wiss., phys.-math. Klasse, XII, (1931) 235.
[3] A. Einstein and W. de Sitter, Proc. Nat. Acad. Sci. **18** (1932) 213.
[4] G. Lemaître, Proc. Nat. Acad. Sci. **20** (1934) 12.
[5] S. Weinberg, Rev. of Mod. Phys. **61** (1989) 1.
[6] J. Solà, *Cosmological constant and vacuum energy: old and new ideas*, J. Phys. Conf. Ser. **453** (2013) 012015. [e-Print: arXiv:1306.1527].
[7] Y. B. Zeldovich, JETP Lett. **6** (1967) 316, Pisma Zh. Eksp. Teor. Fiz. 6 (1967) 883; Sov. Phys. Usp. 11 (1968) 381, republished in Gen. Rel. Grav. **40** (2008) 1557 (edited by V. Sahni and A. Krasinski).
[8] J. Solà, J. de Cruz Pérez, A. Gómez-Valent, *Towards the firsts compelling signs of vacuum dynamics in modern cosmological observations*, arXiv:1703.08218.
[9] J. Solà, A. Gómez-Valent, Int. J. Mod. Phys. D**24** (2015) 1541003.
[10] Macaulay E., Wehus I.K. and Eriksen H.C., Phys. Rev. Lett., **111** (2013) 161301.
[11] A. Gómez-Valent, J. Solà, S. Basilakos, JCAP **01** (2015) 004.
[12] A. Gómez-Valent and J. Solà, *Relaxing the σ_8-tension through running vacuum in the Universe*, arXiv:1711.00692 (accepted in Europhys. Lett. 2018).
[13] J. Solà, A. Gómez-Valent, and J. de Cruz Pérez, Phys. Lett. B**774**, 317 (2017).
[14] P. A. R. Ade *et al*: Planck 2015 results A &A **594** (2016) A13; A &A **594**, A14 (2016); T.M.C. Abbott *et al*. (DES Collab.), arXiv:1708.01530.
[15] S.M. Turner and M. White, Phys. Rev. D**56**, R4439 (1997).

[16] P.J.E. Peebles and B. Ratra, ApJ Lett. **325**, L17 (1988).
[17] J. Solà, A. Gómez-Valent A, J. de Cruz Pérez, Mod. Phys. Lett. **A32** 1750054.
[18] J. Solà, A. Gómez-Valent, and J. de Cruz Pérez, ApJ Lett. **811**, L14 (2015).
[19] J. Solà, A. Gómez-Valent, J. de Cruz Pérez, ApJ **836** (2017) 43.
[20] G.B Zhao et al., Nat. Astron. **1**, 627 (2017).

MAGNETIC FIELD GENERATION IN DENSE GAS OF MASSIVE ELECTRONS WITH ANOMALOUS MAGNETIC MOMENTS ELECTROWEAKLY INTERACTING WITH BACKGROUND MATTER

Maxim Dvornikov [a]

Pushkov Institute of Terrestrial Magnetism, Ionosphere and Radiowave Propagation (IZMIRAN), 108840 Troitsk, Moscow, Russia
Physics Faculty, National Research Tomsk State University, 36 Lenin Avenue, 634050 Tomsk, Russia

Abstract. We describe the generation of the electric current flowing along the external magnetic field in the system of massive charged fermions, possessing anomalous magnetic moments and electroweakly interacting with background matter. This current is shown to result in the instability of the magnetic field leading to its growth. Some astrophysical applications are discussed.

Strong magnetic fields with $B > 10^{15}$ G can be found in some compact stars called magnetars [1]. Recently the elementary particle physics mechanisms, involving the chiral magnetic effect (CME) [2], were applied to model magnetic fields in magnetars [3]. CME is based on the existence of the current $\mathbf{J} = \alpha_{em} (\mu_R - \mu_L) \mathbf{B}/\pi$ in the external magnetic field $\mathbf{B}$. Here $\alpha_{em} = e^2/4\pi$ is the fine structure constant, $e > 0$ is the absolute value of the elementary charge, and $\mu_{R,L}$ are the chemical potentials of right and left chiral fermions,. However CME implies the unbroken chiral symmetry [4], i.e. charged particles should be considered massless. The electroweak chiral phase transition is unlikely in astrophysical media. There is a possibility to restore chiral symmetry in quark matter owing to QCD effects. Thus one can expect the magnetic field amplification in the core of some compact stars where quark matter can be present [5].

In this paper, we summarize our recent findings in [6] on the magnetic field generation in the system of massive electrons with nonzero anomalous magnetic moments. We shall suppose that, besides the external magnetic field, these electrons electroweakly interact with background matter. It turns out that there is a nonzero electric current of these electrons along the magnetic field. The computation of the current is based on the exact solution of the Dirac equation in the considered external fields which was recently found in [7]. If the current $\mathbf{J} \sim \mathbf{B}$ is taken into account in the Maxwell equations, it will result in the magnetic field instability causing the enhancement of a seed field. Finally, we briefly discuss some astrophysical applications.

Let us consider a massive fermion, i.e. an electron, having the anomalous magnetic moment μ. We suppose that this electron electroweakly interacts with background matter under the influence of the external magnetic field. We

---
[a] E-mail: maxdvo@izmiran.ru

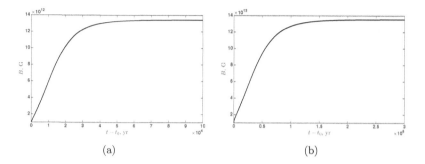

Figure 1: Magnetic field evolution obtained in the numerical solution of Eq. (2) for different length scales. (a) $L = 10^2$ cm, and (b) $L = 10^3$ cm.

found in [6] that, in this situation, there is a nonzero electric current along the magnetic field,

$$\mathbf{J} = \Pi \mathbf{B}, \quad \Pi = -8\mu m V_5 B \frac{\alpha_{\text{em}}}{\pi \tilde{\chi}^3} \sum_{n=1}^{N} \sqrt{\tilde{\chi}^2 - m_{\text{eff}}^2}, \quad (1)$$

where m is the electron mass, $V_5 = (V_L - V_R)/2$, $V_{R,L}$ are the effective potentials of the electroweak interaction for right and left chiral projections, which are given in [8], $m_{\text{eff}} = \sqrt{m^2 + 2eBn}$, $n = 0, 1, \ldots$ is the discrete quantum number which energy levels depend on, $\tilde{\chi} = \chi - \bar{V}$, χ is the chemical potential, $\bar{V} = (V_L + V_R)/2$, and N is maximal integer, for which $\tilde{\chi}^2 - m^2 - 2eBN \geq 0$. In Eq. (1), we consider highly degenerate electron gas, in which $\chi \gg T$, where T is the temperature.

If we take into account the current in Eq. (1) in the Maxwell equations and consider the MHD approximation, the amplitude of the magnetic field obeys the equation,

$$\dot{B} = -\frac{k}{\sigma_{\text{cond}}}(k + \Pi) B, \quad (2)$$

where σ_{cond} is the electric conductivity and the parameter k determines the length scale $L = 1/k$ of the magnetic field. Since Π in Eq. (1) is negative, the magnetic field, described by Eq. (2), can be unstable.

Basing on Eq. (2), we can describe the magnetic field amplification in a dense degenerate matter which can be found in a neutron star. Taking into account the typical density of matter in a neutron star, the time dependence of the electric conductivity [9] and the dependence of μ on the magnetic field [10], in Fig. 1, we plot the time dependence of the magnetic field for different length scales starting from the seed magnetic field $B_0 = 10^{12}$ G, which is typical for a young pulsar.

One can see that the magnetic reaches the saturated magnetic field $B_\text{sat} \approx 1.3 \times 10^{13}$ G. The strength of B_sat and its length scale are close to the values of the corresponding parameters in a magnetic field fluctuation [11], which then can trigger the burst of a magnetar [1]. We have suggested in [6] that the energy source, powering the magnetic field growth presented in Fig. 1, can be the kinetic energy of the stellar rotation.

Finally it is interesting to compare the appearance of the new current along the magnetic field in Eq. (1) with the CME [2]. As shown in [2], only massless fermions at the zero Landau level in an external magnetic field contribute to the generation of the anomalous current along the magnetic field since there is an asymmetry in the motion of these particles with respect to the external magnetic field [8, 9]. In the situation described in the present work, i.e. when massive electrons with nonzero anomalous magnetic moment propagate in the electroweak matter, the particles at the lowest energy level can move in any direction with respect to the magnetic field without an asymmetry. On the contrary, higher energy levels with n > 0 are not symmetric for electrons moving along and opposite **B** [6]. Moreover, the term in the energy spectrum responsible for such an asymmetry is proportional to $\mu B m V_5$ [7]. It is this factor, which **J** in Eq. (1) depends on.

Acknowledgments

I am thankful to the Tomsk State University Competitiveness Improvement Program and to RFBR (research project No. 15-02-00293) for a partial support.

References

[1] R. Turolla, S. Zane, A.L. Watts, Rep. Prog. Phys. **78** (2015) 116901.
[2] A. Vilenkin, Phys. Rev. **D 22** (1980) 3080.
[3] G. Sigl, *"Astroparticle Physics: Theory and Phenomenology"* (Atlantis Press, Paris, 2017).
[4] M. Dvornikov, Phys. Lett. **B 760** (2016) 406, arXiv:1608.04940.
[5] M. Dvornikov, Nucl. Phys. **B 913** (2016) 79, arXiv:1608.04946.
[6] M. Dvornikov, Pis'ma v ZhETF **106** (2017) 741.
[7] I.A. Balantsev, A.I. Studenikin, I.V. Tokarev, Phys. Part. Nucl. **43** (2012) 727.
[8] M. Dvornikov, V.B. Semikoz, *Phys. Rev.* **D 91** (2015) 061301,
[9] M. Dvornikov, V.B. Semikoz, JCAP 05 (2015) 032,
[10] I.M. Ternov, et al., Sov. Phys. JETP **28** (1969) 1206.
[11] M. Dvornikov, Int. J. Mod. Phys. **D** (2017).

VACUUM POLARIZATION IN COSMIC-STRING AND GLOBAL-MONOPOLE BACKGROUND

Yuri V. Grats [a] and Pavel Spirin [b]
Faculty of Physics, Moscow State University, 119991 Moscow, Russia

Abstract. We analyze the gravity-induced effects associated with a massless scalar field in a higher-dimensional spacetime being the tensor product of $(d-n)$-dimensional Minkowski space and n-dimensional spherically/cylindrically-symmetric space with a solid/planar angle deficit. In particular, we revisit the computation of the vacuum polarization effects for a non-minimally coupled massless scalar field in the spacetime of a straight cosmic string. We compute the renormalized vacuum expectation value of the field square $\langle \phi^2(x) \rangle_{\text{ren}}$ and the renormalized vacuum averaged of the scalar-field's energy-momentum tensor $\langle T_{MN}(x) \rangle_{\text{ren}}$. The explicit dependence of the results upon the dimensionalities of both the bulk and conical submanifold and upon the coupling constant to the curvature ξ is discussed.

1 Introduction

The metric under consideration reads:

$$ds^2 = g_{MN}\, dx^M dx^N = -dt^2 + dx_{d-1}^2 + \ldots + dx_{n+1}^2 + \mathrm{e}^{-2(1-\beta)\ln r}\, \delta_{ik} dx^i dx^k\,,$$

with $r^2 = \delta_{ik} x^i x^k$, $i, k, \ldots = 1, \ldots, n$ while $M, N, \ldots = 0, 1, \ldots, d-1$. Here $d \geqslant 3$ and $2 \leqslant n \leqslant d-1$.

This spacetime represents the tensor product of the $(d-n)$-dimensional Minkowski space and the n-dimensional centro-symmetric conformally flat space with a solid angle deficit equal to $\delta\Omega = 2\left(1 - \beta^{n-1}\right)\pi^{n/2}/\Gamma\left(n/2\right)$, if $n \geqslant 3$, or planar angular deficit equal to $\delta\varphi = 2\pi(1-\beta)$, if $n = 2$. These spacetimes are considered as simple models for a multidimensional global monopole (if $n \geqslant 3$) or cosmic string (if $n = 2$) with $(d - n - 1)$ flat extra dimensions.

The corresponding Ricci tensor and the scalar curvature are determined by the conical sector only:

$$R_{ik} = 2\pi(1-\beta)\,\delta^2(\mathbf{r})\,\delta_{ik}\,, \qquad R = 4\pi(1-\beta)\,r^{2(1-\beta)}\,\delta^2(\mathbf{r})\,, \quad n=2;$$

$$R_{ik} = (1-\beta^2)(n-2)\frac{r^2\,\delta_{ik} - x_i x_k}{r^4}\,, \qquad R = (1-\beta^2)\frac{(n-1)(n-2)}{r^{2\beta}}\,, \quad n \geqslant 3.$$

For $\beta \approx 1$ the angle deficit becomes proportional to $\beta' = (1-\beta) \ll 1$, hence we are capable to express all the quantities we need as a series of β'.

---
[a] E-mail: grats@phys.msu.ru
[b] E-mail: salotop@list.ru

2 Green's function: perturbation theory

Feynman propagator for the scalar field in curved background satisfies the equation:

$$\mathfrak{L}\mathcal{G} = -1, \qquad \mathcal{G} = -\mathfrak{L}^{-1}.$$

If the field operator $\mathfrak{L}$ allows to be expressed as $\mathfrak{L} = \mathfrak{L}_0 + \delta\mathfrak{L}$, where $\delta\mathfrak{L}$ is considered as a small perturbation, then representing the solution of eq. above in the form $\mathcal{G} = \mathcal{G}_0 + \delta\mathcal{G}$, with $\mathcal{G}_0 = -\mathfrak{L}_0^{-1}$ being the unperturbed Green's function, one obtains

$$\mathcal{G} = \left[-\mathfrak{L}_0 \left(1 - \mathcal{G}_0 \delta\mathfrak{L}\right) \right]^{-1} = \mathcal{G}_0 + \mathcal{G}_0\,\delta\mathfrak{L}\,\mathcal{G}_0 + \mathcal{G}_0\,\delta\mathfrak{L}\,\mathcal{G}_0\,\delta\mathfrak{L}\,\mathcal{G}_0 + \dots .$$

In the case under consideration $\mathfrak{L}_0$ is determined by the zeroth order in the small quantity β', hence in the coordinate representation

$$\mathcal{L}_0(x,\partial) = \partial^2, \qquad \partial^2 \equiv \eta^{MN}\partial_M\partial_N$$

and the perturbation operator

$$\delta\mathcal{L}(x,\partial) = \partial_M\left(\sqrt{-g}\,g^{MN}\partial_N\right) - \partial^2 - \sqrt{-g}\,\xi R,$$

thus the Green's function is given by momentum representation:

$$G^F(x,x'\,|\,d,n) = G_0^F(x-x') + \int \frac{d^d q}{(2\pi)^d} e^{iqx} \int \frac{d^d p}{(2\pi)^d} \frac{e^{ip(x-x')}}{[p^2 - i\varepsilon][(p+q)^2 - i\varepsilon]} \times$$
$$\times \left[\left(np^2 - 2\mathbf{p}^2 + (n-2)(\mathbf{qp})\right)\mathcal{F}[\alpha](q) - \xi\mathcal{F}[\gamma](q)\right],$$

where $\alpha(r) = \beta' \ln r$ and

$$\gamma(r) = \begin{cases} 4\pi\beta'\delta^2(\mathbf{r}), & n = 2; \\ 2(n-1)(n-2)\beta'/r^2, & n \geqslant 3, \end{cases}$$

Taking into account that formulae for the background curvature differ for cases $n = 2$ and $n \geqslant 3$, we'll start with the the case of a global monopole.

3 Global monopole

From the eqs. above and with the use of the dimensional-regularization method for the first-order correction to the $\langle\varphi^2(x)\rangle_{\text{ren}} = -i\,G^F_{\text{ren}}(x,x)$ we get (for the details see [1,2]):

$$\langle\phi^2(x)\rangle_{\text{reg}} = \beta'\,\frac{n-1}{4\pi^{D/2}}\,\frac{\Gamma(n/2)\,\Gamma^3(D/2)}{\Gamma(D)}\left(\frac{\xi}{\xi_D} - 1\right)\frac{\Gamma\!\left(-\frac{D-2}{2}\right)}{\Gamma\!\left(-\frac{D-n-2}{2}\right)}\,\frac{\mu^{2\varepsilon}}{r^{D-2}},$$

where $D = d - 2\epsilon$ ($\epsilon \to 0+$) and $\xi_D = (D-2)/4(D-1)$. Hereafter an arbitrary parameter μ is introduced to preserve the dimensionality of the regularized expression.

The list of possible cases includes:
- **even d, odd n:**

$$\langle \phi^2(x) \rangle_{\text{ren}} = \frac{\beta'(d-2)\Gamma(d/2)}{(-4\pi)^{d/2}(d-1)} \left[\left(\frac{\xi}{\xi_d} - 1\right) \ln\mu r + \frac{1}{(d-1)(d-2)} \right] \frac{-2}{r^{d-2}};$$

- **odd d and n:** $\langle \phi^2(x) \rangle_{\text{ren}} = 0$;
- **any d, even n:**

$$\langle \phi^2(x) \rangle_{\text{ren}} = (-1)^{n/2} \frac{\beta'}{4\pi^{d/2}} \frac{(n-1)\Gamma(n/2)\Gamma^2(d/2)}{\Gamma(d)} \Gamma\left(\frac{d-n}{2}\right) \left(\frac{\xi}{\xi_d} - 1\right) \frac{1}{r^{d-2}}.$$

The vacuum-averaged energy-momentum tensor is given by

$$\langle T_{MN}(x) \rangle_{\text{reg}} = -i \lim_{x' \to x} D_{MN} G^F_{\text{reg}}(x, x'),$$

where

$$D_{MN} = (1-2\xi)\nabla_M \nabla_{N'} + \frac{1}{2}(4\xi - 1)\nabla_L \nabla^{L'} g_{MN}$$

$$+\xi \left[R_{MN} - \frac{1}{2} R g_{MN} + 2\nabla_L \nabla^L g_{MN} - 2\nabla_M \nabla_N \right].$$

The regularized value of energy-momentum VEV:

$$\langle T_{MN} \rangle_{\text{reg}} = \frac{C\mu^{2\varepsilon}\beta'}{r^D} \left[\left(\frac{8(D-1)(n-1)}{D-n}(\xi - \xi_D)^2 + \frac{1}{D^2-1} \right) \right.$$
$$\left. \times \left(D\frac{\tilde{x}_M \tilde{x}_N}{r^2} - \tilde{\eta}_{MN} - (D-n)\eta_{MN} \right) + \frac{\eta_{MN} - \tilde{\eta}_{MN}}{D+1} \right]$$

with $C = \dfrac{\Gamma(n/2)\,\Gamma^3(D/2)}{4\pi^{D/2}\,\Gamma(D)} \dfrac{\Gamma\left(-\frac{D-2}{2}\right)}{\Gamma\left(-\frac{D-n}{2}\right)}$.

Hereafter the "tilded" quantity with indices means that it equals the corresponding tensor with no tilde for conical-subspace index, and vanishes in the opposite case.

- **d even, n odd:**

$$\langle T_{MN} \rangle_{\text{ren}} = \frac{(-1)^{d/2-1}\Gamma(n/2)\Gamma^2(d/2)}{4\pi^{d/2}\,\Gamma(d)\,\Gamma\left(-\frac{d-n}{2}\right)} \frac{\beta'}{r^d} \left[2\Theta_{MN} \ln\mu r + A_{MN} \right],$$

where

$$\Theta_{MN} \equiv \left(\frac{8(d-1)(n-1)}{d-n}(\xi - \xi_d)^2 + \frac{1}{d^2-1} \right)$$
$$\times \left(d\frac{\tilde{x}_M \tilde{x}_N}{r^2} - \tilde{\eta}_{MN} - (d-n)\eta_{MN} \right) + \frac{\eta_{MN} - \tilde{\eta}_{MN}}{d+1}.$$

- odd d and n: $\langle T_{MN}(x)\rangle_{\text{ren}} = 0$;
- any d, even n:

$$\langle T_{MN}(x)\rangle_{\text{ren}} = (-1)^{n/2-1}\frac{\Gamma(n/2)\,\Gamma^2(d/2)\,\Gamma\left(\frac{d-n+2}{2}\right)}{4\,\pi^{d/2}\,\Gamma(d)}\,\frac{\beta'}{r^d}\,\Theta_{MN}\,.$$

Then, after some straightforward algebra one checks that the constructed renormalized energy-momentum tensor conserves and for conformal coupling its trace anomaly equals its correct value.

4 Straight cosmic strings

Cosmic-string results may be obtained as regular limits $n \to 2$ of the corresponding generic expressions. The cosmic-string results coincide with the known in literature only for minimal and conformal coupling. We refer this discrepancy to the missing of the ξ-correction *inside* the Green's function [1].

5 Conclusions

- Gravity-induced vacuum polarization effects of a massless scalar field are computed;
- Perturbation-theory over angle deficit was used to construct the Green's function; dimensional-regularization method was of usage to reveal and remove divergencies;
- Our expression for the Green function enables one to consider another purely classical problem of gravity-induced self-action of a point static scalar or electric charge [1, 2].

Acknowledgments

Yuri V. Grats thanks prof. A. V. Borisov for fruitful discussions. Work of Pavel Spirin is partially supported by the Russian Foundation of Fundamental Research under the project 17-02-01299a.

References

[1] Y.V. Grats and P. Spirin, Eur. Phys. J. C **77** (2017) no.2, 101.
[2] Yu.V. Grats and P. Spirin, J. Exp. Theor. Phys. **123** (2016) no.5, 807-813 [Zh. Eksp. Teor. Fiz. **150** (2016) no.5, 929-936].

USE OF PSEUDO-HERMITIAN MODELS IN PROMISING RESEARCHES THE POSSIBLE STRUCTURE OF DARK MATTER

Vasily N. Rodionov [a], Galina A. Kravtsova [b], and Arkady M. Mandel [c]
Plekhanov Russian University of Economics, Moscow, Russia
Faculty of Physics, Moscow State University, 119991 Moscow, Russia
MSTU STANKIN, Moscow, Russia

Abstract. The principal positions of non-Hermitian models with γ_5-mass extensions are formulated. A simple estimation for determination maximal permissible fermion mass value M is produced. It is shown that a zone of PT-symmetry includes as ordinary and exotic particles. Paradox distinguish of inertial and gravitation particle's properties for "exotic" particles is shown. It is found that unusual properties of the exotic particles allow to consider their as possible candidates in structure of dark matter.

1 Introduction. Description of the model

We consider non-Hermitian fermion Hamiltonian

$$H = \hat{\vec{\alpha}}\vec{p} + \hat{\beta}(m_1 + \gamma_5 m_2), \qquad (1)$$

which follows from Geometrical Theory with Maximal Mass (see [1] and also [2]) and appears to be PT-symmetrical [3]. Here the physical mass m of a particle is connected with the parameters m_1 and m_2 like that: $m^2 = m_1^2 - m_2^2$. This Hamiltonian corresponds to modified Dirac equations, which are produced due to the ordinary procedure of Dirac "is taking the square root from the Klein-Gordon (KG) equation" [4]. But here we represent the operator KG as a product of two commuting **non-Hermitian** operators of the first order D_1 and D_2 (see [5], and also [6]– [10]): $D_1 = i\partial_\mu \gamma^\mu - m_1 - \gamma_5 m_2$, $D_2 = -i\partial_\mu \gamma^\mu - m_1 + \gamma_5 m_2$. It appears to be the restriction of elementary particles mass values in this model:

$$m = \sqrt{m_1^2 - m_2^2} \leq \frac{m_1^2}{2m_2} = M. \qquad (2)$$

Here M has to be equal to the Maximal Mass in Kadyshevsky' theory [5]. This non-equality is the condition for PT-symmetry (compare to [3]). Now we can express the parameters m_1 and m_2 through the physical parameters: the mass of a particle m and the maximal mass in fermion spectrum M:

$$m_1 = \sqrt{2}M\sqrt{1 \mp \sqrt{1 - \frac{m^2}{M^2}}}; \quad m_2 = M\left(1 \mp \sqrt{1 - \frac{m^2}{M^2}}\right) \qquad (3)$$

---
[a] E-mail: rodyvn@mail.ru
[b] E-mail: gakr@chtc.ru
[c] E-mail: arkadimandel@mail.ru

Two signs in these expressions correspond to two kinds of the fermions: ordinary particles (high sing) and the **exotic** ones (low sing). We will see that the unusual properties of the exotic particles allow to explain many problems in modern physics, in particular, dark matter in the Universe.

2 Non-relativistic limit of modified Dirac equation in the electromagnetic field

In most cases it is no necessity to use the exact solutions of modified Dirac equation with non-Hermitian γ_5-mass extensions. Really one can confine by non-relativistic amendments using expansions v/c and v^2/c^2 [11].

$$\Psi(r,t) = \Psi(r)e^{-i(E+m)t} \qquad (4)$$

$$(E+m)\Psi(r) = \left[\hat{\vec{\alpha}}(\vec{p}-e\vec{A}) + \hat{\beta}(m_1+\gamma_5 m_2) - V\right]\Psi(r), \qquad (5)$$

where $\hat{\vec{\sigma}}$ are matrixes of Pauli, V-potential energy.

Consider representation of $\Psi(r)$ as the four - components function in a form of two component functions and taking into account the standard representation of the matrixes we have

$$\Psi = \begin{pmatrix} \varphi \\ \chi \end{pmatrix} \qquad (6)$$

and expressing

$$\chi = \frac{\hat{\vec{\sigma}}(\vec{p}-e\vec{A}) + m_2}{m+m_1}\varphi \qquad (7)$$

we obtain the Hamiltonian. In a result we have

$$E\varphi(r) = i\frac{\partial}{\partial t}\left[\varphi(r)\exp(-iEt)\right], \qquad (8)$$

$$\hat{H}\varphi(r) = \left[\frac{(\vec{p}-e\vec{A})^2}{2m^*} - \frac{e\hat{\vec{\sigma}}\vec{H}}{2m^*} - V\right]\varphi(r). \qquad (9)$$

Thus, we see that in the obtained equations the parameter m^*

$$m^* = \frac{m+m_1}{2} \geq m \qquad (10)$$

plays the role of mass in the non-relativistic approximation. Note that masses m and m^* are equal only in the usual Dirac limit: $m_2 \to 0$, $m_1 \to m$.

Here we can see that pseudo-Hermitian extension of the usual Dirac equation can lead to not only small differences between usual and extended mass parameters, but sometimes very large ones (if we considered the exotic particles: $M \leq m_1(m_2) \leq 2M$). Really, we can speak about unique properties of particles which were named the "exotic particles". Non-relativistic formulas demonstrate that in the case of exotic particles we have extremely strong decreasing of their electromagnetic interactions ($\sim 1/m_1$) however gravitational interactions stay on the level of ordinary particles $m^2 = m_1^2 - m_2^2$. Perhaps, these extraordinary properties maybe very important under determination **the possible structure of dark matter**.

3 The chiral representation for modified Dirac equation and paradox of two masses

Using the projective operators $P_{(rl)} = \frac{1 \pm \gamma_5}{2}$, we can obtain the system equations of motions which will be differ from the Weyls equations only by the presence of non zero value for the left and right masses

$$i\frac{\partial \xi}{\partial t} - i\vec{\sigma}\vec{\nabla}\xi = 2m_r \eta, \qquad (11)$$

$$i\frac{\partial \eta}{\partial t} - i\vec{\sigma}\vec{\nabla}\eta = 2m_l \xi, \qquad (12)$$

where $m_r = \frac{m_1 + m_2}{2}$, $m_l = \frac{m_1 - m_2}{2}$, and $\xi = \varphi + \chi$, $\eta = \varphi - \chi$. However, the modified Weyl's equations really take the form of a conventional two-component Weyls equations.

Here we can point out the typical mistakes arise when one works with non-Hermitian models. They may be due to the neglect of relationships between the initial parameters m_1 and m_2 (see, for example, [13]). In particular, one argue that when $m_1 = m_2$ the system describes a massless right-wing or left-wing neutrinos (they mean the limit of the original equations). Indeed in this approach we are faced with quite difficult situation. For example let $m_1 = m_2$. Then not only "left mass" but also **relativistic mass** also $m = m_1^2 - m_2^2$ become zero! However, the **non-relativistic mass** $m^* = (m + m_1)/2$ (and hence the rest energy), generally speaking, is not equal to zero, because m_1 can has a nonzero value. Obviously, to search physical meaning in such a situation quite problematically!

But our approach can provide a way out of this difficult situation. We presuppose to put the question about the observed parameters as the key factor. If we suggest that observable variables are mass of particle m and maximal mass values M, we must express parameters m_1 and m_2 through m and M. Thus the solution of paradox of two mass is reduced to the correct choice of initial values of observed variables.

4 Estimation of difference for relativistic and inert masses

It is important, that difference between gravitation and inert mass allow to estimate the value of fermion Maximal Mass M.

According to PDG 2017 [14] all present experimental tests are compatible with the predictions of the current standard theory of gravitation: Einsteins General Relativity.
The universality of the coupling between matter and gravity (Equivalence Principle) has been verified around the level $\sim 10^{-13}$.
If we will consider value of relativistic mass as gravitational component of particle mass but non-relativistic mass will correspond to component of the inert mass, we can estimate values of maximal mass. Using this values on the base of our expressions we can predict the possible value of Maximal mass at the level $M \approx 10^4 Gev \div 10^7 GeV$.
The significant scatter of values characterizing maximal mass may be explained by big difference mass fermions in Standard Model. It is sufficient to note that the mass of t-quark in approximately 300000 times more then the mass of electron.

References

[1] V. G. Kadyshevsky, Nucl. Phys. **B141** (1978) 477; Fermilab-Pub. 78/22-THY (1978); Fermilab-Pub. 78/70-THY (1978).
[2] V. G. Kadyshevsky, M. D. Mateev, V. N. Rodionov, and A. S. Sorin, CERN TH/2007-150 (2007); hep-ph/0708.4205 (2005).
[3] C. M. Bender, H. F. Jones, and R. J. Rivers, Phys. Lett. **B 625** (2005) 333.
[4] P. A. M. Dirac, *The Relativistic Electron Wave Equation* Preprint KFKI-1977-62. - Budapest: Hungarian Academy of Sciences. Central Research Institute for Physics (1977).
[5] V. N. Rodionov, G. A. Kravtsova, Phys. Part. Nucl. **47** (2016) 135.
[6] V. N. Rodionov, Phys. Part. Nucl. **48** (2017) 319.
[7] V. N. Rodionov, Int. J. Theor. Phys. **54** (2015) 3907.
[8] V. N. Rodionov, Physica Scripta **90** (2015) 045302.
[9] V. N. Rodionov, Phys. Rev. **A 75** (2007) 062111.
[10] V. N. Rodionov, arXiv:1603.08425 (2016).
[11] V. B. Berestetskii, E. M. Lifshitz, L. P. Pitaevsky, *Quantum Electrodynamics, Theoretical Physics, v. IV*, M.: Fizmatlit, 2001.
[12] V. N. Rodionov, A. M. Mandel, arXiv:1708.08394v1 (2017).
[13] J. Alexandre, C. M. Bender and P. Millington, arXiv:1703.05251v1 (2017).
[14] Particle Data Group 2017. *The universality of the coupling between matter and gravity (Equivalence Principle) has been verified around the level* 10^{-13}.

COMPOSITION STUDIES WITH THE TELESCOPE ARRAY SURFACE DETECTOR

Mikhail Kuznetsov, Maxim Piskunov, Grigory Rubtsov, Sergey Troitsky, Yana Zhezher for the Telescope Array collaboration [a]
Institute for Nuclear Research of the Russian Academy of Sciences, 60th October Anniversary st. 7a, Moscow 117312, Russia and Faculty of Physics, Moscow State University, 119991 Moscow, Russia

Abstract. The results on ultra-high energy cosmic rays' chemical composition based on the data from the Telescope Array surface detector are reported. The Telescope Array (TA) is an experiment, located in Utah, USA, designed for observation of extensive air showers from the ultra-high energy cosmic rays. TA surface detector (SD) array consists of 507 detector stations, placed in a square grid with 1.2 km spacing with w total area of approximately 700 km². Each station has two layers of 1.2 cm thick plastic scintillator of 3 m² area. This talk is focused on the analysis of the mass composition of primary particles based on the 9 years data of the TA surface detector. The method employs the Boosted Decision Trees (BDT) technique for the analysis of the multiple composition-sensitive shower parameters. The BDT classifier is trained with the two Monte-Carlo training sets: proton, which is considered as background events, and iron, considered as signal events. The classifiers results in a single variable ξ for data and Monte-Carlo test sets. The data to Monte-Carlo comparison results in an average atomic mass of UHECR for energy range $10^{18.0} - 10^{20.0}$ eV. The comparison with TA hybrid composition results and the other experiments is presented.

1 Introduction

Ultra-high-energy cosmic rays (UHECRs) are particles and nuclei of energies more than 10^{18} eV entering the Earth atmosphere. While UHECRs are registered for many years, their origin remains a puzzle for physicists. Knowledge of the UHECR mass composition is crucial for understanding the source mechanism and propagation of cosmic rays.

UHECRs can't be observed directly due to its low flux. Particles with corresponding energies interact with the atmosphere, causing an extensive cascade of the secondary particles — so-called extensive air shower (EAS). Two large-scale EAS facilities: Pierre Auger Observatory [2] at Southern hemisphere and Telescope Array [1] at Northern hemisphere operate in the hybrid mode, measuring both the particle flux on the ground with the surface detectors and the emitted fluorescence light with the fluorescence telescopes.

The Telescope Array is an experiment designed for observation of extensive air showers from high energy cosmic rays, located in Utah, USA. The Telescope Array surface detector (SD) array consists of 507 detector units, placed in a square grid with 1.2 km spacing with total area covered approximately 700

---
[a] E-mail: zhezher.yana@physics.msu.ru

km². Each detector has two layers of 1.2 cm thick plastic scintillator of 3 m² area each.

In this work, a method of composition studies with the Telescope Array surface detector (SD) data is suggested. A common way to obtain UHECR composition is the calculation of the $\langle X_{max} \rangle$, the average value of the atmospheric depth where the shower development reaches its maximum, which requires operation of fluorescence telescopes. The main advantage of SD data usage is that surface detectors obtain much more data than fluorescense telescopes, which can operate only on clear moonless nights (10 % duty cycle). Surface detectors, on the contrary, operate full duty cycle.

2 Method

2.1 Multivariate analysis

Multivatiate analysis is a common name for a variety of statistical techniques, used to analyze data described by more than one variable. Given a set of observables MVA transforms it into a single variable, usually called ξ, which then allows to apply conventional one-dimensional techniques.

In case of EAS data, during the reconstruction procedure a number of observables is obtained, some of which are known to be composition-sensitive. After applying the MVA method ξ disitribution for data may be compared with the corresponding distributions for Monte-Carlo (MC) event sets and as a result, average atomic mass $\langle \ln A \rangle$ as a function of energy is derived.

2.2 Boosted Decision Trees

The Boosted Decision Trees (BDT) technique is used to build a p-Fe classifier based on multiple observables [3]. The BDT classifier works as follows:

- For each variable a splitting value with best separation is found. This value divides the full range of the values of the variable into two ranges, which are called branches. It will be mostly signal in one branch, and mostly background in another. The classifier is trained using the proton MC and iron MC training sets as background and signal respectively;

- Then the algorithm is repeated recursively on each branch;

- The decision tree will iterate until the stopping criterion is reached (for example, number of events in a branch). The terminal node is called a leaf.

The concept of boosting allows one to create a good classifier using a number of weak ones. In this work, "adaptive boosting", or AdaBoost, was implemented

[4]. In AdaBoost, a weak learner is run multiple times on the training data, and each event is weighted by how incorrectly it was classified. An improved tree with reweighted events may be built now, and as a result, averaging over all trees allows to create a better classifier.

The classifier gives a single variable ξ for each event from the data and from the MC sets. This variable ξ is the main parameter which is subsequently used to distinguish between primaries.

In this work, root::TMVA package [5] is used as a stable implementation. The classifier converts the set of observables for an event to a number $\xi \in [-1:1]$: 1 — pure signal (Fe), -1 – pure background (p).

2.3 Composition-sensitive variables

A set of thirteen composition-sensitive variables is used:

1. Linsley front curvature parameter, a.

   Joint fit of lateral distribution function (LDF) and shower front is performed with 7 free parameters: x_{core}, y_{core}, θ, ϕ, S_{800}, t_0, a [6]:

$$t_0(r) = t_0 + t_{plane} + a \times 0.67 \ (1 + r/R_L)^{1.5} \ LDF(r)^{-0.5},$$
$$S(r) = S_{800} \times LDF(r),$$
$$LDF(r) = f(r)/f(800 \text{ m}),$$
$$f(r) = \left(\frac{r}{R_m}\right)^{-1.2} \left(1 + \frac{r}{R_m}\right)^{-(\eta-1.2)} \left(1 + \frac{r^2}{R_1^2}\right)^{-0.6},$$
$$R_m = 90.0 \text{ m}, \ R_1 = 1000 \text{ m}, \ R_L = 30 \text{ m},$$
$$\eta = 3.97 - 1.79(\sec(\theta) - 1),$$

   where t_{plane} is a shower plane delay, S_{800} is a scintillator signal density at 800 m core distance and a is the Linsley front curvature parameter.

2. Area-over-peak (AoP) of the signal at 1200 m and 3. AoP slope parameter [7].

   Given a time resolved signal from a surface station, one may calculate it's peak value and area, which are both well-measured and not much affected by fluctuations.

   $AoP(r)$ is fitted with a linear fit:

$$AoP(r) = \alpha - \beta(r/r_0 - 1.0),$$

   where $r_0 = 1200$ m, α is $AoP(r)$ value at 1200 m and β is it's slope.

4. Number of detectors hit.
5. Number of detectors excluded from the fit of the shower front by the reconstruction procedure [8].
6. $\chi^2/d.o.f.$
7. $S_b = \sum S_i \times r^b$ parameter [9]

$$S_b = \sum_{i=1}^{N} \left[S_i \times \left(\frac{r_i}{r_0} \right)^b \right],$$

where S_i is the signal of i-th detector, r_i is the distance from the shower core to this station in meters and $r_0 = 1000$ m — reference distance. The two S_b for $b = 3$ and $b = 4.5$ are used in the analysis.

8. The sum of the signals of all the detectors of the event.
9. Asymmetry of the signal at the upper and lower layers of detectors.
10. Total number of peaks within all FADC traces.
11. Number of peaks for the detector with the largest signal.
12. Number of peaks present in the upper layer and not in the lower.
13. Number of peaks present in the lower layer and not in the upper.

In addition, zenith angle θ and S_{800} are included in the data set.

2.4 ξ parameter conversion to $\langle \ln A \rangle$

After applying the BDT method, ξ parameter distribution is derived for proton and iron MC and for the data in each energy bin. The range between $\ln A = 0$ (proton) and $\ln A = 4.02$ (iron) is divided into 40 equal parts and at every point a "mixture" of protons and iron (e.g. 5 % p and 95 % Fe) is created.

The Kolmogorov-Smirnov (KS) distance between ξ parameter distribution of the each "mixture" and data is performed, and the case with the smallest KS-distance is chosen, thus allowing to determine $\langle \ln A \rangle$ at a particular energy bin.

$\langle \ln A \rangle$ is estimated as:

$$\langle \ln A \rangle = \epsilon_p \times \ln(1) + (1 - \epsilon_p) \times \ln(56),$$

where ϵ_p is a fraction of protons in the mixture.

3 Data set

In this work, 9-year data collected by the Telescope Array surface detector is used (2008-05-11 — 2017-05-11). Following cuts are applied:

1. quality cuts used for spectral analysis
2. 7 or more detectors triggered
3. $E > 10^{18}$ eV

After the cuts, the data set contains 18077 events.
The analysis uses two Monte-Carlo event sets modeled for the primary p and Fe with CORSIKA and QGSJETII-03 hadronic model. The dethinning procedure is used to reproduce the shower fluctuations while using THINNING procedure to reduce required computing resources. MC sets are split into two equal parts: (I) for training the classifier, (II) for MVA estimator calculation [10].

4 Results

On Fig. 1 the average atomic mass $\langle \ln A \rangle$ is shown in comparison with the Telescope array hybrid results [11]. On Fig. 2 the average atomic mass $\langle \ln A \rangle$ is shown in comparison with the Pierre Auger Observatory surface detector results, based on muon X_{max} and risetime asymmetry calculations, respectively [12].

The mass composition obtained is qualitatively consistent with both the Telescope Array fluorescence detector and the Pierre Auger Observatory results. Within the sensitivity of the method, no energy dependence may be seen in the chemical composition. If one assumes steady composition and the QGSJETII-03 hadronic model, the average atomic mass of primary particles corresponds to $\langle \ln A \rangle = 1.8 \pm 0.1$.

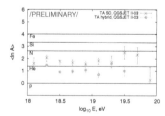
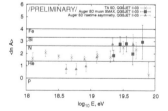

Figure 1: Average atomic mass $\langle \ln A \rangle$ in comparison with the Telescope Array hybrid results [11].

Figure 2: Average atomic mass $\langle \ln A \rangle$ in comparison with the Pierre Auger Observatory surface detector results, based on muon X_{max} and risetime asymmetry calculations [12].

Acknowledgment

The present work is supported by the Russian Science Foundation grant No. 17-72-20291 (INR).

References

[1] Telescope Array Collaboration, Prog. Theor. Phys. Suppl. **151** (2003) 206-210.
[2] Pierre Auger Observatory Collaboration, Nucl. Instrum. Meth. **A798** (2015) 172.
[3] L. Breiman et al., "Classification and Regression Trees" Wadsworth International Group (1984).
[4] Y. Freund, R. E. Schapire, Journal of JSAI, **14(5)** (1999) 771-780.
[5] J. Therhaag, PoS ICHEP **2010** (2010) 510.
[6] Telescope Array Collaboration, Phys. Rev. D **88**, no. 11, 112005 (2013).
[7] Pierre Auger Collaboration, Phys. Rev. Lett. **100** (2008) 211101.
[8] Telescope Array Collaboration, ApJL **768** (2013) L1.
[9] G. Ros, A. D. Supanitsky, G. A. Medina-Tanco et al. Astropart. Phys. **47** (2013) 10.
[10] B. T. Stokes et al., Astropart. Phys. **35** (2012) 759.
[11] W. Hanlon for the Telescope Array Collaboration, UHECR'16.
[12] Pierre Auger Collaboration, Contributions to the 32nd International Cosmic Ray Conference (2011).

PERFORMANCE OF THE MPD EXPERIMENT FOR THE ANISOTROPIC FLOW MEASUREMENT

Petr Parfenov[1,2] [a], Ilya Selyuzhenkov[1,3], Arkadiy Taranenko[1], Anton Truttse[1]

[1] National Research Nuclear University MEPhI (Moscow Engineering Physics Institute), Kashirskoe highway 31, Moscow, 115409, Russia
[2] Institute for Nuclear Research of the Russian Academy of Sciences, Moscow, Russia
[3] GSI Helmholtzzentrum für Schwerionenforschung, Darmstadt, Germany

Abstract. The main goal of the future MPD experiment at NICA is to explore the QCD phase diagram in the region of highly compressed and hot baryonic matter in the energy range corresponding to the highest chemical potential. Properties of such dense matter can be studied using azimuthal anisotropy which is categorized by the Fourier coefficients of the azimuthal distribution decomposition. Performance of the detector response based on simulations with realistic reconstruction procedure is presented for centrality determination, reaction plane estimation, directed and elliptic flow coefficients.

Transverse azimuthally anisotropic flow measurements are one of the key methods to study the time evolution of the strongly interacting medium formed in the nucleus collisions. In the non-central collisions, initial spatial anisotropy results in the azimuthally anisotropic particle emission.

For analysis, particles produced in the heavy-ion collisions were generated using the UrQMD (Ultra-relativistic Quantum Molecular Dynamics) [1]. A million of Au + Au collisions at beam energy $\sqrt{s_{NN}} = 11~GeV$ was generated. Further simulations were carried out using GEANT framework with the MPD detector geometry. Following kinematic cuts were used in the analysis: $|\eta| < 1.5$; $0.2 < p_T < 3~GeV/c$; $N_{hits}^{TPC} > 32$. For this analysis, particle identification (PID) is implemented using the particle's PDG code from GEANT4. For primary particle selection, 2σ DCA cut was used, where DCA is the distance of the closest approach between the reconstructed vertex and a charged particle track. Centrality was determined based on track multiplicity of the emitted charged particles in TPC with impact parameter resolution $5 - 10\%$ in a centrality range $10 - 80\%$.

For the collective flow measurements, event plane method was used [2]. Since the reaction plane was determined based on the finite number of hadrons, measured flow values should be corrected by the resolution correction factor. Estimated resolution factor (see Fig. 1, top) shows the good performance in the wide centrality range ($Res_1\{\Psi_1^{EP}\} \sim 0.9$, $Res_2\{\Psi_1^{EP}\} \sim 0.7$ for centrality $20 - 40\%$). In Fig. 1 (bottom) one can see the directed v_1 and elliptic v_2 flow as a function of p_T. The signal simulated with the UrQMD model is compared with calculations using GEANT simulation and full flow reconstruction procedure (reco) like in the real data analysis. Directed (v_1) and elliptic (v_2)

---

[a] E-mail: petr.parfenov@cern.ch

flow were extracted in simulations using event plane method. Results for the reconstructed (reco) and generated (true) values are in good agreement.

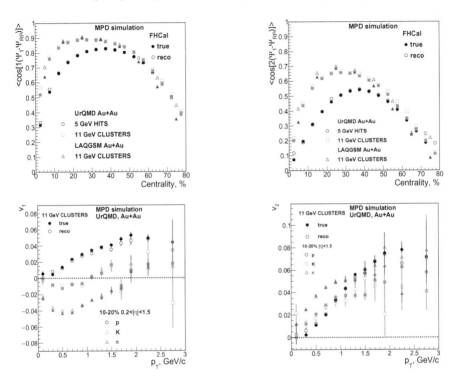

Figure 1: Top: Resolution correction factor as a function of centrality for v_1 (left) and v_2 (right) for the UrQMD and LAQGSM event generators. Bottom: Directed flow v_1 (left) and elliptic flow (right) as a function of p_T. Results from the GEANT simulation marked as true and one from the reconstruction procedure is marked as reco.

Acknowledgments

This work was partially supported by the Ministry of Science and Education of the Russian Federation, grant N 3.3380.2017/4.6, and by the National Research Nuclear University MEPhI in the framework of the Russian Academic Excellence Project (contract No. 02.a03.21.0005, 27.08.2013).

References

[1] S.A. Bass et al., Prog. Part. Nucl. Phys. 41 (1998) 255-369.
[2] A. M. Poskanzer and S. A. Voloshin, Phys. Rev. C **58**, 1671 (1998).

CP Violation and Rare Decays

RECENT RESULTS OF RARE B/D DECAYS FROM BELLE

Min-Zu Wang [a]
on behalf of the Belle Collaboration
Department of Physics, National Taiwan University, Taiwan, R.O.C.

Abstract. We report the recent results of rare B/D decays obtained from the Belle experiment. These include studies of $B \to h\nu\bar{\nu}$, $B^+ \to K^+K^-\pi^+$, $B^0 \to \pi^0\pi^0$ and D^0 to invisible final states.

1 Introduction

Since the discovery of the Higgs boson at the LHC, searches for violations of Standard Model (SM) predictions become more critical for the hints of new physics. Studying rare B/D decays seems to be one of the good strategies. For instance, the deviation from SM predictions might be sizable due to the direct contribution of unknown heavy particles in the FCNC loop or even at the tree level, etc. We use the full data sample, ~ 1 ab^{-1}, collected by the Belle [1] detector at the KEKB asymmetric e^+e^- collider [2] for the study of rare B/D decays. There are about 772 million $B\bar{B}$ pairs being recorded, corresponding to an integrated luminosity of 711 fb^{-1}.

We generated signal Monte Carlo (MC) and background samples in order to optimize the event selection criteria. The final-state particle selections are mainly based on the information obtained from the tracking system (vertex detector, central drift chamber), the electro-magnetic calorimeter, the hadron identification system (central drift chamber, time-of-flight counter, Aerogel Cherenkov counter) and the muon/K_L outer detector. Further selections with sophisticated algorithm using event shape or kinematical information are applied in order to improve the signal to background ratio. The MC-data differences are studied by well measured physics processes and corrected for the determination of the signal efficiency.

2 $B \to h\nu\bar{\nu}$ [3]

There are theoretical predictions [4] for the decay branching fractions of $B \to K^{(*)}\nu\bar{\nu}$ which is about the reach of Belle accumulated data. We use semi-leptonic tagging method to identify the accompanying B and ensure that there is only one extra meson h being found in the detector, where h stands for K^+, K^0_S, K^{*+}, K^{*0}, π^+, π^0, ρ^+ or ρ^0 [5]. Since there are two neutrinos in the final state, we remove the calorimeter clusters matched with the observed particles and take only the rest of the calorimeter energy for signal identification. There is no significant peak found near zero. Therefore, we set the upper limits on the

---

[a] E-mail: mwang@phys.ntu.edu.tw

branching fraction at 90 % confidence level as listed in Table 1 and shown in Fig. 1. It is clear that these will be the golden modes for searches in the future b-factories.

Table 1: Expected (median) and observed upper limits on the branching fraction at 90 % C.L. The observed limits include the systematic uncertainties.

Channel	Efficiency	Expected limit	Observed limit
$K^+\nu\bar{\nu}$	2.16×10^{-3}	0.8×10^{-5}	1.9×10^{-5}
$K^0_S\nu\bar{\nu}$	0.91×10^{-3}	1.2×10^{-5}	1.3×10^{-5}
$K^{*+}\nu\bar{\nu}$	0.57×10^{-3}	2.4×10^{-5}	6.1×10^{-5}
$K^{*0}\nu\bar{\nu}$	0.51×10^{-3}	2.4×10^{-5}	1.8×10^{-5}
$\pi^+\nu\bar{\nu}$	2.92×10^{-3}	1.3×10^{-5}	1.4×10^{-5}
$\pi^0\nu\bar{\nu}$	1.42×10^{-3}	1.0×10^{-5}	0.9×10^{-5}
$\rho^+\nu\bar{\nu}$	1.11×10^{-3}	2.5×10^{-5}	3.0×10^{-5}
$\rho^0\nu\bar{\nu}$	0.82×10^{-3}	2.2×10^{-5}	4.0×10^{-5}

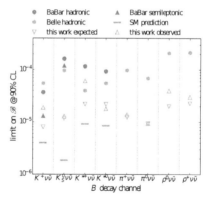

Figure 1: Expected and observed limits for all channels in comparison to previous results for the $BABAR$ measurement with semi-leptonic and hadronic tag , as well as the Belle measurement utilizing hadronic tagging. The theoretical predictions are taken from Ref. [4].

3 $B^+ \to K^+K^-\pi^+$ [6]

It had been reported by $BABAR$ [7] for a peaking structure being found in the low K^+K^- mass spectrum of the charmless $B^+ \to K^+K^-\pi^+$ decay. Later LHCb [8] observed a nonzero inclusive CP asymmetry for the same decay mode and a large unquantified local CP asymmetry in the low K^+K^- mass region. We follow the same analysis procedure. First applying the charm veto to remove the $D^0(J/\psi)$ contaiminated charmful signals and reconstruct the charmless $B^+ \to K^+K^-\pi^+$ events. The signal yield is extracted from the

Table 2: Signal yield, differential branching fraction, and $\mathcal{A}_{CP}$ for individual $M_{K^+K^-}$ bins. The first uncertainties are statistical and the second systematic.

$M_{K^+K^-}$ (GeV/c^2)	N_{sig}	$d\mathcal{B}/dM$ ($\times 10^{-7}$)	$\mathcal{A}_{CP}$
0.8 − 1.1	59.8 ± 11.4 ± 2.6	14.0 ± 2.7 ± 0.8	−0.90 ± 0.17 ± 0.04
1.1 − 1.5	212.4 ± 21.3 ± 6.7	37.8 ± 3.8 ± 1.9	−0.16 ± 0.10 ± 0.01
1.5 − 2.5	113.5 ± 26.7 ± 18.6	10.0 ± 2.3 ± 1.7	−0.15 ± 0.23 ± 0.03
2.5 − 3.5	110.1 ± 17.6 ± 4.9	10.0 ± 1.6 ± 0.6	−0.09 ± 0.16 ± 0.01
3.5 − 5.3	172.6 ± 25.7 ± 7.4	8.1 ± 1.2 ± 0.5	−0.05 ± 0.15 ± 0.01

following two kinematic variables: the beam-energy-constrained mass, $M_{\text{bc}} \equiv \sqrt{E_{\text{beam}}^2 - |\vec{p}_B|^2}$, and the energy difference $\Delta E \equiv E_B - E_{\text{beam}}$. Here, $\vec{p}_B$ and E_B are the momentum and energy of the B-meson candidates, and E_{beam} is the beam energy (all measured in the CM frame). A simutaneous likelihood fit to $B^+ \to K^+K^-\pi^+$ and the corresponding charge conjugated decay is preformed in order to obtain the differential branching fraction as well as the $\mathcal{A}_{CP}$ in bins of K^+K^- mass, where $\mathcal{A}_{CP} = \frac{N(B^- \to K^-K^+\pi^-) - N(B^+ \to K^+K^-\pi^+)}{N(B^- \to K^-K^+\pi^-) + N(B^+ \to K^+K^-\pi^+)}$. The results are shown in Table 2. Our measured inclusive branching fraction and direct CP asymmetry are $\mathcal{B}(B^+ \to K^+K^-\pi^+) = (5.38 \pm 0.40 \pm 0.35) \times 10^{-6}$ and $\mathcal{A}_{CP} = -0.170 \pm 0.073 \pm 0.017$, respectively. Note that the $\mathcal{A}_{CP}$ value in the lowest K^+K^- mass bin is unexpectedly large.

4 $B^0 \to \pi^0\pi^0$ [9]

Previous experimental findings of $B^0 \to \pi^0\pi^0$ [10] showed a factor of 2 larger branching fraction than the theoretical prediction [11] based on the QCD factorization approoch. This decay branching fraction is important for constraining the quark flavor mixing ϕ_2 angle by performing an isospin analysis of the entire $B \to \pi\pi$ system. Due to the large continuum background ($e^+e^- \to q\bar{q}$ ($q \in \{u,d,s,c\}$)), we apply the event shape information to improve the signal to background ratio. A Fisher discriminant (T_c) is constructed with output value between -1 (continuum jet-like) and 1 (signal B-like). This variable together with M_{bc} and ΔE are used in a extended unbinned maximum likelihood fit to extract signal yield. The fit projection plots are shown in Fig. 2. The obtained branching fraction and direct CP asymmetry are $\mathcal{B}(B \to \pi^0\pi^0) = [1.31 \pm 0.19 \text{ (stat.)} \pm 0.19 \text{ (syst.)}] \times 10^{-6}$ and $A_{CP} = +0.14 \pm 0.36$ (stat.) ± 0.10 (syst.), respectively. The signal significance, is 6.4 standard deviations which represents the most significant measurement up-to-date by a single experiment. We combine all Belle $B \to \pi\pi$ measurements to exclude ϕ_2 from $15.5° < \phi_2 < 75.0°$ at 95% confidence level.

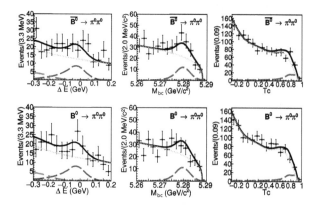

Figure 2: Projections of the fit results onto (left) ΔE, (middle) M_{bc}, (right) T_c are shown in the signal enhanced region: $5.275\text{GeV}/c^2 < M_{bc} < 5.285\text{GeV}/c^2$, $-0.15\text{GeV} < \Delta E < 0.05\text{GeV}$, and $T_c > 0.7$. Each panel shows the distribution enhanced in the other two variables. Data are points with error bars, and fit results are shown by the solid black curves. Contributions from signal, continuum $q\bar{q}$, combined $\rho\pi$ and other rare B decays are shown by the dashed blue, dotted green, and dash-dotted red curves, respectively. The top (bottom) row panels are for events with positive (negative) b tags.

5 D^0 to invisible final states [12]

There are few hundred million D mesons being produced at Belle via the $e^+e^- \to c\bar{c}$ continuum process. We use a charm tagger method introduced in Ref. [13] to construct the missing momentum 4-vector against a positively tagged $D^{(*)}/D_s/\Lambda_c$, extra fragment hadrons and a slow pion in order to form the mass of the targeted D^0 sample. Since the observed yield for $D^0 \to$ invisible is not significant and the unmatched calorimeter energy forms no peak at zero, we calculate a 90% confidence level Bayesian upper limit on the branching fraction which is 9.4×10^{-5} at the 90% confidence level. The fit results are shown in Fig. 3. This is one example to illustrate the probe to the dark sector at the b-factory.

In summary, we are eagerly looking forward to the start of the SuperKEKB collider in 2018.

Acknowledgments

The author wish to thank the local organization committee for making such a wonderful conference. This work is supported by the Ministry of Science and Technology of R.O.C. under the grant MOST-103-2112-M-002-017-MY3.

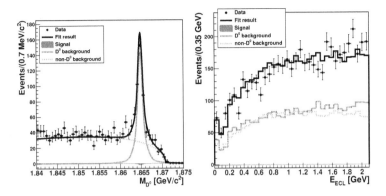

Figure 3: Fit results of $D^0 \to$ invisible decays. The top panel shows the M_{D^0} distribution for $E_{\text{ECL}} < 0.5$ GeV and the bottom one shows E_{ECL} for $M_{D^0} > 1.86$ GeV/c^2.

References

[1] A. Abashian et al. (Belle Collaboration), Nucl. Instrum. Methods Phys. Res., Sect. A **A 479**, (2002) 117; also see the detector section in J. Brodzicka et al., Prog. Theor. Exp. Phys. (2012) 04D001.
[2] S. Kurokawa and E. Kikutani, Nucl. Instrum. Methods Phys. Res. Sect., **A 499**, (2003) 1, and other papers included in this Volume; T. Abe et al., Prog. Theor. Exp. Phys. (2013) 03A001 and following articles up to 03A011.
[3] J. Grygier, et al. (Belle Collaboration), Phys. Rev. **D 96** (2017) 091101.
[4] A. Buras, J. Girrbach-Noe, C. Niehoff, and D. Straub, JHEP **02** (2015) 184.
[5] Throughout the paper, the inclusion of the charge-conjugate decay mode is implied unless otherwise stated.
[6] C.L. Hsu, et al. (Belle Collaboration), Phys. Rev. **D 96** (2017) 031101.
[7] B. Aubert et al. (*BABAR* Collaboration), Phys. Rev. Lett. **99** (2007) 221801.
[8] R. Aaij et al. (LHCb Collaboration), Phys. Rev. **D 90** (2014) 112004.
[9] T. Julius, et al. (Belle Collaboration), Phys. Rev. **D 96** (2017) 032007.
[10] C. Patrignani et al. (Particle Data Group), Chin. Phys. **C 40** (2016) 100001.
[11] H. Li and S. Mishima, Phys. Rev. **D 83** (2011) 034023.
[12] Y. Lai, et al. (Belle Collaboration), Phys. Rev. **D 95** (2017) 011102.
[13] A. Zupanc et al. (Belle Collaboration), J. High Energy Phys. **09** (2013) 139.

SEARCH OF cLFV PROCESSES WITH MUONS AND OTHERS

Kei Ieki [a]

University of Tokyo, 7-3-1 Hongo, Bunkyo-ku, Tokyo

Abstract. Charge lepton flavor violation (cLFV) is a powerful probe to investigate the physics beyond the standard model. There are many cLFV processes searched by the high precision experiments, and they are complementary to the new physics searches in high energy frontier. Several new generation muon cLFV experiments are expected to start in next five years. Status of the cLFV searches, mainly with muons, will be reviewed.

1 Charged Lepton Flavor Violation

Charged Lepton Flavor Violation (cLFV) is forbidden in the standard model (SM). Flavor violation is already discovered for quarks and neutrinos, but not for charged leptons. The branching ratio becomes non-zero by taking into account neutrino oscillation effect in SM. For example, the branching ratio of cLFV decay $\mu \to e\gamma$ is calculated as follows:

$$\mathcal{B}(\mu \to e\gamma) = \frac{3\alpha}{32\pi} \left| \sum_{i=2,3} U^*_{\mu i} U_{ei} \frac{\Delta m^2_{i1}}{M^2_W} \right|^2 \simeq 10^{-54}, \quad (1)$$

where U_{ij} represents the element in Pontecorvo-Maki-Nakagawa-Sakata matrix, Δm^2_{ij} is the squared mass difference of neutrino mass eigenstates and M_W is the mass of W boson. This branching ratio is far below the experimental reach.

On the other hand, some of the well-motivated beyond SM (bSM) models such as SUSY-seesaw model [1] predicts sizable rates of the branching ratio. For example, the branching ratio of $\mu \to e\gamma$ decay is predicted to be $O(10^{-14})$. Therefore, the discovery of cLFV decay would be a clear evidence of bSM physics. The branching ratios predicted by bSM models are within the reach of the experiments which are expected to start in ~5 years.

There are various cLFV decay modes which are searched by the experiments. Some of them are shown in Table 1. Because the different decay modes have different sensitivity to bSM models, it is important to search cLFV in many decay modes. Different decay modes requires different type of beams, and the experimental techniques are also different. In the following chapters, we review the past, ongoing and near future experiments which searches for cLFV decays, mainly focusing on muon experiments.

2 μ LFV

For muons, there are three "golden channels" for the cLFV search: $\mu \to e\gamma$, $\mu \to eee$ and $\mu N \to eN$. Feynman diagrams of these decays are similar to

---
[a]E-mail: iekikei@icepp.s.u-tokyo.ac.jp

Table 1: cLFV decay modes.

Parent particle	μ	τ	H, Z, Z'
Decay modes	$\mu \to e\gamma$	$\tau \to \mu\gamma, e\gamma$	$H \to e\mu, \mu\tau, e\tau$
	$\mu \to eee$	$\tau \to \mu\mu\mu, eee, \mu ee$	$Z \to e\mu, \mu\tau, e\tau$
	$\mu N \to eN$		$Z' \to e\mu, \mu\tau, e\tau$

each other, as shown in Fig. 1. If the photon mediated process is dominant, branching ratio of $\mu \to e\gamma$ is expected to be larger than that of $\mu \to eee$ or $\mu N \to eN$ by a factor of ~400. $\mu \to eee$ and $\mu N \to eN$ are, on the other hand, sensitive to the non-photonic processes.

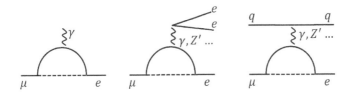

Figure 1: Feynman diagrams of the μ "golden modes".

Figure 2 shows the history of branching ratio upper limits of the μLFV decays. This figure also shows the sensitivity of the near future experiments which are expected to start in ~5 years. $\mu \to e\gamma$ will be searched by MEG II, $\mu \to eee$ will be searched by Mu3e and $\mu N \to eN$ will be searched by COMET, Mu3e and DeeMe. As can be seen in Fig. 2, sensitivities of the near future experiments are within the region predicted by bSM models.

Table 2 summarizes the signals and backgrounds of these experiments. All of the experiments have e^+/e^- in the final state, and the energy of them must be measured precisely. There are accidental backgrounds in $\mu \to e\gamma$ and $\mu \to eee$, while it is not the case for $\mu N \to eN$. With these similarities and differences taken into account, we compare the design of the experiments in the following sections.

3 Choice of beam

Choice of the beam can be either continuous or pulsed. Continuous beam is desired for the experiments which searches for $\mu \to e\gamma$ or $\mu \to eee$, because the accidental backgrounds must be minimized. On the other hand, pulsed beam is used for the $\mu N \to eN$ search. With the pulsed beam, it is possible to reject beam related prompt backgrounds by searching for the signal in a delayed

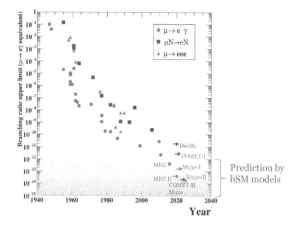

Figure 2: History of the branching ratio upper limit of the μLFV decay modes.

Table 2: Signals and backgrounds of the μLFV experiments.

	$\mu \to e\gamma$	$\mu \to eee$	$\mu N \to eN$
Signal	e and γ $E_e = E_\gamma = 52.8$ MeV back-to-back coincident in time	eee $\sum E = m_\mu, \sum p = 0$ common vertex coincident in time	e $E_e \sim 100$ MeV
BG	accidental e and γ, $\mu \to e\nu\nu\gamma$	accidental $e \times 3$, $\mu \to eee\nu\nu$	μ decay in orbit, prompt beam BG

timing window. Following sections describe the properties of the continuous and pulsed beam for the μLFV experiments.

3.1 Continuous beam for $\mu \to e\gamma$ and $\mu \to eee$

Both MEG II and Mu3e will use the μ^+ beam at Paul Scherrer Institute (PSI). PSI is a unique facility that provides the world's highest intensity continuous μ beam of up to 10^8 muons per second.

A "surface muon" beam is used for these experiments. These muons are generated from stopped pions decaying at near the surface of the production target, and therefore have an advantage of small momentum bite. Contamination of electrons and positrons are removed by Wien filter, and only positive muons are selected.

The beam rate can not be too high for the search of $\mu \to e\gamma$ and $\mu \to eee$ because the rate of accidental backgrounds increases faster. This is not the case for $\mu N \to eN$ search, which requires the beam intensity to be as high as possible.

3.2 Pulsed beam for $\mu N \to eN$

COMET, Mu2e and DeeMe exploits pulsed beam for the search of $\mu N \to eN$. High rate muon beam can be realized by a capture solenoid magnet which surrounds the pion production target (Fig. 3). A long transport solenoid magnet connects the production target to the muon stopping target and the detector. This allows all background pions to decay, and removes the backgrounds with high energy or wrong sign.

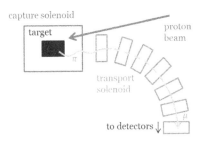

Figure 3: Schematic view of muon beamline for $\mu N \to eN$ search.

4 Detector technologies

Sensitivity of the experiments also strongly depends on the resolution of the detectors because it affects the signal to background ratio. Requirements for the detectors are different for each decay mode, but there are some similarities in the e^+/e^- detectors. Following sections summarizes the similar detectors as well as special detectors for each experiment.

4.1 e^+, e^- detector

Tracker

All of the μLFV experiments described above require precise measurement of e^+/e^- momentum to distinguish signal events from backgrounds. Background e^+ and e^- originates from normal muon decay at extremely high rate. Detector must have low mass in order to minimize the multiple scattering because that limits the momentum resolution.

The MEG II fulfills the low mass requirement by using a cylindrical drift chamber (Fig. 4 left). It will be placed inside a gradient magnetic field coil.

Thanks to this coil, low momentum background positrons will not reach the detector and it will be quickly swiped away, thus makes it possible to operate the drift chamber at high rate.

Similar to MEG II, Mu2e uses cylindrical shape straw tube detector with a hole in the middle so that low momentum electrons cannot reach the detector because of small track radius inside a magnetic field. On the other hand, the COMET experiment will install a second curved solenoid (similar to that in Fig. 3) between muon stopping target and detector so that only high energy electrons can reach the detector.

For the Mu3e experiment, it is not possible to remove high rate low momentum e^-/e^+ by using magnetic field, because the momentum of the signal e^-/e^+ is also low. Instead of using a gas detector which is difficult to operate in high rate, Mu3e developed an ultra thin (50 μm) HV-CMOS sensor detector. Because of its fast response, it is operational in high rate.

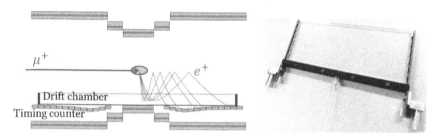

Figure 4: Schematic view of MEG II e^+ detector (left) with a photo of timing counter (right).

Timing detector

Precise timing measurement is also required for $\mu \to e\gamma$ and $\mu \to eee$. Plastic scintillator detectors read out by SiPMs are used in both MEG II and Mu3e for the timing measurement. Right side of Figure 4 shows one scintillator counter from MEG II as an example. Six SiPMs are connected at both ends of the scintillators to collect as much scintillation light as possible and to achieve good timing resolution accordingly. Many counters are installed on the trajectory of the positron track so that the positrons intersect multiple counters and the statistical uncertainty of the timing measurement is reduced.

γ detector

In addition to e detector, high resolution γ detector is required for the search of $\mu \to e\gamma$ decay. MEG II uses 900 l liquid xenon for the γ detector (Fig. 5), which has advantages on high light yield, uniformity, high stopping power and fast response. These properties are important for the precision measurement of

Table 3: Upper bounds of the decay modes.

Decay mode	Upper limit (90% C.L.)	Experiment (year)
$\mu^+ \to e^+\gamma$	4.2×10^{-13}	MEG (2016)
$\mu^+ \to e^+e^-e^+$	1.0×10^{-12}	SINDRUM (1988)
μ^- Au $\to e^-$Au	7×10^{-13}	SINDRUM II (2006)

energy, position and timing of the γ. Because the wavelength of the xenon scintillation light is in vacuum ultraviolet region, UV-sensitive PMTs were developed for MEG. In MEG II, 2-inch PMTs in the γ entrance face are replaced with new 12×12 mm^2 UV-sensitive 4092 SiPMs. Thanks to the improved sensor coverage uniformity and improved granularity, the energy resolution and position resolution are expected to improve by a factor of 2.

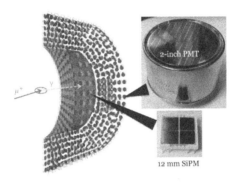

Figure 5: Liquid xenon γ detector for MEG II.

4.2 Status of the μLFV experiments

Current upper limit of μLFV decays are summarized in Table 3. Several μLFV experiments are expected to start in the next few years.

MEG II experiment aims to search $\mu \to e\gamma$ with a sensitivity goal of 4×10^{-14}, which is an order of magnitude better than the current upper limit given by MEG. In order to reach this goal, the μ^+ beam intensity will be increased by more than a factor of two (7×10^7 muons/sec) and both the detector efficiency and resolutions will be improved by a factor of two. The beamline at PSI already provides high intensity μ^+ beam required by MEG II. Construction of the liquid xenon γ detector and e^+ timing counter is already finished, and the complete set of detector including e^+ drift chamber is expected to be ready in 2018.

Mu3e experiment will start from the silicon e tracker only and eventually extend the detector with timing counter. A first physics result is expected around 2021. Final sensitivity goal of $Br(\mu \to eee) < 10^{-16}$ can be reached by an upgraded beamline planned at PSI. COMET at J-PARC and Mu2e at Fermilab will search $\mu N \to eN$ with a final sensitivity goal of $\mathcal{O}(10^{-17})$, while DeeMe at J-PARC is able to start quickly with the sensitivity goal of $10^{-13} \sim 10^{-14}$.

5 Other LFV searches

5.1 τ LFV

Tau LFV decays are similar to that of muons, but it has different sensitivity to physics models. Because of the heavier mass, there are many decay channels available, but it is not possible to have "τ beam" because of its short lifetime. The best upper limits are given by Belle and Babar experiments by looking for τLFV decay in e^+e^- colliders where τ^+ and τ^- are produced in pair. LHCb experiment also searches for τLFV from τ produced in D meson decays. Tau LFV search will be continued by Belle II which starts from 2018 with 40 times higher luminosity beam.

5.2 cLFV at ATLAS and CMS

ATLAS and CMS can do a direct search of cLFV in the decays of H, Z and Z'. LFV decay upper limits provide constraints to these decay modes under assumption of certain models. There are no signature of LFV discovered so far, but we expect to have updated results with new data sets.

6 Summary

There are various experiments aiming to search for cLFV processes, and some of them are expected to start in the next few years with improved sensitivity. Thanks to the improved technology in both beam and detector, sensitivity of the next generation experiment reaches the region where the bSM models predict to have signal.

Reference

[1] S. Antusch, E. Arganda, M.J. Herrero, A.M. Teixeira, J. High Energy Phys. 11 (2006) 090.

STUDY OF $K^+ \to \pi^0 e^+ \nu_e \gamma$ DECAY WITH OKA SETUP

A.Yu. Polyarush [a]

(on behalf of OKA collaboration)

Institute for Nuclear Research RAS, Moscow, Russia.

Abstract. Results of study of the $K^+ \to \pi^0 e^+ \nu \gamma$ decay at OKA setup are presented. 13118 events of this decay have been observed. The branching ratio with cuts $E_\gamma^* > 10$ MeV, $0.6 < cos\Theta_{e\gamma}^* < 0.9$ is calculated $R = \frac{Br(K^+ \to \pi^0 e^+ \nu_e \gamma)}{Br(K^+ \to \pi^0 e^+ \nu_e)} = (0.59 \pm 0.02(stat.) \pm 0.03(syst.)) \times 10^{-2}$. For the asymmetry A_ξ we get $A_\xi = -0.019 \pm 0.020(stat.) \pm 0.027(syst.)$

1 Introduction

The decay $K^+ \to \pi^0 e^+ \nu \gamma$ provides fertile testing ground for the Chiral Perturbation Theory (ChPT) [1, 2]. $K^+ \to \pi^0 e^+ \nu \gamma$ decay branching ratio was evaluated up to $O(p^6)$ [3].

The tree-level amplitude for $K^+ \to \pi^0 e^+ \nu \gamma$ has general structure

$$T = \frac{G_F}{\sqrt{2}} eV_{us} \varepsilon^\mu(q) \bigg\{ (V_{\mu\nu} - A_{\mu\nu}) \bar{u}(p_\nu) \gamma^\nu (1-\gamma_5) v(p_l)$$

$$+ \frac{F_\nu}{2p_l q} \bar{u}(p_\nu) \gamma^\nu (1-\gamma_5)(m_l - \hat{p}_l - \hat{q}) \gamma_\mu v(p_l) \bigg\}. \quad (1)$$

First term of the amplitude for this process describes Bremsstrahlung of kaon and direct emission. The lepton Bremsstrahlung is presented by the second part of Eq. (1).

This decay is particularly interesting, since it is sensitive to T-odd contributions. According to the CPT theorem, observing the violation of it T invariance is equivalent to the observation of CP-violating effects.

An important experimental observation used in the searches for CP violation is the T-odd correlation, for $K^- \to \pi^0 e^- \bar{\nu} \gamma$ decay it is defined as

$$\xi = \frac{1}{M_K^3} p_\gamma \cdot [p_\pi \times p_e]. \quad (2)$$

To establish the presence of a nonzero triple-product correlations, one constructs a T-odd asymmetry of the form

$$A_\xi = \frac{N_+ - N_-}{N_+ + N_-}, \quad (3)$$

where N_+ and N_- are numbers of events with $\xi > 0$ and $\xi < 0$.

---

[a] E-mail: polyarush@inr.ac

2 OKA setup

OKA collaboration works on IHEP Protvino proton synchrotron U-70 of NRC "Kurchatov Institute"-IHEP, Protvino. OKA detector (see Fig. 1) is located in positive RF-separated beam with 12.5% of K-meson. The detailed description of the OKA detector is given in our previous publications [4, 5].

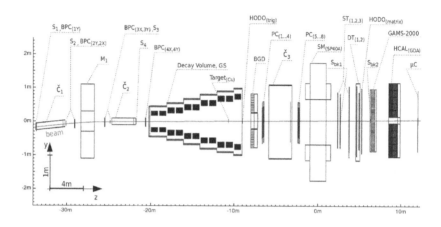

Figure 1: Layout of the OKA detector.

3 $K^+ \to \pi^0 e^+ \nu \gamma$ events selection and background suppression

A study of the $K^+ \to \pi^0 e^+ \nu \gamma$ decay is done with the data set accumulated in November 2012 run with a 17.7 GeV/c beam momentum. Monte-Carlo simulation based on the Geant3 package [6] includes a realistic description of the experimental setup.

To select $K^+ \to \pi^0 e^+ \nu \gamma$ decay channel a set of requirements is applied:

1) One positive charged track detected in tracking system and 4 showers detected in electromagnetic calorimeters GAMS-2000 and EGS.

2) One shower must be associated with charged track.

3) Charged track identified as positron. Identification of an electron is done using the ratio of the shower energy to the momentum of the connected charged track. The particles with $0.8 < E/p < 1.2$ are accepted as electrons.

4) The vertex is located inside the volume of decay.

5) The effective mass $M_{\gamma\gamma}$ for one $\gamma\gamma$-pair is $0.12 < M_{\gamma\gamma} < 0.15$ GeV.

Requires no signals in the veto system above the noise threshold. To suppress the background channels, we used a set of kinematic cuts.

After applying all the cuts, 13118 events were selected, with a background of 1628 events. Background normalization was done by comparison numbers of events for K_{e3} decay in MC and real data samples.

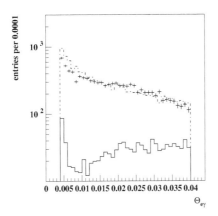

Figure 2: Distribution over $\Theta_{e\gamma}$ — the angle between electron and photon in lab. system. Real data (points with errors), MC background (solid line histogram) and signal plus MC background (dashed line histogram).

4 Results

Distribution over $\Theta_{e\gamma}$ — the angle between electron and photon in laboratory system (see Fig. 2). Reasonable agreement of the data with MC is seen. When generating the signal MC, a generator based on $O(p^2)$ [7] is used.

To obtain the branching ratio of the $K_{\pi^0 e^+ \nu_e \gamma}$ relative to the K_{e3} (R), the background and efficiency corrected number of $K_{e3\gamma}$ events is compared to that of 2812875 K_{e3} events found with the similar selection criteria. Further, the branching ratio with cuts $E_\gamma^* > 10$ MeV, $0.6 < cos\Theta_{e\gamma}^* < 0.9$, chosen for comparability with the previous experiments is calculated.

$$R = \frac{Br(K^+ \to \pi^0 e^+ \nu_e \gamma)}{Br(K^+ \to \pi^0 e^+ \nu_e)} = (0.59 \pm 0.02(stat.) \pm 0.03(syst.)) \times 10^{-2} \quad (4)$$

Systematic errors are estimated by variation of the our cuts and using two different ways of backgrounds normalization.

Statistics more than doubles compared to the previous measurement.
For the asymmetry A_ξ for the same cuts we get

$$A_\xi = -0.019 \pm 0.020(stat.) \pm 0.027(syst.) \quad (5)$$

References

[1] Bijnens J, Echer G and Gasser J Nucl. Phys. B **396** (1993) 81.
[2] Pitch A Rep. Prog. Phys. **58** (1995) 563.
[3] Kubis B.E. *et al.* Eur. Phys. J. **C50** (2007) 557.
[4] Sadovsky AS *et al.* 2017 arXiv:1709.01473 [hep-ex].
[5] Yushchenko O P *et al.* 2017 arXiv:1708.09587 [hep-ex].
[6] Brun R *et al* 1984 Preprint CERN-DD/EE/84-1.
[7] Braguta V.V., Likhoded A A and Chalov A.E Phys. Rev. D **65** (2002) 054038 [hep-ph/0106147].
[8] Ajinenko I V*et al* Yad. Fiz **70** (2007) 2125.
[9] Barmin V V*et al* SJNP **53** (1991) 606.
[10] Bolotov V N *et al* JETP Lett. **42** (1986) 68, Yad. Fiz. **44** (1986) 108.
[11] Romano F *et al* Phys Lett. **36B** (1971) 525.

Hadron Physics

SELECTED RESULTS FROM THE ATLAS EXPERIMENT ON ITS 25th ANNIVERSARY

Farès Djama [a] on behalf of the ATLAS Collaboration
Aix-Marseille Univ, CNRS/IN2P3, CPPM, Marseille, France

Abstract. The Lomonosov Conference and the ATLAS Collaboration celebrated their 25th anniversaries at a few weeks interval. This gave us the opportunity to present a brief history of ATLAS and to discuss some of its more important results.

1 The ATLAS History

The Letter of Intent (LoI) of the ATLAS Collaboration was submitted to the LHCC committee on the 1st of October 1992 [1]. This date has been retained as the date of birth of the ATLAS Collaboration. The 18th Lomonosov Conference was held few weeks before the 25th anniversary of ATLAS, giving us the opportunity to look backward to the ATLAS history and recall a selection of its most important results, while ATLAS is still collecting data during the LHC Run 2. In 1992, the landscape of particle physics was very different from the one of today. Two fermions were still to be observed (top quark and τ-neutrino), CP violation in B mesons and oscillations of neutrinos were not discovered yet, and the Higgs boson was still an hypothesis.

After the LoI acceptation, the ATLAS Technical Proposal was presented in 1994 [2]. ATLAS Experiment was approved in 1996 and the construction of its subdetectors started in 1999. Subdetectors R&D efforts started even before 1992, by testing new ideas and validating them in dedicated collaborations, before adapting them to ATLAS.

The installation in the ATLAS cavern started in 2004 and lasted until 2008. ATLAS recorded its first LHC collision in 2009. During the Run 1 of the LHC, ATLAS recorded 50 pb^{-1} in 2010 and 5 fb^{-1} in 2011, at a centre of mass energy of $\sqrt{s} = 7$ TeV and 21 fb^{-1} in 2012 at $\sqrt{s} = 8$ TeV. The main improvement to ATLAS during the first long LHC shutdown (LS1) in 2013-2014 was the upgrade of its pixel detector with a fourth innermost layer. During Run 2, ATLAS collected 87 fb^{-1} in 2015, 2016 and 2017 at $\sqrt{s} = 13$ TeV.

2 The ATLAS Detector

The ATLAS Detector [3] is made of three main systems: The Inner Tracker called Inner Detector (ID), the calorimetry system and the muon spectrometer.

The innermost components of the ID, made of silicon detectors (four pixel layers followed by eight microstrip layers) deliver high resolution hits for precise tracking and reconstruction of vertices. The outer part is made of thin gazeous

---

[a]E-mail: djama@cppm.in2p3.fr

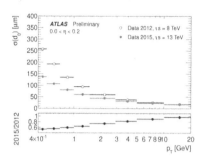
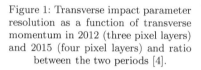
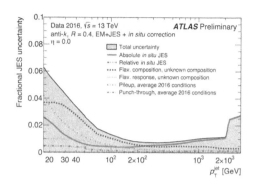

Figure 1: Transverse impact parameter resolution as a function of transverse momentum in 2012 (three pixel layers) and 2015 (four pixel layers) and ratio between the two periods [4].

Figure 2: Fractional jet energy systematic uncertainty as a function of transverse momentum for jets at $\eta = 0$ [6].

straws which provides continuous tracking. The ID is immersed in a 2 T uniform magnetic field created by a supraconductor solenoid magnet for momentum measurements. Precise tracking covers pseudorapidities up to $|\eta| = 2.5$. Figure 1 [4] shows the resolution on transverse impact parameter as a function of the transverse momentum, for Run 1 and Run 2 after the insertion of the fourth innermost layer during LS1 [5].

Electrons and photons are measured by the liquid argon/lead electromagnetic calorimeter. It uses accordion-shaped absorbers and electrodes, allowing an excellent hermeticity and an easy implementation of fine granularity and longitudinal samplings. It is housed in three cryostats (one barrel and two encaps, the barrel one containing also the solenoid magnet). π^0 mesons can be discriminated from photons thanks to the very fine granularity of the first sampling up to $|\eta| = 2.4$, while the acceptance goes up to $|\eta| = 3.2$.

Jets are reconstructed using the ID and the whole calorimeter system, including the hadronic component (scintillators/iron in the barrel, liquid argon/copper in the endcaps, and liquid argon/copper or tungstene in the forward). Forward calorimetry covers pseudorapidities between 3.2 and 4.9, while the two first techniques cover lower values of $|\eta|$. Jet energy is calibrated using both simulation and data-driven methods. Systematic error on the jet energy scale is one of the most important for various measurements and searches. It is shown in Fig. 2 [6] as a function of the jet transverse momentum.

Missing transverse energy is computed by assuming transverse energy conservation using all reconstructed objects without double-counting between ID, calorimeters and muon spectrometer.

The muon spectrometer is based on three air-core toroid supraconductor magnets (one in the barrel and two in the endcaps, each having 8 toroids housed in cryostats). Toroids are instrumented with various kinds of muon chambers, for triggering and tracking, up to $|\eta| = 2.7$. Momentum measurement of muons benefits from both ID and muon spectrometer.

Identification of b-jets (b tagging) is used in a large fraction of physics analysis (top physics, SUSY searches...). b-jets are tagged by exploiting the long lifetime of B hadrons, which gives large impact parameters and secondary vertices. Thus it relies on the performance of the silicon pixel detector. After the insertion of the fourth innermost layer, the light jet rejection has been improved by a factor 4 for the same b-jet efficiency.

3 Selected Results

The discovery of the Higgs boson has been the main goal for LHC physics program. The general purpose detectors at LHC (ATLAS and CMS) were optimised to achieve this goal over all the theoretically allowed mass region, using different channels and reconstructed objects. Simulations of the discovery potential of the Higgs boson were done at each step of ATLAS before the start of LHC. ATLAS (and CMS) discovered the Higgs boson in 2012, using 2011 data at $\sqrt{s} = 7$ TeV and part of 2012 data at $\sqrt{s} = 8$ TeV, resulting in a total integrated luminosity of 10.7 fb^{-1} [7]. The discovery channels were the Higgs boson decay into bosons $H \to \gamma\gamma$, $H \to ZZ^* \to 4l$ and $H \to WW^* \to e\nu\mu\nu$ where 4l stands for two electron pairs, two muon pairs or a pair of each. It is worth noticing that the obtained local significance (5.9 standard deviations at $m_H = 125$ GeV) was not far from the result of the last simulation campaign in 2008 [8], for a Higgs boson mass of 120 GeV and 10 fb^{-1} at $\sqrt{s} = 14$ TeV. Since then, results on two fermionic decays have been published: The observation of $H \to \tau^+\tau^-$ in 2015 [9] and an evidence for $H \to b\bar{b}$ in 2017 [10]. Figure 3 shows the $b\bar{b}$ invariant mass, where the Higgs boson contribution can be clearly seen besides the Z peak.

Since its discovery, measurements of the Higgs boson properties became one of the main LHC topics. ATLAS demonstrated the scalar nature of the discovered particle using the angular distributions of its four lepton decays [11]. The Higgs boson mass has been measured with a 0.2 % precision. More details are given in these proceedings [12].

Studying the top quark is the other important topic at LHC. LHC proton-proton collisions are a copious source of top quarks. It is the heaviest known elementary particle, and it decays before hadronisation, transmitting its properties to its daughter particles, making the properties measurable. The weak production of single top has been measured by ATLAS at all delivered centre of mass energies. Such measurements are sensitive to specific electroweak and QCD observables. In 2017, ATLAS observed the first production of single top associated with a Z boson [13], as it is shown on Fig. 4. This channel features the tZ coupling and is a background for the search for tH production. Details can be found in these proceedings [14].

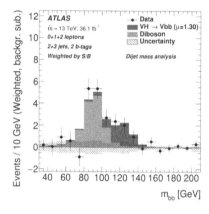 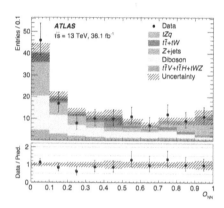

Figure 3: Distribution of bb invariant mass after subtraction of all backgrounds except WZ and ZZ processes [10].

Figure 4: Neural network output after the fit of components contributions. The contribution of tZ production is visible [13].

The recent measurement of the W boson mass m_W by ATLAS [15] is a nice example of electroweak measurements at LHC. A competitive value of m_W has been obtained using $W \to e\nu$ and $W \to \mu\nu$ decays, by fitting the charged lepton transverse momentum and W boson transverse mass. Figure 5 shows the ATLAS measurement together with LEP and Tevatron measurements. Combined with Higgs boson and quark top masses, no deviation from Standard Model predictions is found, as shown in Fig. 6.

Production of W and Z bosons is used to investigate QCD (e.g. multijets production [16]) and electroweak physics (e.g triboson production [17] [18]). More can be found in [19] in these proceedings. Here we show the cross section of $Z + 1$ jet as a function of the jet transverse momentum (Fig. 7) and the cross section of $Z +$ jets production as a function of exclusive jet multiplicity (Fig. 8) [16]. We can see that NLO generators reproduce the jet transverse momentum distribution within uncertainties, providing a test for QCD scaling, while the number of jets produced with a Z boson is well reproduced up to 3 jets, beyond which the fraction of jets produced by parton shower becomes non negligible.

The high energy provided by LHC enables to look for new particles at unprecedented high masses. These searches concern not only particles predicted by the most popular theory beyond the Standard Model, namely SUSY ([20] in these proceedings), but all kind of heavy objects, like illustrated in Fig. 9 which shows two photons invariant mass up to 2.5 TeV, and the absence of any peak [21]. ATLAS searches for dark matter [22] and exotic Higgs boson decays [23] are also summarized in these proceedings. No supersymetric nor any other new particle has been observed so far.

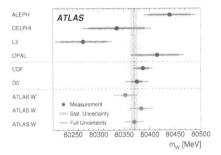
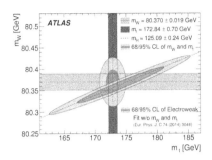

Figure 5: ATLAS m_W measurements compared to LEP and Tevatron results [15].

Figure 6: Comparison between m_W and m_t measurements by ATLAS and a global electroweak fit of both masses using the LHC measurement of the Higgs boson mass [15].

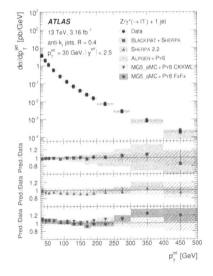
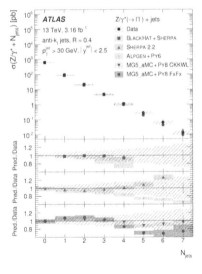

Figure 7: Cross section of $Z + 1$ jet production as a function of the jet transverse momentum, and predictions/data ratio [16].

Figure 8: Cross section of $Z + N$ jets production as a function of N, and predictions/data ratio [16].

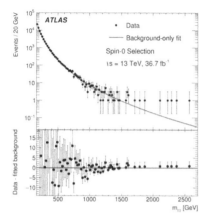
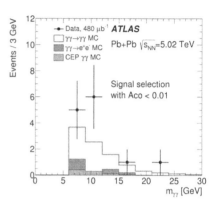

Figure 9: Diphoton invariant mass for the spin-0 hypothesis, with the background-only fit[21].

Figure 10: Diphoton invariant mass after applying the accoplanarity Aco < 0.01 cut [24].

ATLAS pursues also a program of heavy ions physics (lead-lead collisions at $\sqrt{s_{NN}}$ up to 8.2 TeV) where the strong interaction is scrutinized in its quark-gluon plasma regime and its collective sector. Measurement of light by light scattering $\gamma\gamma \to \gamma\gamma$ has been achieved recently [24]. The $\gamma\gamma\gamma\gamma$ vertex is forbidden in the Standard Model, but it occurs via box diagram. The process is rare ($\propto \alpha_{EM}^4$), but the two photons rate in the initial state is enhanced by the atomic number of lead nucleus ($\propto Z^2, Z = 82$). Figure 10 shows the observation of this process.

4 Conclusion and Future

The ATLAS Collaboration is exploiting the LHC potential to extend our knowledge on elementary particles and their interactions. Next steps in the Higgs sector will be the improvement of its measured couplings and measuring $t\bar{t}H$ coupling and Higgs boson self-coupling. Such measurements will validate the mass generation mechanism in the Standard Model. Searches for new particles are ongoing. Top quark, electroweak and QCD measurements are constraining the Standard Model and could give indirect access to new physics in near future.

The immediate future will be the end of the LHC Run 2 by the end of 2018, with an accumulated data sample which will certainly be higher than the expected 120 fb[-1]. During the two years of the following shutdown (LS2), ATLAS will upgrade its calorimetric trigger and replace the inner endcap muon chambers. During the three foreseen years of Run 3 (2021–2023), ATLAS expects 150 fb^{-1} at $\sqrt{s} = 14$ TeV. LS3 will then be devoted to prepare the High Luminosity LHC (HL-LHC) by installing a new tracker, replacing the readout electronics of calorimeters and muon system and upgrading trigger and data acquisition. The HL-LHC era will start in 2026. About 3000–4000 fb^{-1} are expected, which will allow new measurements like Higgs boson self-coupling and rare decays and will extend search limits.

References

[1] ATLAS Collaboration, CERN-LHCC-92-004, 1992.
[2] ATLAS Collaboration, CERN-LHCC-94-43, 1994.
[3] ATLAS Collaboration, JINST 3 S08003, 2008.
[4] ATLAS Collaboration, https://atlas.web.cern.ch/Atlas/GROUPS/PHYSICS/PLOTS/IDTR-2015-007.
[5] ATLAS Collaboration, CERN-LHCC-2010-013, 2010.
[6] ATLAS Collaboration, https://atlas.web.cern.ch/Atlas/GROUPS/PHYSICS/PLOTS/JETM-2017-003.
[7] ATLAS Collaboration, Phys. Lett. B 716 (2012)1-29.
[8] ATLAS Collaboration, CERN-OPEN-2008-20, 2008.
[9] ATLAS Collaboration, JHEP 04 (2015)117.
[10] ATLAS Collaboration, JHEP 12 (2017)24.
[11] ATLAS Collaboration, Eur. Phys. J. C 75 (2015)476.
[12] D. Boumédiene, *Higgs Precision Measurements with ATLAS*, these proceedings.
[13] ATLAS Collaboration, arXiv:1710.03659, 2017.
[14] F. Fabbri, *Top pair and single top production in ATLAS*, these proceedings.
[15] ATLAS Collaboration, arXiv:1701.07240, 2017.
[16] ATLAS Collaboration, Eur. Phys. J. C 77 (2017)361.
[17] ATLAS Collaboration, Eur. Phys. J. C 77 (2017)141.
[18] ATLAS Collaboration, Eur. Phys. J. C 77 (2017)646.
[19] E. Soldatov, *Electroweak precision measurements in ATLAS*, these proceedings.
[20] M. Ronzani, *Searches for supersymmetry with the ATLAS detector*, these proceedings.
[21] ATLAS Collaboration, Phys. Lett. B 775 (2017)105.
[22] P. Czodrowski, *Constraining Dark Matter with ATLAS*, these proceedings.
[23] F. Prokoshin, *Searches for rare and exotic Higgs decays with ATLAS*, these proceedings.
[24] ATLAS Collaboration, Nature Physics 13 (2017)852-858.

REVIEW OF THE RECENT TEVATRON RESULTS

Sergey Denisov [a]

Institute for High Energy Physics of the National Research Centre Kurchatov Institute, 142281, Moscow region, Protvino, pl.Nauki, 1

Abstract. New results obtained by D0 and CDF Collaborations on the top quark and W boson masses, weak mixing angle, and forward-backward asymmetry in $t\bar{t}$ production are presented. These results are used to test CPT invariance, electroweak vacuum stability, and Standard Model self-consistency and predictions. Recent results of D0 Collaboration on the new narrow $X(5568)$ exotic state are discussed.

1 Introduction

The Tevatron is a proton-antiproton collider located at Fermi National Accelerator Laboratory (Batavia, USA). During two long physics runs in 1988–1996 and 2001–2011, two general purpose detectors CDF and D0 collected 10 fb^{-1} of integrated luminosity each. The center-of-mass energy was 1.8 TeV in the Run I and 1.96 TeV in the Run II. The maximum instantaneous luminosity reached in the Run II is equal to 4.3×10^{32} cm^{-2}s^{-1}.

In Section 2, the top-quark mass (m_t) measurements with high precision are discussed and new results of the top-quark mass obtained at the Tevatron and LHC are presented. The 3-rd section deals with CPT-invariance test based on $m_t - m_{\bar{t}}$ mass difference. The stability of the electroweak (EW) vacuum and Standard Model (SM) self-consistency are discussed in the next two sections. New results on the search for exotic particles and on the measurements of forward-backward asymmetry in $t\bar{t}$ production are presented in the sections 6 and 7. A short conclusion is given in the final section.

2 Top-Quark Mass Measurements and Results

The top quark was discovered in 1995 by CDF and D0 collaborations with top quark values of 176±13 GeV (CDF) and 199±30 GeV (D0) [1,2]. Since that time dozens of m_t measurements were performed at the Tevatron and later at the LHC and now we know its value with uncertainty about two orders of magnitude less and much better than for other quarks. Two questions arise: Why do we need a precision value of the top quark mass? Why is it possible to measure m_t with high precision?

There are several arguments for high precision measurements of the top quark mass. Five of them are:

- m_t is a fundamental physical constant (SM does not predict quark masses);
- it allows one to perform CPT-invariance test in the quark sector;

---

[a] E-mail: denisov@ihep.ru

- along with Higgs boson mass provides information on EW vacuum stability;
- it is used to test SM self-consistency via loop corrections to the W mass;
- precise m_t value is important for the background estimates in many new physics searches.

One of the unique top quark properties is short lifetime due to heavy mass. SM predicts $\tau_t \sim 5 \times 10^{-25}$s in agreement with the Tevatron and LHC results [3–5]. This time is much shorter than the time ($\sim 2 \times 10^{-24}$s) required for hadronization and hence top quark decays as a free particle before forming a bound state. This allows one to measure top quark properties:

- directly and hence with much less relative uncertainties than for the lighter quarks characteristics extracted from the parameters of their bound states;
- independently for t and $\bar{t}$ quarks.

Thus the precision m_t measurement is not only important but also achievable.

There are several methods of m_t measurements and corresponding m_t definitions [6]. For example, pole-mass m^{pole} appears in the top quark propagator $1/[p^2 - (m^{pole})^2]$ and can be extracted from the mass dependence of the $t\bar{t}$ production cross-sections (see below). But the most precise top mass results come from the analysis of events with reconstructed top quark decays to Wb with "all jets" (46%), lepton+jets (45%), dileptons (9%) combinations in the final state [7]. Each combination has its advantages and disadvantages for the top quark mass measurements. For example, the highest statistics is available for the "all jets" events but QCD backgrounds are high in this case. Dilepton e, μ- events have low background but their statistics is about an order of magnitude less than that for "all jets". The best m_t values obtained by the CDF and D0 collaborations for different final states are shown in the Table 1. Table 2 presents the combined m_t results.

The top quark pole-masses extracted by the D0 Collaboration from inclusive $t\bar{t}$ production cross-section is equal to $172^{+1.3}_{-0.2}(exp.) \pm 1.1(theor.)$ GeV [18]. The LHC m^{pole} results are the following:
ATLAS: $173.2 \pm 0.9(stat.) \pm 0.8(syst.) \pm 1.2(theor.)$ GeV [19]; CMS: 170.6 ± 2.7 GeV [20].

From the above results, one can conclude:

- the results obtained with different combinations in the final state of the top quark decay are in good agreement,
- the Tevatron and LHC results are in agreement within the errors quoted,

Table 1:

Experiment	Final states	m_t, GeV	Errors		Ref.
			Stat.	Syst.	
CDF	dilepton	171.5	± 1.9	± 2.5	[8]
D0	dilepton	173.50	± 1.31	± 0.84	[9]
CDF	lepton+jets	172.85	± 0.71	± 0.85	[10]
D0	lepton+jets	174.98	± 0.41	± 0.63	[11]
CDF	all jets	175.07	± 1.19	± 1.55	[12]
CDF	MET+jets	173.93	± 1.28	± 1.35	[13]

Table 2:

Combination	Year	m_t, GeV	Errors		Ref.
			Stat.	Syst.	
Tevatron	2016	174.30	± 0.35	± 0.54	[14]
ATLAS	2017	172.51	± 0.27	± 0.42	[15]
CMS	2016	172.44	± 0.13	± 0.47	[16]
Tevatron+LHC	2014	173.34	± 0.27	± 0.71	[17]

- the uncertainties of the Tevatron and LHC measurements are comparable and are mainly due to systematics,

- the total relative errors are about 0.3%, that is much less than for the other quarks,

- within the errors there is no difference between m^{pole} and m_t (according to the recent theoretical studies this difference is less than ∼0.5 GeV).

3 CPT Theorem Test

The fundamental CPT theorem based on the general principles of local relativistic quantum field theory predicts that particle and antiparticle masses must be the same. The CPT symmetry is rigorously conserved in the SM and it was checked with high accuracy for $K_0 - \bar{K}_0$ system: $m(K_0) - m(\bar{K}_0)/average < 6 \times 10^{-19}$ at 90%CL [7]. But some SM extensions permit CPT invariance violation.

The CPT invariance test at the quark level is possible only for the top quarks where t and $\bar{t}$ masses can be measured directly and independently. The first $\Delta m_{t\bar{t}} = m_t - m_{\bar{t}}$ measurements were performed at the Tevatron. To discriminate t against $\bar{t}$ the charge of lepton in $e/\mu+jets$ events was used. The obtained results are the following:

- D0(2011) [21]: 0.80±1.8(stat.)±0.5(syst.) GeV;

- CDF (2013) [22]: -1.95±1.11(stat.)±0.59(syst.) GeV.

Much more precise $\Delta m_{t\bar{t}}$ values were obtained recently at the LHC:

- CMS (2017) [23]: -0.15±0.19(stat.)±0.09(syst.) GeV;
- ATLAS (2017) [24]: -0.67±0.61(stat.)±0.41(syst.) GeV.

Thus CPT holds in the quark sector at the level of $\Delta m_{t\bar{t}}/m_t \sim 10^{-3}$.

4 EW Vacuum Stability Test

The top quark and Higgs boson masses provide information on the EW vacuum stability. As can be seen from Fig.1 experimental data point to the meta-stable vacuum with > 99% CL but the hypothesis that the vacuum is stable and the SM works all the way up to the Planck scale cannot be rejected. Much better precision of the masses (first of all of the top quark mass) is needed for the definite conclusion (> 5σ). That is unlikely possible with existing colliders, but certainly may be achieved at the future lepton colliders.

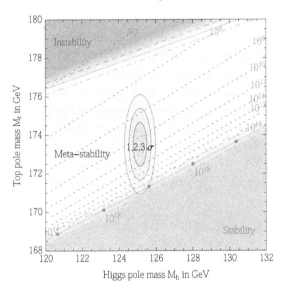

Figure 1: Regions of stability, meta-stability, and instability of the SM vacuum; μ is a renormgroup energy scale (RGE)[25].

5 SM Self-Consistency

There are 6 fundamental EW parameters: $\alpha, G_F/(\hbar c)^3, M_H, M_Z, M_W$ and $\sin^2\theta_W$. First four of them are known from Rydberg constant, muon lifetime and LEP and LHC measurements [7]. Highest precision M_W measurements are performed at the Tevatron. The combined result is $M_W = 80387 \pm 16$ MeV [26] (the world average is equal to 80379 ± 12 MeV [7]).The combined D0 result on the effective weak mixing angle parameter is $\sin^2 \theta_{eff} = 0.23095 \pm 0.00040$ [27]. EW parameters are not independent but related through SM equations:

$$\sin^2 \theta_W = 1 - \frac{M_W^2}{M_Z^2}, M_W^2 \sin^2 \theta_W = \frac{\pi \alpha}{\sqrt{2} G_F}. \quad (1)$$

This allows one to check SM self-consistency. The results are shown in Fig. 2. The horizontal and vertical green bands present the experimental $\sin^2 \theta^l_{eff}$ and M_W values $\pm 1\sigma$. Green ovals show the areas of $\sin^2 \theta^l_{eff}$ and M_W with 1σ and 2σ CL. Contours of blue, yellow and grey areas indicate 1σ and 2σ boundaries for $\sin^2 \theta^l_{eff}$ and M_W obtained from the fit to equations (1) with and without M_H and Z boson width (Γ_Z) measurements. As can be seen from Fig. 2 all areas overlap each other and therefore there is no evidence for the SM non-consistency.

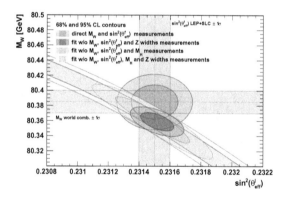

Figure 2: Results of the global fit of SM parameters to relations (1) [28].

Self-consistency of SM can be also tested using the dependence of M_W on mt and M_H via loop corrections:

$$M_W^2 (1 - \frac{M_W^2}{M_Z^2}) = \frac{\pi \alpha}{\sqrt{2} G_F} (1 + \Delta r), \quad (2)$$

where $\Delta r(m_t^2, \ln M_H)$ reflects the loop corrections. The results of the global fit of the precision electroweak data to this relation are presented in Fig. 3 [28]. The vertical and horizontal green belts indicate the $\pm \sigma$ regions for the m_t and M_W direct measurements. The blue and grey areas show 1σ and 2σ regions allowed for m_t and M_W masses, derived from the fit. They correspond to cases when measurements of the Higgs boson mass are included (blue) or excluded (grey) from the fit. The allowed regions coincide well with the green areas indicating $\pm\sigma$ regions for the m_t and M_W experimental values confirming SM self-consistency.

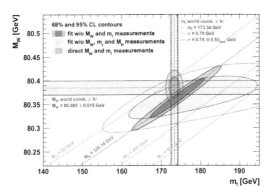

Figure 3: The results of the global fit of electroweak data to relation (2) [28].

6 Searches for Exotic States

In 2016, the D0 Collaboration reported observation of a new narrow $X(5568)$ state, potentially consisting of the b, s, u, and d quarks and decaying into $B_s^0 \pi^\pm$ with $B_s^0 \to J/\psi \phi, J/\psi \to \mu^+\mu^-, \phi \to K^+K^-$ [29]. Thus there are five charged stable particles in the final state as shown in Fig. 4. The following cuts were applied to minimize the background-to-signal ratio:

- two oppositely charged particles identified as muons have $p_T > 1.5$ GeV/c and invariant mass in the range from 2.92 to 3.25 GeV consistent with J/ψ mass;

- two oppositely charged particles assumed to be kaons have $p_T > 0.7$ GeV/c and invariant mass in the range from 1.012 to 1.030 GeV consistent with ϕ mass;

- the fifth charged particle has $p_T > 0.7$ GeV/c and assumed to be a pion;

- $\Delta R = \sqrt{\Delta \eta^2 + \Delta \phi^2} < 0.3$, where $\Delta \eta$ and $\Delta \phi$ are pseudorapidity and azimuthal angle intervals between π and B_s^0 trajectories.

Figure 5 shows the $B_s^0 \pi^\pm$ invariant mass spectrum fitted with the sum of signal and background functions. The signal function is represented by convolution of a relativistic Breit-Wigner function with M_x and Γ_x as free parameters and a Gaussian detector resolution function. The background is well described by the function $F_{bgr}(m_0) = P_4 exp(P_2)$, where $m_0 = m(B_s^0 \pi^\pm) - 5.5$ GeV , P_2 is a second-order polynomial and P_4 is a fourth-order polynomial with a linear term equal to zero. The fit yields the following results:

- $M_x = 5567.8 \pm 2.9(stat.)^{+0.9}_{-1.6}(syst.)$ MeV;

- $\Gamma_x = 21.9 \pm 6.4(stat.)^{+5.0}_{-2.5}(syst.)$ MeV;

- $N_x = 133 \pm 31(stat.) \pm 15(syst.)$,

where N_x is the number of signal events. The ratio $\rho = \sigma(X5568 \to B_s\pi)/\sigma(B_s)$ is measured to be $[8.6 \pm 1.9(stat.) \pm 1.4(syst.)]\%$. The global significance of the signal including Look Elsewhere Effect [30] and systematic uncertainties is estimated to be 5.1σ. The fitted parameters weekly depends on the ΔR cut but without this cut the global significance is reduced to 3.9σ.

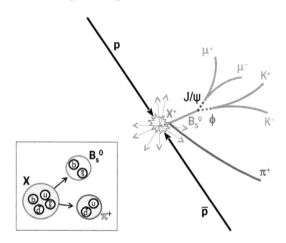

Figure 4: $X(5568)$ decay.

Subsequent analyses performed by LHCb [31] and CMS [32] collaborations in 2016 have not confirmed the existence of the $X(5568)$ in pp interactions at $\sqrt{s}$=7 and 8 TeV. In particular the upper limits of ρ parameter appeared to be equal to 2.4% (LHCb) and 3.9% (CMS) at 95% CL.

In 2017, the D0 collaboration performed a new search for $X(5568)$ using semileptonic B_s^0 decays: $B_s^0 \to D_s^\pm \mu^\pm \nu$, $D_s^\pm \to \phi(1020)\pi^\pm$, $\phi(1020) \to K^+K^-$ [33]. The $m(B_s^0 \pi^\pm)$ distribution for the data is shown in Fig.6 together with the fit results. The fit yields the mass and width of $M_x = 5566.7^{+3.6}_{-3.4}(stat.)^{+1.0}_{-1.0}(syst.)$ MeV, $\Gamma_x = 6.0^{+9.5}_{-6.0}(stat.)^{+1.9}_{-4.6}(syst.)$ MeV and the number of signal events of $N = 139^{+51}_{-63}(stat.)^{+11}_{-32}(syst.)$, which are compatible within the errors with the

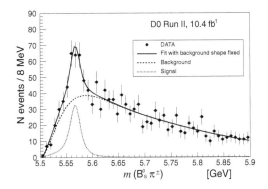

Figure 5: $B_s^0 \pi^\pm$ mass spectra for $B_s^0 \to J/\psi \, \phi$ decays with $\Delta R < 0.3$ cut.

results from the hadronic channel. The local statistical significance from the fit is 4.5σ, the global statistical significance, taking into account the systematic uncertainties, is 3.2σ. The production ratio ρ is measured to be $[7.3^{+2.8}_{-2.4}(stat.)^{+0.6}_{-1.7}(syst.)]\%$ in the semileptonic channel which is in agreement with the hadronic channel. The combined significance for semileptonic and hadronic channels, obtained under the assumptions, that the same object is observed in both channels and the semileptonic and hadronic measurements are independent, is 5.7σ.

There are no reasonable explanations why $X(5568)$ is seen in two different decay modes in $p\bar{p}$ interactions at 2 TeV and not in pp collisions at 7 and 8 TeV. Thus the question about $X(5568)$ existence remains open.

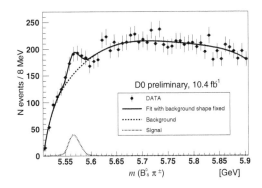

Figure 6: $B_s^0 \pi^\pm$ mass spectra for semileptonic B_s^0 decays.

7 Forward-Backward Asymmetry in $t\bar{t}$ Production

Forward-Backward (FB) asymmetry in $t\bar{t}$ production in the $p\bar{p}$ collisions answers the question: does the top quark prefer the proton direction or the opposite? FB-asymmetry is defined by:

$$A_{FB}^{t\bar{t}} = \frac{N(\Delta y > 0) - N(\Delta y < 0)}{N(\Delta y > 0) + N(\Delta y < 0)}, \quad (3)$$

where $\Delta y = y_t - y_{\bar{t}}$ is the rapidity difference between t and $\bar{t}$ quarks and N is a number of events with Δy above or below zero. FB-asymmetry can also be measured using one or two leptons from top quark decays:

$$A_{FB}^{l} = \frac{N(q_l\eta_l > 0) - N(q_l\eta_l < 0)}{N(q_l\eta_l > 0) + N(q_l\eta_l < 0)}, A_{FB}^{ll} = \frac{N(\Delta\eta > 0) - N(\Delta\eta < 0)}{N(\Delta\eta > 0) + N(\Delta\eta < 0)}, \quad (4)$$

where $\Delta\eta = \eta_{l^+} - \eta_{l^-}$ is the pseudorapidity difference between two leptons and q_l is the sign of the lepton electric charge. QCD does not predict any asymmetry at the leading order. It arises due to high order corrections. Thus FB-asymmetry is a precision probe of SM predictions in the top quark sector.

The results of the first FB-asymmetry measurements at Tevatron showed a deviation from existing next-to-leading order (NLO) QCD predictions by more than 3σ [34,35] . These results stimulated both more precise experimental measurements of $A_{FB}^{t\bar{t}}$ [36–39], A_{FB}^{l} [40–43] and A_{FB}^{ll} [42,43] and more accurate theoretical calculation including next-to-next-to-leading (NNLO) order [44–46]. As a result now there is no contradiction between D0 and CDF measurements and theory. For example, the combined CDF and D0 $A_{FB}^{t\bar{t}}$ value of $0.128 \pm 0.021(stat.) \pm 0.014(syst.)$ [47] is consistent with NNLO QCD + NLO EW prediction of 0.095 ± 0.007 [44] within 1.3σ.

8 Conclusions

Many new important results were obtained by CDF and D0 collaborations recently. Among them the most precise measurements of the top quark and W boson masses: $m_t = 174.30 \pm 0.35(stat.) \pm 0.54(syst.)$ GeV, $M_W = 80387 \pm 16$ MeV. Better precision of these masses is needed for the definite conclusion about SM vacuum stability and SM self-consistency. The significance of the observation of a new exotic state $X(5568)$ is 5.7σ and the corresponding p-value is 5.6×10^{-9}. As LHCb and CMS experiments do not see this state the question about its existence and nature remains open. New D0 and CDF results of the forward-backward asymmetry studies are consistent with the recent NNLO QCD + NLO EW predictions.

Acknowledgments

I gratefully acknowledge the help of T.Z. Gurova, I.A. Razumov, A.A. Shchukin and D.A. Stoyanova in preparation of this manuscript.

References

[1] F. Abe et al. (CDF Collaboration), Phys. Rev. Lett. **74** (1995) 2626.
[2] S. Abachi et al. (D0 Collaboration), Phys. Rev. Lett. **74** (1995) 2632.
[3] V. M. Abazov et al. (D0 Collaboration), Phys. Rev. **D 85**, 091104(R) (2012).
[4] V. Khatcharyan et al. (CMS Collaboration), Phys. Lett. **B 736** (2016) 33.
[5] M. Aaboud (ATLAS Collaboration), arXiv:1709.04207.
[6] E. E. Boos, O. E. Brandt, D. S. Denisov, S. P. Denisov, P. D. Grannis, Physics - Uspekhi **58** (2015) (12) 1133 - 1158.
[7] http://pdg.lbl.gov/
[8] T. Aaltonen et al. (CDF Collaboration), Phys. Rev. **D 92** (2015) no.3, 032003.
[9] V. M. Abazov et al. (D0 Collaboration), Phys. Lett. **B 18** (2016) 752.
[10] T. Aaltonen et al. (CDF Collaboration), Phys. Rev. Lett. **109** (2012) no.3, 152003.
[11] V. M. Abazov et al. (D0 Collaboration), Phys. Rev. Lett. **113** (2014) 032002.
[12] T. Aaltonen et al. (CDF Collaboration), Phys. Rev. **D 90** (2014) no.9, 091101.
[13] T. Aaltonen et al. (CDF Collaboration), Phys. Rev. **D 88** (2013) 011101.
[14] The Tevatron Electroweak Working Group (CDF and D0 Collaboration), arXiv:1608.01881 [hep-ex], 2016.
[15] ATLAS-CONF-2017-071.
[16] CMS Collaboration, Phys. Rev. **D 93** (2016) 072004.
[17] ATLAS, CDF, CMS, D0 Collaborations, arXiv:1403.4427.
[18] V. M. Abazov et al. (D0 Collaboration), Phys. Rev. **D 94** (2016) 092004.
[19] M. Aaboud, The European Physical Journal C November 2017, 77:804.
[20] CMS Collaboration, JHEP **09** (2017) 051.
[21] V. M. Abazov et al. (D0 Collaboration), Phys. Rev. **D 84** (2011) 052005.
[22] T. Aaltonen et al. (CDF Collaboration), Phys. Rev. **D 87** (2013) 052013.
[23] CMS Collaboration, PLB **770** (2017) 17.
[24] G. Aad et al. (ATLAS Collaboration), Phys. Lett. **B 728** (2014) 363.
[25] D. Buttazzo et al., CERN-PH-TH-2013-166, arXiv:1307.3536v4 (2014).
[26] T. Aaltonen et al. (CDF and D0 Collaborations), Phys. Rev. **D 88** (2013) 052018.
[27] V. M. Abazov et al. (D0 Collaboration), submitted to Phys. Rev. Lett., arXiv:1710.03951.

[28] M. Baak et al. Eur. Phys. J. **C74** (2014)3046.
[29] V. M. Abazov et al (D0), Phys. Rev. Lett. **117** (2016) 022003.
[30] E. Gross, O. Vitells, Eur. Phys. J. **C 70** (2010) 525.
[31] R. Aaij et al (LHCb), Phys. Rev. Lett. **117** (2016) 152003.
[32] CMS Collaboration (CMS), CMS-PAS-BPH-16-002 (2016).
[33] https://www-d0.fnal.gov/Run2Physics/WWW/results/prelim/B/B68/.
[34] V. M. Abazov et al. (D0 Collaboration), Phys. Rev. **D 84** (2011) 112005.
[35] T. Aaltonen et al. (CDF Collaboration), Phys. Rev. **D 83** (2011) 112003.
[36] T. Aaltonen et al. (CDF Collaboration), Phys. Rev. **D 87** (2013) 092002.
[37] V. M. Abazov et al. (D0 Collaboration), Phys. Rev. **D 90** (2014) 072011.
[38] T. Aaltonen et al. (CDF Collaboration), Phys. Rev. **D 93** (2016) 112005.
[39] V. M. Abazov et al. (D0 Collaboration), Phys. Rev. **D 92** (2015) 052007.
[40] T. Aaltonen et al. (CDF Collaboration), Phys. Rev. **D 88** (2013) 072003.
[41] V. M. Abazov et al. (D0 Collaboration), Phys. Rev. **D 90** (2014) 072001.
[42] V. M. Abazov et al. (D0 Collaboration), Phys. Rev. **D 88** (2013) 112002.
[43] T. Aaltonen et al. (CDF Collaboration), Phys. Rev. Lett. **113** (2014) 042001.
[44] M. Czakon et al. Phys. Rev. Lett. **115** (2015) 022001.
[45] M. Czakon et al. J.High Energy Physics 05 (2016) 034.
[46] M. Czakon et al., arXiv: 1705.04105 (2017).
[47] T. Aaltonen et al. (CDF and D0 Collaborations), submitted to Phys. Rev. Lett., arXiv:1709.04894 (2017).

TOP PAIR AND SINGLE TOP PRODUCTION IN ATLAS

Federica Fabbri, On behalf of the ATLAS Collaboration [a]
University of Bologna and INFN

Abstract. Measurements of inclusive and differential top-quark production cross sections in proton-proton collisions at a center of mass energy of 8 TeV and 13 TeV at the Large Hadron Collider using the ATLAS detector are presented. The inclusive measurements of top quark pair and single top quark production are compared to the best available theoretical calculations. Differential measurements of the kinematic properties of top quark events are also discussed. These measurements, including results using boosted top quarks, probe our understanding of top quark production up to the TeV scale.

1 Introduction

The physics of top quark still represents one of the most active area in the ATLAS physics program. Due to its large mass it is involved in many theories beyond the standard model (BSM) and is also the only quark decaying before hadronizing, giving direct access to the property of a "bare" quark. The large sample of top quarks collected at the Large Hadron Collider (LHC) and the high precision of the predictions of $t\bar{t}$ and single top production, that reaches the next-to-next to leading order precision (NNLO) on many observables, allow to perform stringent tests of the standard model (SM). The top quark can be produced in pairs through strong interaction or in conjunction to different flavour quarks (t-channel, s-channel) or W bosons (Wt-channel), through electroweak interaction. All the measurements presented in the following are performed on data samples collected by the ATLAS detector [1] in proton-proton collisions at a center of mass energy of $\sqrt{s} = 8$ TeV and $\sqrt{s} = 13$ TeV.

2 $t\bar{t}$ production cross section measurements

The final states of the $t\bar{t} \to WbWb$ are classified depending on the number of leptons originated by the W bosons decay. The channels with higher branching fraction are the single lepton and all-hadronic ones. The presence of leptons allow to easily trigger the events of interest.

The vast majority of the inclusive cross section (σ) measurements are based on the cut-and-count method, where from the event passing the analysis selection (N_{data}) is subtracted the expected background (N_{bkg}) and the σ is obtained as:

$$\sigma(pp \to t\bar{t}) = \frac{N_{data} - N_{bkg}}{BR \times \epsilon \times L} \quad (1)$$

---

[a] E-mail: federica.fabbri@cern.ch

where BR represents the branching ratio in the specific channel, ϵ the efficiency estimated using the Monte Carlo and L is the integrated luminosity. Alternatively, multivariate analysis can be used to improve the separation between signal and background and a likelihood fit can be employed to extract σ, constraining at the same time the systematic uncertainties included in the fit as nuisance parameters. The inclusive $\sigma_{t\bar{t}}$ measurement in the single lepton channel [2] uses both these two techniques to obtain the result with 6% of total uncertainty: $\sigma_{t\bar{t}} = 248.3 \pm 0.7(stat.) \pm 13.4(syst.) \pm 4.7(lumi.)$ pb. All the inclusive measurements show a good agreement with the SM predictions. Depending on the final state, some results can be reinterpreted to set limits on beyond-SM processes, this is the case of the $\sigma_{t\bar{t}}$ measurement in the $\tau + jets$ channel [3] at $\sqrt{s} = 8$ TeV, that sets an upper limit of 22 fb at 95% of confidence level on the production cross section of the BSM process $t \to qH \to q\tau^+\tau^-$.

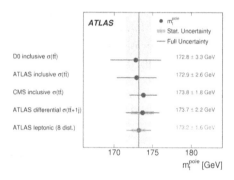

Figure 1: Comparison between the top quark pole mass measurement obtained from the $d\sigma_{t\bar{t}}$ measurement in $e\mu$ channel at $\sqrt{s} = 8$ TeV and previous measurement performed by ATLAS, DØ and CMS [4].

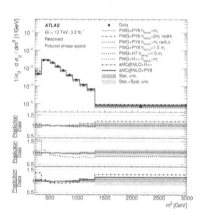

Figure 2: Relative $d\sigma_{t\bar{t}}$ in the single lepton channel, as a function of the $m^{t\bar{t}}$ [6].

With respect to the inclusive measurements, the differential ones (dσ) provide a more stringent test of the perturbative QCD, can be used for the tuning of the Monte Carlo generators and the parton distribution functions (PDF) fitting, and are also sensitive to the possible presence of new physics. The additional challenge is the ability to handle differences between the truth and reconstructed quantities, due to the limited resolution of the detector, resulting in bin-by-bin migrations of the events. In $d\sigma_{t\bar{t}}$ measurement in the $e\mu$ channel [4] at $\sqrt{s} = 8$ TeV the result is obtained as a function of the kinematic variables of the leptons present in the event, compared with the relative predictions. The measured cross section is used to extract a measurement of the top quark pole mass, through a comparison of the results with $d\sigma$ computed using various mass hypothesis. Figure 1 shows that the pole mass obtained with this technique is the most precise single measurement performed until now. The $d\sigma_{t\bar{t}}$

measurement with respect to the $t\bar{t}$ kinematic has been performed at $\sqrt{s} = 13$ TeV both in the dilepton [5] and single lepton [6] channels, as shown in Fig. 2 for the $d\sigma/dm^{t\bar{t}}$. Both analyses observed a good agreement with the standard model expectations.

The high energy and luminosity reached by the LHC allows also to explore the $t\bar{t}$ production at the TeV scale, where the top quark is produced with a high transverse momentum ($p_T > 350$ GeV) and its decay products have a large Lorentz boost and tend to be overlapped in the reconstruction (boosted regime). In case of hadronically decaying top ($t \to Wb \to qq'b$) all the products of top quark which decays before hadronization are contained in a single large-R jet, that can be identified thanks to jet substructure properties. The ATLAS

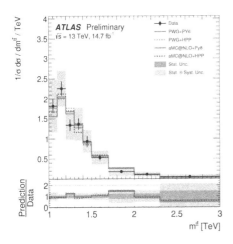
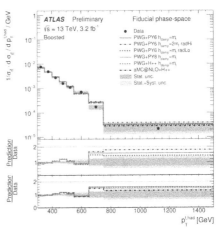

Figure 3: Relative differential $t\bar{t}$ cross section distribution in the all-hadronic channel in boosted regime, as a function of the $m^{t\bar{t}}$ [7].

Figure 4: Relative differential $t\bar{t}$ cross section distribution in the single lepton channel in boosted regime, as a function of the p_T^t [6].

collaboration performed two measurements of the $d\sigma_{t\bar{t}}$ in the boosted regime, one in the single lepton [6] and one in the all-hadronic channel [7], that represent the first cross section measurement in this channel. The results are shown in Figs. 3, 4 and compared with various theoretical predictions. Both analyses observed good agreement with the SM, with some tensions observed only in the high-p_T tails.

3 Single top production cross section measurements

The single top production cross section σ_t is smaller than $\sigma_{t\bar{t}}$, consequently multivariate techniques help in signal/background separation. A neural network

(NN), in particular, has been used in the differential cross section measurement in the t-channel performed at $\sqrt{s} = 8$ TeV [8], that shows a good agreement with the SM expectations. The analysis includes also the measurement of a variable extremely sensitive to the differences between PDFs: $R_t = \sigma_t/\sigma_{\bar{t}}$, shown in Fig. 5. The multivariate approach is used also to extract the inclusive measurement in the Wt-channel [9] $\sigma_{Wt} = 94 \pm 10(stat.)^{+28}_{-22}(syst.) \pm 2(lumi.)$ pb.

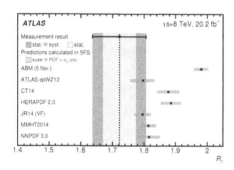
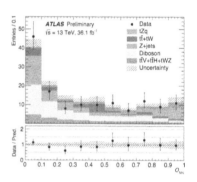

Figure 5: R_t measurement in the t-channel at $\sqrt{s} = 8$ TeV. The PDF uncertainty components are excluded in this comparison [8].

Figure 6: Post-fit NN output in the signal region. The uncertainty band includes statistical and systematic uncertainties [10].

Using the data collected at $\sqrt{s} = 13$ TeV ATLAS obtained the first observation of the tZq process [10], sensitive to tZ and WWZ couplings. The analysis has been performed in the trilepton channel, characterized by a small BR but by high purity. The cross section is extracted applying a likelihood fit on the outcome of a neural network, shown in Fig. 6. The measured cross section $\sigma_{tZq} = 600 \pm 170(stat.) \pm 140(syst.)$ fb corresponds to a significance of 4.2σ and is compatible with the SM expectations.

4 Conclusions

All measurements presented of $t\bar{t}$ and single top production cross sections at $\sqrt{s} = 8, 13$ TeV are in agreement with the SM calculations. The large top quark sample collected by the ATLAS experiment allows to improve the precision of measurements, observe for the first time the process tZq, and extend the $\sigma_{t\bar{t}}$ measurement in the multi-TeV p_T^t region.

References

[1] ATLAS Collaboration, JINST **3** (2008) S08003.
[2] ATLAS Collaboration, ATLAS-CONF-2017-054.

[3] ATLAS Collaboration, Phys. Rev. **D 95** (2017) 072003.
[4] ATLAS Collaboration, Eur. Phys. J. **C 77** (2017) 804.
[5] ATLAS Collaboration, Eur. Phys. J. **C 77** (2017) 299.
[6] ATLAS Collaboration, arXiv:1708.00727.
[7] ATLAS Collaboration, ATLAS-CONF-2016-100.
[8] ATLAS Collaboration, Eur. Phys. J. **C 77** (2017) 531.
[9] ATLAS Collaboration, arXiv:1612.07231.
[10] ATLAS Collaboration, ATLAS-CONF-2017-052.

PHYSICS AT THE COMPACT LINEAR COLLIDER

Ivanka Božović-Jelisavčić[a]
on behalf of the CLICdp collaboration
Vinca Institute of Nuclear Sciences,
University of Belgrade 11000 Belgrade, Serbia

Abstract. This paper (based on an invited talk at the 18th Lomonosov Conference on Elementary Particle Physics) provides an overview of the physics program at CLIC, including updates on the ongoing studies on t-quark precision observables, massive vector-boson scattering and di-photon processes at high energies.

1 Introduction

Highlights from the topical studies ongoing at CLIC in the Higgs, top and Beyond the Standard Model (BSM) sectors are given to illustrate the capabilities of CLIC to address relevant questions in the post-LHC era. An energy-staged approach that includes three running center-of-mass energies (380 GeV, 1.4 TeV and 3 TeV) is optimized to the physics program, including the CLIC unique sensitivity at the highest center-of-mass energy to a possible extensions of the Standard Model (SM). With its full statistics of $\sim 10^6$ Higgs bosons, crucial measurements of the Higgs properties will be possible to indicate the validity scale of the Standard Model.

As at other e^+e^- colliders, the CLIC experimental environment is practically free of QCD background and thus suitable for precision measurements. As discussed in Section 1.1, machine related background can be reduced to the satisfactory level for physics analyses. Staged implementation can provide a long-term precision physics program to complement the LHC searches.

1.1 CLIC accelerator and detector

The CLIC accelerator is based on a novel two-beam acceleration scheme, where a high-intensity beam (drive beam) is used to generate RF power to the main beam. Using normal-conducting accelerator structures, the two-beam accelerations provides gradients of 100 MV/m as has been demonstrated at the CTF3 test facility [1]. Short (σ_x=40 nm, σ_y=1 nm) and dense ($\sim 10^9$ particles) bunches, lead to a strong beamstrahlung induced by the electromagnetic fields of the opposite bunches. Consequently, beamstrahlung photons convert to hadrons causing the occupancy of the central detectors. With a time-stamping window of 10 ns and appropriate cuts on the transverse momentum and reconstructed time of particles in the calorimeter, the average depositions from this type of background can be reduced to the acceptable level [2]. On the other

---

[a]E-mail: ibozovic@vinca.rs

hand, beamstrahlung severely deteriorates the luminosity spectrum, in particular at the highest center-of-mass energies. However, it has been shown in [3] that the effect can be controlled at a permille level in the peak region (>80% of the nominal center-of-mass energy).

A new, post-CDR detector model (CLIC_det) has been proposed [4] with the consideration to the ultimate precision for physics. The detector performance is optimized to physics requirements for hermeticity, flavor tagging, jet energy resolution, p_T reconstruction, etc. Detector comprise all-silicon tracking and compact calorimetry, placed in a 4T magnetic field. High-granularity calorimeters are optimized to the particle flow performance in order to enable a jet energy resolution of ~3.5-5% for jet energies of 10 GeV up to 1.5 TeV. Together with the flavor separation, the above is particularly relevant to distinguish between Z, W or Higgs bosons in their hadronic decay modes.

2 Higgs studies at CLIC

2.1 Combined fit of the Higgs measurements

An extensive program of physics studies based on full simulations of detector and physics processes, together with a full event reconstruction, has been performed in the Higgs sector and documented in [5]. Higgsstrahlung and WW-fusion are dominant production mechanisms of the Higgs boson at low and high center-of-mass energies, respectively. Depending on the production mechanism, appropriate polarization could eventually double the statistics. A combined study of these two processes can be employed to probe the Higgs width and couplings in a model-independent way. This leads to a determination of the Higgs couplings at a level of ~1%, except for the rare decays to light particles such as muons or photons. Assuming that the Higgs total width is constrained by the SM decays, the statistical precision of the Higgs couplings can be improved to a sub-percent level. Both model-independent and model-dependent fits exploit the full statistics available at all CLIC energy stages in a cumulative way. Details of the combined fits can be found in [5] and have already been discussed in [6]. Model-dependent results allow for a comparison to the LHC and HL-LHC experiments where a similar approach is employed. To illustrate the CLIC sensitivity in comparison to the HL-LHC with the full statistics of 3000 fb^{-1}, results of the model-dependent fit are given in Fig. 1 [7]. As can be seen, the CLIC will be able to provide comparable or better precision of the Higgs couplings than HL-LHC.

2.2 Higgs self-coupling

At the high center-of-mass energies of 1.4 TeV and 3 TeV, a double-Higgs production can be used to measure the Higgs boson trilinear self-coupling parameter λ that determines the shape of the Higgs potential. Several processes

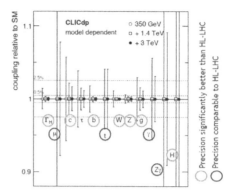

Figure 1: Higgs couplings relative statistical uncertainties in a model-dependent fit. The full statistics of 500 fb^{-1} and 1500 fb^{-1} and 2 ab^{-1} is assumed for 350 GeV, 1.4 TeV and 3 TeV CLIC, respectively, and 3000 fb^{-1} for the HL-LHC.

contribute to the double Higgs production, as illustrated in Fig. 2 (left), where only one Feynman diagram (a) has the sensitivity to λ. Contribution of the non-sensitive Feynman diagrams to the double-Higgs production cross-section is taken into account with the parameterization of the double-Higgs production cross-section $\sigma(e^+e^- \to \nu_e\bar{\nu}_e HH)$ w.r.t. the self-coupling parameter λ:

$$\frac{\Delta\lambda}{\lambda} \sim \kappa \frac{\Delta|\sigma(HH\nu_e\bar{\nu}_e)|}{\sigma(HH\nu_e\bar{\nu}_e)}. \quad (1)$$

Simulation at the generator level gives values of 1.22 and 1.47 for the scaling factor κ, at 1.4 TeV and 3 TeV center-of-mass energy, respectively [5]. Two signal final states are considered at both energies: $HH \to b\bar{b}b\bar{b}$ and $HH \to b\bar{b}W^+W^-$, together with a complete list of relevant background processes. Signal selection ensures that there is no overlap between the different final states of the signal, while a multivariate analysis is employed to separate each signal from background processes.

Combined results for the both Higgs decay channels, including electron polarization of -80%, leads to the relative statistical uncertainty of λ of 19% and 16% at 1.4 and 3 TeV respectively [8]. The achievable precision of λ can be improved further if multivariate analysis method would be employed to separate between λ-sensitive and non-sensitive double-Higgs production channels.

3 Top physics at CLIC

With its energy-staged implementation, CLIC offers a broad program of measurements to study the top-quark. The above includes the top mass determination from a threshold scan with the relative statistical uncertainty of

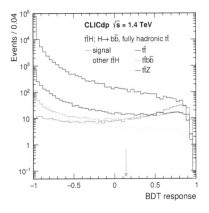 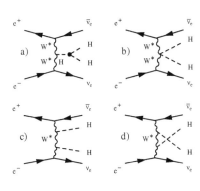

Figure 2: Left: Dominating Feynman diagrams for double-Higgs production at CLIC center-of-mass energies above 1 TeV, where only diagram (a) has a sensitivity to the Higgs self-coupling λ. Right: BDT based separation of the fully hadronic $t\bar{t}H$ signal from various background processes. The arrow denotes the BDT output value corresponding to the maximal significance.

∼15 MeV [9], direct access to the top Yukawa coupling at energies above 500 GeV and the precision studies of the top electroweak (EW) observables providing sensitivity to BSM scenarios. Excellent flavor tagging and jet reconstruction capabilities at CLIC enables precision top Yukawa coupling measurement at 1.4 TeV center-of-mass energy. The relative statistical uncertainty of the $t\bar{t}H$ production cross-section σ translates into a relative statistical uncertainty of the top Yukawa coupling y_t:

$$\frac{\Delta y_t}{y_t} = 0.53 \frac{\Delta \sigma}{\sigma}, \qquad (2)$$

leading to a top Yukawa relative statistical uncertainty of ∼4%, assuming electron polarization of -80% [5]. Analysis has been performed in the Higgs dominant decay channel to $b\bar{b}$, while W bosons from the top-quark decays can either decay leptonicly or hadronically, leading to 6 to 8-jet topology of the final state. The scaling factor in Eq. (2) is determined at the generator level, assuming a Standard Model value of the top Yukawa coupling. Illustration of the signal separation from various background processes, based on the multivariate method (BDT) output value, is given in Fig. 2, right [5]. Multivariate approach allows to maximize a statistical significance of the signal in the presence of various physics backgrounds.

3.1 Top-quark precision EW observables

Top-pair production through the Z,γ exchange in the s-channel, provides direct access to the top EW couplings sensitive to a higher order corrections from

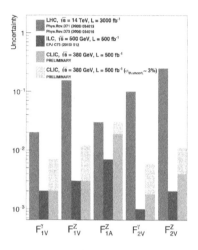 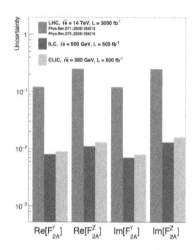

Figure 3: Expected precision on the top-quark EW form factors at different experiments.

BSM processes. The coupling associated to the top-$Z(\gamma)$ vertices is defined through a set of CP conserving and CP violating form factors that can be constrained through measurements of the sensitive EW observables (cross-section, forward-backward asymmetry, helicity angle distribution in top decays). Beam polarization allows to access different form factors independently. Figure 3 [9] gives the relative statistical precision of the top-quark CP conserving and CP violating form factors at different experiments. It demonstrates that future electron-positron colliders (in example of ILC and CLIC) are powerful instruments to study Higgs CP properties.

4 BSM searches

Potential signatures of BSM physics can be searched for at CLIC either through direct reconstruction of new particles, with an approximate mass reach of $\sqrt{s}/2$, or indirectly through the possible deviations in precision observables (cross sections, asymmetries, couplings, etc.). Indirect searches extend the sensitivity for BSM physics beyond the direct kinematic reach of the machine, up to (model-dependent) scales of several tens of TeV. In this paper, two examples of novel CLIC studies are given in Sections 4.1 and 4.2, to illustrate the possible constraints on BSM physics through indirect searches.

4.1 Massive vector boson scattering at high energies

Massive vector boson scattering, and W^+W^- scattering in particular, are sensitive probes to the higher-dimension operators in the Effective Field Theory

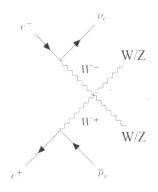
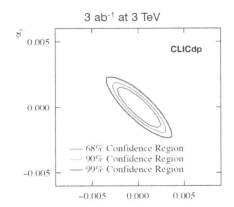

Figure 4: Left: Feynman diagram for W^+W^- scattering. Right: Confidence intervals for the quartic couplings α_4 and α_5, simulated at 3 TeV CLIC with 3 ab^{-1} of the integrated luminosity.

(EFT) approach. Such a scattering process $W^+W^- \to VV, with V = W, Z$ is illustrated in Fig. 4, left. Anomalous couplings at the quartic vertex (i.e. α_4, $\alpha_5 \neq 0$) result in modification of the sensitive observables like the cross-section and invariant mass of the VV system. Fully hadronic final state of the VV system is separated from the background processes in a multivariate analysis approach. A chi squared (χ^2) fit is employed to extract the couplings α_4 and α_5, and the corresponding confidence intervals have been derived. Due to a larger statistics, coupling determination precision at 3 TeV CLIC (Fig. 4, right) is improved for an order of magnitude in comparison to the measurement at 1.4 TeV center-of-mass energy [8]. As for the top-physics studies, this measurement also relies on the excellent jet reconstruction.

4.2 Di-photon production

Di-photon production is a well described QED process that can be exploited to search for the possible deviations of the sensitive observables (cross-section, photon polar angle distribution) from the Standard Model predictions. Depending on the relative uncertainty of the integrated luminosity, various BSM models can be probed at a kinematic limit 10-20 times higher than at LEP. This is illustrated in Table 1 [8]. Forward electron tagging is crucial for this measurement in order to suppress background from the Bhabha scattering.

5 Summary

CLIC is a mature future linear collider option with the physics program combining precision measurements with direct and indirect discovery potential. The

Table 1: 95% CL exclusion limits for various BSM models, obtained from di-photon production at 3 TeV CLIC with 2 ab^{-1} and at LEP.

Scenario	$\Delta L = 0.2\%$	$\Delta L = 0.5\%$	$\Delta L = 0.1\%$	LEP limit
QED cut-off (finite electron size) Λ_{QED}	6.52 TeV	6.33 TeV	6.01 TeV	~ 390 GeV
Contact interactions Λ'	20.7 TeV	20.1 TeV	18.9 TeV	~ 830 GeV
Extra dimensions $M_s/\Lambda^{1/4}$	16.3 TeV	15.9 TeV	15.3 TeV	~ 1 TeV
Excited electron M_{e^+}	5.03 TeV	4.87 TeV	4.7 TeV	~ 250 GeV

CLIC detector model is being optimized to meet the performance requirements from the physics studies, in terms of multi-jet final state reconstruction, photon reconstruction, missing-momentum measurement, forward electron tagging and many more. Staged implementation results in a broad physics program, from precision studies of the Higgs and top-quark physics to BSM probes. At the lowest center-of-mass energy, CLIC features model independence in the Higgs coupling measurements, achievable with a statistical precision at the percent level for most of the couplings. Operation at high energies enables precision measurement of rare Higgs decays, Higgs self-coupling, Higgs mass and CP properties. BSM measurements also benefits from the high-energy operation, with the sensitivity extended far beyond the available center-of-mass energy. Top-quark measurements serve as additional probe to the EW and BSM sectors.

Acknowledgments

Vinca Institute activity at CLIC/CLICdp has been supported by the Ministry of Education, Science and Technological Development of the Republic of Serbia, through the project OI171012.

References

[1] L. Linssen et al. (editors), Physics and Detectors at CLIC: CLIC Conceptual Design Report, ANL-HEP-TR-12-01, CERN-2012-003, DESY 12-008, KEK Report 2011-7, CERN, 2012.

[2] H. Abramowicz et al., Physics at the CLIC e+e- Linear Collider Input to the Snowmass Process 2013, 2013, arXiv:1307.5288.

[3] S. Lukic, I. Bozovic Jelisavcic, M. Pandurovic, I. Smiljanic, Correction of beam-beam effects in luminosity measurement in the forward region at CLIC, JINST **8** (2013) P05008.

[4] M. Weber, The CLIC detector, EPS-HEP 2017, July, 2017, Venice, Italy.
[5] H. Abramowicz et al. CLICdp Collaboration, Higgs physics at the CLIC electron-positron linear collider, Eur. Phys. J. C77 (2017), 475.
[6] I. Bozovic Jelisavcic, CLIC Physics Overview, in the Proceedings of the 17th Lomonosov Conference on Elementary Particle Physics, August 20-26, 2015, Moscow, Russia, Ed. by A. Studenikin, World Scientific, Singapore.
[7] V. Martin, Status of CLIC, 28th International Symposium on Lepton Photon Interactions at High Energy, August, 2017, Guangzhou, China.
[8] F. Simon, CLIC Analyses Highlights, AWLC 2017, June, 2017, Stanford, USA.
[9] A. Filip Zarnecki, Top physics at CLIC and ILC, PoS ICHEP2016 (2016), C16-08-03.

THE TOTEM EXPERIMENT: RESULTS AND PERSPECTIVES

Edoardo Bossini [a], on behalf of the TOTEM collaboration
CERN, Geneva, Switzerland

Abstract. The TOTEM experiment is taking data since 2010 and has produced a large set of measurements on diffractive processes and pp cross sections. In this paper, after a general overview of the detector, the results obtained on total, inelastic and elastic cross section at different energies will be reported. The measurements on the ρ parameter and the recent studies on the Coulomb-Nuclear interference region will be also presented and discussed. Joint results obtained by the TOTEM and CMS experiment on the forward charged particle distribution are reviewed. During the last two years the detector has been upgraded with new timing detectors, opening the way to a new physics program to be carried out with a joint collaboration between the TOTEM and CMS experiments.

1 Introduction

TOTEM [1] is one of the experiments of the LHC, located at the interaction point 5(IP5), together with the CMS experiment. The TOTEM experiment is focused on diffractive processes, which are characterized by the exchange of vacuum quantum numbers, resulting in a colorless interaction. From the kinematic point of view they are identified by the presence of large, non exponentially suppressed, pseudorapidity[b] gaps (regions of space forbidden for the final state particles). In such processes, which include elastic scattering, single, double and central diffraction, one or both hadrons can survive the interaction and be scattered in the very forward region.

The TOTEM experimental setup is made by three different subsystems, two gas tracking detectors T1 and T2 to tag and characterize inelastic events and the Roman Pot (RP) system to tag and measure the scattered protons in the very forward region (Fig. 1).

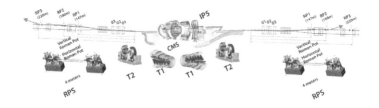

Figure 1: Sketch of the TOTEM experiment detectors.

---

[a] E-mail: edoardo.bossini@cern.ch
[b] pseudorapidity η is defined as $\eta \equiv -ln[\tan\theta/2]$, where θ is the polar angle between the direction of the particle w.r.t the beam direction.

T1 is based on Cathode Strip Chambers installed in two cone-shaped regions centered on the beam and symmetrically placed at a distance of 7.5 m from the IP5. The detectors cover an area in pseudorapidity $3.2 < |\eta| < 4.7$. The T2 telescope is divided in two arms, one on each side of the IP at ∼14 m. Each arm is divided in two semi-cylindrical quarters that are closed on the beam pipe. Each quarter is built with 5 couples of triple Gas Electron Multiplier (GEM) detectors with a radial extension from 42.5 mm to 144 mm from the beam axis (which translates in a coverage $5.3 < |\eta| < 6.5$). The detection of very forward protons is performed in movable beam insertions, called Roman Pots (RP). The RP is a secondary vacuum vessel, hosting some type of detector, that can be moved into the primary vacuum of the machine through vacuum bellows. The detector can thus approach the beam down to few millimeters, which allows to detect protons scattered down to few microradians. One RP station is present on each arm, symmetrically w.r.t the beam interaction point. An RP station is composed by two units, called near and far, about 4 m distant. Each unit is formed by three RPs, two of them approaching the beam vertically and the third one horizontally, with an overlap region with the verticals. All the measurements hereafter reported have been performed with the RP equipped with a set of 10 planes of edgeless planar silicon strip detectors. Thanks to the long lever arm between the two units (4 m) it is possible to reconstruct the local angle of the proton with a resolution of ∼ 10 μrad. Once the position and angle of the track at the RP location is known, it is possible to extrapolate the scattering angle by inverting the transport matrix, which describes the propagation of the proton through the magnetic field of the machine. To reduce the measurement uncertainties and to be able to detect protons with very low (down to ∼ 10^{-4} GeV^2) 4-momentum transferred special LHC optics (magnet configuration) is needed. The optics used is characterized by a large amplitude function β^* (β^* from 90 m up to 2.5 km have been used), larger than the one used for standard LHC run (0.3-3 meters). This grants a lower beam divergence at the collision point, the major source of systematics in the final measurements, at the expense of a lower luminosity.

2 Total, elastic and inelastic cross section

Measurements of elastic, inelastic and total pp cross section were performed at the energies of $\sqrt{s} = 7$ TeV [2, 3] and 8 TeV [4]. New analyses performed at 2.76 and 13 TeV will be published soon. Thanks to the TOTEM experimental apparatus three different strategies can be pursued to compute the total pp cross section σ_{tot}:

- by measuring the elastic differential cross section distribution dN_{el}/dt and performing an extrapolation to t = 0 it is possible to compute the total cross section through the optical theorem together with the machine luminosity $\mathcal{L}$ provided by CMS and ρ (the ratio between real and imaginary part of the scattering amplitude) derived by the COMPETE collaboration [5]: $\sigma_{tot} = 16\pi/(1+\rho^2) \, dN_{el}/dt \big|_{t=0} \mathcal{L}^{-1}$.

- a luminosity independent way, using both the inelastic N_{inel} and elastic N_{el} measurements: $\sigma_{tot} = \frac{16\pi}{1+\rho^2} \frac{dN_{el}/dt|_{t=0}}{N_{el}+N_{in}}$.

- ρ independent way, by the luminosity and the rates: $\sigma_{tot} = 1/\mathcal{L}(N_{el} + N_{inel})$.

At the same time data from the inelastic detectors can be used to compute the inelastic cross section, while from the elastic differential distribution we extrapolate the elastic cross section. Moreover by subtracting the elastic from the total cross section we obtain an inelastic cross section inclusive of the low diffractive masses, main source of uncertainties of such a measurement. The special optics designed for the experiment makes > 90% of the elastic cross section visible, reducing the uncertainties on the extrapolation. For the 7 TeV analysis a total of three datasets was collected, measuring particles with 4-momentum transfer in the range $5 \cdot 10^{-3} < |t| < 2.5$ GeV^2. The position of the diffractive minimum, large-t distribution and the distribution slope B for $|t| < 0.36$ have been measured. In Fig. 2, the results on cross section and the slope B are reported.

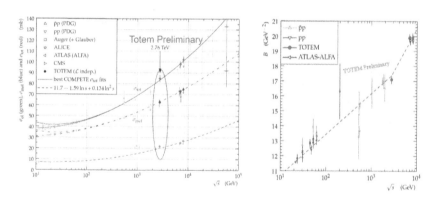

Figure 2: Cross section (left) and differential cross section slope B (right) measured by TOTEM at 7,8 and (preliminary) 2.76 TeV center of mass energy.

3 Very low-t measurements: non-exponential behaviour and CNI region studies

The parametrization of the differential hadronic cross section below the first diffractive minimum has always been parameterized through a simple exponential function $e^{-B|t|}$. Even if several past experiments [6,7] reported about slight deviations from the prediction, such a modelization always proved satisfactory in the description of the experimental data. Also the data used by TOTEM for the result discussed in the previous section were compatible with a pure exponential behaviour. Nevertheless the analysis carried out on a large data sample at the energy of $\sqrt{s} = 8$ TeV and $\beta^* = 90$ m shows that the model is excluded with a significance of 7σ [8]. A new parametrization, with higher order terms at the exponential $d\sigma_{el}/dt = A \cdot exp(\sum_{i=1}^{N_b} b_i t^i)$, has to be used to fit the experimental data. The result has been obtained thanks to large statistics, which leads to a statistical and systematic uncertainty on the differential cross-section below the 1% level in the region $0.027 \leq |t| \leq 0.2$ GeV^2, except for the normalization, which however does not change the shape of the distribution, being an overall scale factor. Previous results of the collaboration remain unchanged even if the new model is applied to the fit performed. Any effect of Coulomb scattering has been neglected in thiese measurements, being negligible for $|t| \geq 0.02$ GeV^2.

Studies on the very low $|t|$ region have been later carried on by merging the large 90 m dataset with another data sample (still at $\sqrt{s} = 8$) collected with a $\beta^* = 1$ km [9]. The 1 km data extend the accessible values of $|t|$ down to $6 \cdot 10^{-4}$ GeV^2. At such low values the pp scattering became affected by the Coulomb interaction. For typical LHC energies of a few TeV Coulomb and nuclear amplitude have similar magnitude at $|t| \sim 5 \cdot 10^4$ GeV^2. The transition region between the $|t|$ range dominated by the hadronic scattering, well described by the phenomenological model of the Regge theory, and the one dominated by the Coulomb interaction, perfectly known from QED, can be thus investigated. This region, called Coulomb-Nuclear Interference (CNI), represents an outstanding source of knowledge. At present multiple models for the CNI exist, none of them experimentally proved (or disproved), which differ both in the interference formula and the hadronic phase used. Moreover the CNI area shows a great sensitivity to the ρ parameter and the data can thus give a direct measurement of the parameter. For each CNI model/hadronic phase taken into account by TOTEM, the ρ parameter has been extrapolated from the 1 km data set and then used to fit the whole $|t|$−range. For the hadronic phase, different central and one peripheral [10] parametrizations have been considered. The interference models under study were the simplified West-Yennie model [11](SWY), Cahn [12] and Kundrát-Lokajíček [13] (KL). The SWY model, based on an exponential hadronic modulus and constant phase (extensively used in past measurements), is ruled out from the experimental measurement ($\chi^2/n.d.f = 180/58 = 3.12$). Moreover purely exponential hadronic module is disfavored for different hypothesis of phase and interference formula. A good interpolation of the experimental data is instead obtained using the Cahn or KL models, either with central of peripheral phases, using a third order ($N_b = 3$) exponential form for the hadronic amplitude, as can be seen in Fig. 3. The measured value of $\rho = 0.12 \pm 0.03$, obtained with the preferred fit (Cahn/KL, $N_b = 3$, either peripheral or central phase), represents the first direct measurement of the ρ at the LHC energy and provides an important constraint to the high energy evolution of the parameter (Fig. 3). A similar analysis is ongoing for the 13 TeV data collected with $\beta^* = 2.5$ km, that will be soon published.

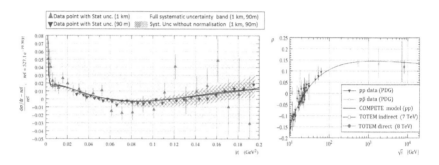

Figure 3: On the left: Differential cross-section plotted as relative difference from a reference exponential with $N_b = 1$ (see vertical axis). The black dots represent data points with statistical uncertainty bars. The coloured continuous curves correspond to the best fits(figure from [9]). On the right: ρ parameter measured at different energies. Only TOTEM has measured it at the LHC energy scale.

4 Pseudorapidity charged particle distribution

Measurements of the forward particle pseudorapidity density have been carried out by TOTEM at energies of $\sqrt{s} = 7$ [14] and 8 TeV [15, 16], using the T2 detector. Diffractive scattering is relevant in the forward region, representing 15-40% of the pp inelastic cross section [17]. The particle multiplicity produced in these processes is modeled in the existing Monte Carlo event generators using non-perturbative phenomenological models, with large uncertainties in the forward region. The experimental data collected can hence be used to tune the MC generators, which play an important role also in the modelization of cosmic-ray showers. While the 7 TeV data analysis has involved only T2 data (covering a region $5.3 < |\eta| < 6.4$), the 8 TeV data has been collected in a joint data taking with the CMS experiment. Combining the data, collected with a minimum bias trigger generated by at least one particle in the TOTEM T2 telescope, which also triggered the readout of CMS, the coverage was extended by the region $|\eta| < 2.2$. A correction to take into account any particle with transverse momentum $P_T > 0$ is also applied. To fill the gap between the central CMS tracker and the T2 detector TOTEM used data collected with displaced collisions, occurring at a distance of 11.25 m from the nominal interaction point.

The two T2 arms were thus able to measure the charged particle density in the regions $3.9 < \eta < 4.7$ and $-6.95 < \eta < -6.9$ (fig.4). Several MC generators have been compared to the data. Their predictions are within the uncertainties, but the need for a tuning in the forward region is evident.

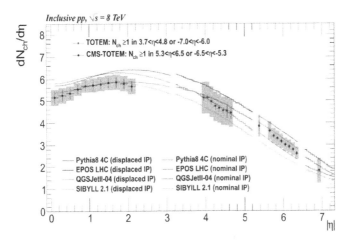

Figure 4: Combined results of charged particle pseudorapidity distributions at 8 TeV. The regions $3.9 < \eta < 4.7$ and $-6.95 < \eta < -6.9$ are accessed using a displaced interaction vertex. The bands show the combined systematic and statistical uncertainties and the error bars represent the η-uncorrelated uncertainties. A comparison with different MC generators is shown. (plot from [16]).

5 Experiment outlook

The capability of performing common data taking with the CMS experiment opens up to the possibility to perform detailed studies of central diffractive (CD) processes [18]. CD is characterized by two intact leading protons (eventually detected in the RP) and a rapidity-isolated system X generated in the central region instrumented by CMS. The possibility of proton tagging leads to an exceptional background reduction. The mass and the transverse momentum of the central system, as well the rapidity gap, can be indeed predicted from the RP measurements and compared with the direct central system reconstruction done within CMS. The physics topics covered in dedicated low-luminosity high-β^* runs have been already tested with a short run performed in 2015:

- Spectroscopy of CD low mass resonances and glueball states. Decay modes, BR and spin can be measured.

- Exclusive production of $c\bar{c}$ states (χ_c), with different states resolved, and J/Ψ production.

- Beyond Standard Model (BSM) physics searches, via missing mass or momentum and rapidity gap violation signature.

- Studies of exclusive ultra-pure gluon-gluon jets, investigating subtle QCD effects(gluon PDF, rapidity gap survival probability,...).

Of particular interest is the search for glueballs, pure bound gluon states predicted by QCD but not yet observed. Candidates predicted through QCD lattice computation have $J^{PC} = 0^{++}, 2^{++}$ which perfectly match with the selection rules of CD processes, and have masses of few GeV ($f_0(1370)$, $f_0(1500)$, $f_0(1710)$ and $f_2(2220)$ as examples). At LHC CD with $M_X \sim$ 1-4 GeV are produced very purely from gg thanks to the unprecedented low-ξ reach.

Studies of CD will be also carried out in standard high-luminosity LHC run, giving access to central masses in the range $300 < M_X < 2$ TeV. Such studies will be done with the CT-PPS (CMS-TOTEM Precision Proton Spectrometer) project, a common detector project done by TOTEM and CMS, whose details and physics program can be found elsewhere [19].

The high luminosity needed to collect all the required data, leads to the problem of pile-up, represented by simultaneous events from multiple pp interactions in the same bunch crossing. Due to their location, far away from the IP, the detectors installed in the RPs are not able to reconstruct the point of origin of the tagged protons with enough precision to assign them to the right pp interaction vertex. To disentangle the detected protons and perform a precise event reconstruction, the RPs need to be equipped with very precise timing detectors (resolution< 50 ps) to measure the proton TOF. This will make it possible to reconstruct the longitudinal coordinate from where the protons were originating by measuring the difference of the proton arrival times in the two detector arms. Two technologies have been exploited by the TOTEM collaboration, ultra-pure single crystal synthetic diamonds [22], and UltraFast Silicon Detectors (UFSD) [21]. Both technologies are actually employed in the CT-PPS project, and future upgrades, like double diamond sensors [20], are foreseen to be used already in 2018, before the LHC long shutdown 2.

6 Conclusion

Totem has produced a large number of scientific results during the last years. A selection of such results has been reported here. Measurements of pp cross sections and elastic differential cross sections have been carried out at energies of $\sqrt{s} = 7$ and 8 TeV. Analysis at 2.76 and 13 TeV are advanced and the results will be published soon. The high statistics and excellent control over the systematics made it possible to investigate the CNI interference region, performing a direct measurement of the ρ parameter at the LHC energy scale (result at 13 TeV ready soon). Moreover, the forward charged particle distribution has been measured at 7 TeV (TOTEM alone) and 8 TeV (with CMS), providing a relevant benchmark for the MC generators in the forward region and for cosmic-ray modelization algorithms. A wide physics program focused on studies of central diffractive processes, in collaboration with the CMS experiment, is actually ongoing, with a further upgrade of the experimental apparatus.

References

[1] G. Anelli et al. (TOTEM coll.), JINST **3** (2008) S08007.
[2] G. Antchev et al. (TOTEM coll.), EPL **101** (2013) 21002.
[3] G. Antchev et al. (TOTEM coll.), EPL **101** (2013) 21004.
[4] G. Antchev et al. (TOTEM coll.), EPL **111** (2013) 21004.
[5] J.R. Cudell et al. (COMPETE coll.), Phys. Rev. Lett. **89** (2002) 201801.
[6] L. Barksay et al., Nucl. Phys. B **148** (1979) 538.
[7] M. Bozzo et al., Nucl. Phys. B **147** (1984) 385.
[8] G. Antchev et al. (TOTEM coll.), Nucl. Phys. B **899** (2015) 527-546.
[9] G. Antchev et al. (TOTEM Coll.), Eur. Phys. J. **C 76** (2016) 12, 661.
[10] V. Kundrát, M. Lokajíček, Z. Phys. C **63** (1994) 619–630.
[11] G.B. West, D.R Yennie, Phys. Rev. **172** (1968) 1413–1422.
[12] R. Cahn, Z. Phys. C **15** (1982) 253–260.
[13] V. Kundrát, M. Lokajíček, Z. Phys. C **63** (1994) 619–630.
[14] G. Antchev et al. (TOTEM Coll.), Europ. Lett. **98** (2012) 3, 31002.
[15] G. Antchev et al. (TOTEM Coll.), Eur. Phys. J. **C 74** (2014) 10, 3053.
[16] G. Antchev et al. (TOTEM Coll.), Eur. Phys. J. **C 75** (2015) 3, 126.
[17] M.G. Ryskin et al., Eur. Phys. J. **C 71** (2011) 1617.
[18] G. Antchev et al. (TOTEM Coll.), CERN TDR, CERN-LHCC-2014-020, TOTEM-TDR-002.
[19] CMS and TOTEM collaborations, CERN Technical Disign Report, CERN-LHCC-2014-021, TOTEM-TDR-003, CMS-TDR-13.
[20] M. Berretti et al., JINST **12** (2017) no.03, P03026.
[21] R. Arcidiacono et al., JINST **12** (2017) no. 03, P03024.
[22] G. Antchev et al. (TOTEM Coll.), JINST **12** (2017) no. 03, P03007.

THE ADAMO PROJECT AND DEVELOPMENTS

V. Caracciolo[a] for the ADAMO collaboration
INFN, Laboratori Nazionali del Gran Sasso, I-67100 Assergi (AQ), Italy

Abstract. The directionality approach to pursue the Dark Matter investigation with $ZnWO_4$ anisotropic crystal scintillator will be addressed and reachable sensitivities - under some given assumptions - in a future pioneer experiment (named ADAMO: Anisotropic detectors for DArk Matter Observation) will be briefly discussed. Moreover, some of the main performances of $ZnWO_4$ crystal scintillator will be briefly summarized.

1 Introduction

In direct Dark Matter (DM) experiments, to obtain a reliable signature for the presence of DM particles in the galactic halo, it is necessary to follow a suitable model independent approach. The most important DM model independent signature is the so-called DM annual modulation signature [1] successfully exploited by the DAMA/NaI and DAMA/LIBRA experiments obtaining, cumulatively, an evidence for the presence of DM particles in our Galaxy at 9.3σ C.L. over 14 annual cycles [2].

Besides the DM annual modulation signature, the directionality approach could be an independent effective strategy to study those DM candidate particles able to induce just nuclear recoils. This strategy is based on the correlation between the arrival direction of the DM candidate particles and the Earth motion in the Galactic rest frame. In this case, the direction of the induced nuclear recoil is strongly correlated with that of the impinging DM particle. Consequently, the observation of an anisotropy in the distribution of nuclear recoil direction could give evidence for such candidates.

In principle Low Pressure Time Projection Chamber (LP-TPC), where the range of recoiling nuclei is of the order of mm (while in solid detectors the range is typically of order of μm) might be suitable to investigate this directionality (see e.g. [3]) through the detection of the tracks directions. However, a realistic experiment with LP-TPCs can be limited e.g. by the necessity of an extreme operational stability, of an extremely large detector size and of a great spatial resolution in order to reach a significant sensitivity. These practical limitations can be overcome by using the anisotropic scintillation detectors. In this case the information on the presence of those candidate particles is given by the variation of the measured counting rate during the sidereal day since both the light output and the pulse shape (PS) vary depending on the direction of the impinging particles with respect to the crystal axes (see the Ref. [4]).

The use of anisotropic scintillators to study the directionality approach was proposed for the first time in Ref. [5] and revisited in [6]. Recently, measurements

---
[a]E-mail: vincenzo.caracciolo@lngs.infn.it

and R&D works by DAMA-Kiev collaboration have shown that the $ZnWO_4$ scintillators can offer suitable features, among others: (i) very good radiopurity [7]; (ii) energy threshold at level of a few keV reachable [8]; (iii) light output and decay time dependence on the direction of nuclear recoil when a heavy particle impinges on the crystal, while the response to γ/β radiation is isotropic. Thus, the $ZnWO_4$ can be an excellent candidate for this type of research and both the anisotropic features can provide two independent ways to exploit the directionality approach.

2 Properties of $ZnWO_4$ Crystal Scintillator

Large volume $ZnWO_4$ single crystals of reasonable quality were grown [9] and studied as scintillators in the eighties [10]. Further development of large volume high quality radiopure $ZnWO_4$ crystal scintillators is described in [4, 11–13]. The first low background measurement with a small $ZnWO_4$ sample (4.5 g) was performed in the Solotvina underground laboratory (Ukraine) to study its radioactive contamination and to search for double beta decay of Zn and W isotopes [14]. More recently, radiopurity and double beta decay processes of Zn and W have been further studied also at LNGS using $ZnWO_4$ detectors with masses 0.1 − 0.7 kg [7, 8, 11–13, 15, 16].

The radioactive contamination of the $ZnWO_4$ crystals, measured in [7], approaches that of specially developed low background NaI(Tl); moreover, $ZnWO_4$ crystals having higher radiopurity could be expected in future realizations.

Measurements with α particles have shown that the light response and the PS of a $ZnWO_4$ scintillator depend on the impinging direction of α particles with respect to the crystal axes [14]. Figure 3 of Ref. [4] shows the dependence in a $ZnWO_4$ crystal of the α/β light ratio (quenching factor) on energy and direction of the α beam relatively to the crystal planes. Instead, the anisotropy of the light response of the $ZnWO_4$ scintillator disappears in case of electron excitation. Moreover, as demonstrated in Ref. [14], the dependence of the PSs on the type of radiation in the $ZnWO_4$ scintillator allows one to discriminate $\gamma(\beta)$ events from those induced by α particles. The PS analysis can be realized by the optimal filter method proposed in Ref. [18] and developed in Ref. [14] for $ZnWO_4$ crystal scintillators. The so-called shape indicator (SI) should be calculated for each signal to obtain a numerical characteristic of scintillation pulse [7]. As an example, Fig. 6 of Ref. [4] shows the dependence of the SI on the energy and on the impinging direction of the α particles obtained with a $ZnWO_4$ scintillator irradiated in the directions 1, 2 and 3 [14]. Also in this case, the anisotropic behaviour of the crystals response is evident. In order to study the response of $ZnWO_4$ to nuclear recoil at keV range for Dark Matter investigation, the DAMA group is presently performing a campaign of measurements by using a 14 MeV neutron beam [17]. As reported in [4], the light output of $ZnWO_4$ largely increase when the crystal scintillator working temperature is decreased.

To profit of this feature a small cryostat has been realized and is currently under test at LNGS.

3 The ADAMO Project

As regards the reachable sensitivity - for the model scenario described in [4] - by the ADAMO project (200 kg of ZnWO$_4$ by 25 detectors, 5 y of data taking and FWHM = $2.4\sqrt{E[keV]}$, the experimental scheme is accurately described in [4]) and is reported in Fig. 1. In particular, two software energy thresholds

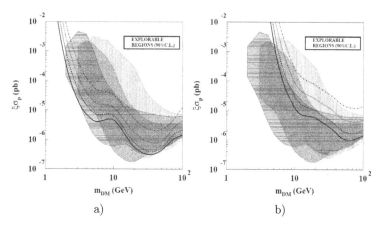

Figure 1: Sensitivity curves at 90% C.L. reachable by the ADAMO project for DM candidates inducing just nuclear recoils in the model scenario given in the text by exploring the directionality approach. Four possible background levels in the low energy region are considered as given in the text as well as two possible software energy thresholds: 2 keVee in (a) and 6 keVee in (b). There m_{DM} is the particle mass, σ_p is the spin-independent cross section on nucleon and ξ is the fraction of the DM local density of the considered candidate. In the figures there are also shown (green, red and blue online) allowed regions obtained in Ref. [20] by performing a corollary analysis of the 9σ C.L. DAMA model independent result in terms of scenarios for DM candidates inducing just nuclear recoils.

have been considered: 2 keVee for Fig. 1 (a) and 6 keVee for Fig. 1 (b). The reachable sensitivity has been calculated considering four possible time independent background levels in the low energy region: 10^{-4} cpd/kg/keV (solid black lines), 10^{-3} cpd/kg/keV (dashed lines), 10^{-2} cpd/kg/keV (dotted lines) and 0.1 cpd/kg/keV (dotted-dashed lines). For the Zn, W and O quenching factors along the three axes of the ZnWO$_4$ crystal we have considered here the values obtained by the method described in Ref. [19] taking into account the data of the anisotropy of α particles. As shown in Fig. 1 the directionality approach can reach - for the DM candidates investigated here and under the given model scenario - a sensitivity to spin-independent cross sections at level of $10^{-5} - 10^{-7}$ pb, for candidate mass between few GeV and hundreds GeV. However, it is worth noting that these plots are model dependent and, thus, always affected by several uncertainties, see Ref. [4] for details. In Fig. 1 the allowed regions (7.5σ from the null hypothesis) obtained by performing a corollary analysis of the DAMA model independent result in term of the scenarios described in Ref. [20] are also reported.

References

[1] K.A. Drukier et al., Phys. Rev. **D 33** (1986) 3495;
K. Freese et al., Phys. Rev. **D 37** (1988) 3388.
[2] R. Bernabei et al., La Rivista del Nuovo Cimento **26** (2003) 1;
R. Bernabei et al., Eur. Phys. J. **C 56**(2008) 333;
R. Bernabei et al., Eur. Phys. J. **C 67** (2010) 39;
P. Belli et al., Phys. Rev. **D 84** (2011) 055014;
R. Bernabei et al., J. Instrum. **7** (2012) P03009;
R. Bernabei et al., Eur. Phys. J. **C 72** (2012) 2064;
R. Bernabei et al., Eur. Phys. J. **A 49** (2013) 64;
R. Bernabei et al., Eur. Phys. J. **C 73** (2013) 2648;
R. Bernabei et al., Int. J. Mod. Phys. **A 28** (2013) 1330022;
R. Bernabei et al., Eur. Phys. J. **74** (C 2014) 2827;
R. Bernabei et al., Int. J. Mod. Phys. **A 31** (2016) 1642004;
R. Bernabei et al., Int. J. Mod. Phys. **A 31** (2016) 1642005;
R. Bernabei et al., Int. J. Mod. Phys. **A 31** (2016) 1642006.
[3] M.J. Lehner et al., in *Heidelberg, DM in astrophy. and parti. physics* (1988) 767-771; G.J. Alner et al., Nucl. Instrum. Meth. **A 555** (2005) 173.
[4] F. Cappella et al., Eur. Phys. J. **C 73** (2013) 2276.
[5] P. Belli et al., Il Nuovo Cim. **D 15** (1992) 475.
[6] R. Bernabei et al., Eur. Phys. J. **C 28** (2003) 203.
[7] P. Belli et al., Nucl. Instrum. Meth. **A 31** (2011) 626-627.
[8] P. Belli et al., J. Phys. **G 38** (2011) 115107.
[9] B.C. Grabmaier, IEEE Trans. Nucl. Sci. **31** (1984) 372.
[10] Y.C. Zhu et al., Nucl. Instrum. Meth. **A 244** (1986) 579; F.A. Danevich et al., Prib. Tekh. Eksp. **5** (1989) 80 [*Instrum. Exp. Tech.* **32** 1059].
[11] L.L. Nagornaya et al., IEEE Trans. Nucl. Sci. **55** (2008) 1469.
[12] L.L. Nagornaya et al., IEEE Trans. Nucl. Sci. **56** (2009) 994.
[13] E.N. Galashov et al., Functional Materials **16** (2009) 63.
[14] F.A. Danevich et al., Nucl. Instrum. Meth. **A 544** (2005) 553.
[15] P. Belli et al., Phys. Lett. **B 658** (2008) 193.
[16] P. Belli et al., Nucl. Phys. **A 826** (2009) 256.
[17] C. Taruggi, ROM2F/2016/06.
[18] E. Gatti, F. De Martini, Nuclear Electronics **2** (1962) 265.
[19] V.I. Tretyak, Astropart. Phys. **33** (2010) 40.
[20] P. Belli et al., Phys. Rev. **D 84** (2011) 055014.

NEW PHYSICS PATTERNS IN $b \to s\ell^+\ell^-$ TRANSITIONS

Bernat Capdevila [a]

Institut de Fisica d'Altes Energies, Universitat Autonoma de Barcelona, E-08193 Bellaterra (Barcelona), Spain

Abstract. In the Standard Model (SM), the rare transitions where a bottom quark decays into a strange quark and a pair of light leptons exhibit a potential sensitivity to physics beyond the SM. In addition, the SM embeds Lepton Flavour Universality (LFU), leading to almost identical probabilities for muon and electron modes. The LHCb collaboration discovered a set of deviations from the SM expectations in decays to muons and also in ratios assessing LFU. Other experiments (Belle, ATLAS, CMS) found consistent measurements, albeit with large error bars. In order to assess the impact of these experimental results, we perform a global fit to all available $b \to s\ell^+\ell^-$ data ($\ell = e, \mu$) in a model-independent way allowing for different patterns of New Physics (NP). In this work we provide a very brief review of the techniques used for the computation of the relevant observables at play in the mentioned transitions and present the latest results of our global fits. For the first time, the NP hypothesis is preferred over the SM by 5σ in a general case when NP can enter SM-like operators and their chirally-flipped partners. LFU violation is favoured with respect to LFU at the 3-4 σ level. Finally, we discuss the implications of the size of LFU violation observed in our fits on the observable P_5' from $B \to K^* \mu^+ \mu^-$.

1 Introduction

In the last few years, the field of Flavour Physics has witnessed the raise of an important amount of deviations from the Standard Model (SM) in various channels mediated by the flavour-changing neutral current (FCNC) transition $b \to s\ell^+\ell^-$, therefore becoming one of the most promising windows for studying possible Beyond the Standard Model (BSM) effects. In this context, the most useful tool to assess the significance and coherence of these deviations is a global model-independent fit.

As stated in [1], the current situation is exceptional since, for the first time, the global analyses of the observed deviations are pointing to several NP explanations that form coherent patterns with very high significances.

This proceeding follows the next structure. In Sec. 2 we present a review of the general statistical and theoretical framework used in our global analyses, paying special attention to the hadronic uncertainties affecting our predictions. This will lead us to Sec. 3, where we display the main outcomes of our global fits to all recent data on $b \to s\ell^+\ell^-$ transitions. Finally, we conclude in Sec. 4.

2 Theoretical Framework

In order to study the mentioned tensions and deviations in a model-independent fashion, we describe $b \to s\ell^+\ell^-$ driven processes by means of an effective

[a] E-mail: bcapdevila@ifae.es

Hamiltonian. In this framework, the heavy degrees of freedom (the top quark, the W and Z bosons, the Higgs field and any possible new heavy particles) are all integrated out in the form of Wilson coefficients C_i, where the short-distance information is encoded, while leaving only a set of effective operators O_i describing the long-distance effects. The relevant operators for the present discussion are,

$$\mathcal{O}_7^{(\prime)} = \frac{\alpha}{4\pi} m_b \left[\bar{s}\sigma_{\mu\nu} P_{R(L)} b\right] F^{\mu\nu},$$
$$\mathcal{O}_{9\ell}^{(\prime)} = \frac{\alpha}{4\pi} \left[\bar{s}\gamma^\mu P_{L(R)} b\right] \left[\bar{\ell}\gamma_\mu \ell\right], \quad (1)$$
$$\mathcal{O}_{10\ell}^{(\prime)} = \frac{\alpha}{4\pi} \left[\bar{s}\gamma^\mu P_{L(R)} b\right] \left[\bar{\ell}\gamma_\mu \gamma_5 \ell\right],$$

where $\ell = e$ or μ, $P_{L,R} = (1 \mp \gamma_5)/2$, m_b stands for the mass of the b quark and $\mu_b = 4.8$ GeV denotes the energy scale. The right-handed (primed) Wilson coefficients are not given since they either vanish or can be neglected in the SM.

The associated Wilson coefficients to the relevant effective operators are $C_7^{(\prime)}$, $C_{9\ell}^{(\prime)}$, $C_{10\ell}^{(\prime)}$. On one hand, $C_7^{(\prime)}$ describe the interaction strength of bottom (b) and strange (s) quarks with photons, while on the other $C_{9\ell}^{(\prime)}$, $C_{10\ell}^{(\prime)}$ encode the interaction strength of b and s quarks with dilepton currents. In the SM $C_{9\ell}^{(\prime)}$ and $C_{10\ell}^{(\prime)}$ are equal for both muons and electrons, however NP can yield very different contributions in muons compared to electrons. To assess the impact of NP on these coefficients, we parametrize C_7, $C_{9\ell,10\ell}$ as their SM value plus a NP contribution, $C_{i\ell} = C_{i\ell}^{\text{SM}} + C_{i\ell}^{\text{NP}}$ (the SM contributions to the chirally-flipped operators are essentially zero), and we estimate the size of these NP contributions by performing a global fit to the available data.

More specifically, our current global fits include up to 175 observables (each bin corresponding to an observable) coming from LHCb, Belle, ATLAS and CMS:

- $B \to K^* \mu^+ \mu^-$ ($P_{1,2}$, $P'_{4,5,6,8}$ [2–5] and F_L in 5 large-recoil bins plus 1 low-recoil bin [6]). All available information on $B \to K^* e^+ e^-$ [7] is included as well. The recent update of the differential branching fraction for $B \to K^* \mu^+ \mu^-$ [8] and the measurement of R_{K^*} [9] by the LHCb collaboration are also important pieces of our fit.

- $B_s \to \phi \mu^+ \mu^-$ (P_1, $P'_{4,6}$ and F_L in 3 large-recoil bins plus 1 low-recoil bin [10]).

- $B^+ \to K^+ \mu^+ \mu^-$ (BR) and $B^0 \to K^0 \ell^+ \ell^-$ (BR) (R_K [11] is implicit).

- $B \to X_s \gamma$ (BR), $B \to X_s \mu^+ \mu^-$ (BR) and $B_s \to \mu^+ \mu^-$ (BR).

- Radiative decays: $B^0 \to K^{*0}\gamma$ (A_I, $S_{K^*\gamma}$ and BR), $B^+ \to K^{*+}\gamma$ (BR) and $B_s \to \phi\gamma$ (BR).

- The new Belle measurements [12] for the isospin averaged but lepton-flavour dependent $B \to K^*\ell^+\ell^-$ observables $P_{4,5}^{\prime e}$ and $P_{4,5}^{\prime \mu}$. See [1] for more details on our treatmeant of the isospin average.

- The new ATLAS measurements [13] on the angular observables P_1, $P'_{4,5,6,8}$ and F_L of the $B^0 \to K^{*0}\mu^+\mu^-$ decay channel in the large-recoil region.

- The new CMS measurements [14] on the angular observables P_1 and P'_5 in $B^0 \to K^{*0}\mu^+\mu^-$, both in the large and low-recoil regions. We also include CMS results on F_L and A_{FB} [15], plus their partial measurements from an earlier analysis [16] at 7 TeV. Regarding CMS latest results, it would be very interesting a check of the stability and consistency of their measurements by performing the extraction of F_L, P_1 and P'_5 altogether, as it is done by the other experiments, and not separately. An independent extraction could imply a non-trivial loss of information regarding their correlations.

For the theoretical calculations, we use a different set of techniques depending on the observable and the corresponding q^2 region under consideration. Inclusive decays are analysed by means of an OPE. On the other hand, for exclusive decays at large-recoil we use heavy quark effective theory and QCD factorisation (QCDF), including factorisable and non-factorisable α_s and Λ/m_b corrections, and lattice QCD and quark-hadron duality at low-recoil. Regarding the $B \to K^*$ form factors at large-recoil, we use the calculation in [17], which presents more conservative errors than the ones in [18], obtained with a different method. For $B_s \to \phi$, the corresponding computation is not available and then we resort to [18], which explains the smaller errors we display in our $B \to \phi\ell^+\ell^-$ predictions.

The statistical procedure used in our analysis follows the lines previously established in [19]. We perform frequentist fits with all known theory and experimental correlations taken into account through the covariance matrix when building the χ^2 function, which is minimised to find best-fit points, pulls, p-values and confidence-level intervals. Depending on the dimensionality of the hypothesis, the minimisation is performed either using a simple scan or a Markov-Chain Monte Carlo Metropolis-Hastings algorithm.

2.1 Some comments about hadronic corrections

In the large-recoil region, we work within the framework called Improved QCDF where four type of corrections are included: i) factorisable α_s corrections computed in QCDF [20, 21], ii) factorisable power-corrections [22], iii) non-factorisable α_s corrections in QCDF [20, 21] and iv) non-factorizable power

Table 1: Main anomalies currently observed in $b \to s\ell^+\ell^-$ transitions, with the current measurements, our predictions for the SM and the NP scenario $C_{9\mu}^{\rm NP} = -1.1$, and the corresponding pulls.

Largest pulls	$P'_{5[4,6]}$	$P'_{5[6,8]}$	$R_K^{[1,6]}$	$R_{K^*}^{[1.1,6]}$
Experiment	-0.30 ± 0.16	-0.51 ± 0.12	$0.745^{+0.097}_{-0.082}$	$0.685^{+0.122}_{-0.083}$
SM	-0.82 ± 0.08	-0.94 ± 0.08	1.00 ± 0.01	1.00 ± 0.01
Pull (σ)	-2.9	-2.9	+2.6	+2.6
$C_{9\mu}^{\rm NP} = -1.1$	-0.50 ± 0.11	-0.73 ± 0.12	0.79 ± 0.01	0.87 ± 0.08
Pull (σ)	-1.0	-1.3	+0.4	-1.3

corrections computed from QCD light-cone sum rules (LCSR) with B-meson distribution amplitudes [17].

In the following lines, we provide a brief description of our treatment of both factorisable and non-factorisable power-corrections. For more elaborated discussions on the nature of such corrections and the controversies they have awakened, we refer the reader to references [22–24], where the state-of-the-art of hadronic uncertainties is discussed in full detail.

The large-recoil symmetries among form factors [25] allow for a decomposition of the seven full form factors in terms of two so-called soft form factors plus $\mathcal{O}(\alpha_s)$ and $\mathcal{O}(\Lambda/m_b)$ symmetry breaking terms. The $\mathcal{O}(\Lambda/m_b)$ breaking terms are known as factorisable power corrections,

$$F(q^2) = F^\infty(\xi_\perp(q^2), \xi_\parallel(q^2)) + \Delta F^{\alpha_s}(q^2) + \Delta F^{\Lambda/m_b}(q^2) \quad (2)$$

To estimate both the size and errors of these corrections we follow the technique first introduced in [26] and shortly after improved in [22]. Numerically, we allow for a generic size of 10% power corrections to the form factor, as suggested by fitting to the particular LCSR form factor calculations from [17, 18].

Non-factorisable power corrections come from insertions of the $\mathcal{O}_2$ effective operator. Unlike the factorisable ones, they cannot be encoded into form factors and yield contributions that enter directly at the amplitude level. These contributions, commonly referred as long-distance charm-loop effects, can be cast as a shift in the Wilson coefficient C_9,

$$C_9^{\text{eff}\,i}(q^2) = C_{9\,\text{pert}}^{\text{eff SM}}(q^2) + C_9^{\rm NP} + C_9^{c\bar{c}\,i}(q^2) \quad (3)$$

where i stands for the different transversity amplitudes ($i = 0, \perp, \parallel$).

Currently, only a partial computation of this effect exists in the literature [17], yielding values $C_{9\,{\rm KMPW}}^{c\bar{c}\,i}$ that tend to enhance the anomalies. In our analysis, we assume that this partial results is representative for the order of magnitude of the total charm-loop contribution and we assign an error to unknown charm-loop effects varying

$$C_9^{c\bar{c}\,i}(q^2) = s_i\, C_{9\,\text{KMPW}}^{c\bar{c}\,i}(q^2), \qquad \text{for } -1 \leq s_i \leq 1 \tag{4}$$

3 Global fit results

In [1], two types of analysis were presented, a complete analysis including the 175 observables mentioned above and a second one with only the subset of LFUV observables ($R_{K^{(*)}}$, $Q_{4,5}$). In both cases the set of radiative and leptonic decays are always included. The SM point yields a χ^2 corresponding to a p-value of 14.6% for the first and 4.4% for the second. The SM pull exceeds 5σ (see [1] for details) in the three main hypothesis of the complete analysis ($C_{9\mu}^{\text{NP}}$, $C_{9\mu}^{\text{NP}} = -C_{10\mu}^{\text{NP}}$ and $C_{9\mu}^{\text{NP}} = -C_{9'\mu}^{\text{NP}}$), even if the last one is unable to explain R_K. But the most interesting outcome is that the 6-D fit (see Tab. 2) has shifted, for the first time, to the 5σ level from the previous 3.6σ. The fit to only LFUV observables exhibits a 4σ significance in several of the NP hypothesis in front of the SM solution, indicating a clear preference in data for a NP contribution to $b \to s\mu^+\mu^-$ but not to $b \to se^+e^-$.

Table 2: 1σ confidence intervals for the NP contributions to Wilson coefficients in the six-dimensional hypothesis for the fit "All". The SM pull is 5.0 σ.

	C_7^{NP}	$C_{9\mu}^{\text{NP}}$	$C_{10\mu}^{\text{NP}}$
Best fit	+0.03	-1.12	+0.31
1σ	$[-0.01, +0.05]$	$[-1.34, -0.88]$	$[+0.10, +0.57]$
	$C_{7'}$	$C_{9'\mu}$	$C_{10'\mu}$
Best fit	+0.03	+0.38	+0.02
1σ	$[+0.00, +0.06]$	$[-0.17, +1.04]$	$[-0.28, +0.36]$

4 Conclusions

Recent experimental results regarding $b \to s\ell^+\ell^-$ transitions are showing significant tensions with respect to SM predictions in several decay modes. Two different global fits, one including 175 $b \to s\ell^+\ell^-$ observables and the other involving only LFUV observables, show several one dimensional and two dimensional NP solutions with significances over 5σ. As it was already noticed in previous analyses, $C_{9\mu}$ represents the strongest NP signal. For the first time, the result of a six-dimensional fit, where all the six relevant effective operators are allowed to recieve NP contributions, is preferred over the SM at the level of 5σ. Mesurements of newly proposed observables will help confirming the signals of NP already observed and signaling out which NP hypothesis finally gets selected.

Acknowledgments

I would like to thank J. Matias, S. Descotes-Genon, J. Virto, A. Crivellin and L. Hofer; this proceeding is based on work made in collaboration with them. This work has received financial support from the grant FPA2014-61478-EXP and the Centro de Excelencia Severo Ochoa SEV-2012-0234.

References

[1] B. Capdevila, A. Crivellin, S. Descotes-Genon, J. Matias and J. Virto, arXiv:1704.05340 [hep-ph].
[2] F. Kruger and J. Matias, Phys. Rev. D **71** (2005) 094009 doi:10.1103/PhysRevD.71.094009 [hep-ph/0502060].
[3] J. Matias, F. Mescia, M. Ramon and J. Virto, JHEP **1204** (2012) 104 doi:10.1007/JHEP04(2012)104 [arXiv:1202.4266 [hep-ph]].
[4] S. Descotes-Genon, T. Hurth, J. Matias and J. Virto, "Optimizing the basis of $B \to K^* ll$ observables in the full kinematic range," JHEP **1305** (2013) 137 [arXiv:1303.5794 [hep-ph]].
[5] S. Descotes-Genon, J. Matias, M. Ramon and J. Virto, "Implications from clean observables for the binned analysis of $B-> K*\mu^+\mu^-$ at large recoil," JHEP **1301** (2013) 048 [arXiv:1207.2753 [hep-ph]].
[6] R. Aaij *et al.* [LHCb Collaboration], "Angular analysis of the $B^0 \to K^{*0}\mu^+\mu^-$ decay using 3 fb^{-1} of integrated luminosity," JHEP **1602** (2016) 104 [arXiv:1512.04442 [hep-ex]].
[7] R. Aaij *et al.* [LHCb Collaboration], JHEP **1504** (2015) 064 doi:10.1007/JHEP04(2015)064 [arXiv:1501.03038 [hep-ex]].
[8] R. Aaij *et al.* [LHCb Collaboration], "Measurements of the S-wave fraction in $B^0 \to K^+\pi^-\mu^+\mu^-$ decays and the $B^0 \to K^*(892)^0\mu^+\mu^-$ differential branching fraction," JHEP **1611** (2016) 047 [arXiv:1606.04731 [hep-ex]].
[9] R. Aaij *et al.* [LHCb Collaboration], "Test of lepton universality with $B^0 \to K^{*0}\ell^+\ell^-$ decays," JHEP **1708** (2017) 055 [arXiv:1705.05802 [hep-ex]].
[10] R. Aaij *et al.* [LHCb Collaboration], "Angular analysis and differential branching fraction of the decay $B_s^0 \to \phi\mu^+\mu^-$," JHEP **1509** (2015) 179 [arXiv:1506.08777 [hep-ex]].
[11] R. Aaij *et al.* [LHCb Collaboration], "Test of lepton universality using $B^+ \to K^+\ell^+\ell^-$ decays," Phys. Rev. Lett. **113** (2014) 151601 [arXiv:1406.6482 [hep-ex]].

[12] S. Wehle et al. [Belle Collaboration], "Lepton-Flavor-Dependent Angular Analysis of $B \to K^*\ell^+\ell^-$," Phys. Rev. Lett. **118** (2017) no.11, 111801 [arXiv:1612.05014 [hep-ex]].

[13] The ATLAS collaboration [ATLAS Collaboration], "Angular analysis of $B_d^0 \to K^*\mu^+\mu^-$ decays in pp collisions at $\sqrt{s} = 8$ TeV with the ATLAS detector," ATLAS-CONF-2017-023.

[14] CMS Collaboration [CMS Collaboration], "Measurement of the P_1 and P_5' angular parameters of the decay $B^0 \to K^{*0}\mu^+\mu^-$ in proton-proton collisions at $\sqrt{s} = 8$ TeV," CMS-PAS-BPH-15-008.

[15] V. Khachatryan et al. [CMS Collaboration], "Angular analysis of the decay $B^0 \to K^{*0}\mu^+\mu^-$ from pp collisions at $\sqrt{s} = 8$ TeV," Phys. Lett. B **753** (2016) 424 [arXiv:1507.08126 [hep-ex]].

[16] S. Chatrchyan et al. [CMS Collaboration], "Angular analysis and branching fraction measurement of the decay $B^0 \to K^{*0}\mu^+\mu^-$," Phys. Lett. B **727** (2013) 77 [arXiv:1308.3409 [hep-ex]].

[17] A. Khodjamirian, T. Mannel, A. A. Pivovarov and Y.-M. Wang, "Charm-loop effect in $B \to K^{(*)}\ell^+\ell^-$ and $B \to K^*\gamma$," JHEP **1009** (2010) 089 [arXiv:1006.4945 [hep-ph]].

[18] A. Bharucha, D. M. Straub and R. Zwicky, "$B \to V\ell^+\ell^-$ in the Standard Model from light-cone sum rules," JHEP **1608** (2016) 098 [arXiv:1503.05534 [hep-ph]].

[19] S. Descotes-Genon, L. Hofer, J. Matias and J. Virto, "Global analysis of $b \to s\ell\ell$ anomalies," JHEP **1606** (2016) 092 [arXiv:1510.04239 [hep-ph]].

[20] M. Beneke, T. Feldmann and D. Seidel, Nucl. Phys. B **612** (2001) 25 doi:10.1016/S0550-3213(01)00366-2 [hep-ph/0106067].

[21] M. Beneke, T. Feldmann and D. Seidel, Eur. Phys. J. C **41** (2005) 173 doi:10.1140/epjc/s2005-02181-5 [hep-ph/0412400].

[22] S. Descotes-Genon, L. Hofer, J. Matias and J. Virto, JHEP **1412** (2014) 125 doi:10.1007/JHEP12(2014)125 [arXiv:1407.8526 [hep-ph]].

[23] B. Capdevila, S. Descotes-Genon, L. Hofer and J. Matias, "Hadronic uncertainties in $B \to K^*\mu^+\mu^-$: a state-of-the-art analysis," JHEP **1704** (2017) 016 [arXiv:1701.08672 [hep-ph]].

[24] R. Aaij et al. [LHCb Collaboration], Eur. Phys. J. C **77** (2017) no.3, 161 doi:10.1140/epjc/s10052-017-4703-2 [arXiv:1612.06764 [hep-ex]].

[25] J. Charles, A. Le Yaouanc, L. Oliver, O. Pene and J. C. Raynal, Phys. Rev. D **60** (1999) 014001 doi:10.1103/PhysRevD.60.014001 [hep-ph/9812358].

[26] S. Jger and J. Martin Camalich, JHEP **1305** (2013) 043 doi:10.1007/JHEP05(2013)043 [arXiv:1212.2263 [hep-ph]].

[27] S. Jäger and J. Martin Camalich, "Reassessing the discovery potential of the $B \to K^*\ell^+\ell^-$ decays in the large-recoil region: SM challenges and BSM opportunities," Phys. Rev. D **93** (2016) no.1, 014028 [arXiv:1412.3183 [hep-ph]].

REVIEW OF NA62 AND NA48 PHYSICS RESULTS

Riccardo Fantechi [a,b]
INFN - Sezione di Pisa and CERN

Abstract. This paper reviews recent results form the NA62 and NA48 experiments at CERN. A subset of the data collected by NA62 in 2016 has been analyzed to validate the analysis strategy for the measurement of the $K \to \pi \nu \bar{\nu}$ decay branching ratio. Improved limits on the coupling of heavy neutrinos are presented as well as NA48 results on the K_{l3} decay form factors.

1 Introduction

1.1 The quest for $K^+ \to \pi^+ \nu \bar{\nu}$ and $K_L \to \pi^0 \nu \bar{\nu}$

The ultra rare decays $K \to \pi \nu \bar{\nu}$ are, among the many rare flavour changing neutral current K and B decays, a powerful tool to search for new physics through underlying mechanisms of flavour mixing. Several circumstances allow to compute the SM branching ratio to very high precision:

- the $O(G_F^2)$ electroweak amplitudes exhibit a power-like GIM mechanism

---

[a] for the NA62 Collaboration: R. Aliberti, F. Ambrosino, R. Ammendola, B. Angelucci, A. Antonelli, G. Anzivino, R. Arcidiacono, M. Barbanera, A. Biagioni, L. Bician, C. Biino, A. Bizzeti, T. Blazek, B. Bloch-Devaux, V. Bonaiuto, M. Boretto, M. Bragadireanu, D. Britton, F. Brizioli, M.B. Brunetti, D. Bryman, F. Bucci, T. Capussela, A. Ceccucci, P. Cenci, V. Cerny, C. Cerri, B. Checcucci, A. Conovaloff, R. Cooper, E. Cortina Gil, M. Corvino, F. Costantini, A. Cotta Ramusino, D. Coward, G. D'Agostini, J. Dainton, P. Dalpiaz, H. Danielsson, N. De Simone, D. Di Filippo, L. Di Lella, N. Doble, B. Dobrich, F. Duval, V. Duk, J. Engelfried, T. Enik, N. Estrada-Tristan, V. Falaleev, R. Fantechi, V. Fascianelli, L. Federici, S. Fedotov, A. Filippi, M. Fiorini, J. Fry, J. Fu, A. Fucci, L. Fulton, E. Gamberini, L. Gatignon, G. Georgiev, S. Ghinescu, A. Gianoli, M. Giorgi, S. Giudici, F. Gonnella, E. Goudzovski, C. Graham, R. Guida, E. Gushchin, F. Hahn, H. Heath, T. Husek, O. Hutanu, D. Hutchcroft, L. Iacobuzio, E. Iacopini, E. Imbergamo, B. Jenninger, K. Kampf, V. Kekelidze, S. Kholodenko, G. Khoriauli, A. Khotyantsev, A. Kleimenova, A. Korotkova, M. Koval, V. Kozhuharov, Z. Kucerova, Y. Kudenko, J. Kunze, V. Kurochka, V.Kurshetsov, G. Lanfranchi, G. Lamanna, G. Latino, P. Laycock, C. Lazzeroni, M. Lenti, G. Lehmann Miotto, E. Leonardi, P. Lichard, L. Litov, R. Lollini, D. Lomidze, A. Lonardo, P. Lubrano, M. Lupi, N. Lurkin, D. Madigozhin, I. Mannelli, G. Mannocchi, A. Mapelli, F. Marchetto, R. Marchevski, S. Martellotti, P. Massarotti, K. Massri, E. Maurice, M. Medvedeva, A. Mefodev, E. Menichetti, E. Migliore, E. Minucci, M. Mirra, M. Misheva, N. Molokanova, M. Moulson, S. Movchan, M. Napolitano, I. Neri, F. Newson, A. Norton, M. Noy, T. Numao, V. Obraztsov, A. Ostankov, S. Padolski, R. Page, V. Palladino, C. Parkinson, E. Pedreschi, M. Pepe, M. Perrin-Terrin, L. Peruzzo, P. Petrov, F. Petrucci, M. Piandani, M. Piccini, J. Pinzino, I. Polenkevich, L. Pontisso, Yu. Potrebenikov, D. Protopopescu, M. Raggi, A. Romano, P. Rubin, G. Ruggiero, V. Ryjov, A. Salamon, C. Santoni, G. Saracino, F. Sargeni, V. Semenov, A. Sergi, A. Shaikhiev, S. Shkarovskiy, D. Soldi, V. Sougonyaev, M. Sozzi, T. Spadaro, F. Spinella, A. Sturgess, J. Swallow, S. Trilov, P. Valente, B. Velghe, S. Venditti, P. Vicini, R. Volpe, M. Vormstein, H. Wahl, R. Wanke, B. Wrona, O. Yushchenko, M. Zamkovsky, A. Zinchenko.
[b] E-mail: fantechi@cern.ch

- the top-quark loops largely dominate the matrix element
- the sub-leading charm-quark contributions have been computed at NNLO order [1]
- the hadronic matrix element can be extracted from the branching ratio of the $K^+ \to \pi^0 e^+ \nu$ decay, well known experimentally [2].

The current predictions for the branching ratios are [3]:

$$BR(K^+ \to \pi^+ \nu \bar{\nu}) = (8.4 \pm 1.0) \times 10^{-11}$$
$$BR(K_L \to \pi^0 \nu \bar{\nu}) = (3.4 \pm 0.6) \times 10^{-11}$$

The uncertainties are dominated by the experimental knowledge of the external inputs. These decays are sensitive to physics beyond the SM, with the largest deviations expected in models with new sources of flavour violation, owing to weaker constraints from B physics [4] [5]. Thanks to the SM suppression and existing constraints from K physics, variations of the $K \to \pi \nu \bar{\nu}$ BRs from the SM predictions induced by new physics at mass scales up to 100 TeV/c^2 can be observed with a 10% precision.

The decay $K^+ \to \pi^+ \nu \bar{\nu}$ has been observed by the experiments E787 and E949 at the Brookhaven National Laboratory and the measured branching ratio is $1.73^{+1.15}_{-1.05} \times 10^{-10}$ [6]. A limit $BR(K_L \to \pi^0 \nu \bar{\nu}) < 2.6 \times 10^{-8}$ at 90%CL has been published by the E391a Collaboration [7].

1.2 Heavy neutrinos

Experimental data on neutrino oscillations cannot be accommodated in the SM, where neutrinos are strictly massless. The simplest renormalizable extensions of the SM, consistent with the neutrino experiments, involve the presence of heavy neutrinos, or heavy neutral leptons (HNLs), which mix with ordinary neutrinos. An example is the neutrino Minimal Standard Model (νMSM) [8], where three massive right-handed neutrinos are introduced.

The mixing between HNLs and the SM neutrinos leads to the production of HNLs in meson decays (i.e. $K^+ \to l^+ N$). The branching fraction is determined by the HNL mass and the mixing parameter $|U_{l4}|^2$ [9]:

$$BR(K^+ \to l^+ N) = BR(K^+ \to l^+ \nu) \cdot \rho_l(m_N) \cdot |U_{l4}|^2$$

where $\rho_l(m_N)$ is a kinematic factor accounting for phase space and helicity suppression.

1.3 Semileptonic kaon decays

The K_{l3} decay width in the absence of radiative corrections can be represented by the Dalitz plot density depending on E_l and E_π [10] and the vector and scalar form factors, $f_+(t)$ and $f_0(t)$. For K_{e3} decays only $f_+(t)$ will contribute. The following form factor parametrizations have been used in the analysis: the quadratic parametrization [11], the pole parametrization [12] and the dispersive parametrization [13]. An improvement on the precision of the form factors measurement contributes to reduce the uncertainty of $|V_{us}|$.

2 The NA62 experiment at CERN

NA62 aims to collect about 100 $K^+ \to \pi^+ \nu \bar{\nu}$ decays in two years of running, to measure the branching ratio with 10% precision, with an acceptance of $\sim 10\%$. This measurement requires a beam line providing at least 10^{13} kaon decays, integrated over the data taking period. In addition a background rejection factor $O(10^{12})$ is required, to extract the signal in the overwhelming abundance of the main K^+ decay modes. Indeed $K^+ \to \pi^+ \pi^0$ (21%) and $K^+ \to \mu^+ \nu$ (63%) should be rejected using respectively photon veto detectors and muon identifiers. The use of a high energy kaon beam (75 GeV/c) has an advantage in the first case, because for $P_{\pi^+} < 35$ GeV/c the energy of the photons from the π^0 is larger than 40 GeV and the efficiency for photon rejection is very high. Additional particle ID is needed to eliminate other backgrounds and to identify the pa rent kaon inside the unseparated beam.

The beam line can provide the required intensity of about 3×10^{12} protons/pulse at an energy of 75 GeV with 1% momentum bite. The beam contains π^+ (70%), protons (23%) and K^+ (6%). The rate seen by the detectors along the beam line, integrated over a surface of 12.5 cm^2, is about 750 MHz. The decay region is defined to be a 80 m volume downstream of the last beam line element. The decay volume is evacuated down to 10^{-6} mbar, to minimize the multiple scattering of the decay products and the number of interactions of the beam with the residual gas. The rate downstream is given mainly by kaon decay products and is about 10 MHz.

The layout of the detector is shown in Fig. 1; the description of the apparatus can be found in [14].

3 The $K^+ \to \pi^+ \nu \bar{\nu}$ analysis

The analysis of 5% of the 2016 dataset corresponding to 2.3×10^{10} kaons is presented here. The complete analysis is done blindly, so that the remaining data are not visible until the analysis procedure is established. The $K^+ \to \pi^+ \nu \bar{\nu}$ signature is one track in the initial and final state with two missing neutrinos. The main kinematic variable is $m_{miss}^2 = (P_K - P_\pi)^2$, where P_K and P_π are the 4-momenta of the K^+ and π^+ respectively. The theoretical shape of the main K^+ background decay modes are compared to the $K^+ \to \pi^+ \nu \bar{\nu}$ on Fig. 2. The analysis is done in the π^+ momentum range between 15 and 35 GeV/c.

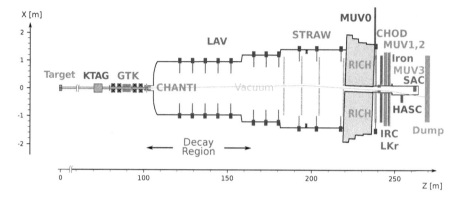

Figure 1: Horizontal view of the NA62 detector. The various labels refers to the detector elements.

Two regions are used: region 1 between $K^+ \to \mu^+\nu_\mu(K_{\mu 2})$ and $K^+ \to \pi^+\pi^0(K_{2\pi})$ and region 2 between $K_{2\pi}$ and $K^+ \to \pi^+\pi^+\pi^-(K_{3\pi})$. The main backgrounds entering those regions are $K_{\mu 2}$ and $K_{2\pi}$ decays through non gaussian resolution and radiative tails; $K_{3\pi}$ through non gaussian resolution; $K^+ \to \pi^+\pi^-e^+\nu_e(K_{e4})$ by not detecting the extra π^-, e^+ particles. Another important source of background is the beam related background coming from upstream decays and beam-detector interactions. Each of the background processes requires a different rejection procedure depending on its kinematics and type of charged particle in the final state. The main requirements for the analysis are: very good kinematic reconstruction to reduce kinematic tails; precise timing to reduce the kaon mis-tagging probability; no extra in time activity in all of the electromagnetic calorimeters (LAV, LKr, IRC, SAC) to suppress $K^+ \to \pi^+\pi^0$ decays with $\pi^0 \to \gamma\gamma$ (photon rejection); clear separation between $\pi/\mu/e$ tracks to suppress decays with μ^+ or e^+ in the final state (particle identification); low multiplicity cuts in the downstream detectors are used to further suppress decays with multiple charged tracks in the final state. The parent K^+ track is reconstructed and time-stamped in the beam tracker (GTK) with 100 ps resolution; the daughter π^+ track is reconstructed in the STRAW tracker, while CHOD and RICH measure π^+ time with resolution below 100 ps. The pion is associated in time to a KTAG kaon signal. The timing and the closest distance of approach between GTK and STRAW tracks allow a precise $K^+ - \pi^+$ spatial matching. The kaon mistagging probability at 40% of nominal intensity is below 2%, with a signal efficiency about 75%. Decays are selected within a 50 m fiducial region beginning 10 m downstream of the last GTK station to reject events originated from interactions of beam particles in GTK and kaon decays upstream of GTK. Figure 3 (left) displays the kinematics of the selected events. The resolution of m^2_{miss} drives the choice of the boundaries of the signal

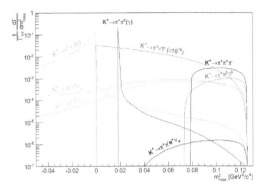

Figure 2: m^2_{miss} distribution for signal (multiplied by 10^{10}) and backgrounds (normalized according to their branching ratio).

regions. Reconstruction tails from $K^+ \to \pi^+\pi^0, K^+ \to \mu^+\nu_\mu, K^+ \to \pi^+\pi^+\pi^-$ set the level of background in signal regions. To reduce it, signal regions are restricted to boxes within a 3D space, defined by the m^2_{miss}, the same quantity computed using the momentum of the particle measured by the RICH under π^+ hypothesis (m^2_{miss}(RICH)) and computed replacing the 3-momentum of the kaon measured by the GTK with the nominal 3-momentum of the beam (m^2_{miss}(No-GTK)). The probability for $K^+ \to \pi^+\pi^0 (K^+ \to \mu^+\nu_\mu)$ to enter the signal regions is $6 \times 10^{-4} (3 \times 10^{-4})$, as measured on data. Calorimeters and RICH separate $\pi^+/\mu^+/e^+$. A multivariate analysis combines calorimetric information and provides $10^5 \mu^+$ suppression for 80% π^+ efficiency. RICH quantities are used to infer particle types, giving $10^2 \mu^+$ suppression for 85% efficiency. The two methods are independent and therefore able to suppress μ^+ by 7 orders of magnitude while keeping ≈65% of π^+. The remaining events after π^+ identification are primarily $K^+ \to \pi^+\pi^0$. Photon rejection exploiting timing coincidences between π^+ and calorimetric deposits suppresses them further. The resulting π^0 rejection inefficiency is $(1.2 \pm 0.2) \times 10^{-7}$, as measured from minimum bias and $K^+ \to \pi^+\nu\bar\nu$-triggered events before and after γ rejection, respectively. Random losses are in the 15-20% range.

A sample of $K^+ \to \pi^+\pi^0$ from a minimum bias trigger sample is used for normalization. About 0.064 $K^+ \to \pi^+\nu\bar\nu$ events are expected from the 2.3×10^{10} decays which represents the 5% of the 2016 data used to validate the analysis strategy. Figure 3 (right) shows the distribution of residual events in the m^2_{miss}(RICH) versus m^2_{miss} plane. Backgrounds from $K^+ \to \pi^+\pi^0, K^+ \to \mu^+\nu_\mu, K^+ \to \pi^+\pi^+\pi^-$ are 0.024, 0.011 and 0.017, respectively. They are estimated from events outside signal regions, with the measured kinematic tails used for extrapolation in signal regions. Simulations studies indicate that background from other processes is lower or negligible. The analysis is still on-going

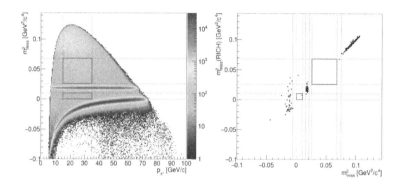

Figure 3: Left: Distribution of m^2_{miss} vs track momentum for minimum bias data. the rd signal regions are drawn for reference. Right: Distribution in the $(m^2_{miss}(\text{RICH}), m^2_{miss})$ plane of $K^+ \to \pi^+ \nu \bar{\nu}$ events passing the selection, except the cut on $m^2_{miss}(\text{No-GTK})$; signal regions (red tick boxes) and lines defining background regions (light dashed lines) are drawn; the event in region 1 has $m^2_{miss}(\text{No-GTK})$ outside the signal region.

together with an optimization of the selection to further reduce backgrounds and increase signal acceptance. No events are observed in signal regions.

4 Heavy neutrino analysis

Production of HNLs in $K^+ \to l^+ N$ decays can be searched for by looking for peaks in the missing mass spectra of the K_{l2} candidates. NA62 has performed two searches: one using the $K_{\mu 2}$ spectrum with data collected in 2007 with the NA48 detector [17] and the other with data collected by the NA62 detector in 2015, for which we present a preliminary result from the K_{e2} spectrum.

The total number of kaons in the fiducial zone is $(5.977 \pm 0.015) \times 10^7$ for the 2007 $K_{\mu 2}$ analysis [15] and $(3.010 \pm 0.011) \times 10^8$ for the 2015 K_{e2} data [16].

The signal region for the search are defined as $300 < m_{miss} < 375$ MeV/c^2 for the 2007 data and $170 < m_{miss} < 448$ MeV/c^2 for the 2015 analysis. The search is performed with steps of 1 MeV/c^2, with the additional condition for the missing mass to be within a resolution window, computed with MC simulation. The statistical analysis is performed using the Rolke-Lopez method to find 90% confidence intervals, with input from the number of observed events and the estimate of the background. No statistically significant HNL production is observed and upper limits are established. These limits are converted into upper limits on the branching fraction $BR(K^+ \to l^+ N)$ for each HNL mass hypothesis. The upper limits on the mixing $|U_{l4}|^2$, computed from the limits on the branching fraction are shown in Fig. 4: the NA62 results improve the existing limits both on $|U_{\mu 4}|^2$ and $|U_{e4}|^2$ in the analyzed signal regions.

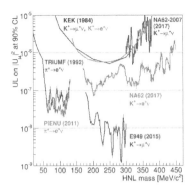

Figure 4: Upper limits on $|U_{l4}|$ from NA62 compared to the previous world limits.

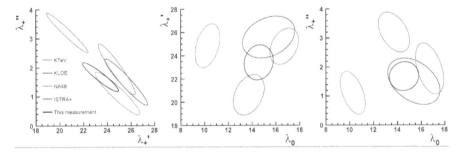

Figure 5: Confidence levels for the quadratic parametrization of the joint form factors. The ellipses correspond to 39%CL (one sigma contour).

5 Form factors of $K^{\pm} \to \pi^0 l^{\pm} \nu$ decays

The total number of events on the final samples of data taken with the NA48 detector [17] are 4.278×10^6 K_{e3} and 2.907×10^6 $K_{\mu 3}$.

The measurements of the form factors are obtained by minimizing the difference between data (background subtracted) and simulated Dalitz plots in 5×5 MeV^2 bins. The semileptonic radiative Monte Carlo samples have been simulated with a specific generator [18].

NA48/2 is the first experiment measuring the form factors using both K^+ and K^-. The $K_{\mu 3}$ result is dominated by the statistical error, the K_{e3} by the uncertainty on background. The $K_{\mu 3}$ and K_{e3} are in agreement within each other and the combined results are competitive with the current world average. To avoid partially correlated systematic uncertainties in the averaging of the $K_{\mu 3}$ and K_{e3} results, the complete analysis has been repeated considering the two decay modes as a unique data set, containing two Dalitz plots simultaneously fitted with a common set of form factor parameters. Figure 5 shows the 39% confidence contours for the joint form factors using the quadratic parametrization, together with earlier measurements.

6 Conclusions

NA62 is taking data to search for physics beyond the SM through the ultrarare decay $K^+ \to \pi^+ \nu \bar{\nu}$. The experiment is performing as expected. A subsample of 5% of the 2016 statistics, corresponding to 2.3×10^{10} kaon decays, has been processed to validate the analysis strategy. The full analysis of the 2016 dataset will reach the SM sensitivity, while from the 2017 and 2018 data some tens of $K^+ \to \pi^+ \nu \bar{\nu}$ events can be detected.

The search for heavy neutrinos in the $K_{\mu 2}$ and K_{e2} channels has improved the existing limits on both couplings. The measurement of the $K^\pm \to \pi^0 l^\pm \nu$ form factors has produced precise measurements competitive with the world average.

References

[1] A. J. Buras, M. Gorbahn, U. Haisch and U. Nierste, JHEP **0611** (2006) 002.
[2] C. Amsler et al., Phys. Lett. **B667** (2008) 1.
[3] A.J. Buras, D. Buttazzo, J. Girrbach-Noe and R. Knegjens, JHEP **1511**, 33 (2015).
[4] M. Blanke, A.J. Buras and S. Recksiegel, Eur. Phys. J. **C76** (2016) 182.
[5] M. Blanke, A.J. Buras, B. Duiling, K. Gemmler and S. Gori, JHEP **0903** (2009) 108.
[6] S. Adler et al., Phys. Rev. **D 79** (2009) 092004.
[7] J.K. Ahn et al., Phys. Rev. **D 81** (2010) 072004.
[8] T. Asaka and M. Shaposhnikov, Phys. Lett. **B620** (2005) 17.
[9] R.E. Shrock, Phys. Rev. **D24** (1981) 1232.
[10] L.M. Chounet, J.M. Gaillard, M. Gaillard, Phys.Rept. **4**, (1972) 199.
[11] C. Patrignani et al. (Particle Data Group), Chin. Phys. **C 40**, (2016) 100001.
[12] P. Lichard, Phys. Rev. D **D 55**, (1997) 5385.
[13] V. Bernard, M. Oertel, E. Passemar, J. Stern, Phys. Rev. **D 80**, (2009) 034034.
[14] E. Cortina Gil et al. [NA62 Collaboration], JINST **12** (2017) P05025.
[15] C. Lazzeroni et al. [NA62 Collaboration], Phys. Lett. **B772** (2017)712.
[16] E. Cortina Gil et al. [NA62 Collaboration], CERN-EP-2017-311, arXiv:1712.00297.
[17] V. Fanti et al., Nucl.Instrum. Meth. **A574**, (2007) 433.
[18] C. Gatti, Eur. Phys. J. **C45** (2006) 417.

STATUS OF THE KLOE-2 EXPERIMENT AT DAΦNE

Paolo Fermani [a] on behalf of the KLOE-2 collaboration
Laboratori Nazionali di Frascati dell'INFN, Via Enrico Fermi 40 (00044), Frascati, Italy

Abstract. The KLOE-2 experiment at the INFN Laboratori Nazionali di Frascati (LNF) is currently taking data at the upgraded e^+e^- DAΦNE collider, collecting until now an integrated luminosity of more than 4 fb^{-1}, with the aim of reaching 5 fb^{-1} for the end of March 2018. The KLOE detector undergone several upgrades including an innovative "state of the art" cylindrical GEM detector, the Inner Tracker, to improve its vertex reconstruction capabilities near the interaction region, and a pair of electron/positron taggers to study the gamma-gamma interactions. An overview of the KLOE-2 experiment will be given including present status and achievements together with future prospects.

1 Introduction

The KLOE-2 experiment is operating at the DAΦNE e^+e^- collider at the Laboratori Nazionali di Frascati (LNF) in Italy [1]. It is the continuation of the KLOE (KLongExperiment) experiment [2] upgraded with a new series of "state of the art" detectors in order to improve its discovery performances and to extend the KLOE physics program [3]. The latter is mainly devoted to Kaon physics [4] (neutral kaon interferometry; discrete simmetry and QM tests; rare K_S decays; CKM test) and high precision $\gamma\gamma$ physics (study of $\Gamma(\pi^0 \to \gamma\gamma)$, Transition Form Factors (TFFs) of pseudoscalar mesons) but it is also focusing on hadron physics [5] (Spectroscopy of light mesons; hadron cross-section at low energy) and dark matter mediators' search [6].

KLOE-2 is collecting data since November 2014 with the intent to acquire at least 5 fb^{-1} of integrated luminosity by the scheduled end of data-taking in March 2018. Currently the total luminosity acquired is roughly more than 4.7 fb^{-1} just a little step below the fixed goal.

2 The KLOE-2 Detector

The KLOE-2 detector is positioned at the interaction point (IP) of the DAΦNE collider; it consists of a large Drift Chamber (DC) and an ElectroMagnetic Calorimeter (EMC), both immersed in a 0.5 T axial magnetic field generated by an outer superconductive coil, that have been inherited from the KLOE experiment. A series of new sub-detectors have also been installed in order to enhance the tracks and vertices reconstruction and the photons, electrons and positrons detection: an innovative Inner Tracker detector (IT) [7]; a system of taggers [8] called LET and HET; two couples of new calorimeters [9] called QCALT and CCALT.

---
[a] E-mail: paolo.fermani@lnf.infn.it

2.1 The KLOE Drift Chamber and Electromagnetic Calorimeter

DC. The main tracker of the KLOE-2 experiment is a cylindrical Drift Chamber [10] with a diameter of 4 m and a length of 3.7 m, whose dimensions are determined by the decay path of the K_L ($\lambda_L \approx 3.5$ m). It works with a Helium (90%) and Isobuthane (10%) gas mixture in order to minimize the effect of multiple scattering, the conversion of photons and the K_L regeneration. The DC has an all stereo geometry (with 12582 cells) that provides $\simeq 150$ μm spatial resolution in the bending plane, $\simeq 2$ mm along the beam axis, $\simeq 3$ mm on decay vertices inside the DC fiducial volume and $\simeq 6$ mm on decay vertices close to the interaction point. Tracks are reconstructed with a momentum resolution of $\sigma_p/p = 0.4\%$.

EMC. The KLOE electromagnetic calorimeter [11] is a very fine sampling lead scintillating fiber calorimeter (with 88 modules of 23 cm thickness for $15X_0$ radiation lengths) with a photomultiplier (PMs) read-out. It is composed by a barrel and two end-caps to increase hermeticity and ensure a 98% of a solid angle coverage. The EMC allows to reconstruct clusters with excellent time ($\sigma_t = 54$ ps/$\sqrt{E(GeV)} \oplus 100$ ps) and good energy ($\sigma_E/E = 5.7\%/\sqrt{E(GeV)}$) resolutions. DC and EMC performances are very stable in time despite the five times higher machine background with respect of the operational conditions of the past KLOE run.

2.2 The new calorimeters

QCALT. Two new Quadrupole CALorimeters with Tiles (QCALT) [12] are installed around the DAΦNE low-β quadrupoles at both sides of the IP, with the aim of improving by a factor 5 the rejection efficiency of photons from $K_L \to 3\pi^0$ background events in CP-violating $K_L \to 2\pi^0$ decays. Each calorimeter, 1 m long, consists of a dodecagonal structure, arranged as a sampling of 5 layers of 5 mm thick scintillator plates, alternated with 3.5 mm thick tungsten plates, for a total $\approx 5X_0$ thickness. The active part of each plane is composed by 20 tiles of 5×5 cm^2 area with 1 mm diameter WaveLength Shifter fibers. Each fiber is then optically connected to a Silicon PhotoMulitplier (SiPM), for a total of 2400 channels. The QCALTs have a ~ 2 cm resolution along the beam axis and a time resolution of ~ 1 ns.

CCALT. Two Crystal CALorimeters with Timing (CCALT) [13] are installed very close to the IP, near the first focusing quadrupoles of DAΦNE, in order to enlarge from 20° to 11° the angular acceptance for particles coming from the IP and also improve multi-photon detection in rare decays such as $K_S \to \gamma\gamma$, $\eta \to \pi^0\gamma\gamma$ and $K_S \to 3\pi^0$. Each CCALT module is made of 4 aluminum shells containing 4 LYSO crystals, customly shaped, readout by SiPMs. Tests performed at the Frascati Beam Test Facility allowed to measure an energy resolution better than 5% and a time resolution of about 49 ns at an enegy of 100 MeV.

2.3 The new system of taggers

A tagging system is installed to detect scattered electrons and positrons in the gamma-gamma interactions: $e^+e^- \to e^+e^-\gamma^*\gamma^* \to e^+e^- + X$ that deviate from the equilibrium orbit, during the propagation along the machine lattice, since their energy is lower than 510 MeV. The KLOE-2 tagging system is composed by a Low Energy Tagger (LET) and a High Energy Tagger (HET).

LET. Its aim [14] is to detect scattered electrons and positrons in the energy range $150 \leq E(MeV) \leq 350$ that form a mean angle $\theta = 11°$ with the beam axis. It consists of two calorimeters symmetrically placed at 1 m from both the sides of the IP, between the beam-pipe outer support structure and the inner wall of the DC. Calorimeters have been chosen because simulations show that there is a weak correlation between the energy and the trajectory of scattered $e^\pm$: so there is no need of an accurate reconstrution of their positions. Conversely LET has a good energy resolution in order to improve the invariant mass resolution of $\gamma\gamma$ interaction products, together with a good time resolution (~ 1 ns) to allow the correlation between the bunch crossing and the detected events and reject accidental particles coming from the machine background. The LET calorimeters are made of 20 LYSO crystals and readout by 3″ SiPMs. The energy resolution is $\sigma_E/E = 2.4\%/\sqrt{E(GeV)} \oplus 6.5\% \oplus 5MeV/E(GeV)$ where the last term accounts for the electronic noise.

HET. Designed [15] to detect scattered electrons and positrons with an energy $E > 420\ MeV$, whose detection is very important for the $\gamma\gamma \to \pi^0$ search: in case of π^0 production the photons from the decay have to be detected in the central KLOE apparatus in association with the HET signal. The two HET detectors are placed in symmetrical positions 11 m away from the IP, on both electrons and positrons sides. Each HET station is an hodoscope consisting of a set of 28 $3 \times 5 \times 6$ mm^3 plastic scintillators (EJ-228), placed at different distances from the beam-line and one additional scintillator covering all the others that is used for coincidence purposes. So, knowing which scintillator has been fired, it is possible to immediately measure the distance between the impinging particle and the beam. The light emitted by each of the 28 scintillators is read out through a plastic light guide by high ($\simeq 35\%$) quantum efficiency photomultipliers. The HET has a time resolution of 550 ps that allows one to correctly determine the bunches which generated the interaction. Figure 1 shows how, using this information, one can reproduce the DAΦNE bunch structure ($T_{RF} = 2.7$ ns) to which the KLOE-2 apparatus is synchronized [16].

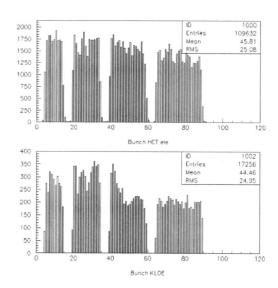

Figure 1: The DAΦNE bunch structure as measured by the electron HET station (top) and by the KLOE central detector (bottom) [16].

2.4 The Inner Tracker

To improve the resolution on decay vertices close to the IP, the Inner Tracker (IT) has been inserted in the free space between the beam pipe and the DC inner wall [17]. The IT is the first ever built and operated cylindrical GEM (Gas Electron Multiplyer) detector produced with a single-mask etching, following a technique developed at LNF, with a total material budget below $2\%X_0$ [18]. This allows to minimize the multiple scattering of low-momentum tracks and the probability of photon conversions. The IT is composed by four co-axial layers with radii from 13 cm (to preserve the $K_S K_L$ quantum interference region) to 20.5 cm (due to the limit coming from the DC inner wall). Each layer has a total active length of 70 cm and is made of 5 concentric cylindrical electrodes: one cathod, three GEM foils that work as multiplication stages and one anode with a stereo read-out made of two views: X longitudinal strips and V diagonal strips, both with a 650 μm pitch, connected through conductive vias and common backplane at an angle of $\simeq 27°$. The readout results in more than 25000 front end electronic channels read by 64-channel GASTONE ASIC chips specifically designed for KLOE-2 [19]. The IT operates in an Argon (90%) and Isobuthane (10%) gas mixture to limit the discharge probability. The efficiency with cosmic-ray muon tracks reconstructed also by the DC is 94% for the single-view and $\approx 85\%$ for the two-views [20].

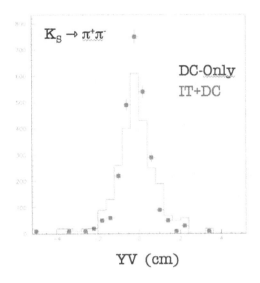

Figure 2: Comparison of the distributions of the Y coordinate of the reconstructed vertices positions for $K_S \to \pi^+\pi^-$ decays in the DC-only case (solid black line) and in the integrated IT+DC reconstruction (red points) [20].

In order to measure efficiency and resolutions and perform IT calibration and alignment, it has been developed a procedure that profits of the excellent DC track reconstruction performance. We used samples of both cosmic-ray muons and Bhabha scattering events, taking into account several effects including one involving non-radial tracks events and another one acounting for the effect of the KLOE-2 magnetic field on the reconstructed tracks.

A first set of alignment and calibration parameters has been validated with a result for the resolution of $\simeq 400\,\mu$m for Bhabha scattering events. This first set of IT calibrations together with the DC information have been applied for the integrated tracking IT+DC (using a Kalman filter technique) and for the reconstruction of the vertices of $K_S \to \pi^+\pi^-$ events. The preliminary results, that can be seen in figure 2, (considering the K_S lifetime contribution and the negligible beam size) give a resolution on reconstructed vertices positions of $\sim$0.7 cm with the IT+DC data, instead of the $\sim$1 cm with DC only data [20].

3 The DAΦNE Collider and KLOE-2 Data Quality

The DAΦNE collider is a Φ factory [1] in which electrons and positrons collide at the energy of the Φ peak ($\sqrt{s}$=1019.4 MeV) mainly producing Kaons. DAΦNE has also been upgraded with sextupoles and a new *crab-waiste* beams interaction scheme [21](with 25 mrad beam crossing angle) with a topping-up injection that allowed to reach a record luminosity of $L_{peak} = 2 \times 10^{32}$ $cm^{-2}s^{-1}$ and a maximum daily rate of $L_{max} = 13$ pb^{-1} with a collection efficiency of 80%. The KLOE-2 detector provides instantaneous luminosity measurement, by counting the large angle Bhabha events, and machine backgrounds monitoring via three different observables: the current seen by the IT, the current seen by the DC and the counting rate of EMC end-caps. These background probes are continuously monitored and threshold values are defined such to allow the detector to operate safely and to collect good quality data.

Some benchmark physics channels are selected and the results are in good agreement with previous KLOE data distributions, despite the increased levels of the machine background. For instance $K_S \to \pi^{\pm}$ (from which the lifetime of K_S can be measured obtaining $t_S = 0.968 \pm 0.034$ with respect of the PDG value [22]) and $\phi \to \eta\gamma$ (with $\eta \to \gamma\gamma$ and $\eta \to 3\pi^0$).

To precisely monitor the energy of the ϕ peak an energy scan is performed by shifting the central value of the DAΦNE radiofrequency. The event rates in different ϕ decay channels are measured and normalized to the large angle Bhabha event rate (including a radiative correction and beam energy spread of 300 keV): they are well fitted by the ϕ lineshape [23].

4 The $K_S \to 3\pi^0$ Preliminary Analysis

The $3\pi^0$ is a pure CP$= -1$ state, so an observation of the $K_S \to 3\pi^0$ decay would be an unambiguous signal of CP violation. The standard model prediction is $BR(K_S \to 3\pi^0) \sim 1.9 \times 10^9$ with an accuracy better than 1%. The KLOE experiment set a limit using 1.7 fb^{-1} of integrated luminosity, collected in the period 2004-2005, obtaining a $BR(K_S \to 3\pi^0) < 2.6 \times 10^8$ at 90% C.L. [24].

This here reported is the first preliminary analysis performed with 300 pb^{-1} of KLOE-2 data. The analysis technique is simple: one looks for a very clean K_S tag that is provided by the K_L interaction in the EMC (K_L-crash events) together with the observation of 6 prompt-photons (neutral particles that travel with $\beta = 1$ from the IP to the EMC) coming from the K_S decay that constitutes the signal. The main background is due to wrongly reconstructed $K_S \to 2\pi^0 \to 4\gamma s$ decays (used also for normalization). The analysis is then based on the photons counting and kinematic fit both in the $3\pi^0$ and $2\pi^0$ hypothesis.

After some cleanings of the events a kinematic fit is performed imposing energy and momentum conservation, the kaon mass and the velocity of the 6 γs in the final state. To improve the rejection of events with split clusters, two χ^2-like variables are used: $\zeta_{2\pi}$ (selecting the best 4 out of 6 clusters satisfying the kinematic constraints, in the $2\pi^0$ hypothesis) and $\zeta_{3\pi}$ (verifying the signal $3\pi^0$ hypothesis by looking at the reconstructed masses of the 3 pions). Those functions are used to select the signal box in which the signal-to-background ratio is maximized. Only events in this region are selected. Then other cuts on the γs energy and on the minimum distance between reconstructed EMC clusters are applied, resulting in a 19.2% efficiency after all cuts and in a 10 times better background rejection compared to the old analysis. The work is still ongoing, the procedure has been veryfied and we expect to reach a sensitivity of the order of $BR(K_S \to 3\pi^0) < 10^{-8}$ at 90% C.L. with an increased statistics.

5 Conclusions

The KLOE-2 experiment, upgraded with a new series of sub-detectors (among which there is the IT: the first cylindrical GEM detector used in high-energy physics experiment), is fully operational and running since 2014 and several physics analysis have started with its data. The goal will be to reach 5 fb^{-1} of acquired luminosity by the end of data-taking in March 2018 with DAΦNE that is delivering luminosity at $\sim$10 pb$^{-1}/day$, in line with the KLOE-2 plan.

References

[1] C. Milardi et al., JINST, **7** (2012) T03002.
[2] F. Bossi et al., Nuovo Cimento, **30** (2008) 10.
[3] G. Amelino Camelia et al., Eur. Phys. J. C, **68** (2010) 619.
[4] A. Di Domenico et al., PoS (EPS-HEP), **314** (2017) 213.
[5] A. Kupsc et al., PoS (EPS-HEP), **314** (2017) 385.
[6] G. Mandaglio et al., PoS (EPS-HEP), **314** (2017) 073.
[7] G. Bencivenni et al., JINST, **12** (2017) C07016.
[8] D. Babusci et al., Acta Phys. Pol. B, **46** (2015) 81.
[9] F. Happacher et al., Nucl. Phys. Proc. Suppl., **197** (2009) 215.
[10] M. Adinolfi et al., Nucl. Instrum. Meth. A, **488** (2002) 51.
[11] M. Adinolfi et al., Nucl. Instrum. Meth. A, **482** (2002) 364.
[12] M. Cordelli et al., Nucl. Instrum. Meth. A, **617** (2010) 105.
[13] M. Cordelli et al., Nucl. Instrum. Meth. A, **718** (2013) 81.
[14] D. Babusci et al., Nucl. Instrum. Meth. A, **617** (2010) 81.
[15] F. Archilli et al., Nucl. Instrum. Meth. A, **617** (2010) 266.
[16] F. Curciarello et al., JINST, **12** (2017) C06037.
[17] A. Di Cicco et al., Acta Phys. Pol. B, **46** (2015) 73.
[18] G. Bencivenni et al., Nucl. Instrum. Meth. A, **581** (2007) 581.
[19] A. Balla et al., Nucl. Instrum. Meth. A, **721** (2013) 523.
[20] E. De Lucia et al., PoS (EPS-HEP), **314** (2017) 491.
[21] M. Zobov et al., Phys. Rev. Lett. 104, **174801** (2010) 1.
[22] C. Patrignani et al. (PDG), Chin. Phys. C, **40** (2016) 100001.
[23] A. Passeri et al., J. Phys. Conf. Ser., **800** (2017) 012038.
[24] D. Babusci et al., Phys. Lett. B, **723** (2013) 54.

LATEST TESTS OF HARD QCD AT HERA

Oleksandr Zenaiev [a] on behalf of the H1 and ZEUS Collaborations
DESY, Notkestraße 85, D-22609, Hamburg, Germany

Abstract. New results from HERA are presented. H1 and ZEUS measurements of open beauty and charm production cross sections in deep inelastic scattering (DIS) are combined. The new combined data are used to determine the charm and beauty quark masses. H1 measurements of jet production cross sections in DIS are presented and compared for the first time to next-to-next-to-leading order (NNLO) predictions. The strong coupling constant is determined from these data using NNLO predictions to be $\alpha_S(M_Z) = 0.1157(20)_{\rm exp}(29)_{\rm th}$. ZEUS measurements of prompt photons production in DIS provide a precise test of perturbative QCD.

1 Introduction

Measurements of heavy-quark, jet and prompt photon production in neutral-current (NC) deep inelastic scattering (DIS) at HERA provide stringent tests of perturbative QCD. They can also serve as input to extract QCD parameters. In particular, measurements of charm and beauty production are used to extract the heavy-quark masses, while studies of jet production allow precise determination of the strong coupling constant, $\alpha_S(M_Z)$, and provide constraints on the gluon distribution in the fits of parton distribution functions (PDFs).

2 Combination and QCD analysis of beauty and charm production measurements

At HERA various flavour tagging methods were applied for beauty and charm cross section measurements [1–13], thus allowing for a significant reduction of statistical and systematic uncertainties when combining different measurements. The reduced charm, $\sigma_{\rm red}^{c\bar{c}}$, and beauty, $\sigma_{\rm red}^{b\bar{b}}$, cross sections are combined to create one consistent set of charm and beauty cross sections in the kinematic range of photon virtuality $2.5 \leq Q^2 \leq 2000$ GeV^2 and Bjorken scaling variable $3 \times 10^{-5} \leq x_{\rm Bj} \leq 5 \times 10^{-2}$. This analysis is an extension of the previous H1 and ZEUS combination [14] of charm measurements in DIS [1–8] with new charm and beauty data [1, 9–13].

The reduced cross sections are obtained from the visible cross sections, defined as the D-, μ-, e- or jet-production cross sections in a particular kinematic range, using next-to-leading order (NLO) theoretical predictions by the HVQDIS program [15]. Only the shape of these theory predictions in terms of kinematic variables is relevant for the corrections, while their normalisation cancels. The combination of reduced cross sections is based on the procedure

---
[a] E-mail: oleksandr.zenaiev@desy.de

described elsewhere and used in previous HERA combinations [14, 16–19], accounting for all correlations in the uncertainties.

In total, 209 charm and 57 beauty data points are combined simultaneously to 52 reduced charm and 27 beauty cross-section measurements, respectively. A total χ^2 of 149 for 187 degrees of freedom is obtained in the combination indicating consistency of input data and conservative estimates of the uncertainties. The individual datasets as well as the results of the combination are shown in Fig. 1, while Fig. 2 present a comparison of the NLO QCD predictions in the fixed-flavour-number scheme (FFNS) to the combined data. The predictions are calculated using HERAPDF2.0 FF3A [16] and ABM11 PDFs [20] at NLO, and ABMP16 PDFs [21] at approximate NNLO as implemented in the OPENQCDRAD program [22] interfaced in the xFitter framework [23]. All calculations yield a similar reasonable description of the data, within the uncertainties, in the whole kinematic range.

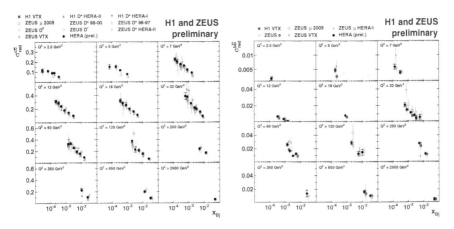

Figure 1: Combined reduced charm (left) and beauty (right) cross sections (full circles) as a function of x_{Bj} for different values of Q^2. For presentation purposes each individual measurement is shifted in x_{Bj}.

The combined beauty and charm data are included in a QCD analysis at NLO, performed using XFITTER [23], together with the combined HERA inclusive DIS data [16]. The methodology follows closely the approach of HERAPDF2.0 FF3A [16], employing the fixed-flavour scheme with three active flavours at all scales. The heavy-quark masses are left free in the fit. The uncertainties are estimated as in the general approach of HERAPDF2.0 [16] in which the fit, model, and parametrisation uncertainties are taken into account, and the total uncertainty is obtained by adding the fit, model, and parametrisation uncertainties in quadrature.

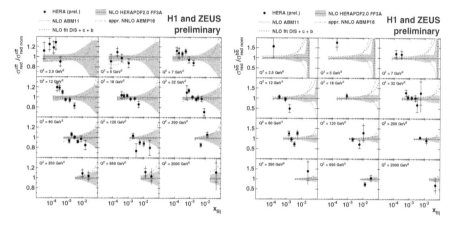

Figure 2: Combined reduced charm (left) and beauty (right) cross sections, compared to the NLO and approximate NNLO QCD predictions obtained using various PDFs, normalised to HERAPDF2.0 FF3A.

The results for the fitted heavy-quark masses extracted are:

$$m_c(m_c) = 1290^{+46}_{-41}(\text{fit})^{+62}_{-14}(\text{mod})^{+7}_{-31}(\text{par}) \text{ MeV},$$
$$m_b(m_b) = 4049^{+104}_{-109}(\text{fit})^{+90}_{-32}(\text{mod})^{+1}_{-31}(\text{par}) \text{ MeV}. \quad (1)$$

The model uncertainties are dominated by theoretical uncertainties arising from the scale variations for heavy-quark production. The resulting theoretical predictions are shown in Fig. 2. A cross check was performed using the Monte Carlo method [24, 25] which yielded consistent results. These data can be also used to evaluate the heavy-quark mass running to test QCD expectations [26].

3 Measurements of jet cross sections in DIS and determination of $\alpha_S(M_Z)$

New measurements of jet production cross sections in NC DIS have been performed in the kinematic region $5.5 < Q^2 < 80$ GeV^2 and inelasticity $0.2 < y < 0.6$ by H1 [27]. Inclusive jet cross sections were measured as a function of Q^2 and jet transverse momentum, P_T^{jet}, and dijet and trijet cross sections measured as functions of Q^2 and the average P_T^{jet} of the two or three leading jets. Furthermore, the kinematic range of an earlier measurement of inclusive jet cross sections [28] at higher values of Q^2 was extended to lower values of P_T^{jet}. In Fig. 3 the data are compared to NLO, approximate NNLO and full NNLO predictions [29], whenever available. The predictions describe the data well within the experimental and theoretical uncertainties. The NNLO predictions improve the description of inclusive jet and dijet cross sections as compared to lower order theoretical calculations. Measurements of jet cross sections normalised to the inclusive NC DIS cross section in the respective Q^2 range further improve the experimental precision, owing to partial cancellation of experimental uncertainties.

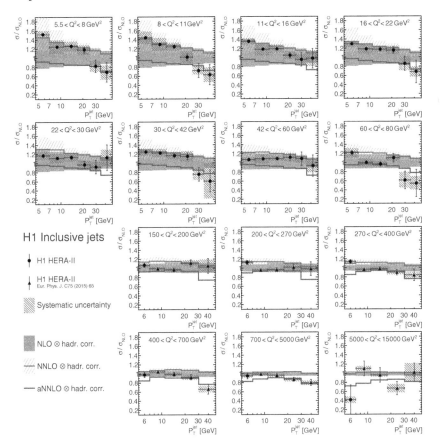

Figure 3: Inclusive jet cross sections in NC DIS compared to NLO, approximate NNLO and full NNLO predictions.

The measured jet production cross sections are exploited for a determination of $\alpha_S(M_Z)$ using the full NNLO calculations. The determined value is $\alpha_S(M_Z) = 0.1157(20)_{\rm exp}(29)_{\rm th}$ [30], where the jet data are restricted to high scales above 28 GeV. Uncertainties due to the input PDFs or the hadronisation corrections are found to be small, and the largest source of uncertainty is from scale variations of the NNLO calculations. Values of $\alpha_S(M_Z)$ determined from inclusive jet data or dijet data alone are found to be consistent with the main result. All these results are found to be consistent with each other and with the world average value of $\alpha_S(M_Z)$ (see Fig. 4, left). As an alternative approach, a combined determination of PDF parameters and $\alpha_S(M_Z)$ in NNLO accuracy was performed, yielding consistent results. The inclusion of H1 jet data into such a simultaneous PDF and $\alpha_S(M_Z)$ determination provides stringent constraints on $\alpha_S(M_Z)$ and the gluon density. This is the first precision extraction of $\alpha_S(M_Z)$ from jet data at NNLO involving a hadron in the initial state.

The running of the strong coupling constant, i.e. the renormalisation scale dependence $\alpha_S(\mu_R)$, is tested in the range of approximately 7 to 90 GeV by dividing the jet data into ten subsets of approximately constant scale [30] (see Fig. 4, right). The scale dependence of the coupling is found to be consistent with the expectation.

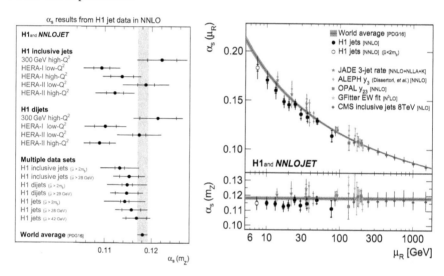

Figure 4: Summary of $\alpha_S(M_Z)$ values obtained from fits to individual and multiple H1 jet data sets (left), and results for $\alpha_S(M_Z)$ and $\alpha_S(\mu_R)$ for fits to data points arranged in groups of similar μ_R (right).

4 Measurements of isolated photon production in DIS

The production of isolated photons accompanied by jets has been measured in DIS by ZEUS, extending earlier ZEUS results [31] which studied single-particle distributions. In the new measurements, differential cross sections have been evaluated as functions of pairs of measured variables in combination: the fraction of the incoming photon energy and momentum that is transferred to the outgoing photon and the leading jet, x_γ; the fraction of the incoming proton energy taken by the parton that interacts with the exchanged photon, x_p; the differences in azimuthal angle and pseudorapidity between the outgoing photon and the leading jet, $\Delta\phi$ and $\Delta\eta$, respectively, and between the outgoing photon and the scattered electron, $\Delta\eta_{e,\gamma}$ and $\Delta\phi_{e,\gamma}$, respectively.

The PYTHIA prediction for the quark-radiated photon component plus the DJANGOH–HERACLES calculation for the lepton-radiated component describes all the distributions, if the PYTHIA prediction is scaled up by a factor of 1.6. Furthermore, predictions from two theoretical models were also compared to the data (see Fig. 5): the BLZ model based on the k_T-factorisation method [32] gives fair description of the data but does not describe well the overall normalisation or the shape of some of the distributions, while the AFG predictions at NLO based on the $\overline{MS}$ scheme [33] gives a good description of almost all the distributions.

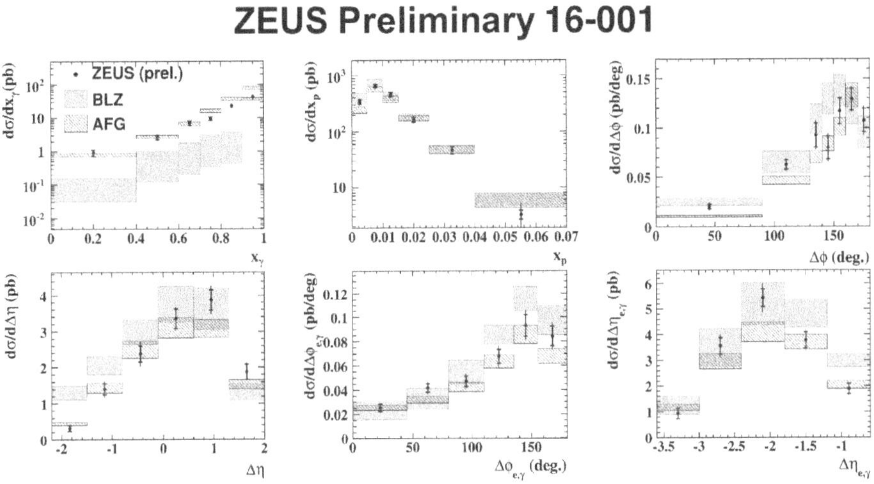

Figure 5: Differential cross sections for selected variables compared to the AFG and BLZ theoretical predictions.

5 Summary

Ten years after the end of data taking at the HERA accelerator the H1 and ZEUS experiments continue providing new measurements and also combinations of previously published results. Measurements of beauty and charm production cross sections were combined, accounting for their statistical and systematic correlations. This improved the previously published charm cross sections, and the beauty cross sections have been combined for the first time. The running charm and beauty masses were extracted in the QCD analysis using the new combined data. The first precision extraction of $\alpha_S(M_Z)$ from new jet data at NNLO involving a hadron in the initial state has been performed by H1, opening a new chapter of precision QCD measurements at hadron colliders. New measurement of isolated photon production by ZEUS complements earlier measurements of this process and provides further tests of perturbative QCD.

References

[1] F. D. Aaron et al., Eur. Phys. J. **C65**, (2010) 89 [arXiv:0907.2643].
[2] A. Aktas et al., Eur. Phys. J. **C51**, (2007) 271 [hep-ex/0701023].
[3] F. D. Aaron et al., Eur. Phys. J. **C71**, (2011) 1769 [arXiv:1106.1028].
[4] F. D. Aaron et al., Phys. Lett. **B686**, (2010) 91 [arXiv:0911.3989].
[5] J. Breitweg et al., Eur. Phys. J. **C12**, (2000) 35 [hep-ex/9908012].
[6] S. Chekanov et al., Phys. Rev. **D69**, (2004) 012004 [hep-ex/0308068].
[7] S. Chekanov et al., Eur. Phys. J. **C63**, (2009) 171 [arXiv:0812.3775].
[8] S. Chekanov et al., Eur. Phys. J. **C65**, (2010) 65 [arXiv:0904.3487].
[9] H. Abramowicz et al., JHEP **05**, (2013) 023 [arXiv:1302.5058].
[10] H. Abramowicz et al., JHEP **05**, (2013) 097 [arXiv:1303.6578].
[11] H. Abramowicz et al., JHEP **09**, (2014) 127 [arXiv:1405.6915].
[12] H. Abramowicz et al., Eur. Phys. J. **C71**, (2011) 1573 [arXiv:1101.3692].
[13] H. Abramowicz et al., Eur. Phys. J. **C69**, (2010) 347 [arXiv:1005.3396].
[14] F. D. Aaron et al., Eur. Phys. J. **C73**, (2013) 2311 [arXiv:1211.1182].
[15] B. W. Harris and J. Smith, Phys. Rev. **D57**, (1998) 2806 [hep-ph/9706334].
[16] H. Abramowicz et al., Eur. Phys. J. **C75**, (2015) 580 [arXiv:1506.06042].
[17] A. Glazov, AIP Conf. Proc. **792**, (2005) 237.
[18] A. Atkas et al., Eur. Phys. J. **C63**, (2009) 625 [arXiv:0904.0929].
[19] F. D. Aaron et al., JHEP **01**, (2010) 109 [arXiv:0911.0884].
[20] S. Alekhin, J. Blümlein and S. Moch, Phys. Rev. **D86**, (2012) 054009 [arXiv:1202.2281].
[21] S. Alekhin, J. Blümlein, S. Moch and R. Placakyte, [arXiv:1701.05838].
[22] S. Alekhin, J. Blümlein, S. Moch, Phys. Rev. **D86**, (2012) 054009 [arXiv:1202.2281].
[23] S. Alekhin et al., Eur. Phys. J. **C75**, (2015) 304 [arXiv:1410.4412], www.xfitter.org.
[24] W. T. Giele and S. Keller, Phys. Rev. **D58**, (1998) 094023 [hep-ph/9803393].
[25] W. T. Giele, S. A. Keller and D. A. Kosower, hep-ph/0104052.
[26] A. Gizhko et al., Phys. Lett. **B775**, (2017) 233 [arxiv:1705.08863].
[27] V. Andreev et al., Eur. Phys. J. **C77**, (2017) 215 [arxiv:1611.03421].
[28] V. Andreev et al., Eur. Phys. J. **C75**, (2015) 65 [arxiv:1406.4709].
[29] J. Currie et al., Phys. Rev. Lett. **117**, (2016) 042001 [arxiv:1606.03991].
[30] V. Andreev et al., Eur. Phys. J. **C77**, (2017) 791 [arxiv:1709.07251].
[31] H. Abramowicz et al., Phys. Lett. **B715**, (2012) 88.
[32] S. Baranov, A. Lipatov and N. Zotov, Phys. Rev. **D81**, (2010) 094034.
[33] P. Aurenche, M. Fontannaz and J.Ph. Guillet, Eur. Phys. J. **C44**, (2005) 395.

xFITTER

Oleksandr Zenaiev [a] on behalf of the xFitter team
DESY, Notkestraße 85, D-22609, Hamburg, Germany

Abstract. An accurate knowledge of the parton distribution functions (PDFs) plays a critical role for the precision tests of the Standard Model (SM) and impact the theory predictions of beyond SM production. The xFitter project provides a unique open-source software framework for the determination of the proton PDFs and the interpretation of experimental measurements in the context of QCD. The latest xFitter software release includes many new features and improvements. We also highlight recent results obtained using xFitter, as well as novel studies performed by the xFitter team.

The interpretation of the measurements in hadron collisions relies on the concept of the factorisation in QCD, when inclusive cross sections are given by:

$$\sigma = \sum_{a,b} \int_0^1 dx_1\ dx_2 f_a(x_1, \mu_f^2) f_b(x_2, \mu_f^2) \times \hat{\sigma}^{ab}(x_1, x_2; \alpha_s(\mu_r^2), \mu_r^2, \mu_f^2).$$

Here the cross section σ is expressed as a convolution of parton distribution functions (PDFs) f_a and f_b with the partonic cross section $\hat{\sigma}^{ab}$. At leading order (LO) in the perturbative expansion of the strong coupling constant α_s, the PDFs represent the probability of finding a specific parton a (b) in the first (second) hadron carrying a fraction x_1 (x_2) of its momentum. The PDFs depend on the factorisation scale, μ_f, while the partonic cross sections depend on α_s, and the factorisation and renormalisation scales, μ_f and μ_r, respectively. The parton cross sections $\hat{\sigma}^{ab}$ are calculable in perturbative QCD whereas PDFs are determined in global fits to a variety of experimental data.

Several groups are doing the extraction of PDFs, using different input data, theoretical assumptions and fit strategies. Their recent reviews can be found e.g. in Refs. [1–3]. The rapid flow of data from the LHC experiments and the corresponding theoretical developments, which are providing predictions for more complex processes at increasingly higher orders, has motivated the development of tools to combine them together in a fast, efficient, open-source framework.

The open-source QCD fit framework xFitter (former HERAfitter) [4,5] has been developed for the determination of PDFs and the extraction of fundamental parameters of QCD such as the heavy-quark masses and the strong coupling constant. It also provides a common framework for the comparison of different theoretical approaches. Furthermore, it can be used to test the impact of new experimental data on the PDFs and on the SM parameters.

The diagram in Fig. 1 gives a schematic overview of the xFitter structure:

---

[a] E-mail: oleksandr.zenaiev@desy.de

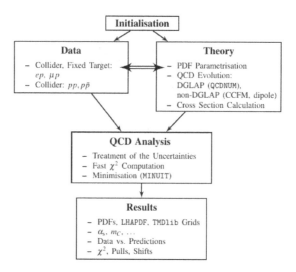

Figure 1: Schematic overview of the xFitter program.

- **Data:** Measurements from various processes in pp, $p\bar{p}$, ep and μp scattering are provided in the xFitter package including the information on their uncorrelated and correlated uncertainties. In total 48 data sets are publicly available as of December 2017 [6]. It is easily possible to add new custom data sets.

- **Theory:** The PDFs are parametrised at a starting scale, Q_0^2, using a functional form and a set of free parameters (several parametrisation forms are available) and evolved to the scale of the measurements Q^2, $Q^2 > Q_0^2$ as predicted by QCD. The calculation of the cross section for a particular process is obtained by the convolution of the evolved PDFs with the corresponding parton scattering cross section. A fast evaluation of cross sections for various processes is possible via an interface to fast grid computations (APPLGRID [7], fastNLO [8–10] and APFELgrid [11]).

- **QCD Analysis:** The free parameters of a QCD analysis are determined in a χ^2 fit. Various choices are available for the treatment of experimental uncertainties in the χ^2 definition. Besides the χ^2 minimisation, alternative approaches to PDF studies are available, such as reweighting (probability distribution based PDFs are updated with new data inputs) and profiling (the individual PDF eigenvector sets of the input PDFs are constrained taking into account the new data). A simple calculation of the χ^2 accounting for data and PDF uncertainties, provided in the LHAPDF format, is available as well, to easily assess the description of the data by the predictions.

- **Results:** The resulting PDFs are provided in a format ready to be used by the LHAPDF library [12, 13] or by TMDlib [14]. xFitter drawing tools can be used to display the PDFs with their uncertainties at a chosen scale, as well as the comparison of data to theoretical predictions, pulls, shifts of nuisance parameters etc.

The latest version of the xfitter program, 2.0.0 Frozen Frog, includes a number of new developments, improvements for the users and bug fixes (for more details see [5]). We foresee further considerable code improvements in the next release aiming to facilitate interfacing new theoretical calculations.

The xfitter program has been used in a number of experimental and theoretical analyses (for the full list see [5]), performed by xfitter developers, theory groups and experimentalists. Latest studies performed by the xfitter team include a new determination of the photon PDF from a fit of HERA inclusive DIS structure functions supplemented by ATLAS data on high-mass Drell-Yan cross sections [15] and a study of the impact of the heavy flavour matching scales on a PDF fit to the combined HERA data set [16].

References

[1] Jon Butterworth et al. *J. Phys.*, G43:023001, 1510.03865.
[2] A. Accardi et al. *Eur. Phys. J.*, C76(8):471, 1603.08906.
[3] J. Gao, L. Harland-Lang, and J. Rojo. 1709.04922.
[4] S. Alekhin et al. *Eur. Phys. J.*, C75(7):304, 2015, 1410.4412.
[5] xFitter web site, https://www.xfitter.org.
[6] Hepforge web site, http://xfitter.hepforge.org/data.html.
[7] T. Carli et al. *Eur. Phys. J.*, C66:503–524, 0911.2985.
[8] T. Kluge, K. Rabbertz, and M. Wobisch. In *Deep inelastic scattering. Proceedings, 14th International Workshop, DIS 2006, Tsukuba, Japan, April 20-24, 2006*, pages 483–486, hep-ph/0609285.
[9] M. Wobisch et al. 1109.1310.
[10] D. Britzger et al. In *Proceedings, 20th International Workshop on Deep-Inelastic Scattering and Related Subjects (DIS 2012)*, pages 217–221, 2012, 1208.3641.
[11] V. Bertone, S. Carrazza, and N. P. Hartland. *Comput. Phys. Commun.*, 212:205–209, 1605.02070.
[12] M. R. Whalley, D. Bourilkov, and R. C. Group. In *HERA and the LHC: A Workshop on the implications of HERA for LHC physics. Proceedings, Part B*, 2005, hep-ph/0508110.
[13] LHAPDF web site, http://lhapdf.hepforge.org.
[14] F. Hautmann et al. *Eur. Phys. J.*, C74:3220, 1408.3015.
[15] F. Giuli et al. *Eur. Phys. J.*, C77(6):400, 1701.08553.
[16] V. Bertone et al. 1707.05343.

LOW ENERGY NUCLEAR REACTIONS IN NORMAL AND EXPLODING BATTERIES

Y. Srivastava[a], A. Widom, J. Swain
Physics Department, Northeastern University, Boston, MASS USA
Georges de Montmollin, L. Rosselli
Lenr-Cities Suisse Sàrl, Rue Charles Knapp 29, CH-2000 Neuchâtel CH

Abstract. Acceleration of electrons and ions in materials under severe stress is known to cause neutron and other particle production in nature [earthquakes, thunderstorms, lightning]. We present theoretical and laboratory evidence of the same through fracture and shredding of electrodes and Coulomb explosion in batteries. Our considerations show that a necessary condition for stable (Lithium battery pack) operation (avoiding Alfvén plasma instability) is that Lithium ions should not be transferred between two battery terminals at a rate exceeding two moles per minute.

1 Introduction

While studies about fracture of materials under stress are legion that are described in standard texts, only recently have they been theoretically studied [1-6] with an eye towards neutron and other particle production when brittle materials undergo fracture and shredding under extreme stress. Experimentally, radioactive emissions from geophysical fracturing of rocks in the earth's crust just prior to large earthquakes have been observed [7]. Neutrons have also been observed from bolts of lightning [8] and from thunderstorms [9]. An excellent review of the experimental situation about neutron and other particle production from fracture, both in nature and in the laboratory, can be found in [10] where the observed fracto-emissions are successfully described through applications of the electro-weak and electro-strong theory [4-6].

Another experimentally well established path, actively followed over the past two decades, leading to neutron and other particle creation (called LENR, low energy nuclear reactions) is through Coulomb explosions [11]. In the pioneering experiments [11], electrons were removed from a beam of deuterium atoms resulting in clusters of positively charged nuclei, deuterons. Once the charge density of a cluster exceeded a critical value, the cluster exploded and the deuterons flying out acquired quite large kinetic energies [$\geq$ several KeV's]. When two such deuterons (from different clusters) "met", they could easily overcome their mutual Coulomb repulsion and cause a nuclear reaction. The legendary dd fusion reaction observed through such Coulomb explosions

$$d + d \rightarrow {}^3\text{He} + n, \qquad (1)$$

has since then been confirmed and refined by several groups and other reactions studied.

---
[a]E-mail: yogendra.srivastava@gmail.com

But deuterons are not mandatory since, Coulomb explosions can also be arranged for highly reactive alkali atoms, such as lithium, sodium and potassium, as well as nickel, all materials we have experimented with, and successfully exploded. As described in [13–15] in some detail, the observed lithium Coulomb explosions are strikingly similar to lithium battery explosions, raising therefore the exciting possibility that the "usual" lithium battery explosions are not entirely chemical. We find [14, 15] that under well defined conditions, indeed there is a nuclear component.

1.1 Battery Cell experiments

The first set of our experiments and their analysis described in [13] concerns the fracture and shredding of both copper and tungsten electrodes in an electrolytic cell; i.e. in a battery. During the fracturing process, a bright plasma glow may be observed on the tip of the cathode, however the plasma is by no means a requirement for the production of new materials in the cell [17]. Many nuclear reactions in the plasma glow mode have been observed by Mizuno [18] and by Cirillo [19] employing scanning electron microscope (SEM) & energy dispersive spectroscopy (EDX) probing product atoms left on the cathode.

An alternative route to the pioneering plasma cell experiments [18, 19], focussing on the energetics and dynamics of shredding and fracture of electrodes through *electro-weak* and *electro-strong* theories has been presented in [13].

The final nuclear reaction products may often be described by reactions of the form

$$e^- + p \to n + \nu_e; \quad \text{[electron capture]}$$
$$n + {}^A X_Z \to {}^{A+1} X_Z; \quad \text{[neutron absorption or halo nuclei]}$$
$${}^{A+1} X_Z \to {}^{A+1} X_{Z+1} + e^- + \bar{\nu}_e; \quad \text{[Fermi beta decay process]}. \quad (2)$$

In our experiments, the dc voltage was about 400 Volts and when employing an electrolyte, the current was about 5 Amperes, depending upon the state of disrepair of the cathode. The plasma lasted for about 5 minutes. A similar experiment with distilled water at the same 400 Volt source, yielded a small current of about 4 milli-Amperes. At such small currents there was no visible plasma near the cathode tip [17]; however, high voltage shredding of the cathode occurred over a long time scale (of ten hours). Evidence of new elements were found in both cases [13].

Comparative current-voltage & energy considerations in lightning and shredding of electrodes in batteries are displayed in the Table below. Of particular note are (i) the mean current (35 Kilo-Amps, twice the Alfvén current) in lightning and (ii) the energy density in our battery set up being thousands of time larger than that in a lightning.

Parameters	Lightning	Battery
Voltage	5×10^8 Volts	4×10^2 Volts
Peak Current	3.5×10^4 Amperes	5 Amperes
Duration	10^{-2} sec	3×10^2 sec.
Peak Power	1.5×10^{13} Watts	2×10^3 Watts
Peak Energy	1.5×10^{11} Joules	6×10^5 Joules
Fiducial Volume	$1.5 \times 10^2 \ m^3$	$3 \times 10^{-7} m^3$
Peak Energy/Volume	$10^9 \ Joules/m^3$	$2 \times 10^{12} \ Joules/m^3$

Very recently, nuclear transmutations following our *electro-strong* theory have been reported in thunderstorms [12]. Since lightning produces neutrons, the considerably enhanced energy density (proportional to the square of the electric field that accelerates electrons and ions) opens the way for LENR transmutations to occur in our battery setups [13].

1.2 Necessary safety conditions for Lithium battery packs

Our rule [14] involves the Alfvén current $I_A \approx 17.3$ kiloAmps and an Alfvén time $\tau_A = N_A|e|/I_A \approx 5.66$ sec. (N_A is the Avagadro's number). Our safety rule asserts that for safe operation of a battery, the current $I \leq 0.2 \ I_A$, yielding a Li^+ transfer rate of less than two moles per minute, for each battery in the pack.

1.3 Tensile and explosive properties of current carrying wires

A sufficiently violent explosion from a strong current pulse is a conventional manufacturing method for fabricating metallic nano-particles. Such high current fracture explosions are known to often be accompanied by LENR [15]. Neutron emission from exploding wire discharge was discovered in [20].

Through a detailed theoretical analysis [15], we have shown that in a wire carrying a current I, the relativistic energy of an electron is given by

$$\mathcal{E} = mc^2 \gamma = mc^2 \sqrt{1 + \eta^2 \left(\frac{I}{I_A}\right)^2}, \quad (3)$$

wherein η is a dimensionless measure of the inductance per unit length of the wire.

The chemical engineering thermodynamic laws involving nuclear reactions are clearly the same as those involving any other reactions. The electron capture nuclear chemical reaction Eq.(2) occurs if the electron current I is above a threshold current

$$I > I_{(e^-+p^+ \to n+\nu)} = \frac{I_A}{\eta} \sqrt{\left(\frac{M_n - M_p}{m}\right)^2 - 1} \approx \left(\frac{39.63 \text{ kilo ampere}}{\eta}\right). \quad (4)$$

The threshold current for the reaction in Eq.(4) is clearly exceeded in strong lightning bolts wherein neutrons have been observed [8]. We note that the threshold current may be decreased in wire environments provided that $\eta \gg 1$. Threshold currents for other electroweak electron capture events in nuclei may be found in [15]. Once the supply of neutrons from electroweak processes is established and becomes substantial, neutron fission reactions are then also kinematically allowed. The particular nuclear reaction obtained depends on the details of the chemical and isotopic composition of the metal wire.

References

[1] G. Preparata, *Il Nuovo Cimento*, **104**, 1289 (1991).
[2] A. Widom and L. Larsen, *Eur Phy. J.* **C46** 107 (2006).
[3] Y. Srivastava, A. Widom, L. Larsen, *Pramana - J. Phys.* **75** 617 (2010).
[4] A. Widom, J. Swain, Y. Srivastava, *J. Phys. G. Nucl. Part. Phys.* **40**, 015006 (2013).
[5] Y. N. Srivastava et. al., *Key Engineering Materials* **543**, 68 (2013).
[6] A. Widom, J. Swain and Y. N. Srivastava, *Meccanica*, **50** 1205 (2015); arXiv: 1306.6286 [phys. gen-ph].
[7] W. Plastino et al., *J. Radioanal. Nucl. Chem.* **282** 809 (2009); *J. Environ. Radioactivity* **101** 45 (2010).
[8] G. Shah, H. Razdan, C. Bhat and Q. Ali, *Nature* **313**, 773 (1985).
[9] A. Gurevich et al., *Phys. Rev. Lett.* **108**, 125001 (2012).
[10] *Acoustic, Electromagnetic, Neutron Emissions from Fracture and Earthquakes*, Editors, A. Carpinteri, G. Lacidogna and Amedeo Manuello; Springer Berlin (2015); A. Carpinteri and O. Borla, *Engineering Fracture Mechanics*, **177** (2017) 230.
[11] J. Zweiback et al, *Phys. Rev. Lett.*, **84**, 2634(2000);*Phys. Rev.* **E74**, 016403 (2006).
[12] T. Enoto et al, *Nature*, http://dx.doi.org/10.1038/nature24630 (2017).
[13] A. Widom, Y. Srivastava, J. Swain, G. de Montmollin, L. Rosselli, *Reaction products from electrode fracture and Coulomb explosions in batteries*, Engineering Fracture Mechanics, **184**, 88 (2017).
[14] A. Widom, J. Swain, Y. Srivastava, G. de Montmollin, *Lithium Ion Battery Packs and the Alfvén Current Instability*, (under submission).
[15] A. Widom, J. Swain, Y. Srivastava, G. de Montmollin, *Tensile and Explosive Properties of Current Carrying Wires*, (under submission).
[16] Some pictures and videos of the experiments can be found at: http://www.lenr-cities.ch/Article.
[17] J. Dash, M. Zhu and J. Solomon, MIT Colloquium, March 22 (2014).
[18] T. Mizuno et al., *Jpn. J. Appl. Phys.*, **39** (2000) 6055; **40** (2001) L989.
[19] D. Cirillo et. al., Key Engineering Materials **495** 104, **495** 124 (2012).
[20] E. Stephanakis, et al, *Phys. Rev.* **29**, 568 (1972).

CALCULATION OF THE R–RATIO OF $e^+e^- \to$ HADRONS AT THE HIGHER–LOOP LEVELS

A.V. Nesterenko [a]

BLTPh JINR, Dubna, 141980, Russian Federation

Abstract. The calculation of the R–ratio of electron–positron annihilation into hadrons is discussed. The method, which enables one to properly account for all the effects due to continuation of the spacelike perturbative results into the timelike domain at an arbitrary loop level, is delineated.

The theoretical description of a variety of the strong interaction processes is inherently based on the hadronic vacuum polarization function $\Pi(q^2)$, which is defined as the scalar part of the hadronic vacuum polarization tensor

$$\Pi_{\mu\nu}(q^2) = i \int d^4 x\, e^{iqx} \langle 0|\, T\, \{J_\mu(x)\, J_\nu(0)\}|0\rangle = \frac{i}{12\pi^2}(q_\mu q_\nu - g_{\mu\nu} q^2) \Pi(q^2), \quad (1)$$

the related Adler function [1]

$$D(Q^2) = -\frac{d\,\Pi(-Q^2)}{d \ln Q^2}, \quad (2)$$

and the function $R(s)$

$$R(s) = \frac{1}{2\pi i} \lim_{\varepsilon \to 0_+} \bigl[\Pi(s + i\varepsilon) - \Pi(s - i\varepsilon)\bigr] = \frac{\sigma(e^+ e^- \to \text{hadrons}; s)}{\sigma(e^+ e^- \to \mu^+ \mu^-; s)}, \quad (3)$$

which is identified with the so–called R–ratio of electron–positron annihilation into hadrons. The functions $\Pi(q^2)$ (1) and $D(Q^2)$ (2), being the functions of the spacelike kinematic variable $Q^2 = -q^2 > 0$, can be directly accessed within QCD perturbation theory, whereas the R–ratio (3), being the function of the timelike kinematic variable $s = q^2 > 0$, can be described only by making use of the relevant dispersion relations. Specifically, the relation, which expresses the R–ratio (3) in terms of the theoretically calculable Adler function and provides a native way to properly account for the effects due to continuation of the spacelike perturbative results into the timelike domain, can be obtained by integrating Eq. (2) in finite limits, that yields [2,3]

$$R(s) = \frac{1}{2\pi i} \lim_{\varepsilon \to 0_+} \int_{s+i\varepsilon}^{s-i\varepsilon} D(-\zeta)\, \frac{d\zeta}{\zeta}. \quad (4)$$

The integration contour in this equation lies in the region of analyticity of the integrand in the complex ζ–plane.

---

[a] E-mail: nesterav@theor.jinr.ru, nesterav@gmail.com

It is necessary to outline that the dispersion relations, which express the functions $\Pi(q^2)$, $R(s)$, and $D(Q^2)$ in terms of each other, rely only on the kinematics of the process on hand and involve neither model–dependent phenomenological assumptions nor additional approximations. In turn, such relations impose a number of strict physical intrinsically nonperturbative constraints on the functions $\Pi(q^2)$, $R(s)$, and $D(Q^2)$, that should certainly be accounted for when one comes out of the limits of applicability of the QCD perturbation theory. It is worthwhile to note also that these nonperturbative restrictions have been merged with corresponding perturbative input in the framework of dispersively improved perturbation theory (DPT) [4–6] (its preliminary formulation was discussed in Ref. [7]). In particular, the DPT enables one to overcome some inherent difficulties of the QCD perturbation theory and extend its applicability range towards the infrared domain, see book [4] and references therein.

This study is primarily focused on the theoretical description of the R–ratio of electron–positron annihilation into hadrons (3) at moderate and high energies, so that the nonperturbative aspects of the strong interactions will be disregarded hereinafter. For this purpose the effects due to the masses of the involved particles can be safely neglected (the impact of such effects on the low–energy behavior of the functions on hand was discussed in, e.g., Refs. [4–6,8]). In the massless limit the relation (4) can be represented as

$$R^{(\ell)}(s) = 1 + r^{(\ell)}(s), \quad r^{(\ell)}(s) = \int_s^\infty \rho^{(\ell)}(\sigma) \frac{d\sigma}{\sigma}, \tag{5}$$

where

$$\rho^{(\ell)}(\sigma) = \frac{1}{2\pi i} \lim_{\varepsilon \to 0_+} \left[d^{(\ell)}(-\sigma - i\varepsilon) - d^{(\ell)}(-\sigma + i\varepsilon) \right] \tag{6}$$

stands for the spectral function and $d^{(\ell)}(Q^2)$ denotes the ℓ–loop strong correction to the Adler function (2). As mentioned above, only perturbative contributions will be retained in Eq. (6) in what follows, that makes Eq. (5) identical to that of the "Analytic perturbation theory" [9] (for some of its applications see Refs. [10–19]). A discussion of the nonperturbative terms in the spectral density $\rho^{(\ell)}(\sigma)$ can be found in, e.g., Refs. [20–23].

The perturbative expression for the Adler function (2) takes the form of the power series in the so–called QCD couplant $a_s^{(\ell)}(Q^2) = \alpha_s^{(\ell)}(Q^2)\,\beta_0/(4\pi)$

$$D_{\text{pert}}^{(\ell)}(Q^2) = 1 + d_{\text{pert}}^{(\ell)}(Q^2), \quad d_{\text{pert}}^{(\ell)}(Q^2) = \sum_{j=1}^{\ell} d_j \left[a_s^{(\ell)}(Q^2) \right]^j. \tag{7}$$

Here ℓ specifies the loop level, $d_1 = 4/\beta_0$, $\beta_0 = 11 - 2n_f/3$, n_f is the number of active flavors, the common prefactor $N_c \sum_{f=1}^{n_f} Q_f^2$ is omitted throughout, $N_c = 3$ denotes the number of colors, and Q_f stands for the electric charge of

f–th quark. The QCD couplant entering Eq. (7) can be represented as

$$a_s^{(\ell)}(Q^2) = \sum_{n=1}^{\ell} \sum_{m=0}^{n-1} b_n^m \frac{\ln^m(\ln z)}{\ln^n z}, \tag{8}$$

where $z = Q^2/\Lambda^2$, b_n^m is the combination of the β function perturbative expansion coefficients ($b_1^0 = 1$, $b_2^0 = 0$, $b_2^1 = -\beta_1/\beta_0^2$, etc.), and Λ denotes the QCD scale parameter. The Adler function perturbative expansion coefficients d_j were calculated up to the four–loop level [24], whereas the β function perturbative expansion coefficients β_j are available up to the five–loop level [25].

Since the calculation of the spectral function $\rho^{(\ell)}(\sigma)$ (6) becomes rather cumbrous beyond the one–loop level (the explicit expressions for $\rho^{(\ell)}(\sigma)$ at first four loop levels can be found in Ref. [26]), one commonly re–expands the strong correction $r^{(\ell)}(s)$ (5) at high energies, that eventually leads to [27, 28]

$$r^{(\ell)}(s) = \sum_{j=1}^{\ell} d_j \left[a_s^{(\ell)}(|s|) \right]^j - \sum_{j=1}^{\ell} d_j \sum_{n=1}^{\infty} \frac{(-1)^{n+1}}{(2n+1)!} \pi^{2n}$$

$$\times \sum_{k_1=0}^{\ell-1} \cdots \sum_{k_{2n}=0}^{\ell-1} \left(\prod_{p=1}^{2n} B_{k_p} \right) \left[\prod_{t=0}^{2n-1} \left(j + t + k_1 + k_2 + \ldots + k_t \right) \right]$$

$$\times \left[a_s^{(\ell)}(|s|) \right]^{j+2n+k_1+k_2+\ldots+k_{2n}}, \qquad \sqrt{s}/\Lambda > \exp(\pi/2). \tag{9}$$

It is necessary to emphasize that the re–expansion (9) is valid only for $\sqrt{s}/\Lambda > \exp(\pi/2) \simeq 4.81$, and it converges rather slowly when the energy scale approaches this value. If the number of terms retained on the right–hand side of Eq. (9) is large enough, then it can provide quite accurate approximation of the strong correction to the R–ratio (5). However, one usually truncates the re–expansion (9) at the order ℓ, thereby neglecting all the higher–order π^2–terms (though, the latter may not necessarily be negligible, see Ref. [28]), that results in the expression commonly employed in the practical applications:

$$R_{\text{appr}}^{(\ell)}(s) = 1 + r_{\text{appr}}^{(\ell)}(s), \quad r_{\text{appr}}^{(\ell)}(s) = \sum_{j=1}^{\ell} r_j \left[a_s^{(\ell)}(|s|) \right]^j, \quad r_j = d_j - \delta_j. \tag{10}$$

Here d_j denote the Adler function perturbative expansion coefficients (7) and δ_j embody the contributions of relevant π^2–terms (9), see Refs. [4, 27–31].

At the same time, the explicit expression for the perturbative spectral function entering Eq. (5) can be calculated at an arbitrary loop level (it is assumed that the involved perturbative coefficients d_j and β_j are known) by making use of the method developed in Ref. [28], namely

$$\rho^{(\ell)}(\sigma) = \sum_{j=1}^{\ell} d_j \sum_{k=0}^{K(j)} \binom{j}{2k+1}(-1)^k \pi^{2k}$$
$$\times \left[\sum_{n=1}^{\ell}\sum_{m=0}^{n-1} b_n^m\, u_n^m(\sigma)\right]^{j-2k-1} \left[\sum_{n=1}^{\ell}\sum_{m=0}^{n-1} b_n^m\, v_n^m(\sigma)\right]^{2k+1}. \qquad (11)$$

In this equation

$$u_n^m(\sigma) = \begin{cases} u_n^0(\sigma), & \text{if } m=0, \\ u_n^0(\sigma)u_0^m(\sigma) - \pi^2 v_n^0(\sigma)v_0^m(\sigma), & \text{if } m \ge 1, \end{cases} \qquad (12)$$

$$v_n^m(\sigma) = \begin{cases} v_n^0(\sigma), & \text{if } m=0, \\ v_n^0(\sigma)u_0^m(\sigma) + u_n^0(\sigma)v_0^m(\sigma), & \text{if } m \ge 1, \end{cases} \qquad (13)$$

$$u_n^0(\sigma) = \frac{1}{(y^2+\pi^2)^n} \sum_{k=0}^{K(n+1)} \binom{n}{2k}(-1)^k \pi^{2k} y^{n-2k}, \qquad (14)$$

$$v_n^0(\sigma) = \frac{1}{(y^2+\pi^2)^n} \sum_{k=0}^{K(n)} \binom{n}{2k+1}(-1)^k \pi^{2k} y^{n-2k-1}, \qquad (15)$$

$$u_0^m(\sigma) = \sum_{k=0}^{K(m+1)} \binom{m}{2k}(-1)^k \pi^{2k} \bigl[L_1(y)\bigr]^{m-2k} \bigl[L_2(y)\bigr]^{2k}, \qquad (16)$$

$$v_0^m(\sigma) = \sum_{k=0}^{K(m)} \binom{m}{2k+1}(-1)^{k+1} \pi^{2k} \bigl[L_1(y)\bigr]^{m-2k-1} \bigl[L_2(y)\bigr]^{2k+1}, \qquad (17)$$

$$L_1(y) = \ln\sqrt{y^2+\pi^2}, \qquad L_2(y) = \frac{1}{2} - \frac{1}{\pi}\arctan\!\left(\frac{y}{\pi}\right), \qquad (18)$$

$K(j) = [(j-2) + (j \bmod 2)]/2$, $y = \ln(\sigma/\Lambda^2)$, and $n \ge 1$ is assumed. In turn, Eq. (11) enables one to properly account for the effects due to continuation of the spacelike perturbative results into the timelike domain at an arbitrary loop level, that plays a valuable role in the studies of a variety of the strong interaction processes, see paper [28] and references therein for the details.

References

[1] S.L. Adler, Phys. Rev. D **10**, 3714 (1974).
[2] A.V. Radyushkin, report JINR E2-82-159 (1982); JINR Rapid Commun. **78**, 96 (1996); arXiv:hep-ph/9907228.
[3] N.V. Krasnikov and A.A. Pivovarov, Phys. Lett. B **116**, 168 (1982).

[4] A.V. Nesterenko, *"Strong interactions in spacelike and timelike domains: Dispersive approach"* (Elsevier, Amsterdam, 222 p., 2016).
[5] A.V. Nesterenko and J. Papavassiliou, J. Phys. G **32**, 1025 (2006).
[6] A.V. Nesterenko, Phys. Rev. D **88**, 056009 (2013); J. Phys. G **42**, 085004 (2015).
[7] A.V. Nesterenko and J. Papavassiliou, Phys. Rev. D **71**, 016009 (2005); Int. J. Mod. Phys. A **20**, 4622 (2005); Nucl. Phys. B (Proc. Suppl.) **152**, 47 (2005); **164**, 304 (2007).
[8] A.V. Nesterenko, Nucl. Phys. B (Proc. Suppl.) **186**, 207 (2009); **234**, 199 (2013); Nucl. Part. Phys. Proc. **258**, 177 (2015); **270**, 206 (2016); SLAC eConf C0706044, 25 (2008); C1106064, 23 (2011); PoS ConfinementX, 350 (2012); AIP Conf. Proc. **1701**, 040016 (2016); EPJ Web Conf. **137**, 05021 (2017).
[9] D.V. Shirkov and I.L. Solovtsov, Phys. Rev. Lett. **79**, 1209 (1997); Theor. Math. Phys. **150**, 132 (2007); K.A. Milton and I.L. Solovtsov, Phys. Rev. D **55**, 5295 (1997); Phys. Rev. D **59**, 107701 (1999).
[10] G. Cvetic and C. Valenzuela, Phys. Rev. D **74**, 114030 (2006); **84**, 019902(E) (2011); Braz. J. Phys. **38**, 371 (2008); G. Cvetic and A.V. Kotikov, J. Phys. G **39**, 065005 (2012); A.P. Bakulev, Phys. Part. Nucl. **40**, 715 (2009); N.G. Stefanis, *ibid.* **44**, 494 (2013).
[11] G. Cvetic, A.Y. Illarionov, B.A. Kniehl, and A.V. Kotikov, Phys. Lett. B **679**, 350 (2009); A.V. Kotikov, PoS (Baldin ISHEPP XXI), 033 (2013); PoS (Baldin ISHEPP XXII), 028 (2015); A.V. Kotikov and B.G. Shaikhatdenov, Phys. Part. Nucl. **44**, 543 (2013); Phys. Atom. Nucl. **78**, 525 (2015); A.V. Kotikov, V.G. Krivokhizhin, and B.G. Shaikhatdenov, J. Phys. G **42**, 095004 (2015); A.V. Kotikov, B.G. Shaikhatdenov, and P. Zhang, arXiv:1706.01849 [hep-ph]; C. Ayala, G. Cvetic, A.V. Kotikov, and B.G. Shaikhatdenov, arXiv:1708.06284 [hep-ph].
[12] G. Cvetic and C. Villavicencio, Phys. Rev. D **86**, 116001 (2012); C. Ayala and G. Cvetic, *ibid.* **87**, 054008 (2013); P. Allendes, C. Ayala, and G. Cvetic, *ibid.* **89**, 054016 (2014); C. Ayala, G. Cvetic, and R. Kogerler, J. Phys. G **44**, 075001 (2017); C. Ayala, G. Cvetic, R. Kogerler, and I. Kondrashuk, arXiv:1703.01321 [hep-ph].
[13] M. Baldicchi and G.M. Prosperi, Phys. Rev. D **66**, 074008 (2002); AIP Conf. Proc. **756**, 152 (2005); M. Baldicchi, G.M. Prosperi, and C. Simolo, *ibid.* **892**, 340 (2007); M. Baldicchi, A.V. Nesterenko, G.M. Prosperi, D.V. Shirkov, and C. Simolo, Phys. Rev. Lett. **99**, 242001 (2007); M. Baldicchi, A.V. Nesterenko, G.M. Prosperi, and C. Simolo, Phys. Rev. D **77**, 034013 (2008).
[14] K.A. Milton, I.L. Solovtsov, and O.P. Solovtsova, Phys. Lett. B **439**, 421 (1998); Phys. Rev. D **60**, 016001 (1999); R.S. Pasechnik, D.V. Shirkov, and O.V. Teryaev, *ibid.* **78**, 071902 (2008); R.S. Pasechnik, J. Soffer, and O.V. Teryaev, *ibid.* **82**, 076007 (2010).

[15] K.A. Milton, I.L. Solovtsov, and O.P. Solovtsova, Phys. Lett. B **415**, 104 (1997); K.A. Milton, I.L. Solovtsov, O.P. Solovtsova, and V.I. Yasnov, Eur. Phys. J. C **14**, 495 (2000).
[16] A. Bakulev, K. Passek–Kumericki, W. Schroers, and N. Stefanis, Phys. Rev. D **70**, 033014 (2004); **70**, 079906(E) (2004); A.P. Bakulev, A.V. Radyushkin, and N.G. Stefanis, *ibid.* **62**, 113001 (2000); A.P. Bakulev, A.I. Karanikas, and N.G. Stefanis, *ibid.* **72**, 074015 (2005); A. Bakulev, A. Pimikov, and N. Stefanis, *ibid.* **79**, 093010 (2009); N. Stefanis, Nucl. Phys. B (Proc. Suppl.) **152**, 245 (2006).
[17] G. Cvetic, R. Kogerler, and C. Valenzuela, Phys. Rev. D **82**, 114004 (2010); J. Phys. G **37**, 075001 (2010); G. Cvetic and R. Kogerler, Phys. Rev. D **84**, 056005 (2011).
[18] O. Teryaev, Nucl. Phys. B (Proc. Suppl.) **245**, 195 (2013); A.V. Sidorov and O.P. Solovtsova, Nonlin. Phenom. Complex Syst. **16**, 397 (2013); Mod. Phys. Lett. A **29**, 1450194 (2014); J. Phys. Conf. Ser. **678**, 012042 (2016); Phys. Part. Nucl. Lett. **14**, 1 (2017).
[19] C. Contreras, G. Cvetic, O. Espinosa, and H.E. Martinez, Phys. Rev. D **82**, 074005 (2010); C. Ayala, C. Contreras, and G. Cvetic, *ibid.* **85**, 114043 (2012).
[20] A.V. Nesterenko, Phys. Rev. D **62**, 094028 (2000); **64**, 116009 (2001).
[21] A.V. Nesterenko, Int. J. Mod. Phys. A **18**, 5475 (2003); Nucl. Phys. B (Proc. Suppl.) **133**, 59 (2004).
[22] A.V. Nesterenko, Mod. Phys. Lett. A **15**, 2401 (2000); A.V. Nesterenko and I.L. Solovtsov, *ibid.* **16**, 2517 (2001).
[23] A.C. Aguilar, A.V. Nesterenko, and J. Papavassiliou, J. Phys. G **31**, 997 (2005); Nucl. Phys. B (Proc. Suppl.) **164**, 300 (2007).
[24] P.A. Baikov, K.G. Chetyrkin, and J.H. Kuhn, Phys. Rev. Lett. **101**, 012002 (2008); **104**, 132004 (2010); P.A. Baikov, K.G. Chetyrkin, J.H. Kuhn, and J. Rittinger, Phys. Lett. B **714**, 62 (2012).
[25] P.A. Baikov, K.G. Chetyrkin, and J.H. Kuhn, Phys. Rev. Lett. **118**, 082002 (2017); F. Herzog, B. Ruijl, T. Ueda, J.A.M. Vermaseren, and A. Vogt, JHEP **1702**, 090 (2017).
[26] A.V. Nesterenko and C. Simolo, Comput. Phys. Commun. **181**, 1769 (2010); **182**, 2303 (2011).
[27] A.V. Nesterenko and S.A. Popov, Nucl. Part. Phys. Proc. **282**, 158 (2017).
[28] A.V. Nesterenko, arXiv:1707.00668 [hep-ph].
[29] J.D. Bjorken, report SLAC–PUB–5103 (1989).
[30] A.L. Kataev and V.V. Starshenko, Mod. Phys. Lett. A **10**, 235 (1995).
[31] G.M. Prosperi, M. Raciti, and C. Simolo, Prog. Part. Nucl. Phys. **58**, 387 (2007).

LOCAL GROUPS OF INTERNAL TRANSFORMATIONS ISOMORPHIC TO LOCAL GROUPS OF SPACETIME TETRAD TRANSFORMATIONS

Alcides Garat [a]

Former Professor at Universidad de la República, Instituto de Física, Facultad de Ingeniería, J. Herrera y Reissig 565, 11300 Montevideo, Uruguay.

Abstract. A new tetrad is introduced within the framework of geometrodynamics for non-null electromagnetic fields in curved four-dimensional Lorentzian spacetimes. This tetrad diagonalizes the electromagnetic stress-energy tensor locally and covariantly and enables the maximum simplification of the electromagnetic field. The Einstein-Maxwell equations will also be simplified. The tetrad vectors have a construction structure that allows for the proof of new group isomorphisms. The local group of electromagnetic gauge transformations is proved to be isomorphic to the new group LB1. LB1 is the group of local tetrad transformations comprised by SO(1,1) plus two different kinds of discrete transformations. The local group of electromagnetic gauge transformations is also isomorphic to the local group of tetrad transformations LB2. LB2 is SO(2). Therefore, we proved that LB1 is isomorphic to LB2. When studying the vectors that locally and covariantly diagonalize the stress-energy tensor, we find that it is the stress-energy tensor the object that determines the local gauge structure of space-time. These group results amount to proving that the no-go theorems of the sixties like the S. Coleman- J. Mandula, the S. Weinberg or L. ORaifeartagh versions are incorrect. Not because of their internal logic, but because of the assumptions made at the outset of all these versions. The local group of "internal" electromagnetic gauge transformations is proved to be isomorphic to a local subgroup of the Lorentz group. In addition, the local LB1 isomorphic to LB2 which is SO(2) means that the boosts plus two discrete transformations can be put in a one to one relation to SO(2) which also contradicts the group assumptions made at the outset of the no-go theorems. The explicit and manifest isomorphic link between the Abelian local "internal" electromagnetic gauge transformations and the local tetrad transformations on special orthogonal unique planes of eigenvectors of the Einstein-Maxwell stress-energy tensor is manifest evidence of these incorrect no-go results as it will be proved. Simply because the Lorentz transformations on a local plane in a four-dimensional curved Lorentzian spacetime do not commute with Lorentz transformations on a different local plane in general, element of contradiction with the no-go theorems. These new tetrads are useful in astrophysics spacetime evolution algorithms since they introduce maximum simplification in all relevant objects, specially in stress-energy tensors.

1 Diagonalization of the stress-energy tensor

Throughout the paper we use the conventions of Ref. [1]. In particular we use a metric with sign conventions $-+++$. The only difference in notation with paper [1] will be that we will call our geometrized electromagnetic potential A_μ, where $f_{\mu\nu} = A_{\nu;\mu} - A_{\mu;\nu}$ is the geometrized electromagnetic field $f_{\mu\nu} = (G^{1/2}/c^2) F_{\mu\nu}$. The stress-energy tensor according to equation (14a) in Ref. [1],

---

[a] E-mail: garat.alcides@gmail.com

can be written as, $T_{\mu\nu} = f_{\mu\lambda} f_\nu{}^\lambda + *f_{\mu\lambda} * f_\nu{}^\lambda$, where $*f_{\mu\nu} = \frac{1}{2} \epsilon_{\mu\nu\sigma\tau} f^{\sigma\tau}$ is the dual tensor of $f_{\mu\nu}$. The local duality rotation given by equation (59) in paper [1] $f_{\mu\nu} = \xi_{\mu\nu} \cos\alpha + *\xi_{\mu\nu} \sin\alpha$, allows us to express the stress-energy tensor in terms of the extremal field $T_{\mu\nu} = \xi_{\mu\lambda} \xi_\nu{}^\lambda + *\xi_{\mu\lambda} * \xi_\nu{}^\lambda$. We can express the extremal field as, $\xi_{\mu\nu} = e^{-*\alpha} f_{\mu\nu} = \cos\alpha f_{\mu\nu} - \sin\alpha * f_{\mu\nu}$. Extremal fields are local gauge invariants in the electromagnetic sense as it can be noticed from its definition. Extremal fields satisfy the equation $\xi_{\mu\nu} * \xi^{\mu\nu} = 0$. This a condition imposed on extremal fields in order to find a local scalar named the complexion α. The explicit expression for the complexion, which is also a local electromagnetic gauge invariant, can be given when imposing the given condition, by $\tan(2\alpha) = -f_{\mu\nu} * f^{\mu\nu}/f_{\lambda\rho} f^{\lambda\rho}$. Through the use of the general identity, $A_{\mu\alpha} B^{\nu\alpha} - *B_{\mu\alpha} * A^{\nu\alpha} = \frac{1}{2} \delta_\mu{}^\nu A_{\alpha\beta} B^{\alpha\beta}$ which is valid for every pair of antisymmetric tensors in a four-dimensional Lorentzian spacetime [1], when applied to the case $A_{\mu\alpha} = \xi_{\mu\alpha}$ and $B^{\nu\alpha} = *\xi^{\nu\alpha}$, it can be proved that condition $\xi_{\mu\nu} * \xi^{\mu\nu} = 0$ yields the equivalent condition, $\xi_{\alpha\mu} * \xi^{\mu\nu} = 0$. The extremal field $\xi_{\mu\nu}$ and the scalar complexion α have been previously defined through equations (22)–(25) in Ref. [1]. It is our purpose to find a tetrad in which the stress-energy tensor is diagonal. This tetrad would simplify the analysis of the geometrical properties of the electromagnetic field. There are four tetrad vectors that at every point in spacetime diagonalize the stress-energy tensor in geometrodynamics,

$$V_{(1)}^\alpha = \xi^{\alpha\lambda} \xi_{\rho\lambda} X^\rho \tag{1}$$

$$V_{(2)}^\alpha = \sqrt{-Q/2}\, \xi^{\alpha\lambda} X_\lambda \tag{2}$$

$$V_{(3)}^\alpha = \sqrt{-Q/2}\, *\xi^{\alpha\lambda} Y_\lambda \tag{3}$$

$$V_{(4)}^\alpha = *\xi^{\alpha\lambda} * \xi_{\rho\lambda} Y^\rho , \tag{4}$$

where $Q = \xi_{\mu\nu} \xi^{\mu\nu} = -\sqrt{T_{\mu\nu} T^{\mu\nu}}$ according to equations (39) in manuscript [1]. Q is assumed not to be zero, because we are dealing with non-null electromagnetic fields. We are free to choose the vector fields X^α and Y^α, as long as the four vector fields (1)–(4) are not trivial. For vectors (1)–(2) the stress-energy eigenvalue is $\frac{Q}{2}$ while for vectors (3)–(4) the eigenvalue is $-\frac{Q}{2}$. We would like to make a convenient and particular choice of the vector fields X^α and Y^α. In geometrodynamics, the Maxwell equations, $f^{\mu\nu}{}_{;\nu} = 0$ and $*f^{\mu\nu}{}_{;\nu} = 0$ are telling us that two potential vector fields exist [2], $f_{\mu\nu} = A_{\nu;\mu} - A_{\mu;\nu}$ and $*f_{\mu\nu} = *A_{\nu;\mu} - *A_{\mu;\nu}$. For instance, in the Reissner-Nordstrom geometry the only non-zero electromagnetic tensor component is $f_{tr} = A_{r;t} - A_{t;r}$ and its dual $*f_{\theta\phi} = *A_{\phi;\theta} - *A_{\theta;\phi}$. The symbol ";" stands for covariant derivative with respect to the metric tensor $g_{\mu\nu}$ and the star in $*A_\nu$ is just a name, not the dual operator, meaning that $*A_{\nu;\mu} = (*A_\nu)_{;\mu}$. The vector fields $X^\alpha = A^\alpha$ and $Y^\alpha = *A^\alpha$ represent a possible choice in geometrodynamics for the vectors X^α and Y^α.

2 Gauge geometry. Gauge transformations on blades one and two

Once we make the choice $X^\alpha = A^\alpha$ and $Y^\alpha = *A^\alpha$ the question about the geometrical implications of electromagnetic gauge transformations of the tetrad vectors (1)–(4) arises. We first notice that a local electromagnetic gauge transformation of the "gauge vectors" $X^\alpha = A^\alpha$ and $Y^\alpha = *A^\alpha$ can be just interpreted as a new choice for the gauge vectors $X_\alpha = A_\alpha + \Lambda_{,\alpha}$ and $Y_\alpha = *A_\alpha + *\Lambda_{,\alpha}$. When we make the transformation, $A_\alpha \to A_\alpha + \Lambda_{,\alpha}$, $f_{\mu\nu}$ remains invariant, and the transformation, $*A_\alpha \to *A_\alpha + *\Lambda_{,\alpha}$, leaves $*f_{\mu\nu}$ invariant, as long as the functions Λ and $*\Lambda$ are scalars. It is valid to ask how the tetrad vectors (1-2) will transform under $A_\alpha \to A_\alpha + \Lambda_{,\alpha}$ and (3)–(4) under $*A_\alpha \to *A_\alpha + *\Lambda_{,\alpha}$. Schouten defined what he called, a two-bladed structure in a spacetime [4]. These local blades or planes are the planes determined by the pairs $(V^\alpha_{(1)}, V^\alpha_{(2)})$ and $(V^\alpha_{(3)}, V^\alpha_{(4)})$. Given the space constraint in these proceedings we will limit ourselves to show a few illustrative results as far as tetrad transformations for gauge vector choice given by electromagnetic gauge transformations. The whole analysis is given in manuscript [3]. In order to simplify the notation we will write $\Lambda_{,\alpha} = \Lambda_\alpha$. First we study the change in (1)–(2) under $A_\alpha \to A_\alpha + \Lambda_{,\alpha}$. Using the following notation, $C = (-Q/2) V_{(1)\sigma} \Lambda^\sigma / (V_{(2)\beta} V^\beta_{(2)})$ and $D = (-Q/2) V_{(2)\sigma} \Lambda^\sigma / (V_{(1)\beta} V^\beta_{(1)})$, several cases arise on blade one. We would like to calculate the norm of the transformed vectors $\tilde{V}^\alpha_{(1)}$ and $\tilde{V}^\alpha_{(2)}$, $\tilde{V}^\alpha_{(1)} \tilde{V}_{(1)\alpha} = [(1+C)^2 - D^2] V^\alpha_{(1)} V_{(1)\alpha}$ and $\tilde{V}^\alpha_{(2)} \tilde{V}_{(2)\alpha} = [(1+C)^2 - D^2] V^\alpha_{(2)} V_{(2)\alpha}$, where the relation $V^\alpha_{(1)} V_{(1)\alpha} = -V^\alpha_{(2)} V_{(2)\alpha}$ has been used and $V^\alpha_{(1)}$ assumed timelike for simplicity. In order for these transformations to keep the timelike or spacelike character of $V^\alpha_{(1)}$ and $V^\alpha_{(2)}$ the condition $[(1+C)^2 - D^2] > 0$ must be satisfied. The condition $[(1+C)^2 - D^2] > 0$ allows for two possible situations, $1 + C > 0$ or $1 + C < 0$. For the particular case when $1 + C > 0$, the transformations of the normalized (1)–(2) are telling us that an electromagnetic gauge transformation on the vector field A^α, that leaves invariant the electromagnetic field $f_{\mu\nu}$, generates a boost transformation on the normalized tetrad vector fields $\left(\frac{V^\alpha_{(1)}}{\sqrt{-V^\beta_{(1)} V_{(1)\beta}}}, \frac{V^\alpha_{(2)}}{\sqrt{V^\beta_{(2)} V_{(2)\beta}}} \right)$. We show as an example for the normalized

(1)–(2) $\frac{\tilde{V}^\alpha_{(1)}}{\sqrt{-\tilde{V}^\beta_{(1)} \tilde{V}_{(1)\beta}}} = \frac{(1+C)}{\sqrt{(1+C)^2 - D^2}} \frac{V^\alpha_{(1)}}{\sqrt{-V^\beta_{(1)} V_{(1)\beta}}} + \frac{D}{\sqrt{(1+C)^2 - D^2}} \frac{V^\alpha_{(2)}}{\sqrt{V^\beta_{(2)} V_{(2)\beta}}}$ and

$\frac{\tilde{V}^\alpha_{(2)}}{\sqrt{\tilde{V}^\beta_{(2)} \tilde{V}_{(2)\beta}}} = \frac{D}{\sqrt{(1+C)^2 - D^2}} \frac{V^\alpha_{(1)}}{\sqrt{-V^\beta_{(1)} V_{(1)\beta}}} + \frac{(1+C)}{\sqrt{(1+C)^2 - D^2}} \frac{V^\alpha_{(2)}}{\sqrt{V^\beta_{(2)} V_{(2)\beta}}}$. The case $1 + C < 0$, represents the composition of two transformations. An inversion of the normalized tetrad vector fields $\left(\frac{V^\alpha_{(1)}}{\sqrt{-V^\beta_{(1)} V_{(1)\beta}}}, \frac{V^\alpha_{(2)}}{\sqrt{V^\beta_{(2)} V_{(2)\beta}}} \right)$, and a boost. If the case is that $[(1+C)^2 - D^2] < 0$, the vectors $V^\alpha_{(1)}$ and $V^\alpha_{(2)}$ will change their timelike or spacelike character, $\tilde{V}^\alpha_{(1)} \tilde{V}_{(1)\alpha} = [-(1+C)^2 + D^2] (-V^\alpha_{(1)} V_{(1)\alpha})$ and

$(-\tilde{V}^\alpha_{(2)}\,\tilde{V}_{(2)\alpha}) = [-(1+C)^2 + D^2]\,V^\alpha_{(2)}\,V_{(2)\alpha}$. These are improper transformations on blade one. They have the property of being a composition of boosts and a discrete transformation given by $\Lambda^o{}_o = 0$, $\Lambda^o{}_1 = 1$, $\Lambda^1{}_o = 1$, $\Lambda^1{}_1 = 0$. We notice that this discrete "switch" transformation is not a Lorentz transformation. They might also be composed with an inversion, see Ref. [3] for the whole analysis. LB1 is the group comprised by $SO(1,1)$ plus minus the identity 2×2, (the full inversion) and the switch. On blade or plane two, the choice $Y_\alpha = *A_\alpha + *\Lambda_{,\alpha}$ induces just local spatial tetrad vector transformations. We reiterate that local tetrad electromagnetic gauge transformations can be interpreted as new or different gauge choices $X_\alpha = A_\alpha + \Lambda_{,\alpha}$ and $Y_\alpha = *A_\alpha + *\Lambda_{,\alpha}$.

3 Group isomorphisms

We will just limit ourselves to state these new theorems proved in detail in Ref. [3].

Theorem 1 *The mapping between the group of electromagnetic gauge transformations and the group LB1 defined above is isomorphic.*

Theorem 2 *The mapping between the group of electromagnetic gauge transformations and the group LB2, which is SO(2), defined above is isomorphic.*

4 Conclusions

We can state the following conclusions as a summary.

- New orthonormal tetrad for non-null electromagnetic fields in four-dimensional curved Lorentzian spacetimes [3]. Maximum simplification of relevant tensors and field equations. This tetrad diagonalizes locally and covariantly the Einstein-Maxwell stress-energy tensor. Astrophysical applications in spacetime evolution [5,6].

- Isomorphisms between the local electromagnetic gauge group and the local groups of tetrad spacetime transformations LB1 and LB2. There is an isomorphism between kinematic states and gauge states of the gravitational fields, locally. Extension or generalization to non-Abelian theories with gauge group $SU(2) \times U(1)$, see Ref. [7].

- New tetrads encode gravitational and electromagnetic gauge information. The Einstein-Maxwell stress-energy tensor determines the local gauge structure of spacetime.

- We are introducing an explicit "link" between the "internal" and the "spacetime", so far detached from each other. Hypotheses made at the outset of the no-go theorems [8–10] proved incorrect. Therefore, the no-go theorems are incorrect.

Acknowledgments

The author acknowledges and is grateful to the organizers for the financial help that allowed his participation in the 18th Lomonosov Conference on Elementary Particle Physics.

References

[1] C. Misner and J. A. Wheeler, Annals of Physics **2**, 525 (1957).
[2] N. Cabibbo and E. Ferrari, Nuovo Cim. **23**, 1147 (1962).
[3] A. Garat, J. Math. Phys. **46**, 102502 (2005). A. Garat, Erratum: Tetrads in geometrodynamics, J. Math. Phys. **55**, 019902 (2014). gr-qc/0412037.
[4] J. A. Schouten, *Ricci Calculus: An Introduction to Tensor Calculus and Its Geometrical Applications* (Springer, Berlin, 1954).
[5] A. Garat, Euler observers in geometrodynamics, Int. J. Geom. Methods Mod. Phys., Vol. **11** (2014), 1450060.
[6] A. Garat, Covariant diagonalization of the perfect fluid stress-energy tensor, Int. J. Geom. Methods Mod. Phys., Vol. **12**, No. 3 (2015), 1550031.
[7] A. Garat, Tetrads in Yang-Mills geometrodynamics, Gravitation and Cosmology, (2014) Vol. **20** No. 1, pp. 116-126. Pleiades Publishing Ltd. arXiv:gr-qc/0602049.
[8] S. Weinberg, Phys. Rev. **139**, B597 (1965).
[9] L. O' Raifeartaigh, Phys. Rev. **139**, B1052 (1965).
[10] S. Coleman, J. Mandula, Phys. Rev. **159**, 1251 (1967).

New Developments in Quantum Field Theory

HADRONIZATION VIA GRAVITATIONAL CONFINEMENT

Constantinos G. Vayenas[1,2,a] and Dimitrios Grigoriou[1,b]
[1] LCEP, 1 Caratheodory St., University of Patras, Patras GR 26504, Greece
[2] Division of Natural Sciences, Academy of Athens, 28 Panepistimiou Ave.,
GR-10679 Athens, Greece

Abstract. We examine the thermodynamics of hadronization using the rotating lepton model in which the strong force is modeled as the gravitational force between ultrarelativistic leptons, i.e. neutrino triplets in the case of baryons, neutrino pairs in the case of mesons, and neutrino-$e^{\pm}$ combinations, in the cases of $W^{\pm}$, Z^o and H^o bosons. The model contains no adjustable parameters and allows for the computation of the hadron and boson masses within typically 2%. The model also allows for the computation of the energies of formation of hadrons and bosons and shows that hadronization is an extremely exoergic process which could provide energy for gravitational waves.

1 Introduction

The potential importance of gravitational interactions between neutrinos was first discussed by Wheeler in his geon models [1], long before it was known that neutrinos have masses [2–4]. In a book [5] and several recent papers [6–13] we have shown that, when accounting for special relativistic and the equivalence principle, gravitational interactions between ultrarelativistic (>100 MeV) neutrinos become very strong and lead to the formation of stable rotational neutrino states with the masses and other properties of baryons. This has been shown by formulating simple Bohr-type models which comprise only two equations as shown in Table 1, i.e. the special relativistic equation of motion for the rotating particles, e.g. quarks, which have been shown in [5, 6] to have the same rest mass as electron neutrinos [3, 4], and the de Broglie wavelength equation which is the basis of quantum mechanics. The gravitational force in the equation of motion is expressed via Newton's gravitational Law but with the important difference of using gravitational masses rather than rest masses. Alternatively the same results can be obtained [6, 12] when solving the model using the Schwarzschild geodesics of general relativity coupled with the generalized uncertainty principle [12, 14].

According to the equivalence principle, the gravitational mass of a particle, m_g, equals its inertial masses, m_i, and the latter is given by

$$m_i = dF/d\mathbf{a} = d\left[(dp/dt)/(d\mathbf{v}/dt)\right] = d\left[(\gamma m_o \mathbf{v})/dt\right]/(d\mathbf{v}/dt) = \gamma^3 m_o, \quad (1)$$

where $\mathbf{a}$ is the acceleration, F is the force, p is the momentum, m_o is the rest mass and $\gamma (= (1 - \mathbf{v}^2/c^2)^{-1})$ is the Lorentz factor. This equation was originally

---
[a] E-mail: cgvayenas@upatras.gr
[b] E-mail: dgrigoriou@chemeng.upatras.gr

Table 1: The two equations of the rotating neutrino model for baryons [5, 6].

Equation of motion	De Broglie equation
Neutrino as *particle*	Neutrino as *wave*
$\gamma_\nu m_{o,\nu} \dfrac{v^2}{r} = \dfrac{Gm_{o,\nu}^2 \gamma^6}{\sqrt{3}\, r^2}$ (I)	$\dfrac{\hbar}{\gamma_\nu m_{o,\nu} v} \approx \dfrac{\hbar}{\gamma_\nu m_{o,\nu} c} = r$ (II)
↑ a	↑ b

[a] Relativistic equation of motion for circular motion.
[b] Newton's gravitational law accounting for special relativity ($m_i = \gamma^3 m_{o,\nu}$) and for the equivalence principle ($m_{g,\nu} = m_{i,\nu}$); $m_{o,\nu}$ is the neutrino rest mass.

derived for linear motion by Einstein in his pioneering 1905 special relativity paper [15] and was shown recently, using instantaneous reference frames [16], to remain valid for arbitrary particle motion [5,6]. Accordingly, Newton's universal gravitational law takes the form

$$F = \frac{Gm_{1,o}m_{2,0}\gamma_1^3\gamma_2^3}{d^2}, \quad (2)$$

where d is the particle distance. Thus denoting by $m_{o,\nu}$ the rest mass of a neutrino and by γ_ν the corresponding Lorentz factor, one obtains

$$F_{\nu\nu} = \frac{Gm_{o,\nu}^2 \gamma_\nu^6}{d^2}, \quad (3)$$

for the force between two neutrinos with the same speed with respect to an observer.

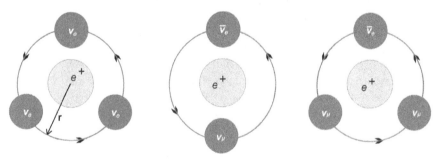

Figure 1: Rotating neutrino model geometry for a proton (a) [11], for a muon μ^+ (b) and for a pion π^+ (c). The central positron is at rest with respect to the observer ($\gamma = 1$) and thus adds little ($0.511\ MeV/c^2$) to the total mass of the composite state.

Noting that for the circular geometry of Figure 1a the parameter d^2 is replaced by $\sqrt{3}r^2$ [5,6], where r is the rotational radius, one obtains the first equation (I), of Table 1. Solving this equation together with the de Broglie equation (II) of Table 1, one obtains that, for large γ_ν values, it is

$$\gamma_\nu = 3^{1/12}(\hbar c/G)^{1/6} m_{o,\nu}^{-1/3} = 3^{1/12} m_{Pl}^{1/3} m_{o,\nu}^{-1/3}, \quad (4)$$

where $m_{Pl}(=\hbar c/G)^{1/2}$ is the Planck mass. Thus the mass, m_c, and radius, r_c, of the composite confined state are given by

$$m_c = 3\gamma_\nu m_{o,\nu} = 3^{13/12}\left(m_{Pl} m_{o,\nu}^2\right)^{1/3} \quad ; \quad r_c = \hbar/\gamma_\nu m_{o,\nu} c, \quad (5)$$

Using the value of 0.0437 eV/c^2 for the rest mass of neutrinos [3, 5, 6] which is in good agreement with the recent results of the superKamiokande measurements [3], one obtains, surprisingly, that

$$m_c = 939.5 \ MeV/c^2, \quad r_c = 0.63 \ fm. \quad (6)$$

The m_c value differs less than 0.2% from the neutron and proton masses (939.56 MeV/c^2 and 938.27 MeV/c^2 respectively) and the r_c value is also in good agreement with the experimental values [17, 18]. It is also worth noting that, using equation (4), the neutrino gravitational mass, $\gamma_\nu^3 m_{e,\nu}$, is given by

$$m_{g,\nu} = \gamma_\nu^3 m_{o,\nu} = 3^{1/4} m_{Pl}. \quad (7)$$

This result implies that the gravitational force $Gm_{Pl}^2/3^{1/2}r^2$, between two such neutrinos equals the Strong Force value of $\hbar c/3^{1/4}r^2$ [5, 18]. Consequently the Strong Force appears to be a relativistic gravitational force. Electrostatic interactions between e^+ or e^- and polarized neutrinos [19] can also have a minor role [8, 12]. Gravitational collapse is prevented via the generalized uncertainty principle [5, 6, 12].

Here we first use the same methodology to obtain expressions similar to eq. (5) for the masses of muons and pions, and then we derive a simple equation and corresponding general graphical method for determining the masses of gravitationally confined hadron structures. We then discuss one of their key thermodynamic properties, i.e. their energy of formation.

2 Results and discussion

2.1 Masses of muons and pions

Although muons are known as leptons, they are also known to decay according to $\mu \to e\nu_\mu \bar{\nu}_e$ [17, 18]. It is thus tempting to model them according to the geometry of Figure 1b: In this case, for $v \approx c$, the equations of Table 1 become:

$$\frac{\gamma_{\nu_\mu} m_{o,\nu_\mu} c^2}{r} = \frac{\gamma_{\nu_e} m_{o,\nu_e} c^2}{r} = \frac{G m_{o,\nu_\mu} m_{o,\nu_e} \gamma_{\nu_\mu}^3 \gamma_{\nu_e}^3}{r^2}. \quad (8)$$

$$\frac{\hbar}{\gamma_{\nu_\mu} m_{o,\nu_\mu} c} = \frac{\hbar}{\gamma_{\nu_e} m_{o,\nu_e} c},\qquad(9)$$

and thus

$$m_\mu = m_c/2 = \gamma_{\nu_\mu} m_{o,\nu_\mu} = \gamma_{\nu_e} m_{o,\nu_e} = 2^{1/3}(m_{Pl} m_{o,\nu_\mu} m_{o,\nu_e})^{1/3}.\qquad(10)$$

The muon neutrino mass eigenstate is at this point unknown, but substituting in equation (10) the experimental m_μ value of 105.66 MeV/c² [17,18] and using $m_{\nu,e}$=0.0437 eV/c² [5–8] we find

$$\mu_{o,\nu_\mu} = m_\mu^3/(2 m_{Pl} m_{o,\nu_e}) = 1.105 \cdot 10^{-3} \text{ eV}/c^2.\qquad(11)$$

In order to check the accuracy of this value we proceed to also model the pion $\pi^\pm$ structure, which is known to have a mass of 139.57 MeV/c² and to decay according to $\pi^\pm \to \mu^\pm \nu_\mu$ [17, 18]. These decay products suggest the model structure of figure 1c. In this case the first equation of Table 1 becomes

$$F = \gamma_{\nu_\mu} m_{o,\nu_\mu} c^2/r = \gamma_{\nu_e} m_{o,\nu_e} c^2/r$$

which implies

$$F = (\gamma_{\nu_\mu} \gamma_{\nu_e})^{1/2} (m_{o,\nu_\mu} m_{o,\nu_e})^{1/2} c^2/r\qquad(12)$$

Hence the synchronization condition ($v_{\nu_\mu} = v_{\nu_e}$, thus $\gamma_{\nu_\mu} = \gamma_{\nu_e}$) implies that the three particles get hybridized [9] with a mass

$$m_{o,\nu_{\mu e}} = (m_{o,\nu_\mu} m_{o,\nu_e})^{1/2} = 6.95 \cdot 10^{-3} eV.\qquad(13)$$

and a Lorentz factor, $\gamma_{\nu_{\mu e}}$, given by $\gamma_{\nu_{\mu e}} = (\gamma_{\nu_\mu} \gamma_{\nu_e})^{1/2}$.

Consequently, as in the case of the proton, equations (I) and (II) of Table 1 remain valid with $m_{\mu e}$ replacing $m_{o,\nu}$ and $\gamma_{\mu e}$ replacing γ_ν. Thus it follows

$$m = 3\gamma_{\nu_{\mu e}} m_{o,\nu_{\mu e}} = 3^{13/12}(m_{Pl} m_{o,\nu_\mu} m_{o,\nu_e})^{1/3}\qquad(14)$$

Upon decomposition of the hybridized state, two isoenergetic fragments are produced, each with mass

$$m_\pi = m/2 = (1/2)3^{13/12}(m_{Pl} m_{o,\nu_\mu} m_{o,\nu_e})^{1/3} = 137.82\ MeV/c^2\qquad(15)$$

which is in close agreement with the masses of the neutral pion (134.98 MeV/c²) and of the charged pions $\pi^\pm$ (139.56 MeV/c² [17,18]. This supports the accuracy of the m_{o,ν_μ} value extracted from the muon model equation (11).

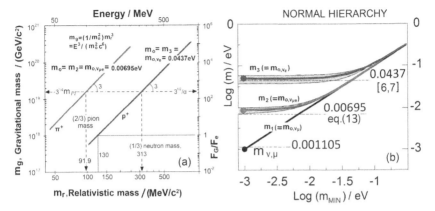

Figure 2: (a) Plot of equations (17) for electron neutrinos (eigenstate 3, $m_{o,\nu_e} = 0.0437$ eV and for muon neutrinos, eigenstate 2, $m_{o,\nu_\mu} = 6.95 \cdot 10^{-3} eV/c^2$ [3–6]) showing that protons and pions form from neutrinos when their gravitational mass, m_g, reaches $3^{1/4}$ times the Planck mass. At these points the relativistic neutrino masses reach the effective mass of quarks and the F_G/F_e ratio reaches $3^{1/2}/\alpha (\approx 237)$ respectively; $\alpha = e^2/\epsilon c\hbar \approx 1/137.035$. (b) The three neutrino eigenstate masses as a function of the lightest mass for the normal hierarchy, [3]; comparison with the m_{o,ν_e} and m_{o,ν_μ} values computed from the rotating neutrino model of baryons [5, 6] and bosons [7–9] and from the present work.

2.2 Gravitational mass dependence on particle velocity and total energy

Combining the weak equivalence principle of Einstein and Eötvös, i.e. $m_g = m_i$ which has been confirmed by the most modern torsion balance measurements to at least 1 part in 10^{12} [5], and equation (1), it follows $m_g = m_i = \gamma^3 m_o$, where m_o is the rest mass. Also, from Einstein's special relativity [15, 16], the relativistic mass, m_r, and the total (i.e. rest plus kinetic) energy, E, of a particle are related to the rest mass m_o via

$$m_r = \gamma m_o; \qquad E = m_r c^2 = \gamma m_o c^2. \qquad (16)$$

Upon eliminating γ between (1) and (16) one obtains

$$m_g = (1/m_o^2) m_r^3; \qquad m_g = (1/m_o^2 c^6) E^3. \qquad (17)$$

These simple equations are found to be quite useful, since they express directly the gravitational mass of a particle in terms of its total energy.

Figure 2a presents plots of equation (17) for two different m_o values, $m_{o,\nu_e} = 0.0437\ eV/c^2$ and $m_{o,\nu_{\mu e}} = 6.95 \cdot 10^{-3}$ eV/c^2, extracted above from the modeling of protons-neutrons and muons-pions, respectively. As shown in Figure 2b, these values lie quite close to the neutrino mass eigenstates 3 and 2, respectively, of the normal hierarchy [3, 5, 9]. Figure 2a shows that when the gravitational mass, m_g, which varies with E^3, eq. (17), reaches $3^{1/4}m_{Pl}$, then the relativistic mass of the composite state equals the effective quark masses, i.e. 1/3 and 2/3 respectively of the proton and muon masses. Consequently, for given m_o, Figure 2a allows for the determination of the rest mass of the composite three-particle confined state.

The right hand of axis of Figure 2a shows the dependence on $m_{r,\nu}$ and E of the ratio of the gravitational force, F_G, between two neutrinos with rest mass $m_{o,\nu}$ each, computed from equation (3), and of the Coulombic electrostatic force, F_e, between a positron, e^+, and an electron, e^-, at the same distance. This force is computed from Coulomb's law, i.e. from $F_e = e^2/\epsilon r^2$, with $\epsilon = 4\pi\varepsilon_o = 1.11 \times 10^{-10}$ C^2/Nm, where ε_o is the permittivity of vacuum.

Surprisingly, as shown in Figure 2, the ratio, F_G/F_e computed from Coulomb's law and equation (3), i.e. from

$$\frac{F_G}{F_e} = \frac{\epsilon G m_{o,\nu}^2}{e^2}\gamma_\nu^6 = \frac{\epsilon G m_{o,\nu}^2}{e^2}\left[\frac{E}{m_{o,\nu}c^2}\right]^6 = \frac{\epsilon G E^6}{e^2 m_{o,\nu}^4 c^{12}}, \quad (18)$$

exceeds unity for relativistic neutrino masses above 130 MeV/c^2. Consequently, for neutrino energies above 130 MeV, Newtonian gravity with gravitational masses is stronger than Coulombic attraction or repulsion. As shown in Figure 2a, for a neutrino energy of 313 MeV (corresponding closely to the effective mass of u and d quarks [17,18]) the F_G/F_e ratio takes the value $3^{1/2}/\alpha \approx 237$. Clearly, for such neutrino energies, their gravitational attraction is as strong as the Strong Force itself.

As also shown by equation (18), for fixed particle energy E, the ratio F_G/F_e increases dramatically with decreasing neutrino rest mass $m_{o,\nu}$. This in turn shows that gravitational confinement of neutrinos can take place with moderate (e.g. less than 500 MeV) energies (Fig. 2) because the rest mass of neutrinos is so small. Equation (16) written for electrons, which have a rest mass $1.17 \cdot 10^7$ times larger than that of neutrinos, leads to $F_G/F_e \approx 1$ for $E = 14.3$ TeV. This shows why lighter particles such as neutrinos, are ideal for building composite particles such as hadrons.

On the other hand, integration of the rhs of the force equation (I) of Table 1, using the r vs γ dependence expressed by the same equation (I), leads [5,6,20] to the potential energy, V_G, equation

$$V_G(r_c) = -5\gamma_{\nu,c}m_{o,\nu}c^2 = -(5/3)m_p c^2, \quad (19)$$

where $m_p = 3\gamma_{\nu,c}m_{o,\nu}$, is the proton mass.

This equation shows that the ΔH of the hadronization reaction

$$3\nu_e + e^+ \to p \quad (20)$$

which is analogous to hadronization via quark-gluon plasma condensation [14] is very negative (-1560 MeV/c^2) [20] and that heavier neutrinos create more stable baryons, as experimentally observed, since protons contain the heaviest of all neutrino eigenstates [5,6]. The ΔH of (20) could also provide evergy for gravitational waves.

3 Conclusions

The same Bohr-type rotating neutrino model methodology used already for computing the mass of the proton, can also be used without any adjustable parameters, for computing the masses of muons and pions. A simple expression for the gravitational mass of elementary particles in terms of their rest mass and their total energy shows that relativistic gravity between two neutrinos is stronger than the Coulombic attraction between a positron and an electron at the same distance and explains why neutrinos are ideal as hadronization components.

Acknowledgments

We thank Professor A. Fokas for numerous very helpful discussions.

References

[1] R.B. Dieter, J.A. Wheeler, Rev. Mod. Phys. **29(3)** (1957) 465.
[2] T. Kajita, Int. J. Mod. Phys. A, **24** (2009) 3437.
[3] R.N. Mohapatra et al, Rep Prog Phys **70** (2007) 1757.
[4] H. Fritzsch, Modern Physics Letters A, **30(16)** (2015) 1530012(1-6).
[5] C.G. Vayenas, S. Souentie *Gravity, special relativity and the strong force: A Bohr-Einstein-de-Broglie model for the formation of hadrons* (Springer, New York, 2012).
[6] C.G. Vayenas, S. Souentie & A. Fokas, Physica A **405** (2014) 360.
[7] C.G. Vayenas, A.S. Fokas, D. Grigoriou, Physica A **450** (2016) 37.
[8] A.S. Fokas, C.G. Vayenas, Physica A **464** (2016) 231.
[9] A.S. Fokas, C.G. Vayenas, D. Grigoriou, Physica A, https://doi.org/10.1016/j.physa.2017.11.003 (2017).
[10] C.G. Vayenas, A.S. Fokas, D. Grigoriou, J. Phys. **738** (2016) 012080.
[11] C.G. Vayenas, A.S. Fokas, D. Grigoriou, arXiv:1606.09570 (2016).
[12] C.G. Vayenas, D. Grigoriou, J. Phys.: Conf. Ser. **574** (2015) 012059.
[13] C.G. Vayenas, A.S. Fokas, D. Grigoriou, Appl. Catal. B **203** (2017) 582.
[14] S. Das, E.C. Vagenas, Can. J. Phys. **87** (2009) 233.
[15] A. Einstein, Ann. der Physik. **17** (1905) 891.
[16] A.P. French, *Special relativity* (W.W. Norton and Co., New York, 1968)
[17] D. Griffiths, *Introduction to Elementary Particles*. (2nd ed. Wiley-VCH Verlag GmbH & Co. KgaA, Weinheim, 2008).

[18] C.G. Tully, *Elementary Particle Physics in a Nutshell* (Princeton University Press 2011).
[19] C. Giunti, A. Studenikin, Rev. Mod. Phys. **87** (2015) 531.
[20] C.G. Vayenas, A.S. Fokas, D. Grigoriou, J. Phys.:Conf. Ser. **888** (2017) 012174.
[21] P. Braun-Munzinger, J. Stachel, Nature **448** (2007) 302.

SWITCHING-ON AND -OFF EFFECTS OF EXTERNAL FIELDS ON THE VACUUM INSTABILITY

T. C. Adorno[1 a], R. Ferreira[4 b], S. P. Gavrilov[1,3 c] and D. M. Gitman[1,2,4 d]

[1] Department of Physics, Tomsk State University, Lenin Prospekt 36, 634050, Tomsk, Russia;
[2] P. N. Lebedev Physical Institute, 53 Leninskiy prospekt, 119991, Moscow, Russia;
[3] Department of General and Experimental Physics, Herzen State Pedagogical University of Russia, Moyka Embankment 48, 191186 St. Petersburg, Russia;
[4] Instituto de Física, Universidade de São Paulo, Caixa Postal 66318, CEP 05508-090, São Paulo, S.P., Brazil.

Abstract. In the framework of QED with t-electric potential steps, the influence of switching-on and -off processes of a time-dependent electric field on the vacuum instability are considered. The effect is discussed for a composite electric field, characterized by a switching-on, a switching-off and a constant interval. Differential and total quantities are compared with a T-constant electric field, in which such processes are absent.

1 Introduction

Vacuum instability by strong external backgrounds predicted by quantum theories, such as the quantum electrodynamics (QED), ranks among the most important nonperturbative effects in physics. The effect have been recognized since the seminal works of Klein [1], Sauter [2], Heisenberg and Euler [3] and later by Schwinger [4]. Since then, it has been extensively studied over the years. For example, particle scattering, creation and annihilation by t-electric and x-electric potential steps were considered in the framework of the relativistic quantum mechanics in [5–7], particle creation by strong gravitational fields were considered in several works and textbooks, e.g., [8–12]. The general formulation of QED with t-electric potential steps was developed in Refs. [13–15] and, more recently, by the so-called x-electric potential steps [16]. Nowadays, there is a number of reviews that discuss the subject in detail, for instance, [17–20].

In this proceedings contribution, we address to a particular aspect of particle creation, namely, on how switching-on and switching-off processes of a time-dependent electric field affect vacuum instability. To this end we consider an electric field composed by three independent intervals, in which it grows exponentially during the switching-on interval I $= (-\infty, t_1)$, remains constant in the intermediate interval II $= [t_1, t_2]$ and decreases exponentially during the switching-off interval III $= (t_2, +\infty)$. It is homogeneous, positively oriented along a fixed direction $\mathbf{E}(t) = \left(E^i(t) = \delta^i_1 E(t), i = 1, \ldots, D\right)$

---

[a] E-mail: tg.adorno@gmail.com, tg.adorno@mail.tsu.ru
[b] E-mail: rafaelufpi@gmail.com
[c] E-mail: gavrilovsergeyp@yahoo.com, gavrilovsp@herzen.spb.ru
[d] E-mail: gitman@if.usp.br

and described by a vector along the same direction, $A^\mu = \left(A^0 = 0, \mathbf{A}\left(t\right)\right)$, $\mathbf{A}\left(t\right) = \left(A^i\left(t\right) = \delta_1^i A_x\left(t\right)\right)$, whose explicit forms are

$$E\left(t\right) = E \begin{cases} e^{k_1(t-t_1)}, & t \in \mathrm{I}, \\ 1, & t \in \mathrm{II}, \\ e^{-k_2(t-t_2)}, & t \in \mathrm{III}, \end{cases} \quad (1)$$

$$A_x\left(t\right) = E \begin{cases} k_1^{-1}\left(-e^{k_1(t-t_1)} + 1 - k_1 t_1\right), & t \in \mathrm{I}, \\ -t, & t \in \mathrm{II}, \\ k_2^{-1}\left(e^{-k_2(t-t_2)} - 1 - k_2 t_2\right), & t \in \mathrm{III}. \end{cases} \quad (2)$$

Here $t_1 < 0$, $t_2 > 0$ are fixed time instants and $(E, k_1, k_2) > 0$.

Since the electric field (2) is a t-electric potential step [20], Dirac spinors can be conveniently presented as[e]

$$\psi_n\left(x\right) = \exp\left(i\mathbf{p}\mathbf{r}\right)\psi_n\left(t\right), \quad n = \left(\mathbf{p}, \sigma\right),$$
$$\psi_n\left(t\right) = \left\{\gamma^0 i\partial_t - \left[p_x - U\left(t\right)\right] - \gamma\mathbf{p} + m\right\}\varphi_n\left(t\right) v_{\chi,\sigma}, \quad (3)$$

where $v_{\chi,\sigma}$ denotes a set of constant and orthonormalized spinors (χ and σ are spin quantum numbers), $\varphi_n\left(t\right)$ is a scalar function and $U\left(t\right) = -eA_x\left(t\right)$ is the potential energy of an electron ($e > 0$). The scalar function $\varphi_n\left(t\right)$ satisfy the second-order ordinary differential equation

$$\left\{\frac{d^2}{dt^2} + \left[p_x - U\left(t\right)\right]^2 + \pi_\perp^2 - i\chi\dot{U}\left(t\right)\right\}\varphi_n\left(t\right) = 0, \quad \pi_\perp = \sqrt{\mathbf{p}_\perp^2 + m^2}, \quad (4)$$

whose exact solutions for each one of the intervals above have been discussed by us previously in [20–23]. For example, at the switching-on I and -off III intervals, the solutions are expressed in terms of Confluent Hypergeometric Functions (CHF) $\Phi\left(a, c; \eta\right)$ [25],

$$\varphi_n^j\left(t\right) = b_2^j y_1^j\left(\eta_j\right) + b_1^j y_2^j\left(\eta_j\right), \quad (5)$$

with $y_1^j\left(\eta_j\right) = e^{-\eta_j/2}\eta_j^{\nu_j}\Phi\left(a_j, c_j; \eta_j\right)$, $y_2^j\left(\eta_j\right) = e^{\eta_j/2}\eta_j^{-\nu_j}\Phi\left(1 - a_j, 2 - c_j; -\eta_j\right)$ (the index j distinguish quantities associated to the switching-on $j = 1$ from the switching-off $j = 2$ intervals). As for the constant interval II, they are expressed in terms of Weber's Parabolic Cylinder Functions (WPCF) $D_\nu\left(z\right)$ [25],

$$\varphi_n\left(z\right) = b^+ u_+\left(z\right) + b^- u_-\left(z\right), \quad (6)$$

---

[e]$\psi(x)$ is a $2^{[d/2]}$-component spinor ($[d/2]$ stands for the integer part of the ratio $d/2$, $d = D + 1$), m denotes the electron mass and γ^μ are γ-matrices in d dimensions. Hereafter we use the relativistic system of units ($\hbar = c = 1$ and $\alpha = e^2/4\pi$), except when indicated otherwise.

where $u_+(z) = D_{\beta+(\chi-1)/2}(z)$, $u_-(z) = D_{-\beta-(\chi+1)/2}(iz)$. In these solutions, a_j, c_j, ν_j and β are parameters

$$a_1 = \frac{1+\chi}{2} + i\Xi_1^-, \quad a_2 = \frac{1+\chi}{2} + i\Xi_2^+, \quad \Xi_j^\pm = \frac{\omega_j \pm \Pi_j}{k_j},$$

$$c_j = 1 + 2\nu_j, \quad \nu_j = \frac{i\omega_j}{k_j}, \quad \beta = \frac{i\lambda}{2}, \quad \lambda = \frac{\pi_\perp^2}{eE},$$

$$\omega_j = \sqrt{\Pi_j^2 + \pi_\perp^2}, \quad \Pi_j = p_x - \frac{eE}{k_j}\left[(-1)^j + k_j t_j\right], \quad (7)$$

z and η_j are time-dependent functions

$$\eta_1(t) = ih_1 e^{k_1(t-t_1)}, \quad \eta_2(t) = ih_2 e^{-k_2(t-t_2)}, \quad z(t) = (1-i)\frac{eEt - p_x}{\sqrt{eE}}, \quad (8)$$

where $h_j = 2eE/k_j^2$ and $b_{1,2}^j$, $b^\pm$ are constants, fixed by initial conditions.

In virtue of asymptotic properties of the CHF, one can classify exact solutions of the switching-on and -off intervals as

$$_+\varphi_n(t) = {}_+\mathcal{N}\exp(i\pi\nu_1/2)\, y_2^1(\eta_1),$$
$$_-\varphi_n(t) = {}_-\mathcal{N}\exp(-i\pi\nu_1/2)\, y_1^1(\eta_1),\, t \in \mathrm{I};$$
$$^+\varphi_n(t) = {}^+\mathcal{N}\exp(-i\pi\nu_2/2)\, y_1^2(\eta_2),$$
$$^-\varphi_n(t) = {}^-\mathcal{N}\exp(i\pi\nu_2/2)\, y_2^2(\eta_2),\, t \in \mathrm{III}, \quad (9)$$

to learn that at the infinitely remote past $t \to -\infty$ and future $t \to +\infty$, the set above behaves as plane-waves,

$$_\zeta\varphi_n(t) = {}_\zeta\mathcal{N} e^{-i\zeta\omega_1 t}, \quad t \to -\infty, \quad {}^\zeta\varphi_n(t) = {}^\zeta\mathcal{N} e^{-i\zeta\omega_2 t}, \quad t \to +\infty, \quad (10)$$

where ω_1 denotes the energy of initial particles at $t \to -\infty$, ω_2 denotes the energy of final particles at $t \to +\infty$ and ζ labels electron ($\zeta = +$) and positron ($\zeta = -$) states. The normalization constants $_\zeta\mathcal{N}$, $^\zeta\mathcal{N}$ are calculated with respect to the usual inner product for Fermions $(\psi, \psi') = \int d\mathbf{x}\psi^\dagger(x)\psi'(x)$ [24] and have the form $_\zeta\mathcal{N} = \left(2\omega_1 q_1^\zeta V_{(d-1)}\right)^{-1/2}$, $^\zeta\mathcal{N} = \left(2\omega_2 q_2^\zeta V_{(d-1)}\right)^{-1/2}$, in which $q_j^\zeta = \omega_j - \chi\zeta\Pi_j$.

With the help of the solutions (9), one can use Eq. (3) to introduce IN $_\zeta\psi(x)$ and OUT $^\zeta\psi(x)$ sets of Dirac spinors. Through relations between them

$$^\zeta\psi_n(x) = g\left(_+|^\zeta\right)\, _+\psi_n(x) + g\left(_-|^\zeta\right)\, _-\psi_n(x).$$

One can evaluate Bogoliubov coefficients $g\left(_\zeta|^{\zeta'}\right)$ from the inner product $\left(_\zeta\psi_n, {}^{\zeta'}\psi_{n'}\right) = \delta_{nn'} g\left(_\zeta|^{\zeta'}\right)$. These coefficients can be equivalently calculated by requiring conditions of continuity of $_\zeta\varphi_n(t)$ and $^\zeta\varphi_n(t)$ and their time derivatives at t_1, t_2. It yields, in particular, the coefficient $g\left(_-|^+\right)$

$$g\left(_-|^+\right) = \sqrt{\frac{q_1^-}{8eE\omega_1 q_2^+ \omega_2}} e^{\frac{i\pi}{2}(\nu_1-\nu_2+\beta+\chi/2)} \left[f_1^-(t_2) f_2^+(t_1) - f_1^+(t_2) f_2^-(t_1)\right]$$

$$f_k^\pm(t_j) = \left[(-1)^j k_j \eta_j \frac{dy_k^j(\eta_j)}{d\eta_j} + y_k^j(\eta_j)\partial_t\right] u_\pm(z)\bigg|_{t=t_j}, \qquad (11)$$

that is the most important coefficient for the study of pair creation, since it provides a direct identification with the differential mean number of pairs created N_n^{cr}, the total number N^{cr} and vacuum vacuum transition probability P_v through the formulae,

$$N_n^{\mathrm{cr}} = \left|g\left(_-|^+\right)\right|^2, \quad N^{\mathrm{cr}} = \sum_n N_n^{\mathrm{cr}}, \quad P_v = \exp\left[\kappa \sum_n \ln\left(1 - \kappa N_n^{\mathrm{cr}}\right)\right]. \qquad (12)$$

In Fig. 1, we present some plots of the mean number of Fermions created from the vacuum N_n^{cr} (12) as a function of p_x for different values of $k = k_1 = k_2$ and $\mathcal{T}$ for a fixed amplitude E of the composite field. For simplicity, we set $\mathbf{p}_\perp = 0$ and select the reduced system of units, in which^f $\hbar = c = m = 1$. In these plots, p_x, $\mathcal{T}$ and k are relative to electron's mass m, corresponding to dimensionless quantities, i. e., p_x/m, $m\mathcal{T}$ and k_j/m, respectively.

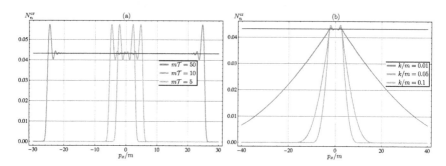

Figure 1: Differential mean number of Fermions created from the vacuum N_n^{cr} by a symmetrical composite field, with $k_1 = k_2 = k$ and amplitude $E = E_c = m^2/e = 1$ fixed. In (a), $k/m = 1$ is fixed while in (b) $m\mathcal{T} = 5$ is fixed. The horizontal dashed line denotes the uniform distribution $e^{-\pi\lambda}$ which, in the reduced system of units and $\mathbf{p}_\perp = 0$, is $e^{-\pi}$.

---

^{f}In this system, the Compton wavelength correspond to one unit of length $\lambda_e = \hbar/mc =$ 1 [ru] $\simeq 3.8614 \times 10^{-14}$ m, one unit of time is $\lambda_e/c = 1$ [ru] $\simeq 1.3 \times 10^{-21}$ s and electron's rest energy is one unit of energy $mc^2 = 1$ [ru] $\simeq 0.511$ MeV.

All formulas above can be generalized to consider particle creation of Klein-Gordon particles. For example, exact solutions for the Klein-Gordon equation $\phi_n(x)$ can be presented as $\phi_n(x) = \exp(i\mathbf{p}\mathbf{r})\varphi_n(t)$, in which $\varphi_n(t)$ satisfies Eq. (4) with $\chi = 0$ and $n = \mathbf{p}$. The corresponding Bogoliubov coefficient $g(_-|^+)$ for this case can be obtained from Eq. (11) by setting $\chi = 0$ and substituting $\sqrt{q_1^-/(\omega_1 q_2^+ \omega_2)} \to -\sqrt{1/(\omega_1\omega_2)}$. We refer the reader to the Refs. [14, 20, 21] for further details.

2 Differential and total quantities

The most favorable conditions for pair creation from the vacuum are characterized by strong fields acting over a sufficiently large time duration. For the electric field in consideration, it means small values for the switching-on k_1 and -off k_2 phases and large $\mathcal{T} = t_2 - t_1$,[g] so that the following condition is satisfied

$$\sqrt{eE}\mathcal{T} \gg \max\left(1, \frac{m^2}{eE}\right), \quad \min\left(eEk_1^{-2}, eEk_2^{-2}\right) \gg \max\left(1, \frac{m^2}{eE}\right). \quad (13)$$

Within the conditions above, one has to study how N_n^{cr} depends on the parameters above and compare them with parameters involving momenta of pairs created. In this direction, it is known that a t-electric potential step with large time duration does not create a significant number of pairs with large $\mathbf{p}_\perp$. This is meaningful as long as charged pairs are accelerated along the direction of the electric field. Thus, a wider range of values of p_x instead $\mathbf{p}_\perp$ is expected. As a result, only a restricted range of values to $\mathbf{p}_\perp$ is of importance, namely, $\sqrt{\lambda} < K_\perp$ in which $\min\left(\sqrt{eE}\mathcal{T}, eEk_1^{-2}, eEk_2^{-2}\right) \gg K_\perp^2 \gg \max(1, m^2/eE)$. Moreover, the most significant range of values to the longitudinal momentum is $-\sqrt{eE}\left(\mathcal{T}/2 + k_1^{-1}\right) \leq p_x/\sqrt{eE} \leq \sqrt{eE}\left(\mathcal{T}/2 + k_2^{-1}\right)$, in which it can be shown that the mean number of pairs created acquires three main asymptotic forms,

$$N_n^{\mathrm{cr}} \sim \begin{cases} \exp\left(-2\pi\Xi_1^-\right), & \text{for } p_x/\sqrt{eE} < -\sqrt{eE}\mathcal{T}/2, \\ e^{-\pi\lambda}, & \text{for } |p_x|/\sqrt{eE} \leq \sqrt{eE}\mathcal{T}/2, \\ \exp\left(-2\pi\Xi_2^+\right), & \text{for } p_x/\sqrt{eE} > +\sqrt{eE}\mathcal{T}/2, \end{cases} \quad (14)$$

as $k_i \to 0$, $\mathcal{T} \to \infty$, $\min(k_1\mathcal{T}/2, k_2\mathcal{T}/2) \gg 1$. Comparing these results with the differential mean numbers of pairs created by a T-constant electric field [20, 21] and by a Peak electric field [20, 23], one may conclude that the asymptotic forms given by Eq. (14) coincides with asymptotic forms of the electric fields mentioned above. In virtue of that, we refer the electric field in the present configuration as a slowly varying field configuration.

---

[g]Without loss of generality, we select from now on a symmetrical interval II, in which $t_1 = -\mathcal{T}/2 = -t_2$.

By definition, the total number of pairs created N corresponds to the summation of the differential mean numbers N_n^{cr} over the momenta $\mathbf{p}$, and spin degrees-of-freedom

$$N^{\text{cr}} = V_{(d-1)} n^{\text{cr}}, \quad n^{\text{cr}} = \frac{J_{(d)}}{(2\pi)^{d-1}} \int d\mathbf{p} N_n^{\text{cr}}, \tag{15}$$

which, in fact, is reduced to the calculation of the density of pairs created from the vacuum n^{cr}. Here the summation over $\mathbf{p}$ was transformed into an integral and $J_{(d)} = 2^{[d/2]-1}$ denotes the total number spin projections in a d-dimensional space. Using the asymptotic forms given by Eq. (14), the total density of pairs created takes the form[h]

$$n^{\text{cr}} = \left[\mathcal{T} + \left(k_1^{-1} + k_2^{-1}\right) G\left(\frac{d}{2}, \frac{\pi m^2}{eE}\right)\right] r^{\text{cr}}, \quad r^{\text{cr}} = \frac{J_{(d)} (eE)^{d/2}}{(2\pi)^{d-1}} \exp\left(-\frac{\pi m^2}{eE}\right), \tag{16}$$

where r^{cr} is the rate of pairs created from the vacuum. Here $G(\alpha, z) = e^z z^\alpha \Gamma(-\alpha, z)$, is proportional to the incomplete Gamma function $\Gamma(-\alpha, z)$ [25]. From Eq. (12), the vacuum-vacuum transition probability P_v reads

$$P_v = \exp\left(-\mu V_{(d-1)} n^{\text{cr}}\right), \quad \mu = \sum_{l=0}^{\infty} \frac{(-1)^{l(1-\kappa)/2} \epsilon_{l+1}}{(l+1)^{d/2}} \exp\left(-\frac{\pi m^2}{eE} l\right), \tag{17}$$

where

$$\epsilon_l = \frac{\left[\mathcal{T} + \left(k_1^{-1} + k_2^{-1}\right) G\left(d/2, \pi m^2 l/eE\right)\right]}{\left[\mathcal{T} + \left(k_1^{-1} + k_2^{-1}\right) G\left(d/2, \pi m^2/eE\right)\right]}. \tag{18}$$

Comparing the total density of pairs created by a T-constant electric field [21] and by a Peak electric field [23] in the slowly varying regime, one concludes that Eq. (16) correspond to a sum of densities, one of pairs created during the switching-on and -off intervals and another of pairs created during the constant interval; see Ref. [26] for an universal approximate representation.

To analyze the influence of switching-on and -off effects on vacuum instability, we compare densities of pairs created by the composite field (2) and by the T-constant electric field [21] in the slowly varying regime. In this case, the densities are given by Eq. (16) and by $n^{\text{cr}} = \mathcal{T} r^{\text{cr}}$, respectively, so that we study the ratio

$$\mathcal{R} = \frac{\mathcal{T}}{\mathcal{T} + \left(k_1^{-1} + k_2^{-1}\right) G\left(d/2, \pi m^2/eE\right)}, \tag{19}$$

---

[h]In computing integrals over the perpendicular momentum $\mathbf{p}_\perp$, we incorporate exponentially small contributions by extending the integration domain to the infinity, i. e., $\int_{\sqrt{\lambda} < K_\perp} d\mathbf{p}_\perp \to \int_{\sqrt{\lambda} < \infty} d\mathbf{p}_\perp$.

between them for the special case $T = \mathcal{T}$. This allow us to separate particle creation effects associated with the interval where the field is constant from those exclusively associated with switching-on and -off processes. In Fig 2, we present the ratio $\mathcal{R}$ for some specific values of $\mathcal{T} = T$, k_j, the amplitude E and the space-time dimensions.

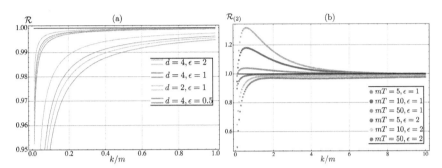

Figure 2: (a): Ratio $\mathcal{R}$ between densities of pairs created from the vacuum by a T-constant electric field and the composite field in the slowly varying regime, as a function of k/m ($k_1 = k_2$), for $m\mathcal{T} = mT = 1000$ (solid lines) and $m\mathcal{T} = mT = 100$ (dashed lines) for some specific values for the amplitude E and space-time dimensions d. Here $k_{\min}/m = 0.01$ and $k_{\max}/m = 1$. (b): Ratio $\mathcal{R}_{(2)}$ between the total number of particles created by a T-constant field (with time duration T) and a symmetrical composite field ($k_1 = k_2$ and $\mathcal{T} = T$), as a function of k/m, for some values of $\epsilon = E/E_c$ and $\mathcal{T}$. The distance between adjacent points is $\Delta k_i/m = k_{i+1}/m - k_i/m = 0.05$. The leftmost point is $k/m = 0.1$ while the rightmost one is $k/m = 10$. All numerical values refer to the reduced system of units.

According to Fig. 2, the ratio $\mathcal{R}$ decreases as k/m decreases and tends to the unity as k/m increases, a common feature to any amplitude E, space-time dimensions d or time duration T. This is due to the fact that the difference between the pulse shapes (time dependence) of a T-constant field and the composite field decreases as k/m increases. This means that a significant number of pairs is created during the switching-on an -off processes when k/m is small and decreases as k/m increases. In addition, for k and T fixed, the influence of switching-on and -off processes decreases as E increases (compare, for example, orange lines ($E/E_c = \epsilon = 2$) with pink lines ($E/E_c = \epsilon = 0.5$), in the graph (a)). This means that a larger number of pairs is created during the constant interval if E increases. Moreover, irrespectively of the value of k/m, the influence of switching-on and -off processes decreases as $\mathcal{T}$ increases. At last, but not least, for k/m, E and T fixed, one can compare results corresponding to $d = 4$ (red lines, plot (a)) with those for $d = 2$ (light blue lines, plot (a)) to conclude that the switching-on and -off processes are more significant for low dimensional systems. The total density given by Eq. (16) depends on the space-time dimensions via an incomplete Gamma function that, inevitably, decreases as d increases.

To extend the comparison above to cases where $\mathcal{T} = T$ is also finite, we compute numerically the ratio

$$\mathcal{R}_{(2)} = \frac{\int dp_x \, |g(_-|^+)_T|^2}{\int dp_x \, |g(_-|^+)|^2}, \tag{20}$$

between the exact density of pairs created by the T-constant electric field[i] over the total density of pairs created by the composite field, given by Eq. (11), in the case of a two-dimensional space-time $d = 2$ (see plot (b) in Fig. 2). Besides the same features discussed above, the ratio $\mathcal{R}_{(2)}$ exceeds the unity for some specific values of E/E_c within a certain range for k/m, as it can be seen from the results corresponding to $E/E_c = 1$ (green/blue/red dots). By definition, $\mathcal{R}_{(2)} > 1$ means that the T-constant field (with total time duration $T = \mathcal{T}$) produces more particles from the vacuum than the composite field with the same duration $\mathcal{T}$ and certain values to the switching-on and -off phases k. These results show that the existence of switching-on and -off processes imply, in certain cases, in a particle creation rate slower than the case where such processes are absent.

Supported by Russian Science Foundation, research project No. 15-12-10009.

References

[1] O. Klein, Z. Phys. **41**, 407 (1927); **53**, 157 (1929).
[2] F. Sauter, Z. Phys. **69**, 742 (1931); **73**, 547 (1932).
[3] W. Heisenberg and H. Euler, Z. Phys. **98**, 714 (1936); [arXiv:physics/0605038].
[4] J. Schwinger, Phys. Rev. **82**, 664 (1951).
[5] A.I. Nikishov, Zh. Eksp. Teor. Fiz. **57**, 1210 (1969) [Transl. Sov. Phys. JETP **30**, 660 (1970)].
[6] A. I. Nikishov, in Proc. P.N. Lebedev Phys. Inst. **111**, 153 (Nauka, Moscow, 1979).
[7] A.I. Nikishov, Nucl. Phys. **B21**, 346 (1970).
[8] S. W. Hawking, Comm. Math. Phys. **43**, 199 (1975); Phys. Rev. D **14**, 2460 (1976).
[9] B. S. DeWitt, Phys. Rev. **160**, 1113 (1967); Phys. Rev. **162**, 1195 (1967); Phys. Rev. **162**, 1239 (1967).
[10] L. Parker, Phys. Rev. Lett. **21**, 562 (1968).
[11] Y. B. Zeldovich, A. A. Starobinsky, Sov. Phys. JETP **34**, 1159 (1972).
[12] N.D. Birrell and P.C.W. Davies, *Quantum Fields in Curved Space* (Cambridge University Press, Cambridge, 1982).

---

[i] $|g(_-|^+)_T|^2$ denotes the mean number of pairs created by a T-constant field in $d = 2$. We refer the reader to Refs. [20, 21] for its explicity form.

[13] A. A. Grib, S. G. Mamaev, and V. M. Mostepanenko, *Vacuum Quantum Effects in Strong Fields* (Friedmann Laboratory, St. Petersburg, 1994).
[14] E.S. Fradkin, D.M. Gitman, and S.M. Shvartsman, *Quantum Electrodynamics with Unstable Vacuum* (Springer-Verlag, Berlin, 1991).
[15] D.M. Gitman, Izw. VUZov Fizika **19**, No. 10, 86 (1976); S.P. Gavrilov and D.M. Gitman, Izw. VUZov Fizika **20**, No. I, 94 (1977) [Sov. Phys. Journ. **20**, 75 (1977)]; D. M. Gitman, J. Phys. A: Math. Gen. **10**, 2007 (1977).
[16] S.P. Gavrilov and D.M. Gitman, Phys. Rev. D **93**, 045002 (2016).
[17] R. Ruffini, G. Vereshchagin, and S. Xue, Phys. Rep. **487**, 1 (2010).
[18] F. Gelis and N. Tanji, Prog. Part. Nucl. Phys. **87**, 1 (2016).
[19] G.V. Dunne, in I. Kogan Memorial Volume, *From fields to strings: Circumnavigating theoretical physics*, Eds. M Shifman, A. Vainshtein and J. Wheater, World Scientific, 2005.
[20] T. C. Adorno, S. P. Gavrilov and D. M. Gitman, Int. J. Mod. Phys. **32**, 1750105 (2017).
[21] S.P. Gavrilov and D.M. Gitman, Phys. Rev. D **53**, 7162 (1996).
[22] T. C. Adorno, S. P. Gavrilov, and D. M. Gitman, Phys. Scr. **90**, 074005 (2015).
[23] T. C. Adorno, S. P. Gavrilov, and D. M. Gitman, Eur. Phys. J. C **76**, 447 (2016).
[24] S. Schweber, *An Introduction to Relativistic Quantum Field Theory* (Harper & Row, New York, 1961).
[25] *Higher Transcendental functions* (Bateman Manuscript Project), edited by A. Erdelyi et al. (McGraw-Hill, New York, 1953), Vols. 1 and 2.
[26] S. P. Gavrilov and D.M. Gitman, Phys. Rev. D **95**, 076013 (2017).

THE THREE-LOOP CONTRIBUTION TO β-FUNCTION QUARTIC IN THE YUKAWA COUPLINGS FOR THE $\mathcal{N} = 1$ SUPERSYMMETRIC YANG-MILLS THEORY WITH THE HIGHER COVARIANT DERIVATIVE REGULARIZATION

Vikentii Shakhmanov [a]
*Department of theoretical physics, Faculty of Physics, Moscow State University,
119991 Moscow, Russia*

Abstract. We calculate terms quartic in the Yukawa couplings in the three-loop β-function and in the two-loop anomalous dimension of the matter superfields for the $\mathcal{N} = 1$ SYM theory regularized by higher covariant derivatives. Comparing the results, we obtain that these contributions to the renormalization group functions defined in terms of the bare couplings satisfy the NSVZ relation independently of the subtraction scheme. For the simplest form of the higher derivative regularizing term we also calculate terms quartic in the Yukawa couplings in the renormalization group functions defined in terms of the renormalized couplings. These terms do not satisfy the NSVZ relation in the considered approximation for a general renormalization prescription. However, we demonstrate that in the case of using the recently proposed renormalization prescription in the non-Abelian case the NSVZ relation between the terms quartic in the Yukawa couplings is really valid.

It is known that the β-function is connected with the anomalous dimension of the matter superfields in $\mathcal{N} = 1$ supersymmetric theories by a relation, which is claimed to be valid exactly in all orders of the perturbation theory. This relation (1) is called the exact NSVZ β-function [1] and has the form

$$\beta(\alpha, \lambda) = -\frac{\alpha^2 \Big(3C_2 - T(R) + C(R)_i{}^j (\gamma_\phi)_j{}^i (\alpha, \lambda)/r\Big)}{2\pi(1 - C_2\alpha/2\pi)}. \tag{1}$$

Eq. (1) can be rewritten as the relation between the β-function and the anomalous dimensions of the matter superfields, of the quantum gauge fields and of the Fadeev-Popov ghosts [2],

$$\frac{\beta(\alpha, \lambda)}{\alpha_0^2} = -\frac{1}{2\pi}\Big(3C_2 - T(R) - 2C_2\gamma_c(\alpha, \lambda)$$
$$-2C_2\gamma_V(\alpha, \lambda) + C(R)_i{}^j(\gamma_\phi)_j{}^i(\alpha, \lambda)/r\Big). \tag{2}$$

Assuming that this relation is valid for the renormalization group functions defined in terms of the bare couplings one can obtain the renormalization scheme in which the NSVZ β-function is valid in all orders of perturbation theory similar to the Abelian case [3]. The renormalization prescription [2] is to use the higher covariant derivative regularization and to put some restrictions (3) on the renormalization constants, where x_0 is a fixed value of $\ln \Lambda/\mu$.

$$Z_\alpha(\alpha, \lambda, x_0) = 1; \; Z_\phi(\alpha, \lambda, x_0)_i{}^j = \delta_i{}^j; \; Z_c(\alpha, \lambda, x_0) = 1; \; Z_V = Z_\alpha^{1/2} Z_c^{-1}. \tag{3}$$

[a] E-mail: shakhmanov@physics.msu.ru

We will investigate the general massless $\mathcal{N}=1$ SYM theory with matter and regularize it by the BRST invariant version of the higher covariant derivative regularization [4]. The corresponding expression for the regularized action can be found in [5].

Here we consider the supergraphs presented in Fig. 1. These supergraphs have

Figure 1: The considered supergraphs.

no external lines. (It is possible to demonstrate that the second one does not contribute to the β-function.) We attach to the supergraths two external lines of background gauge field by all possible ways. Then we have got a set of diagrams contributing to the β-function. The expression for a sum of these diagrams can be presented as an integral of double total derivative in the momentum space. This allows us to calculate one of three loop integrals. After calculating these diagrams we obtained the following result (4) for the contribution to the β-function defined in terms of the bare couplings

$$\frac{\Delta\beta}{\alpha_0^2} = \frac{1}{\pi r} C(R)_i{}^j \frac{d}{d\ln\Lambda}\left[-\lambda_0^{imn}\lambda_{0jmn}^*\int\frac{d^4k}{(2\pi)^4}\frac{1}{k^4 F_k^2} + \lambda_0^{iab}\lambda_{0kab}^*\lambda_0^{kcd}\lambda_{0jcd}^*\right.$$
$$\left.\times\int\frac{d^4k d^4l}{(2\pi)^8}\frac{1}{k^4 F_k^2 l^4 F_l^2} + \lambda_0^{iab}\lambda_{0jac}^*\lambda_0^{cde}\lambda_{0bde}^*\int\frac{d^4k d^4l}{(2\pi)^8}\frac{4}{k^4 F_k^3 l^2 F_l(k+l)^2 F_{k+l}}\right], \quad (4)$$

where λ_0^{ijk} denotes the bare Yukawa couplings. The corresponding contribution to the anomalous dimension of the matter superfields (also defined in terms of the bare couplings) is given by the following expression:

$$(\Delta\gamma_\phi)_j{}^i = \frac{d}{d\ln\Lambda}\left(\lambda_0^{iab}\lambda_{0jab}^*\int\frac{d^4k}{(2\pi)^4}\frac{2}{k^4 F_k^2} - \lambda_0^{iab}\lambda_{0kab}^*\lambda_0^{kcd}\lambda_{0jcd}^*\int\frac{d^4k d^4l}{(2\pi)^8}\right.$$
$$\left.\times\frac{2}{k^4 F_k^2 l^4 F_l^2} - \lambda_0^{iab}\lambda_{0jac}^*\lambda_0^{cde}\lambda_{0bde}^*\int\frac{d^4k d^4l}{(2\pi)^8}\frac{8}{k^4 F_k^3 l^2 F_l(k+l)^2 F_{k+l}}\right). \quad (5)$$

We see that the expressions for the β-function and for the anomalous dimension both defined in the terms of the bare couplings with the higher covariant derivative regularization satisfy the NSVZ relation:

$$\frac{\Delta\beta(\alpha_0,\lambda_0)}{\alpha_0^2} = -\frac{1}{2\pi r}C(R)_i{}^j\Delta\gamma_\phi(\lambda_0)_j{}^i. \quad (6)$$

However, the standard RG functions are defined in terms of the renormalized couplings. We get the β-function defined in terms of the renormalized couplings using the simplest example of regulator function for explicit calculations:

$$\frac{\tilde{\beta}(\alpha,\lambda)}{\alpha^2} = -\frac{1}{2\pi}(3C_2 - T(R))$$

$$+ \frac{1}{2\pi r} C(R)_i{}^j \left[-\frac{1}{4\pi^2} \lambda^{iab}\lambda^*_{jab} + \frac{1}{16\pi^4} \lambda^{iab}\lambda^*_{kab}\lambda^{kcd}\lambda^*_{jcd}(b_2 - g_1) \right.$$

$$\left. + \frac{1}{16\pi^4} \lambda^{iab}\lambda^*_{jac}\lambda^{cde}\lambda^*_{bde}(1 + 2b_2 - 2g_1) \right] + O(\alpha) + O(\lambda^6). \tag{7}$$

This expression for the contribution to the β-function depends on some arbitrary constants b_2 and g_1. That is why the expression is scheme dependent. Thus, the NSVZ relation is not valid for an arbitrary renormalization prescription. However, if we impose the boundary conditions (3) which possibly define the NSVZ scheme, these conditions give the values of arbitrary constants $b_2 = g_1 = -x_0$. We see that for these values of finite constants the β-function defined in terms of the renormalized coupling constant coincides with the β-function defined in terms of the bare coupling constant and the NSVZ relation is valid:

$$\frac{\tilde{\beta}(\alpha,\lambda)}{\alpha^2} = -\frac{1}{2\pi}(3C_2 - T(R)) - \frac{1}{2\pi r} C(R)_i{}^j \tilde{\gamma}_\phi(\alpha,\lambda)_i{}^j + O(\alpha) + O(\lambda^6). \tag{8}$$

Thus, we have verified that the boundary conditions (3) really produce the NSVZ scheme in the considered approximation.

Acknowledgments

The author is very grateful to K.V.Stepanyantz and A.E.Kazantsev for valuable discussions.

References

[1] V. A. Novikov, M. A. Shifman, A. I. Vainshtein and V. I. Zakharov, Nucl. Phys. B **229** (1983) 381.
[2] K. V. Stepanyantz, Nucl. Phys. B **909** (2016) 316.
[3] A. L. Kataev and K. V. Stepanyantz, Nucl. Phys. B **875** (2013) 459.
[4] A. A. Slavnov, Nucl. Phys. B **31** (1971) 301.
[5] V. Y. Shakhmanov and K. V. Stepanyantz, Nucl. Phys. B **920** (2017) 345.

QUANTUM FIELD THEORY EFFECTS ON THE HORIZONS OF STATIC METRICS

Michael Fil'chenkov[a] and Yuri Laptev[b]

Institute of Gravitation and Cosmology, Peoples' Friendship University of Russia, 117198 Moscow, Russia

Abstract. Kerr–Newman and Kottler's metrics with two horizons are considered. Evolution of their horizons is analysed in terms of an effective temperature. The results are applied to black hole physics and cosmology.

1 Introduction

In spaces with a nontrivial topology there arises vacuum polarization and particle creation [1]. The horizons lead to a loss of information, i. e. a gain in entropy, and hence the presence of an effective temperature. Evaporation and decay of the horizons are considered for two static metrics with two horizons.

2 Kerr–Newman Metric Horizons

Kerr–Newman's metric [2] is defined as

$$ds^2 = \frac{\Delta^2}{\rho^2}\left(cdt - \frac{a}{c}\sin^2\theta d\varphi\right)^2$$
$$-\frac{\sin^2\theta}{\rho^2}\left[\left(r^2 + \frac{a^2}{c^2}\right)d\varphi - adt\right]^2 - \frac{\rho^2}{\Delta^2}dr^2 - \rho^2 d\theta^2, \quad (1)$$

where

$$\Delta^2 = r^2 - \frac{2GMr}{c^2} + \frac{a^2}{c^2} + \frac{GQ^2}{c^4}, \quad (2)$$

$$\rho^2 = r^2 + \frac{a^2}{c^2}\cos^2\theta, \quad (3)$$

a is the specific angular momentum, Q the charge, M the mass of a black hole. In case $a = 0$ we obtain Reissner–Nordström's metric [3,4], and for $Q = 0$ (1) reduces to Kerr's metric [5]. The horizons of Kerr–Newman's metric

$$r_\pm = \frac{GM}{c^2}\left(1 \pm \sqrt{1 - \frac{a^2c^2 + GQ^2}{G^2M^2}}\right) \quad (4)$$

are roots of the equation $\Delta^2 = 0$, where r_+ is the event horizon, r_- is the Cauchy horizon.

---
[a] E-mail: fmichael@mail.ru
[b] E-mail: yplaptev@rambler.ru

The cases with $Q \neq 0$ and $a \neq 0$ are possible only if $M \neq 0$, i.e. there cannot exist massless charged or rotating black holes. Hence the relation between M, Q and a as follows is assumed to be valid:

$$Q^2 + \frac{a^2 c^2}{G} = \alpha^2 M^2, \qquad (5)$$

where $\alpha = const$, $Q = \alpha M$ for $a = 0$, and $a = \frac{\alpha \sqrt{G} M}{c}$ for $Q = 0$. Then the horizons take the form:

$$r_\pm = \frac{GM}{c^2}\left(1 \pm \sqrt{1 - \frac{\alpha^2}{G}}\right), \qquad (6)$$

where $r_+ = r_g$ is Schwarzschild's horizon.

The effective temperatures of Kerr–Newman's horizons are defined as

$$kT_+ = \frac{\hbar c^3}{8\pi GM}, \qquad (7)$$

$$kT_- = \frac{2\hbar c^3}{\pi \alpha^4 M}, \qquad (8)$$

which transforms into the corresponding formulae for Reissner–Nordström's metric for $Q = \alpha M$.

The power of thermal radiation from the horizons

$$P_\pm = 4\pi r_\pm^2 \sigma T_\pm^4, \qquad (9)$$

where is Stefan–Bolzmann's constant.

The corresponding horizon lifetimes are defined as

$$\tau_+ = c^2 \int \frac{dM}{P_+}, \qquad (10)$$

for the event horizon and

$$\tau_- = \int \frac{dE_-}{P_-}, \qquad (11)$$

where

$$E_- = \frac{\alpha^2 M^2}{r_-} = 2Mc^2, \qquad (12)$$

for the Cauchy horizon.

The evaporation from Schwarzschild's horizon is Hawking's effect [6] of pairs being created on it. The evaporation from the Cauchy horizon is followed by a loss of charge Q and specific angular momentum a of the black hole converting it into a Schwarzschild one [7,8]. Really, $dM < 0$ in Hawking's effect, whereas

$Q = \alpha M$, hence $dQ < 0$ for $a = 0$, and $a = \frac{\alpha \sqrt{G} M}{c}$, hence $da < 0$ for $Q = 0$. As to the specific angular momentum, its loss is due to Penrose's process of extracting a rotational energy from the ergosphere of Kerr's black hole [9].

For the horizon lifetimes we obtain the expressions

$$\tau_+ = 5 \cdot 2^{10} \pi \left(\frac{M}{m_{pl}}\right)^3 t_{pl}, \qquad (13)$$

$$\tau_- = \frac{5\pi}{2} \left(\frac{\alpha}{\sqrt{G}}\right)^{12} \left(\frac{M}{m_{pl}}\right)^3 t_{pl}, \qquad (14)$$

where m_{pl} and t_{pl} are the Planckian mass and time respectively.

The Cauchy horizon evaporates much faster than the event horizon. Thus the ratio

$$\frac{\tau_-}{\tau_+} = 2 \left(\frac{\alpha}{2\sqrt{G}}\right)^{12} \ll 1, \qquad (15)$$

since $\frac{\alpha}{\sqrt{G}} \ll 1$, which means that a charged and rotating black hole first loses its charge, angular momentum and part of its mass, converting into a Schwarzschild one, and then evaporates, losing the remaining mass.

3 Evolution of Kottler Metric Horizons

Kottler's metric [10] is defined by its temporal component

$$g_{00} = 1 - \frac{r_g}{r} - \frac{r^2}{r_0^2}, \qquad (16)$$

where r_0 is de Sitter's horizon satisfying the relationship:

$$\frac{1}{r_0^2} = \frac{8\pi G \varepsilon_0}{3c^2}, \qquad (17)$$

where ε_0 is the energy density of de Sitter's vacuum.

If $r_0 \gg r_g$, the equation $g_{00}(r_h) = 0$ has two roots: $r = r_g$ and $r = r_0$. The effective temperature of de Sitter's vacuum reads

$$kT_{dS} = \frac{\hbar c}{2\pi r_0}, \qquad (18)$$

and de Sitter's horizon lifetime has the form

$$\tau_{dS} = \int \frac{dE_{dS}}{P_{dS}}, \qquad (19)$$

where

$$E_{dS} = \varepsilon_0 r_0^3 \qquad (20)$$

is de Sitter's vacuum energy.

Schwarzschild's horizon lifetime with $\tau_H = \tau_+$ is given by formula (13), and de Sitter's lifetime has the form

$$\tau_{dS} = \frac{30 r_0^3}{l_{pl}^2 c}. \tag{21}$$

The de Sitter–Schwarzschild lifetime ratio reads

$$\frac{\tau_{dS}}{\tau_H} = \frac{3}{2^6 \pi} \left(\frac{r_0}{r_g}\right)^3. \tag{22}$$

De Sitter's vacuum of dark energy proves to be more stable than that in the early Universe, since $\frac{\tau_{dS}}{\tau_H} < 1$ for $\frac{r_0}{r_g} < 4\sqrt[3]{\frac{\pi}{3}}$.

4 Conclusion

Kerr–Newman's black holes convert into Schwarzschild's by losing their charge and specific angular momentum, when a particle, with the sign of charge opposite to that of the black hole, falls under the event horizon, whereas the other particle of the pair carries its charge and angular momentum to infinity. Kottler's metric horizon evolution results in a decay of the de Sitter horizon in the Big Bang and in an evaporation of massive black holes against the background of dark energy.

References

[1] A.A. Grib, S.G. Mamaev, and V.M. Mostepanenko, *"Vacuum Quantum Effects in Strong Fields"* (Friedmann Laboratory Publ., St. Petersburg, 1994).
[2] E.N. Newman et al., J. Math. Phys. **6** (1965) 918.
[3] H. Reissner, Ann. Phys. **50** (1916) 15.
[4] G. Nordström, Proc. Kon. Ned. Akad. Wt. **20** (1918) 1238.
[5] R.P. Kerr, Phys. Rev. Lett. **11** (1963) 237.
[6] S.W. Hawking, Comm. Math. Phys. **43** (1975) 199.
[7] D.V. Gal'tsov, *"Particles and Fields in the Vicinity of Black Holes"* (Moscow University Press, Moscow, 1986, in Russian).
[8] I.D. Novikov and V.P. Frolov *"Black Hole Physics"* (Nauka, Fizmatlit, Moscow, 1986, in Russian).
[9] R. Penrose, Rev. Nuovo Cim., Ser. 1, **1** (1969) 252.
[10] F. Kottler, Ann. Phys. **56** (1918) 401.

Problems of Intelligentsia

A FUTURE FOR THE INTELLIGENTSIA

John Kuhn Bleimaier [a]

15 Witherspoon Street, Princeton, New Jersey 08542 USA

The first Moscow University conference on the intelligentsia which I attended took place in 1997. It lasted for five days and there were twenty-five papers presented. The topic back in 1997 was, the Intelligentsia and Power. It was the consensus of the presenters that the Intelligentsia had an inextricable relationship with the exercise of power in the society. Here we are in 2017 at the 11th conference on the intelligentsia and out topic is the future of the Intelligentsia. I am standing before you as the first and last speaker. Is there a message here? Is the Intelligentsia as a central component of society withering away and destined for extinction?

My answers to these two questions are respectively yes and no. Yes, the declining participation in this conference should send a message to our organizing committee for purposes of planning the 2019 conference. No, the Intelligentsia is not withering away.

At a perfectly prosaic level one of the reasons for the popularity of the 1997 conference was the fact that the proceedings were presented in two languages, Russian and English. At this venue the presentation of papers and the conduct of discussions in the Russian language significantly broadened the base of participation. The term, Intelligentsia was, after all, first coined in Russia in the 19th Century. The Russian Intelligentsia played a significant role in the winding up of the communist leviathan and the reemergence of a Russian nation state at the end of the 20th Century. In 1997, with the easygoing Boris Yeltsin as the head of state, the Intelligentsia seemed to have a legitimate place, close to the halls of power. Discussion about the Intelligentsia and power was lively and voluble.

In 2017 the Russian Federation is no longer a work in progress, it has become a stable and rigid sovereign entity with a popular and assertive leader. The Intelligentsia will always be an important component of Russian society and speak for the conscience of the people. However, it seems that the Russian Intelligentsia does not have, or seek to have, a decisive role in directing the ship of state at the present time.

It is axiomatic that the future of our civilization is in the hands of sentient individuals who are capable of analyzing the status quo, planning for development and following through with implementation. It seems to me that people with a broad understanding, a consciousness of the past and possessed of ideals are the best qualified to shepherd the way into the future. Such a description of the ideal leadership cadre may define the Intelligentsia. If that is the case, the Intelligentsia in Russia, Europe, Asia and the Americas and across the globe has a pivotal role to play in the future of humankind.

---

[a] B.A., Columbia College of Columbia University; Master of International Affairs, Columbia University School of International Affairs; Juris Doctor, Saint John's University School of Law; member of the New York, New Jersey United States Supreme Court bars; formerly Fellow of Mathey College of Princeton University.

1

The Intelligentsia is that social stratum comprised of individuals for whom intellectual activity is a passion. To a certain extent all human enterprise has an intellectual component, being based on mans sentient nature. In a technologically advanced society most employment requires advanced education and the application of complex reasoning to problem solving. However, the members of the Intelligentsia engage in analytical and creative thinking for its own sake. The Intelligentsia probes the theoretical underpinnings and seeks transcendent understanding. This entails attempting to discover fundamental principles.

There was a time when it was assumed that all educated people were members of the Intelligentsia. The nature of formal higher education entailed exposure to multiple disciplines and it seemed clear that an intelligent persons intellectual curiosity would be stimulated resulting in de facto membership in the Intelligentsia. Ironically, in recent times the proliferation of specialized knowledge has resulted in ever more extensive professional education and ever declining general education. The personnel needs of the marketplace have put a premium on applied knowledge skills and has devalued the importance of liberal arts and pure science. Thus today many individuals with advanced degrees and academic positions are yet unqualified to be called members of the Intelligentsia. They do not possess the expansive world view required to analyze the big picture.

A highly trained technocrat may function well as an advanced cog in the contemporary economy and yet lack the overarching intellectual curiosity and interdisciplinary awareness characteristic of the Intelligentsia. Perhaps we can posit that the difference between the technical professional and the intellectual is akin to the difference between an artisan and an artist. The artisan performs a discrete function. The artist seeks to understand and influence the world.

The world as we know it can continue to function possessed exclusively of the services of technical professionals who maintain the various levels of the infrastructure. However, progress and advancement require input from the Intelligentsia, the creative innovators who can bring together concepts from diverse disciplines. Membership in the Intelligentsia is not expanding in proportion to the expansion in higher education. However, the future of society as a positively evolving human community depends on the continuing existence and preeminence of abstract thinkers, members of the Intelligentsia.

2

2017 marked the 500th anniversary of the event which brought the Intelligentsia to the forefront of European civilization. The Protestant Reformation, dating from 1517, was the pivotal moment for our civilization. Martin Luthers concept of the priesthood of all believers made it possible for the individual to question the teachings and actions of those in positions of authority. The Reformation established the freedom of conscience. This notion, more than any scientific observation of the position of the sun in the solar system, put man at the center of the universe and made him the master of his own destiny. The Reformation made possible the Intelligentsia as we know it. The freedom of conscience inherent in the priesthood of all believers opened the possibility to question authority and engage in free intellectual enquiry.

There appears to have been an Intelligentsia prior to the Reformation, but its existence is of purely academic interest. With the Reformation the ideas of intellectuals were transformed into political and social action, into scientific inquiry and discovery. Power structures were attacked, hierarchies were set at naught, merely on the basis of the force of an idea. The members of the Intelligentsia transformed the social structure and the world of natural science. The freedom of scientific inquiry made possible the theoretical and practical advancements which ultimately brought about the industrial revolution.

The Treaty of Westphalia and the Counterreformation within the Roman Catholic Church brought establishment recognition to the new social order with its preeminent Intelligentsia. This in turn made the Enlightenment possible. The history of western civilization during the last 500 years is intimately intertwined with the Intelligentsia.

3

The Russian Intelligentsia enjoys a unique position in the history of western culture. First of all the term Intelligentsia was initially coined in 19th Century Russia where a self-conscious group of Renaissance men and women came to recognize that their intellectual pursuits set them apart as a distinct grouping within the society at large. It has long been posited that there is something in the Russian national character which is predisposed to introspection and abstract contemplation. Be that as it may. Ultimately, the Russian Intelligentsia came to occupy a historically central position because of the Russian Revolution and subsequent developments which had global significance.

When we are celebrating the 500th anniversary of the Reformation we simultaneously observed the 100th anniversary of the Russian Revolution in 2017. The Russian Intelligentsia played a critical role in the events which lead to the Revolution. Russian thinkers, theorists, teachers and propagandists paved the way for an enormous and unprecedented social turmoil. Members of the Russian Intelligentsia facilitated the demise of the Russian imperial system. They established the provisional government after the removal of the czar. They destabilized and toppled the provisional government paving the way for the October Bolshevik Revolution. Members of the Russian Intelligentsia held important posts in the successor Soviet state from its inception. Subsequently, the Russian Intelligentsia established a dissident parallel power structure during the waning years of the Soviet Union and the members of the Intelligentsia ultimately toppled the Marxist leviathan.

The Russian Intelligentsia continues a lively existence in todays Russian Federation. Perhaps, it has not as yet coalesced into a formidable force for or against the current governmental system. However, it is certain that whether the current Russian body politic either crumbles or solidifies, the Intelligentsia will play a decisive role in the process. This optimistic assessment of the future of the Russian Intelligentsia today leads to the inescapable conclusion that there is also an important role for the global Intelligentsia at large in the future of the entire human community. Whether as a source of stability or instability the significant role of the Intelligentsia is vouched safe.

4

In the 21st Century there has been a crisis of the Intelligentsia associated with the transformation of the leisure class and a shift in the nature of education. The traditional class structure of society and a reexamination of the purpose of education have changed the composition and challenged the regeneration of the Intelligentsia.

In the 19th Century a significant component of the Intelligentsia was recruited from the leisure, rentier, class, that social stratum composed of people whose economic situation insulated them from the need to engage in conventional work, who had time to cultivate contemplation, the arts and to study pure science. The great depression of the 1930s dislodged the leisure class and this fertile recruiting ground for membership in the Intelligentsia began to dramatically shrink. The super rich and oligarchs of the 21st Century constitute a unique nouveau riche: rooted in financial speculation, unselfconscious and culturally indifferent. Todays oligarchy has not proven to be a fertile field for the development of introspective, contemplative intellectuals.

The educational system has also gradually shifted from an emphasis on inculcation of values and introduction to civilization to the preparation of specialized technicians. There was an earlier concept that individuals having been educated in the arts and sciences were intellectually qualified for responsible positions in the private and public sectors. Now it is generally accepted that higher education should efficiently prepare technical specialists who are immediately equipped to fill niche occupations in the military industrial complex. Introduction to the heritage of western civilization has been relegated to a leisure time activity of self enrichment for those so inclined.

In a world of television, video entertainments and other pervasive electronic media there is little time for the average professional to devote to disinterested intellectual pursuits. If formal education has failed to engender an Intelligentsia mentality it is doubtful that people will stumble upon the pillars of western culture on their own in their quest for diversion. For this reason the Intelligentsia has not grown in proportion to the growth of higher education in society.

5

In the mid 20th Century there was a sense that the future of democracy as a political institution was at risk. The rise of fascism and communism at various times appeared to jeopardize the continued existence of liberal democratic society. Both Aldus Huxley and George Orwell conceived frightening future societies respectively in their novels BRAVE NEW WORLD and 1984. It seemed possible that human culture would be reshaped into an anthill of featureless, well disciplined existence.

In a world dominated by either the new socialist man or the Nietzscheian superman there would seem to be no room for introspective intellectuals. Just as the Intelligentsia as we know it had its genesis in the emancipating movements of the Reformation and the Enlightenment, so the rise of national socialism or Marxist socialism could have signaled the death knell of the Intelligentsia as a social class. We intuitively realize that the continuation of free intellectual inquiry requires a modicum of individual political liberty. However, the 20th Century witnessed both the rise and the fall of both, national socialist and Marxist socialist ideology. Today the nightmares conceived by both Huxley and Orwell seem remote and no longer immediately forbidding.

Thus, the 20th Century authoritarian threats to the continued existence of the Intelligentsia seem to have evaporated along with the imminent violent ethos which brought them into existence.

6

No discussion of the future of the Intelligentsia would be complete without consideration of Oswald Spenglers magnum opus, THE DECLINE OF THE WEST. 2018 marks the one hundredth anniversary of the first publication of Spenglers work. Oswald Spengler studied and absorbed the historic political and intellectual rise and fall of several mighty civilizations, or cultures. With great sensitivity he recognized their strengths and weaknesses. Because the temporal world exists, by definition, in linear time, Spengler saw the cycle of birth, maturation, decline and death in the cultures which he studied. From his research, Spengler synthesized a theory of history which claimed the existence of immutable cycles in the life of every civilization. He claimed to have discovered an inevitable law of nature which seals the fate of every culture.

The Intelligentsia which we have examined supra is largely the product of western civilization. Thus if Spengler is correct that the decline of the West is inevitable, we can assume that the decline of the western intelligentsia is concomitant. But was Spengler correct? Is the decline of the West inevitable?

Oswald Spenglers genius was his extraordinary sensitivity. He perceived trends which were only in their infancy during his lifetime. On the basis of these delicate perceptions he assumed the role as philosophical soothsayer and predicted the future of western culture and civilization. However, once we recognize that Spenglers philosophy did not actually foretell future developments then we can free ourselves from the notion that the long term future, as he foresaw it, is inevitable. Spenglers pessimistic view of the outlook for western civilization is inconsistent with the spirit of optimism and faith in progress, which has formed the foundation of western Christian culture.

As all great thinkers, Oswald Spengler has provoked close examination and intense debate on the subject of the human condition. His model of cyclical historic development can be a useful point of departure in contemplating the past and planning for the future. However, the need to reestablish the fundamental verities of Western Christian culture as the basis of social organization is abundantly obvious. In the international community today the savagery of Islamic fundamentalism puts the lie to the concept that all religions exercise a positive, normative influence over human conduct. The Israeli/Palestinian conflict shows the futility of settling disputes based on force and in the absence of fairness and equity. The precipitous decline in public morality in various countries shows that moral relativism is bankrupt. As politicians go corrupt, and many common people go wild, it becomes crystal clear that a successful society must espouse morality. Christian morality remains unsurpassed in its utility as it conjoins the strictures of law with the moderation of mercy and forgiveness. The western Intelligentsia has been struggling with the application of our cultural dynamism to the problems of social organization for five hundred years.

Let us postulate that this is the Intelligentsias answer to the prediction of the decline of the west. The end of our civilization is not a mythic, predestined curse as seen by romanticists like Spengler. It is not a phase of social evolution in the Marxist idiom. It is not the product of the survival of the fittest and the precursor of the age of the superman. The decline of the west is a malady which engages the interest of the Intelligentsia and harnesses its intellectual strength for the task of working out solutions for societys problems. In view of the centenary of the first publication of Oswald Spenglers historic theory we will discuss the concept of the decline of the west in greater detail separately in the Appendix, infra.

7

The Resurgence of Islam in the 21st Century is a phenomenon with which the Intelligentsia must contend. Not only in the Middle East, but in Europe and North America the influence of renewed Muslim fundamentalism has serious repercussions. In an interdependent global community the development of any societal movement has transnational impact. The presence of petroleum natural resources in the Islamic middle east gives additional weight to developments in that region. Furthermore the presence of Israel, the Jewish national state, squarely in the center of the hotbed of Islamic militancy tends to further focus international attention.

Historic Islam was associated with advances in mathematics, science, astronomy and architecture from the 9th to the 15th Centuries. It would be logical to postulate the existence of a Muslim Intelligentsia at that distant time when the Islamic world was at the cutting edge of scientific inquiry. However, in our time an Islamic Intelligentsia seems to exist exclusively as a component of the greater, international, western oriented Intelligentsia. Empirical evidence suggests that today Islamic scientists and intellectuals are components of western society on the basis of training and world view. They are part of the generalized Intelligentsia.

The 21st Century resurgence of Islam has not, by and large, sprung from the community of well educated Islamic intellectuals. It appears to be the creature of parochial theological education centered on the study of the Koran and associated pious writings. The leaders of the new Islamic fundamentalism have not been members of anything we could call an Intelligentsia. They are not introspective or even broadly educated. Their narrow world view might be considered the antithesis of that of an Intelligentsia. The new Islamic fundamentalism destroys historic landmarks; eschews the arts; and insists on doctrinal rigidity to the exclusion of any independent inquiry.

Because it is a broad based, youth oriented movement, resurgent Islam is a real challenge to the Intelligentsia. However, it is a challenge for which the Intelligentsia should be well prepared. The answer to fundamentalist Islam is education. Not the education of the Taliban Koranic academy, but the education of liberal arts and sciences which opens the mind for intellectual pursuits.

8

China is the new major league player on the global stage. An ancient culture which was largely eclipsed as a result of 19th Century western imperialism, a stultifying bureaucracy and the ubiquitous presence of opium, China was quiescent in the global arena for a long time, through the mid 20th Century. Militant Maoist communism, demographics and the western hunger for inexpensive labor in the production of industrial goods have cast China in a pivotal role today. The radical dislocations wrought by Mao Zedong in a heretofore moribund social structure helped to unleash great potential. Today, the presence of more than 1.5 billion people has presented both a problem and an opportunity. Western consumerism coupled with environmentalism gave the Chinese economy a mighty jolt. From the time of the Sun Yat-sen revolution at the beginning of the 20th Century there has been a tiny western oriented Intelligentsia in China. The subsequent Sino-Japanese conflict, the communist revolution and the so called cultural revolution, rent asunder the fabric of Chinese society. At the dawning of the 21st Century a unique, nominally communist and economically capitalist structure has taken shape. The scale of the Chinese presence on the planet makes its potential contribution to the future of the Intelligentsia significant. Broadly speaking, 95More significantly, there is a substantial Chinese presence at all centers of higher education and scientific inquiry. Chinese students are absorbing the heritage of western technology, science and civilization at a voracious pace. This cannot but have an impact on the future of the Intelligentsia. Many of todays Chinese students in the West will surely be members of the Intelligentsia in the future. The expansion in the size of the Intelligentsia worldwide will undoubtedly have an impact on the demand for works of art and literature. This impact is already being felt in the sphere of collector art. The Chinese cultural traditions will likely impact the evolution of the Intelligentsias perspectives and tastes. While the ultimate impact of the expanding Chinese component of the Intelligentsia is as yet unclear, there is no reason to believe that the Chinese influence will be negative. From the point of view of the Intelligentsia the expansion of learning, creativity and intellectual contemplation is never a bad thing.

9

Is the Intelligentsia a monolithic, integrated social entity with a single unified perspective? The answer appears clearly to be in the negative. Shared interests and occupations do not entail common values and beliefs. While it is true that the members of the Intelligentsia seem to agree on the importance of preserving the natural environment and advancing world peace, they do not agree as to the means of achieving these broad goals. This has ever been the case and will likely continue in the future. The 19th Century Russian Intelligentsia, our historic prototype, was distinctly fragmented. Liberal western oriented members of the Intelligentsia gravitated toward Count Leo Tolstoy, flirted with socialism and despised the tsarist autocratic structure. On the other hand conservative Slavophils extolled the pronouncements of Fyodor Dostoyevsky and the Orthodox theology of Father John of Kronstadt, supporting the Russian monarchy as the backbone of the nation. Both groups included rigorous thinkers and articulate spokesmen. There were other philosophical movements and groupings within the Intelligentsia. It remains the same today. The Intelligentsia is not grouped on any particular end of the political spectrum. However, in the 21st Century the diversity within the Intelligentsia is not always easily discerned. That is because the phenomena of mass electronic media has provided an unprecedented bully pulpit to those exponents of the Intelligentsia who share the ideology of the economic oligarchy. Because mega mass media often speaks with one voice there is a tendency to accept that there is but a single unified stand taken by the members of the Intelligentsia. However, one need only look to the internet to realize that there are educated, thoughtful and articulate spokesmen for alternative belief systems at large in society,

10 Conclusion

The future of the human species and of the planet which we inhabit may well be uncertain. However, it seems clear that as long as there are human beings there will be study, analysis, planning, creativity, dialogue, dispute and faction. The presence of individuals with an intellectual bent of mind guarantees that there will always be an Intelligentsia in human society. Some things occur apparently randomly in the fate of civilization. But advances occur as a result of careful observation and meticulous planning. This is the work of the Intelligentsia. This ensures that the Intelligentsia will continue to play an important role in the future of humankind.

Appendix: The Intelligentsia, Christianity and Progress in Relation to the Decline of the West

We are living in the backwash of the cataclysmic events of the 20th Century. The First and Second World Wars and the New World Order have completely restructured the planet, politically, economically and socially. The seemingly unshakable dominance of European Christian culture has been put in question. Historians try to make sense of the present by elucidating the past. Whether this is a useful pastime or not, it is human nature to seek understanding.

No society can emerge unscathed from a century of civil war with the attendant loss of blood and treasure. If we postulate that Europe together with North America have constituted a single social entity with shared history and morality, together with interlocking economy, then we can realize that the 20th Century was a period of uninterrupted civil war. As a result of the massive death and destruction, experienced during the last hundred years, European society has nearly destroyed itself. The erstwhile Euro-American civilization has not been overpowered by a competitor. It has effectively flirted with mass suicide.

Because this European Christian culture has always been characterized by a belief in progress and attendant optimism, I, for one, am prepared to believe that a resurgence is at least possible. For this reason and for entertaining diversion, I am committed to studying and understanding history.

A

The decline of the West, as used in the title of this appendix, is intended to have two distinct meanings. First, the decline of which I speak is the perceived retreat of western hegemony on the global stage during the 20th Century. The second sense in which I refer to The Decline of the West, is as the title of Oswald Spenglers magnum opus wherein he expounds his theory of history as an inevitably cyclical phenomenon. It is my thesis that the former is at most a retrenchment and perhaps only a misperception. As for the latter, that is, Spenglers philosophy of historiography, I find it to be an elegant but flawed construct, a romantic folly.

In the context of this analysis I will use various terms such as culture and civilization, in accordance with their conventional dictionary definition, not necessarily according to Spenglers usage. In accordance with Christian doctrine, In the beginning was the Word, and the Word was with God, and the Word was God. John 1:1 Language is sacred. Coining new terms is the essence of creativity. However, redefining existing terms and ascribing altogether new meanings to them is highly suspect. The propagandist distorts language in order to redirect the thinking of his audience. It is appropriate to praise and to critique Oswald Spenglers philosophy without complacent resort to his own redefinitions.

At various points in this appendix I make reference to western Christian civilization. By this term I mean to include the civilization of Russia, eastern Europe and Australia as well as the other communities espousing values with there roots in Christian doctrine, regardless of their geographic location.

B

On June 28, 1914 the world was enjoying unprecedented economic prosperity. Industries were working at capacity producing capital and consumer goods. The automobile, the airplane, sound recording, radio and electric lighting were transforming human existence. Scientific advances were improving the lot of all peoples. Medicine, hygiene and sanitation were materially decreasing infant mortality and increasing life expectancy. Social consciousness was breaking down artificial barriers between classes, redistributing the wealth of nations and combating poverty. Oswald Spengler was sitting at his desk, at work on the manuscript of The Decline of the West.

In 1914, nearly a century had passed since the Congress of Vienna had terminated the Napoleonic wars. The European continent had enjoyed a prolonged period of relative peace. Christendom had reached its first high watermark. The Doctrine of the Golden Rule, do unto others as you would have them do unto you, Matthew 7:12 and the ideal of loving ones neighbor as one loved ones self, had been spread to the far corners of the earth. By no means had a utopian state of affairs been attained, however, the teachings of Christ were universally recognized as representing an ideal. The exponents of Christian theology had taken their message to all peoples. The imperative of the Great Commission (Matthew 28:16-20) had been fulfilled.

At a secular level, the moral consciousness of the planet had been transformed. Even among non-Christians the valuation of human life, concepts of justice and equity, notions of fairness had been radically impacted. The naked force of the western idea coupled with the tangible strength of the western societies made their influence inescapable, from pole to pole, form cape to cape.

Shortly before the fateful year 1914, Admiral Nikolaus von Horthy traveled around the world with the Austro-Hungarian naval squadron (sic). From the Mediterranean, to the Atlantic, to the Pacific; from Europe, to Africa, to Asia, to the New World, everywhere the Habsburg vessels traveled on their grand tour they encountered resplendent western culture. Over a period of centuries North and South America had been transformed into component parts of the respective Anglo-Saxon and Iberian cultures. Most of Africa and Asia were experiencing colonial dependency being integrated into the European way of life and value system.

Yes, this was the age of imperialism. European empires stretched their influence around the planet. We have become accustomed to thinking of colonialism in negative terms. The word, imperialist, has become a form of abuse. However, a dispassionate historic assessment will ultimately conclude that the age of imperialism was a time of generalized enlightenment. Advanced science, technology and value systems based on the Golden Rule were spread abroad by the Europeans. In the most extreme cases Stone Age life styles were propelled forward into the industrial age. Feudal social structures were wrenched upward toward modernity. The human condition is inherently imperfect. But a high watermark had been attained.

The epoch of European empires may be evaluated variously by the colonized societies. However, there is no denying that the positive advances attained by all peoples in the 21st Century were made possible by the colonial dissemination of ideas and concepts during the 19th Century and the first decades of the 20th. Certainly the colonial experience was diverse, ranging from benign to exploitative. But from the disinterested perspective of future historians imperialism on the whole will be judged to have been a very effective tool in the service of progress.

This was the state of affairs a century ago, when Oswald Spengler began writing his definitive volume. Evidences of progress were everywhere to be seen. In the Old World itself prosperity was becoming generalized. The social order was generally democratized. Absolute monarchies were transforming into constitutional monarchies.

On June 28, 1914 the heir to the Habsburg thrown was assassinated by a Serbian terrorist in the Balkan province of Herzegovina. Neither the Archduke Franz Ferdinand, nor the aspirations of Serbian nationalists, nor the Balkan region appear to have been overpoweringly important from our perspective in the 21st Century. Nevertheless, the event which took place on that long ago day in that far away place was destined to plunge Christendom into a fratricidal war whose consequences nearly destroyed the underpinnings of Western civilization during the course of the ensuing century.

As we find ourselves at the centennial of the outbreak of the Great War it behooves us to not only study the history of a bygone epoch but to contemplate the ways in which we might bridge the gap between the past and the present in order to build a better future on worthy foundations. It is not possible to return the evils released from Pandoras box over the course of a hundred years. However, the essence of Western Christian civilization has been the unshakable confidence in the possibility of progress. If we can recapture that confidence in the improvability of the future we can take up the reins of destiny and put the tram car of civilization back on track.

What caused the outbreak of the world conflagration? It was clearly not the person of the victim of the Balkan crime, nor the ideology of the perpetrator of the act. Critical examination of the events, which preceded the Great War, shows that the western powers were on a collision course and the death of the Archduke was merely the chance spark, which lit the fateful flame.

C

The notion of the decline of the west is inextricably intertwined with 19th Century romanticism. The doomed heroic struggle is the essence of Wagnerian opera. The fixation on ruins of classical civilization is central to the plastic arts starting with Giovanni Battista Piranesi. The demographic paradigm called the yellow peril forms the backdrop for the projections into the future. The massive destruction brought about by the First World War provides the justification for universal pessimism. The environmental degradation of the planet justifies the sense of the finite. The loss of faith in aesthetics, disillusionment in relation to objective beauty and the pervasive anarchy in the arts seem to presage a culmination. All in all, the intellectual tendencies of the last century have been a self-fulfilling prophecy.

However, the fixation on decline and the glamorization of defeat are the antithesis of traditional western culture. They are an aberration, a solecism, inconsistent with the overall thrust of historical development. The cult of decline does foster regression. However, it will inevitably consume itself before it can take civilization down the path to destruction.

The essence of Western Christian civilization is the belief in progress. The improvability of man and the advancement of the human condition are at the center of two thousand years of history. This established the preeminence of Western Christian culture and guarantees its survival long after the romanticism of decline has slipped off the charts.

Unquestionably a genius, Oswald Spengler studied and absorbed the historic political and intellectual rise and fall of several mighty civilizations, or cultures. With great sensitivity he recognized their strengths and weaknesses. Because the temporal world exists, by definition, in linear time, Spengler saw the cycle of birth, maturation, decline and death in the cultures, which he studied. From his research, Spengler synthesized a theory of history, which claims the existence of immutable cycles in the life of every civilization. He claims to have discovered an inevitable law of nature which seals the fate of every culture.

Interestingly, by positing the ineluctable decline of western culture, Spengler places himself and his philosophy outside the mainstream of the western intellectual tradition. Western Christian civilization is premised upon optimism, improvement and progress. The spirit of the West is infused with the notion of progress for humankind and advancement of society. Western Christian culture perceives life as the cosmic struggle between good and evil. However, the long-term outcome is never in doubt. Good will triumph over evil. God will banish Satan to the bottomless pit. The omnipotence of good will totally overcome the preposterous presumption of evil. An overall decline of Western Christian culture is, thus, impossible.

D

From the perspective of Western Christian culture the historic philosophy of Oswald Spengler represents the culmination of an anti Christian solecism of the 19th Century a romantic, spiritual movement exemplified by Darwin, Nietzsche, Marx and Wagner. Spenglers hypothesis of the decline of the West is the natural consequence of a worldview, which divorces the doctrine and revelations of Christianity from the equation of mans existence. But the philosophy of decline can be a self-fulfilling prophecy, just as the belief in progress engenders itself.

Charles Darwins theory on the origin of species, just as the entire notion of biologic evolution from the microbe to the human being, is premised on a certain progression. To this extent it is consistent with the general thrust of Western Christian thought. Creation moves forward progressively in response to natural selection from the simple and primitive to the complex and exalted. Genesis1:1-2:3. However, since Darwins theory divorces God from the equation, and thus is devoid of notions of good and evil, there is no place for moral progress. It is precisely moral progress, which is at the fulcrum of Western Christian thought.

Friedrich Nietzsche applies the Darwinian theory of evolution and natural selection in the context of human civilization. Nietzsche proposes the evolution of man into superman. He transforms the concept of the survival of the fittest into a pseudo moral imperative in its own right. He thus argues that it is the duty of the fit to survive and prosper without regard for the plight of the less fit. Like Darwin, Nietzsche envisions the world devoid of Christian doctrine. He is the exponent of a new, anti Christian morality where might makes right.

Karl Marx, on the other hand, fashioned an anti Christian ideology premised on evolving social organization, from feudalism, to capitalism, to socialism, to communism. The Marxist worldview is also somewhat colored by the Christian value system envisioning the lowly inheriting the earth. However, the centrality of materialism to Marxist thinking places this ideology forever at odds with the Christian ideal of spirituality and its inherent rejection of the material world.

It is difficult to overestimate the position of the arts in the life of civilization. It has been said that perception is reality. How we perceive the world around us is molded by the genius of creative artists. Richard Wagner believed that opera was the most exalted of the arts as it combined music, poetry, the plastic and visual arts. Wagner made himself the master of this genre, which, in an age before mass media, spoke to an enormous audience across boundaries of nation and class. The unique romanticism of Richard Wagner incorporated a neo-paganism with a nostalgic attachment to the collapse of a doomed order. The fascination of 19th Century romanticists with ruins of bygone cultures was given palpable force in Wagners opera. Is it not inevitable that the twilight of the gods should be transformed into the decline of the West?

E

The notion of progress is only possible in the context of a value system, which defines good and bad. In the absence of a basis for determining what is desirable and what is undesirable, the concept of progress or regression have no meaning. Christianity presents revealed truth. This makes it possible to have a belief system that stands apart from myth and superstition. The revealed truth also establishes a baseline for the extrapolation of philosophical concepts and the making of value judgments. Using the revealed truths as a point of departure it is not necessary to resort to the Cartesian cognito ergo sum as a starting point for the reasoning process. Revealed truth removes the necessity for arguing about fundamentals and provides for development with self-confidence.

Western Christian culture has believed in itself as the bearer of ineluctable verities. It has believed in the predestined ultimate triumph of its message. Of enormous importance is the Wests confidence in the universality of its truth. Christianity is not a tribal religion extolling the preeminence of our gods versus your gods. It has believed in its mission in relation to all humanity. For these reasons Western Christian society is incapable of decline by its own terms.

Oswald Spengler has provided an elegant intellectual structure for the analysis of historic development; for the examination of the contemporary socio-political environment; and for the contemplation of the future. His erudition is formidable. Spenglers writings establish the syllabus for an advanced academic seminar. While his reasoning is impeccable his conclusions do not possess the inevitable force of nature, which he would have us believe.

Spenglers thesis, in the final analysis, possesses the stultifying quality of the doctrine of predestination. He would have us accept the notion that the cyclical rise and fall of civilizations is inescapable. Taken to its logical conclusion this worldview justifies passivity and inaction. If the cyclical rise and fall of societies is unavoidable there is no reason for any person or group of persons to take any action. Indeed, was Spenglers motivation in putting pen to paper merely to help us to accept our dismal fate?

This is the antithesis of the Western Christian tradition, which is premised on the individuals freedom of action, obligation to pursue good works and potential for the achievement of improvement.

F

The essence of eastern philosophy is the attainment of inner peace. Thus, the wise man is the individual who is able to estrange himself from the cares of his physical environment and to look within himself for a source of balance and repose. Western Christian intellectual history leads to the opposite conclusion. While spiritual salvation is attained by faith and grace, the people of God are known by their works. Titus 2:14 . Indeed, God has chosen for Himself a special people zealous of good works. Thus, the history of the Western Christian world is a catalogue of societal edifices designed to achieve the good result.

The history of the Christian Church holds within itself the key to the dynamism of Western Christian society. The original church of biblical times was pacifist, anarchic and egalitarian in accordance with the teachings of Christ which emphasized turning the other cheek, giving away your shirt when asked for your mantle and loving ones enemies. All this changed dramatically when Christianity became an official religion of the Roman Empire at the time of Constantine. The opportunity to come out of the shadows and to become the spiritual core of the greatest political entity on the planet was irresistible. The leaders of the 4th Century Church fashioned a theology that alloyed the anarchic New Testament teachings of Jesus with the strictures of the Old Testament Mosaic Law. In the practical interests of administering a vast empire the ideal of disinterested love was modified. However, by recognizing the centrality of Christ, the classical theologians enshrined the basis of the well-formed conscience, independent of temporal considerations.

This seeming contradiction between Christ and the Law is at the heart of the dynamism of Western Christian culture. Romans 10:4. There is a perpetual pendulum effect with Western Christian jurisprudence ever swinging between forgiveness and retribution.

G

Oswald Spenglers genius was his extraordinary sensitivity. He perceived trends, which were only in their infancy during his lifetime. On the basis of these delicate perceptions he assumed the role as philosophical soothsayer and predicted the future of western culture and civilization. Soon after the publication of his magnum opus the subtle tremors of change dramatically increased in intensity. In the 21st Century we see clearly that much of what Spengler foretold has, indeed, come to pass. However, it is my firmly held belief that Spengler was not an intellectual clairvoyant, but rather an extremely astute observer and social scientist. He recognized the societal trends before his contemporaries were aware that radical change was afoot.

Once we recognize that Spenglers philosophy has not actually foretold future developments then we can free ourselves from the notion that the long-term future, as he foresaw it, is inevitable. Spenglers pessimistic view of the outlook for Western civilization is inconsistent with the spirit of optimism and faith in progress, which has formed the foundation of Western Christian culture. The acuity of his observation in his own time ought not give us pause to doubt the fundamental verities upon which our civilization is grounded. We need not give up our optimism. Progress is still possible. Indeed, if we remain true to the underlying principles of Western Christian civilization, progress is inevitable.

As does every creative theorist, Spengler established a historical paradigm, which forms the core of his philosophy. The spring, summer, autumn and winter of the historic civilizations upon which he bases his theory of cyclical history are only a loose fit in relation to the real history of the world. Just as Marx realized before Spengler, the study of the past lends itself to division into various epochs following one another. But these epochs are not repetitive cycles although human history contains certain recurring leitmotivs. This is because human nature is a constant. However, there are no ineluctable cycles which seal mans destiny in advance.

Philosophical constructs, just like ideological worldviews, represent helpful tools for understanding the past, analyzing the present and extrapolating into the future. However, when we become convinced that a theory of history is in fact an unalterable prediction of what will transpire down the road, we deceive ourselves and hobble our potential. While I greatly admire Spengler for his encyclopedic knowledge and for his creative insight, I refuse to accept him as a mystical prophet or unerring fortuneteller.

Oswald Spengler identified Christianity as the grandmother of Bolshevism, an ideology which he profoundly rejected. While Spengler is not generally considered as an enemy of Christianity as were both Marx and Nietzsche, a careful analysis of Spenglers philosophy of history reveals that he rejected the fundamental Christian message. His view of religion as an inevitable component in the rise and fall of cultures has no place for the acceptance of revealed Truth and a Divine role in the creation and evolution of human society. For Spengler, the role of Christianity in our, what he calls Faustian, culture is not materially different from the role of their polytheistic beliefs in the ancient Egyptian civilization. He perceives any religion as just one of the characteristics of every civilization, destined to ascend, experience metamorphosis and decline.

In this regard I would argue that Spengler errs. Independently of its theological claim to universality and timelessness, Christianity has proven to possess a dynamism which has allowed it to successfully transcend the rise and fall of whole societies. The alliance of the anarchic, pacifism of Christs teaching with the rigors of the law of the Old Testament has created an extraordinary resiliency. There is an ever present pendulum effect as society fluctuates between the extremes of mercy and retribution. The practical legal structure established on the basis of the Old Covenant is effectively leavened by free dissemination of mercy. The well formed Christian conscience prevents excesses.

H

The Common Era: is a term used in an anti Christian society to refer to the historical epoch since the birth of Jesus Christ. Traditionally the historic calendar has referred to years before His birth as: BC, before Christ. By the same token events taking place since Jesus nativity have been referred to as taking place in a year: Anno Domini, in the year of our Lord, such and such. A militantly secular society wishing to divorce itself from the Christ centered view of history has coined a new term. In doing so, however, the anti Christian pundits have actually effectively affirmed the primacy of the Saviors birth in the history of the world.

There is no gainsaying that the year one of our common number line is the year of Christs birth. By referring to our time as the Common Era and to the pre Christian past as, Before the Common Era, they have precisely underlined the significance of that single historic birth as a historical turning point. The contemporary era common to all humankind dates from the birth of Jesus Christ. It is He who declared the universal message for all peoples. And it is His followers who have spread a unique gospel of progress across the planet to cultures old and new. Truly, the coming of Jesus Christ heralded the dawning of a common era for all the species.

When we recognize the commencement of this common era some two thousand years ago we contradict the fundamental premise of Oswald Spenglers cyclical view of history. When there is, or can be, a world changing event there is no possibility for the existence of immutable cycles of rise and fall. The progress made possible by the Christian message has taken place in fits and starts with an uneven progression toward a brighter future. But the presence of an inescapable upward trend toward progress for all humanity makes redundant the quest for a cyclical pattern incorporating an inevitable ebb and flow.

Having posited the folly of the Spenglerian historic scenario it behooves us to explain how it is that some of his predictions have indeed come true. What is the significance of the signs of decadence and regression to be seen in our time? If we are not living through an inescapable autumn and winter of our culture what facts explain where we find ourselves today? Let us trace the malaise du jour to its roots in the 19 th Century. While the 19 th Century was unquestionably a Christian century from the perspective of human progress, the seeds of present problems were sown in the period following the French Revolution.

The Protestant Reformation established the freedom of conscience and broke the Roman Churchs virtual stranglehold on the moral perceptions of western man. The thesis of Martin Luther was that individual believers had the inherent right and ability to make moral choices, independent of a church structure. The American Revolution enshrined the supremacy of conscience when Thomas Jefferson stated that man has the right, if not the duty, to set at naught the dictates of the sovereign if they are in contravention of the law of God or nature

Both the Reformation and the American republican experiment are important components of the progress attained by Western Christian culture. However, both pivotal events brought with them the increased risk that anti Christian tendencies might make inroads in the social structure. By placing the individual well formed conscience on a pedestal Martin Luther challenged the unified hierarchy of Western Christian values. When Jefferson penned the American Declaration of Independence he once and for all terminated the paternalistic, authoritarian structure of governance in the West. Significantly it was a pious monk who struck the blow to the churchs monopoly. And it was a prosperous land owner and lawyer who undercut the foundation of the establishment. No Godless, sans-culotte radicals these.

Freedom of conscience and civil liberty played central roles in attaining the scientific and societal progress which have characterized Western Christian civilization. Yet unfettered individualism provided fertile soil not just for a bountiful harvest but for the proliferation of tares and weeds which have sapped the body politic. The genius of Western Christian culture and the verity of the Christian revealed truth render inevitable the ultimate resumption of progress and the quest for the ideal. Both fascism and communism have fallen by the wayside. Yet, not long ago in the historic past both of these ideologies showed enormous promise in the competition to overthrow the Western Christian ideal. Had either fascism or communism attained preeminence we might well long for the promised decline of the west prophesized by Oswald Spengler. Indeed, we might anxiously await the rise of the next culture destined to cycle forward to civilization. However, the ideological antagonists of Western Christendom have fallen of their own accord. My prognosis is that their fate will be shared by other politico-social constructs which oppose the mighty Christian juggernaut.

I

No analysis of Western Christian culture from the Renaissance to the present could be complete in the absence of a discussion of the Jewish question. Jewish people have been present in the west in substantial numbers at least since the Crusades. The significance of Jewish influence in society prior to the 18th Century is subject to dispute. From our 21st Century perspective we may be tempted to greatly overestimate the importance of the Jewish people in the epochs prior to the French Revolution. However, from the storming of the Bastille to the establishment of the State of Israel, the involvement of Jewish people in the social, political and economic life of the west has been enormous.

As we have seen, the Protestant Reformation and the American Revolution challenged the centralized authority of the church and the state. Both these social upheavals promoted the development of individualism and initiative. From science to industry this free and liberated environment fostered great progress. It helped to fulfill the positive destiny of Western Christian culture.

The Jews have traditionally been outsiders in Western Christian culture. Their religion, customs and appearance set them apart from the majority of the population of the countries where they have lived. They have performed an economic and social function as they engaged in activities for which demand existed, but from which host populations estranged themselves.

Outsiders always have a diminished stake in the establishment. Quite naturally outsiders have an interest in diminishing the power of the establishment and loosening the restraints imposed by the social structure. It should therefore hardly be found unusual that the Jews should have been a destabilizing force. Outsider Jews supported the French Revolution, the risings of 1848, the Paris Commune and the Russian Revolution. Jewish intellectuals attacked the teachings of the church and questioned the legitimacy of the divine right of kings.

From the time of the Napoleonic Wars international finance has played a central role in the economic life and political destiny of western nations. The involvement of Jewish people in the banking industry and their transnational mobility as a group has provided them with a position of importance vastly disproportionate to their size as a component part of the population of the west. From the world of finance Jewish people came to exercise an important influence in the entire economic life of the west.

The crux of the Jewish question centers around the significant role played by Jews in western societies coupled with their perpetual status as outsiders. Thus, Jews have had one foot in the establishment with the other foot outside, inclined to trip.

The Jewish people have tended to support the secularization of Western Christian society, apparently so as to render their position as outsiders less strident. The risk presented by this position is that it is precisely the Christian nature of society which has allowed strangers the opportunity to enjoy substantial freedom of activity and to attain real material success.

Zionism, the construct of Theodore Hertzl may well owe its ascendancy to societal anti-Semitism. However, it effectively divided the world into them and us. This is never a positive state of affairs. The impact of Zionism has been to deflect the universal nature of the Christian message and, ironically, to stoke the fires of anti-Semitism.

One of the tragedies of the 20 th Century is the emergence of National Socialism as a movement one of whose principal tenets was virulent anti Semitism. The systematic depredations visited upon the Jewish people in Germany and the lands occupied by National Socialist Germany during the Second World War has seared itself into western consciousness and colored all political dialogue since the middle of the 20 th Century. The National Socialist experience has, quite naturally, had the greatest impact on the socio-political consciousness of Jewish people. It has increased their sense of isolation and enhanced the attraction of Hertzls Zionist ideology. It would be impossible to understand the creation and the continuing policies of the emergent state of Israel without analysis of the Nazi context.

It is critical to realize that National Socialism was, ab initio, a fundamentally non Christian movement. The elements of neo paganism along with a Nietzscheian fascination with Zoroastrianism and eastern mysticism were central to the Nazi mythos. The National Socialist manifestation was as alien to the Western Christian civilization as was Marxism. For this reason the persistent phenomenon of western guilt feeling, particularly among the peoples who expended prodigious amounts of blood and treasure to defeat Nazi Germany, is difficult to explain. Even within the bounds of the Third Reich, the Confessing Lutheran Church , consistently maintained steadfast opposition to the anti Christian elements of the National Socialist state.

The publication of Alexander Solzhenitsyns history of the interaction between Jews and Christians in Russia in 2002 marked a new chapter in the study of Jewish influence in Western Christian societies. Solzhenitsyns point of departure was avowedly Christian. His perspective was opposed to anti Semitism. His stated intention was to effect reconciliation. Neither an apology nor an attack, the Nobel laureate novelists text apparently attempts to divorce history from emotion.

One of the towering figures of Western Christian culture was Felix Mendelssohn, a Jew. His contributions to western sacred music are enormously important. His magnificent 5th Symphony is a virtual anthem of Christian militancy. Music occupies the contentious middle ground between the intellect and the emotions. The impact of Mendelssohns compositions can be nicely contrasted to that of Richard Wagner. The former is the standard-bearer of Christendom in the mold of Johann Sebastian Bach, while the latter pushed the envelope in the direction of paganism in the same sense that Nietzsche did in the world of philosophy.

J

The fact that Christianity is a universal religion explains its continuing appeal in diverse cultural environments. The dramatic spread of Christianity in Korea during the 20 th Century is a classic example. An ancient Asian culture has maintained its identity and fidelity to its historic roots while orienting its spiritual life to the Christian way. This is not altogether unprecedented. Christian missionaries in Japan enjoyed great success in the early 16 th Century before the brutal repression of the church under the later shogunate. Millions of Chinese also turned to Christ despite many obstacles including the martyrdoms associated with the Boxer Rebellion and the subsequent communist repressions.

Ironically Maoism can be seen as the secular implementation of Christian values in China. Mao Tse-tung turned China from the stultifying tradition of ancestor worship and passive acceptance of rigid structure to revolution, egalitarianism and western notions of fairness. Thus Chinese communism, while avowedly anti Christian, is philosophically closer to Christianity than its predecessor Confucianism.

The successes in spreading Christianity outside Europe show the continuing viability of Western Christian civilization. If the west were entering its preordained winter stage of cultural development the spread of the religion associated with the wests ascendance would surely be counterintuitive. Oswald Spengler had failed to realize that the advent of Christs teaching was a game changer. Western Christian culture is not doomed to cycle out of the historic continuum.

K

Islam has been the dark universal. The religious movement begun by Mohammed strives to universality, as does Christianity. It is not a tribal or national religion but sees itself as appropriate for all peoples. It is a proselytizing system of ideas which compels adherents to spread the teachings of the Koran. However, Islam has traditionally condoned its expansion by brute force. Conversion to Islam on the basis of intimidation is acceptable. Acceptance of the Muslim faith is first overt and formulaic, rather than based on internal spiritual transformation.

The present dimensions of the Islamic world have been established exclusively through conquest. Traditional Islam is violent Islam. The 21 st Century phenomenon of Muslim extremism is nothing more that a reversion to traditional Islamic methods. The movement known as ISIS (Islamic State in Iraq and Syria) or ISIL (Islamic State in Iraq and the Levant) is not a radical deviation from Muslim teaching, but rather the opposite. Those who would spread and implement the strict interpretation of Sharia, or Islamic law, are behaving in a fashion which is perfectly consistent with historic Muslim precedent.

Indeed, moderate Islam seems to represent the deviation from Koranic teaching. Moderate Muslims are those who have come under the influence of Christianity and have, consciously or unconsciously, been swayed by the pacifist teachings of Christ. One of the products of 19 th and early 20 th Century imperialism was the fact that millions of Muslims came to live under Western Christian suzerainty. The brutal strictures of Islamic law were suppressed in the western colonial possessions, even if the Muslim religion as a whole was nominally tolerated.

The resurgence of violent Islam in the 21 st Century is precisely the result of the political withdrawal of the erstwhile western colonial powers from the Islamic societies. It is also the result of a Christian retreat from doctrinal militancy and a loss of faith in the absolute virtue of Christs basic teaching. Islam, on its own, is inherently an ideology premised on the unabashed supremacy of the sword. For this reason, if, in the future, the physical authority of Western Christian civilization should further recede globally, the virulence of Islam will increase. There is no secular obstacle for a religion which blends mans innate spiritual yearning with his equally innate dark lust for violence and dominion.

The inherent truth of Christs peaceful message and the inherent strength of the culture which blended the very different messages of the Old and New Testaments, constitute the only hope for containing the Islamic ambition for dominance. The Western Christian culture, which at one point over spread the planet, owed its unparallel strength to the alloy of New Covenant love and forgiveness with an Old Covenant, strict moral code. The swinging pendulum between the seemingly contradictory doctrines has allowed for adaptability and flexibility. Thus in the long term Western Christian civilization is not destined to join its predecessors in the dustbin of history, 20 th Century aberrations, notwithstanding.

L

The organized church is itself responsible for some of the anti Christian sentiment which has its roots in the 19 th Century. The situation of Count Leo Tolstoy presents a classic example. In his later years Tolstoy promulgated a temporal philosophy explicitly premised of the teachings of Jesus Christ, characterized by humility, egalitarianism and pacifism. Tolstoys teachings were, however, at variance with the official dogma of the Russian Orthodox Church on various theological matters. Instead of accepting the mighty novelist into its fold and trying to minimize apparent discord, with the objective of winning over the errant on the controversial issues, the institutional church declared Tolstoys writings anathema. In so doing the church branded the Tolstoyans as opponents of Christianity. Indeed this was self-fulfilling. A significant component of the Russian intelligentsia turned against the institutional church with disastrous consequences.

The decline of the west is only possible if the west estranges itself from the Christian roots which made its ascendance possible. Morbid romanticism characterized by its fascination on decay and decline fixed its attention on pre Christian civilization. Social Darwinism, the philosophy of Friedrich Nietzsche, as well as fascism have certainly contributed to the regression of Western Christian culture in the 20 th Century. Marxism and Zionism have unleashed conflicts, which have unraveled the fabric of global harmony.

Thus, the perception that the west is in decline is directly in proportion to the diminished influence of Christianity in western society. It is not an inherent, unavoidable cycle of destiny. When we recognize this we should cease to passively accept a pessimistic prediction. On the contrary, we should once again strive to be a people zealous of good works and convinced of the attainability of progress. This will prove to be a positive self-fulfilling prophecy.

M

The year 1900, on the eve of the 20 th Century, witnessed two events which could well have presaged a great leap forward for Western Christian culture. The establishment of the predecessor of the International Court of Justice at The Hague was intended to banish internecine conflict forever from the civilized world. The signatory nations, encompassing virtually all of Christendom, committed themselves to the peaceful resolution of disputes in accordance with legal and equitable principles. The international law which was to form the basis of the Hague courts jurisprudence was precisely that body of doctrine first assembled by the great Christian philosopher, Hugo Grotius. If the sovereigns of the western world had been true to their enunciated principles and had allowed the disputes of the 20th Century to have been resolved by an impartial tribunal on the basis of Christian values, we might today live in a veritable paradise.

Absent the First World War, the Second World War and the Cold War the natural and human resources of the planet would be immeasurably more abundant than they are today. It is impossible to accurately calculate the extent to which the patrimony of humanity has been squandered in the context of the 20th Century mega conflicts.

The second event which took place in 1900 and which potentially had great significance for future developments was the international effort exercised for the suppression of the Boxer Rebellion. The Boxer Rebellion was an anti Christian development in China, largely orchestrated by the Chinese dowager empress in order to exterminate foreign and particularly Christian influence in Imperial China. In the context of the Boxer Rebellion hundreds of thousands of Chinese Christians along with foreign missionaries were massacred. The Boxers, as the Chinese insurgents were called, ultimately besieged the foreign legations in Beijing and threatened the annihilation of the diplomatic and foreign commercial community in the Chinese capitol. In response to the Boxer Rebellion an unprecedented international coalition undertook the relief of the legations at Beijing and the suppression of the Boxers. Soldiers of Britain, France, Germany, Russia, the United States, Italy, Austria-Hungary and Japan participated in the action. The alliance against the Boxers seemed to promise the dawning of a new era of international cooperation. Countries who had conflicting national ambitions and who were competitors in the international economic arena were able to undertake a successful joint effort in the defense of Christendom. It is of great significance that the Japanese participated in the international coalition against the Boxer Rebellion. This proves that it was an alliance not based on race or religion in a dogmatic sense, but a grouping premised on a shared allegiance to the principles of Western Christian culture.

Alas the great promise exemplified by the establishment of the international court at the Hague and the forging of an alliance against the barbarism of the Boxer Rebellion has remained unfulfilled. The same sovereigns who established the court and collaborated in the suppression of anti Christian outrages precipitated the World War which brought about a century of misery. However, the gross misdeeds of 1914 and their aftermath are not the result of a mystic historic cycle of ascendance and decline. The predicament of the 20 th Century is the product of the confluence of multiple, trifling, unfortunate events and the overall waning of faith in Christian doctrine. This was not inevitable and it is not irreversible.

N

I personally dispute Oswald Spenglers final conclusion while respecting his scholarly erudition. In 1936, in the context of Oswald Spenglers obituary, the New York Times eloquently assessed his mighty contribution to the world of the intellect. Whatever may be the final reputation of Oswald Spengler, whatever the fate of his philosophy or predictions whether or not his theories correspond to reality, he painted a world panorama that, like a great play or a great symphony, is its own justification for existence.

From my perspective, the need to reestablish the fundamental verities of Western Christian culture as the basis of social organization is clear. Christian morality remains unsurpassed in its utility as it combines the rigidity of law with the moderation of mercy, equity and forgiveness.

This is the answer to the decline of the West. It is not a mythic, predestined curse as seen by romanticists. It is not a phase of social evolution in the Marxist idiom. It is not the product of the survival of the fittest and the precursor of the age of the superman. The decline has resulted from a retreat from Christian values. This diagnosis renders the selection of prophylaxis obvious.

EIGHTEENTH LOMONOSOV CONFERENCE ON ELEMENTARY PARTICLE PHYSICS

TWELFTH INTERNATIONAL MEETING ON
PROBLEMS OF INTELLIGENTSIA:
"The Future of the Intelligentsia"

Moscow State University, Moscow, 24–30 August, 2017

Under the Patronage of the Rector
of Moscow State University
Victor Sadovnichy

Dedicated to the 25th Anniversary
of the Lomonosov Conferences

International Advisory Committee

E.Akhmedov (Max Planck, Heidelberg), V.Belokurov (MSU), V.Berezinsky (GSSI/LNGS, INFN), S.Bilenky (JINR, Dubna), J.Bleimaier (Princeton), M.Danilov (Lebedev Phys. Inst., Moscow), A.Dolgov (Novosibirsk State Univ. & ITEP, Moscow), N.Fornengo (University of Torino & INFN), C.Giunti (University of Torino & INFN), M.Itkis (JINR, Dubna), L.Kravchuk (INR, Moscow), A.Masiero (INFN, Padua), V.Matveev (JINR, Dubna), M.Panasyuk (MSU), K.Phua (World Scientific, Singapore), V.Rubakov (MSU & INR, Moscow), J.Silk (Univ. of Oxford), A.Skrinsky (INP, Novosibirsk), A.Slavnov (MSU & Steklov Math. Inst.), A.Smirnov (Max Planck, Heidelberg), P.Spillantini (INFN, Florence), A.Starobinsky (Landau Inst., Moscow), N.Sysoev (MSU), Z.-Z.Xing (IHEP, Beijing)

Organizing Committee

V.Bagrov (Tomsk Univ.), I.Balantsev (MSU), V.Bednyakov (JINR, Dubna), Yu. Dubasova (Piligrim Center of Moscow Patriarchate), A.Egorov (ICAS), D.Galtsov (MSU), A.Grigoriev (MSU), A.Kataev (INR, Moscow), Yu.Kudenko (INR, Moscow), K.Kouzakov (MSU), A.Lokhov (INR, Moscow) – Scientific Secretary, N.Nikiforova (MSU), A.Nikishov (Lebedev Phys. Inst., Moscow), S.Ovchinnikov (MSU), A.Popov (MSU), Yu.Popov (MSU), P.Pustoshny (MSU), V.Ritus (Lebedev Phys. Inst., Moscow), V.Savrin (MSU), K.Stankevich (MSU), M.Sumin(MSU), V.Tsaturyan (MSU), V.Yakemenko (MSU), A.Studenikin (MSU & ICAS) – Chairman

Organizers and Sponsors

Faculty of Physics, Moscow State University
Joint Institute for Nuclear Research, Dubna
Ministry of Education and Science of Russia
Russian Foundation for Basic Research
Institute for Nuclear Research of Russian Academy of Sciences
Bruno Pontecorvo Neutrino and Astrophysics Laboratory (MSU)
Interregional Center for Advanced Studies

EIGHTEENTH LOMONOSOV CONFERENCE ON
Programme of the 18th Lomonosov Conference on Elementary Particle Physics
/duration of 25 (20/15) min talks includes 5 (3) minutes for discussion/

24 August, Thursday

08.00 – 09.00 **Registration** (Hall in front of Central Physics Auditorium, Faculty of Physics, Moscow State University)

09.00 – 09.30 **Opening** (Conference Hall)

Chair: **Alexander Studenikin**, Chairman of Organizing Committee, MSU & JINR

Grigory Trubnikov, Vice Minister, Ministry of Education and Science of Russia
Andrey Fedyanin, Vice Rector of Moscow State University
Grigory Rubtsov, Vice Director of Institute for Nuclear Research RAS
Andrey Slavnov, Head of Department of Theoretical Physics of MSU

09.30 – 13.20 **MORNING SESSION** (Conference Hall)

Chair: **Ivanka Bozovic Jelisavcic**

09.30 M. Giammarchi (INFN, Milan) *The Lomonosov Conference and physics in the last 25 years* (25 min)
09.55 S. Kostromin (JINR) *NICA accelerator complex* (20 min)

10.15 F. Djama (CPPM Marseille) *Overview of the ATLAS experiment at the LHC* (25 min)
10.40 S. Swain (National Institute of Science Education and Research) *CMS highlight overview* (25 min)

11.05 – 11.40   Tea break

Chair: **Valery Rubakov**

11.40 E. Coccia (GSSI & INFN) *The detection of gravitational waves: 50 years of experimental efforts* (25 min)
12.05 S. Goswami (Physical Research Laboratory, India) *Neutrino phenomenology* (25 min)
12.30 C. Ott (CalTech) *Core-collapse supernova mechanisms* (25 min)
12.55 K. Long (Imperial College London) *Accelerator-based neutrino physics; past, present and future* (25 min)

13.20 – 15.00   Lunch

15.00 – 18.30   **AFTERNOON SESSION** (Conference Hall)

Chair: **Yogendra Narain Srivastava**

15.00 V. Berezinsky (GSSI) *Protons in UHECR* (25 min)
15.25 P. Serpico (LAPTh) *Entering the cosmic ray astrophysics precision era* (25 min)
15.50 I. Maris (Univ. Libre de Bruxelles) *Results from the Pierre Auger Observatory* (25 min)

16.15 – 16.45   Tea break

Chair: **Karol Lang**

16.45 I. Bozovic Jelisavcic (Univ. of Belgrade) *Physics at the Compact Linear Collider (CLIC)* (25 min)
17.10 B. Gavela (Univ. Autonoma de Madrid) *On axions and ALPs* (25 min)

17.35 A. Starobinsky (Landau Inst.) *Recent developments in the inflationary scenario* (25 min)
18.00 S. Mihara (KEK) *Muon particle physics program at J-PARC* (20 min)

25 August, Friday

09.00 – 13.35   MORNING SESSION (Conference Hall)

Chair: **Yury Kudenko**

09.00 M. Ackermann (DESY) *Astrophysics and particle physics with IceCube neutrinos* (25 min)
09.25 S.-H. Seo (Seoul Nat. Univ.) *New Results form RENO & future prospects with T2HKK* (25 min)
09.50 M. Smy (Univ. of California, Irvine*) T2K neutrino oscillation results and constraints on delta CP* (25 min)
10.15 C. Hagedorn (Univ. of Southern Denmark) *Leptonic CP violation* (25 min)
10.40 L. Ludhova (Forschungszentrum Jülich) *Low energy neutrino physics with liquid scintillator detectors* (25 min)

11.05 – 11.30   Tea break

Chair: **Seon-Hee Seo**

11.30 M. Yokoyama (Univ. of Tokyo) *Hyper-Kamiokande* (25 min)
11.55 A. Kouchner (Univ. Paris Diderot - Laboratoire APC) *High-energy neutrino searches in the Mediterranean Sea* (25 min)
12.20 A.G. Cocco (INFN, Naples) *Relic neutrino detection* (25 min)
12.45 A. Guglielmi (INFN, Padua) *ICARUS: from CNGS to Booster beam* (25 min)
13.10 J. Evans (Univ. of Manchester) *The DUNE experiment* (25 min)

13.35 – 15.00   Lunch

15.00 – 19.10   AFTERNOON SESSION (Conference Hall)

Chair: **Konstantin Kouzakov**

15.00 I. Drachnev (PINP, Gatchina) *Review on solar neutrino studies with Borexino* (20 min)
15.20 Y. Li (IHEP, Chinese Academy of Sciences) *Status of light sterile neutrinos* (20 min)
15.40 S. Zhou (IHEP, Chinese Academy of Sciences) *Neutrino mass ordering and neutrinoless double-beta decays* (20 min)
16.00 M. Yoshimura (Okayama Univ.) *Neutrino mass spectroscopy* (25 min)
16.25 Zh. Xu (BNL/SDU) *Recent Results from STAR at RHIC* (25 min)

16.50 – 17.20   Tea break

SESSION   25.08. A   (Neutrinos)

Chair: **Justin Evans**

17.20 S. Nasri (United Arab Emirates Univ.) *Neutrino Mass, Dark Matter, and Baryon Asymmetry of the Universe* (20 min)
17.40 M. Gromov (SINP MSU) *The SOX project* (20 min)
18.00 D. Chiesa (Univ. & INFN Milano-Bicocca) *The CUORE experiment at LNGS* (20 min)
18.20 M. Ghosh (Tokyo Metropolitan Univ.) *Sensitivity of the T2HKK experiment to the non-standard interactions* (15 min)
18.35 S. Gariazzo (IFIC-CSIC & Univ. de Valencia) *Neutrino clustering in the Milky Way* (15 min)
18.50 S. Qian (IHEP Chinese Academy of Sciences) *The R&D of the large area MCP-PMT for the neutrino Experiments* (20 min)

SESSION   25.08. B   (Physics at Colliders)

Chair: **Fares Djama**

17.20 U. Schneekloth (DESY) *Electroweak fits to HERA data* (20 min)

17.40 D. Boumediene (CNRS/LPC) *Higgs precision measurements with ATLAS* (15 min)
17.55 S. Kiselev (ITEP) *Hadronic resonance production with ALICE at the LHC* (20 min)
18.05 O. Zenaiev (DESY) *xFitter* (15 min)
18.20 N. Sasao (Okayama Univ.) *Intense gamma radiation by accelerated quantum ions* (20 min)
18.40 M. Ronzani (CERN) *Searches for supersymmetry with the ATLAS detector* (15 min)

SESSION 25.08. C (Gravitation, cosmology, astrophysics)

Chair: **Yogendra Narain Srivastava**

17.20 V. Berezin (INR RAS) *Conformal invariance and phenomenology of cosmological particle production* (20 min)
17.40 E. Alvarez (Univ. Autonoma de Madrid) *Weyl invariant, first order quadratic gravity* (20 min)
18.00 V. Dokuchaev (INR RAS) *Global geometry of the Vaidya space-time* (20 min)
18.20 E. Pozdeeva (MSU) *Cosmological attractors in multifield inflation* (15 min)

19.30 – 22.30 Sight-seeing bus excursion in Moscow

26 August, Saturday

09.00 – 13.30 MORNING SESSION (Conference Hall)

Chair: **Eugenio Coccia**

09.00 I. Bilenko (MSU) *LIGO project: status, results and future* (25 min)
09.25 W. Del Pozzo (Univ. of Pisa) *Fundamental physics from gravitational waves observations* (25 min)
09.50 R. Ciolfi (INAF, Padua) *Gravitational wave sources and multimessenger astronomy* (25 min)

10.15 R. Sparvoli (Univ. of Rome Tor Vergata) *The CSES/Limadou mission* (20 min)
10.35 M. Panasyuk (SINP MSU) *Astrophysical space observatory "Lomonosov": the first results* (25 min)

11.00 – 11.30   Tea break

Chair: **Alexander Dolgov**

11.30 P. Spillantini (INAF) *Direct measurements of cosmic rays* (25 min)
11.55 Y. Srivastava (Northeastern Univ.) *Critical indices and limits on space-time dimensions from cosmic rays* (20 min)
12.15 V. Choutko (MIT) *Latest AMS results on high energy cosmic ray measurements* (25 min)
12.40 R. Bernabei (Univ. of Rome Tor Vergata) *Recent results and perspectives of DAMA/LIBRA* (25 min)
13.05 A. Drukier (Stockholm Univ.) *Ultimate dark-matter detection* (25 min)

13.30 – 15.00   Lunch

15.00 – 18.50   AFTERNOON SESSION (Conference Hall)

Chair: **Shun Zhou**

15.00 Ch. Tunnell (Univ. of Chicago) *First results from the XENON1T Experiment* (20 min)
15.20 J. Sola (Univ. de Barcelona) *Current state of dynamical dark energy versus observations* (25 min)
15.45 T. Kobayashi (SISSA) *Spontaneous Baryogenesis and Geometric Baryogenesis* (25 min)
16.10 C. Kouvaris (Univ. of Southern Denmark) *Probing light dark matter* (25 min)
16.35 P. Salucci (SISSA) *The distribution of dark matter in galaxies and its nature* (25 min)

17.00 – 17.25   Tea break

Chair: **Matthew Malek**

17.25 R. Rangarajan (Physical Research Laboratory, India) *Current status of warm inflation* (25 min)
17.50 L. Yang (Tsinghua Univ.) *Recent status of CDEX and CJPL* (20 min)
18.10 B. Capdevila (Institut de Física d'Altes Energies, Barcelona) *Patterns of New Physics in $b \to sl^+l^-$ transitions in the light of recent data* (20 min)
18.30 M. Kirsanov (INR RAS) *Search for dark sector physics in missing energy events with NA64* (20 min)

27 August, Sunday

9.00 – 19.00   Bus excursion to Sergiev Posad

28 August, Monday

09.00 – 13.35   MORNING SESSION (Conference Hall)

Chair: **Michael Smy**

09.00 T. Matsuo (TOHO Univ.) *Results from the OPERA experiment* (20 min)
09.20 T. Nakadaira (KEK) *Future J-PARC neutrino beam upgrade and T2K-II experiment* (20 min)
09.40 K.T. Knoepfle (Max Planck Institute, Heidelberg) *Status and prospects of the search for neutrinoless double beta decay of Ge-76* (25 min)
10.05 D. Zaborov (CPPM, Marseille) *The KM3NeT neutrino*

telescope and the potential of a neutrino beam from Russia to the Mediterranean Sea (25 min)
10.30 J. Xu (IHEP of Chinese Academy of Sciences) *Recent results from Daya Bay* (20 min)
10.50 C. Vayenas (Univ. of Patras) *Gravitationally confined relativistic neutrinos* (25 min)

11.15 – 11.50   Tea break

Chair: **Srubabati Goswami**

11.50 J. Zalesak (Institute of Physics, Czech Academy of Sciences) *Recent results from the NOvA Experiment* (25 min)
12.15 G. Salamanna (Roma Tre Univ.) *Status and physics potential of the JUNO experiment* (20 min)
12.35 T. Matsubara (Tokyo Metropolitan Univ.) *Recent results from Double Chooz reactor neutrino experiment* (20 min)
12.55 V. Belov (ITEP, Kurchatov Institute) *From EXO-200 to nEXO* (20 min)
13.15 A. Konovalov (ITEP, Kurchatov Institute) *Observation of coherent elastic neutrino-nucleus scattering at SNS* (20 min)

13.35 – 15.00   Lunch

15.00 – 19.00   AFTERNOON SESSION (Conference Hall)

SESSION 28.08. A   (Neutrinos)

Chair: **Yufeng Li**

15.00 Z. Zhao (Liaoning Normal Univ.) *Breakings of the neutrino reflection symmetry* (20 min)
15.20 L.V. Maret (Univ. of Geneva) *Cross-section measurements in the T2K experiment* (20 min)
15.40 M. Rebelo (Univ. of Lisbon) *Neutrino physics and leptonic weak basis invariants* (25 min)
16.05 R. Soualah (Univ. of Sharjah) *Probing the majorana neutrinos nature at actual and future colliders* (15 min)

16.20 A. Esmaili Taklimi (Pontificia Univ. Catholica do Rio de Janeiro) *High energy neutrinos and dark matter* (20 min)

SESSION 28.08. B   (BSM, Theory)

Chair: **Anatoly Borisov**

15.00 A. Nesterenko (JINR) *Dispersive approach to QCD and its applications* (20 min)
15.20 M. Butenschoen (Univ. of Hamburg) *Theory of heavy quarkonium production* (20 min)
15.40 V. Kopeliovich (INR RAS) *Rescaling of quantized skyrmions: from nucleon to baryons with heavy flavor* (20 min)
16.00 R. Sultanov (St. Cloud State Univ.) *Influence of the $\bar{p} - p$ nuclear interaction on the rate of the low-energy $\bar{p} + H_\mu \to (\bar{p}p)_\alpha + \mu^-$ reaction* (20 min)
16.20 Y. Srivastava (Northeastern Univ.) *Low energy nuclear reactions in normal and exploding batteries* (20 min)

SESSION 28.08. C   (Physics at Colliders)

Chair: **Sanjay Swain**

15.00 O. Zenaiev (DESY) *Latest tests of hard QCD at HERA* (25 min)
15.25 M. Tokarev (JINR) *Fractality of strange particle production in pp collisions at RHIC* (20 min)
15.45 S. Poslavsky (IHEP, Provtino) *PANDA physics* (20 min)
16.05 F. Fabbri (INFN, Bologna) *Top pair and single top production in ATLAS* (20 min)

SESSION 28.08. D   (Dark matter searches)

Chair: **Raghavan Rangarajan**

15.00 P. Czodrowski (CERN) *Constraining dark matter with ATLAS* (20 min)
15.20 V. Caracciolo (LNGS, INFN) *The ADAMO project and developments* (20 min)
15.40 N. Ferreiro Iachellini (Max Planck Inst. for Physics, Munich)

Light dark matter search with CRESST-III (20 min)
16.00 A. Alexandrov (INFN, Naples) *Directional detection of dark matter with a nuclear emulsion based detector* (20 min)
16.20 V. Solovov (LIP) *LUX Dark Matter Experiment: the latest results* (20 min)

16.40 - 17.10   Tee break

SESSION 28.08. E   (Neutrinos)

Chair: **Giuseppe Salamanna**

17.10 V. Semikoz (IZMIRAN, Troitsk) *Non-conservation of the lepton current and asymmetry of relic neutrinos* (20 min)
17.30 Ch. Das (JINR) *Effects of BSMs on θ_{23} determination* (15 min)
17.45 T. Karkkainen (Univ. of Helsinki) *Constraining the nonstandard interaction parameters in long baseline neutrino experiments* (15 min)
18.00 A. Borisov (MSU) **& P. Sizin** (NUST MISiS) *Nonstandard neutrino interactions in a dense magnetized medium* (15 min)
18.15 S. Mayburov (Lebedev Institute of Physics) *Search for periodical variations of Fe-55 nuclide weak decay parameters* (15 min)

SESSION 28.08. F   (Theory: cosmic rays, cosmology)

Chair: **Raghavan Rangarajan**

17.10 V. Belokurov (MSU) *Evolution of the Universe caused by the averaged potential of the quantum scalar field* (15 min)
17.25 M. Dvornikov (IZMIRAN, Troitsk & Tomsk State Univ.) *Chiral magnetic effect and the generation of astrophysical magnetic fields* (15 min)
17.40 W. Chang (National TsinHua Univ.) *Making case for the singlet majoronic dark radiation* (20 min)
18.00 O. Antipin (Institut Rudjer Boskovic) *Revisiting the decoupling effects in the running of the Cosmological Constant* (15 min)

18.15 O. Zenaiev (DESY) *Astrophysical applications of PROSA PDFs* (20 min)
18.35 E.J. Chun (Korea Institute for Advanced Study) *Baryogenesis and n-n̄ oscillation from R-parity violation (neutron-antineutron oscillations)* (20 min)
18.55 V. Rodionov (REU, Moscow) *Use of pseudo-Hermitian models in promising researches the possible structure of dark matter* (20 min)

SESSION 28.08. G   (Hadronic, Theory)

Chair: **Stanislav Poslavsky**

17.10 A. Issadykov (JINR) *The study of b-s anomaly decays in covariant quark model* (15 min)
17.25 A. Kataev (INR RAS) *The generalized Crewther relation in $SU(N\_c)$: the current status of investigations* (15 min)
17.40 M. Fil'chenkov (Peoples' Friendship Univ. of Russia) *Quantum field theory effects on the horizons of static metrics* (20 min)
18.00 Yu. Grats (MSU), **P.A. Spirin** (MSU) *Vacuum polarization in cosmic-string and global-monopole backgrounds* (20 min)
18.20 A. Garat (Univ. de la Republica, Montevideo) *New tetrads in Riemannian geometry and new ensuing results in group theory, gauge theory and fundamental physics, in particle physics, general relativity and astrophysics* (20 min)

SESSION 28.08. H   (Physics at colliders)

Chair: **Sanjay Swain**

17.10 E. Soldatov (NRNU MEPhI) *Electroweak precision measurements in ATLAS* (15 min)
17.25 S. Poslavsky (IHEP, Provtino) *Heavy quarkonia production at LHC* (15 min)
17.40 F. Prokoshin (Univ. Tecnica Federico Santa Maria, Chile) *Searches for rare and exotic Higgs decays with ATLAS* (15 min)

17.55 A. Krutenkova (ITEP, Kurchatov Institute) *Production of nuclear fragments in the $^{12}C+^7Be$ collisions at intermediate energies* (15 min)
18.10 A. Gridin (JINR) *Search for exotic charmonium-like states at COMPASS* (20 min)

29 August, Tuesday
09.00 – 13.30   MORNING SESSION (Conference Hall)

Chair: **Marc Knecht**

09.00 S. Pokorski (Univ. of Warsaw) *Naturalness* (25 min)
09.25 D. Jeans (KEK) *The International Linear Collider project* (25 min)
09.50 P. Fermani (LNF, INFN) *Status of the KLOE-2 experiment at DAFNE* (25 min)
10.15 A. Taranenko (Moscow Phys. Eng. Inst.) *Recent results from PHENIX experiment at RHIC* (25 min)
10.40 X. Shi (IHEP of Chinese Academy of Sciences) *Overview of the CEPC Project* (20 min)

11.00 – 11.30   Tea break

Chair: **Marc Knecht**

11.30 P. Lewis (Univ. of Hawaii) *The Belle II/SuperKEKB project* (25 min)
11.55 S. Levonian (DESY) *Recent results on diffraction at HERA* (20 min)
12.15 E. Bossini (CERN & INFN, Pisa) *The TOTEM experiment: Results and perspectives* (25 min)
12.40 I. Selyuzhenkov (MEPhI) *Recent ALICE highlights* (25 min)
13.05 U. Schneekloth (DESY) *The OLYMPUS experiment - hard two-photon contribution to elastic lepton-proton scattering* (20 min)

13.30 – 15.00   Lunch

15.00 – 19.00   AFTERNOON SESSION (Conference Hall)

Chair: **Sergey Levonian**

15.00 S. Denisov (IHEP, Protvino) *Review of the Tevatron results* (25 min)
15.25 R. Fantechi (INFN, Pisa & CERN) *Review of NA62 and NA48 physics results* (25 min)
15.50 W. Gohn (Univ. of Kentucky) *The Muon g-2 Experiment at Fermilab* (20 min)
16.10 T. Mibe (KEK) *J-PARC muon g-2/EDM experiment* (20 min)
16.30 S. Serednyakov (INP-Novosibirsk) *Recent measurements of exclusive hadronic cross sections at BABAR and the implication for the muon g-2 calculation* (20 min)
16.50 A. Dzyuba (Petersburg Nuclear Physics Institute) *Charm physics at LHCb* (20 min)

17.10 – 17.40   Tea break

SESSION 29.08. A   (Neutrinos)

Chair: **Paolo Salucci**

17.40 A. Segarra (IFIC-Valencia) *The coherent weak charge of matter* (15 min)
17.55 D. Grigoriou (Univ. of Patras) *Comparison of the special relativistic and of the Schwarzschild geodesics treatments of the formation of bound rotational neutrino states stabilized via the uncertainty principle* (15 min)
18.10 N. Agafonova (INR RAS) *Search for supernova neutrino bursts with the LVD experiment* (15 min)
18.25 O. Kharlanov (MSU) *Signatures of neutrino magnetic moment in collective oscillations of supernova neutrinos* (15 min)
18.40 S. Demidov (INR RAS) *Influence of NSI on neutrino signal from dark matter annihilations in the Sun* (15 min)
18.55 S. Luchuk (INR RAS) *Study of the ν_μ charged-current quasi-elastic-like interactions in the NOvA near detector* (10 min)

SESSION 29.08. B   (Colliders)

Chair: **Oleg Rogachevsky**

17.40 P. Parfenov (MEPhI) *Azimuthal anisotropic flow performance for MPD experiment at NICA* (15 min)
17.55 A. Polyarush (INR RAS) *Study of* $K^+ \to \pi^0 e^+ \nu$ γ-*decay with OKA setup* (15 min)
18.05 N. Ermakov (MEPhI) *Transverse mass dependence of kaon Bose-Einstein correlations in p+p collisions at energy* $\sqrt{s} = 200$ *and* 510 *GeV with STAR* (15 min)

SESSION 29.08. C   (Theory)

Chair: **Alexander Nesterenko**

17.40 A. Kulikova (Tomsk State Univ.) *Relativistic radiation theory with Hertz tensor fields* (15 min)
17.55 T. Adorno (Tomsk State Univ.) *Preliminary aspects of switching-on and -off effects on vacuum instability* (15 min)
18.10 A. Shitova (Yaroslavl State Univ.) *A decay of an ultra-high energy neutrino* $\nu_e \to e^- W^+$ *in an extremely high magnetic field* (15 min)
18.25 V. Shakhmanov (MSU) *The three-loop contribution to beta-function quartic in the Yukawa couplings for the N=1 supersymmetric Yang-Mills theory with the higher covariant derivative regularization* (15 min)

SESSION 29.08. D   (Astrophysics, cosmic rays)

Chair: **Joan Sola**

17.40 N. Tomassetti (Univ. of Perugia & INFN) *Results from the AMS experiment on low-energy cosmic rays: time-dependence of cosmic-ray fluxes and solar modulation* (20 min)

18.00 Y. Zhezher (INR RAS) *Chemical composition of ultra-high energy cosmic rays from the Telescope Array experiment surface detector data* (15 min)
18.15 G. Bigongiari (INFN, Pisa) *Calocube: a novel approach for a homogeneous calorimeter for high-energy cosmic rays detection in space* (20 min)
18.35 Y. Khyzhniak (MEPhI) *Measurement of cosmic rays in the STAR experiment* (15 min)

30 August, Wednesday

09.00 – 12.55 MORNING SESSION (Conference Hall)

Chair: **Vladimir Petrov**

09.00 P. Lukin (Budker Inst. of Nuclear Physics and Novosibirsk State Univ.) *Overview of the CMD-3 recent results at e+e collider VEPP-2000* (20 min)
09.20 K. Suzuki (Nagoya Univ.) *Belle II Physics Prospects* (25 min)
09.45 Min-Zu Wang (National Taiwan Univ.) *Recent results of rare B/D decays from Belle* (20 min)
10.05 K. Ieki (Univ. of Tokyo) *Search of cLFV processes with muons and others* (25 min)

10.30 – 11.00 Tea break

Chair: **Alexander Zakharov**

11.00 G. Gratta (Stanford Univ.) *Tests of gravity at very large and very short distance* (25 min)
11.25 A. Dolgov (Novosibirsk State Univ. & ITEP Kurchatov Institute) *New problems in cosmology and primordial black holes* (25 min)
11.50 E. Arbuzova (Dubna State Univ. and Novosibirsk State Univ.) *Problems of spontaneous and gravitational baryogenesis* (20 min)
12.10 A. Taketa (Univ. of Tokyo) *Geophysics using neutrino oscillations* (20 min)
12.30 M. Shirchenko (JINR, Dubna) *Status and recent results of the MAJORANA DEMONSTRATOR* (20 min)
12.50 M. Knecht (CNRS) *Theoretical aspects of the anomalous magnetic moment of the charged leptons* (20 min)

13.10 – 14.30 Lunch

ROUND TABLE DISCUSSION
"Frontiers in Particle Physics"
(Conference Hall)

Chair: **Marco Giammarchi**

14.30 G. Chiarelli (INFN, Pisa) *Legacy measurements at the Tevatron* (25 min)
14.55 J. Pinfold (Univ. of Alberta) *The MoEDAL Experiment at the LHC - a new light on the high energy frontier* (25 min)
15.20 M. Malek (Univ. of Sheffield) *Neutrinos and cosmology* (25 min)
15.45 L. Stanco (INFN, Padua) *Neutrino mass hierarchy from oscillation* (25 min)
16.10 M. Smy (Univ. of California, Irvine) *Future Prospects of Solar Neutrino Measurements* (25 min)

16.35 – 17.00   Tea break

Chair: **Marco Giammarchi**

17.00 S. Popov (MSU) *Fast radio bursts: new exotic puzzle in astrophysics* (20 min)
17.20 D. Prokhorov (Wits Univ.) *The spectrum of the core of Centaurus A with H.E.S.S. and Fermi-LAT* (20 min)
17.40 G. Marsella (Univ. of Salento) *Recent Results from the DAMPE experiment* (20 min)
18.00 V. Petrov (IHEP, Protvino) *QUO VADIS? Strong Interactions at High Energies in Historical Prespective* (25 min)

TWELFTH INTERNATIONAL MEETING ON PROBLEMS OF INTELLIGENTSIA:
"The Future of the Intelligentsia"
(Conference Hall)

Chair: **Alexander Studenikin**

18.30 J. Bleimaier (Princeton) *The future of the intelligentsia* (25 min)

Closing of the 18[h] Lomonosov Conference
on Elementary Particle Physics
and the 12th International Meeting on Problems of Intelligentsia

SPECIAL SESSION (40^0)

POSTER SESSION

N. Tomassetti (Univ. of Perugia & INFN) *Astrophysical background for dark matter search with cosmic-ray antinuclei*
N. Tomassetti (Univ. of Perugia & INFN) *Predicting solar modulation of cosmic-ray particles and antiparticles in the Heliosphere*
V. Koryukin (Mari State Univ) *On the role of quantum chromodynamics in the theory of the Universe matter evolution*
M. Organokov (Institut Pluridisciplinaire Hubert CURIEN) *Netrino entanglement with magnetic field*
J. Gornaya (MEPhI) *Analysis of flow in the experiment NA49*
O. Golosov (MEPhI) *Analysis of anisotropic transverse flow in Pb-Pb collisions at 40AgeV in the NA49 experiment at CERN SPS using Qn-Corrections Framework*

List of participants of the 18th Lomonosov Conference on Elementary Particle Physics and the 12th International Meeting on Problems of Intelligentsia

Ackermann Markus	Deutsches Elektronen Synchrotron	markus.ackermann@desy.de
Adorno Tiago	Tomsk State University	tg.adorno@gmail.com
Agafonova Natalia	INR RAS	agafonova@inr.ru
Alexandrov Andrey	Istituto Nazionale di Fisica Nucleare sezione di Napoli	andrey.alexandrov@na.infn.it
Alvarez Enrique	Universidad Autonoma de Madrid	enrique.alvarez@uam.es
Andrzej Drukier	Department of Physics OCS Stockholm University	adrukier@gmail.com
Antipin Oleg	Institut Rudjer Boskovic.	oantipin@irb.hr
Arbuzova Elena	Dubna State University and Novosibirsk State University	al.arbuzova@gmail.com
Bell Ailec	Cuban Isotopes Center	ailec.caridad@gmail
Belokurov Vladimir	Lomonosov Moscow State University	vvbelokurov@yandex.ru
Belov Vladimir	SSC RF ITEP NRC KI	belov@itep.ru
Berezin Victor	INR RAS	berezin@inr.ac.ru
Berezinsky Veniamin	Gran Sasso Science Institute	berezinsky@lngs.infn.it
Bernabei Rita	University of Rome Tor Vergata	rita.bernabei@roma2.infn.it
Bernat Capdevila	Facultat de Ciencies, Edif. Cn, IFAE	bcapdevila@ifae.es
Bigongiari Gabriele	Istituto Nazionale di Fisica Nucleare - sezione di Pisa	gabriele.bigongiari@pi.infn.it
Bilenko Igor	MSU	igorbilenko@phys.msu.ru
Bleimaier John	Princeton	bleimaier@aol.com
Bordovitsyn Vladimir	National Research Tomsk State Univ. MSU	vabord@sibmail.com
Borisov Anatoly		borisov@phys.msu.ru
Bossini Edoardo	Centro Studi e Ricerche E.Fermi & INFN-Pisa	edoardo.bossini@pi.infn.it
Boumediene Djamel	Centre National de Recherche Scientifique	djamel.boumediene@cern.ch
Bozovic Jelisavcic Ivanka	Vinca Institute of Nuclear Sciences	ibozovic@vinca.rs
Braguta Victor	ITEP	victor.v.braguta@gmail.com
Butenschoen Mathias	University of Hamburg	mathias.butenschoen@desy.de
Caracciolo Vincenzo	National Laboratory of the Gran Sasso	vincenzo.caracciolo@lngs.infn.it

Chang We-Fu	National TsinHua University	wfchang@phys.nthu.edu.tw
Chiarelli Giorgio	Istituto Nazionale di Fisica Nucleare	giorgio.chiarelli@pi.infn.it
Chiesa Davide	INFN of Milano-Bicocca	davide.chiesa@mib.infn.it
Choutko Vitali	Massachusetts Institute of Technology	vitali.choutko@cern.ch
Chun Eung Jin	Korea Institute for Advanced Study	ejchun@kias.re.kr
Ciolfi Riccardo		riccardo.ciolfi@oapd.inaf.it
Coccia Eugenio	Osservatorio Astronomico di Padova	eugenio.coccia@lngs.infn.it
Cocco Alfredo		alfredo.cocco@na.infn.it
Czodrowski Patrick	Gran Sasso Science Institute	czodrows@cern.ch
Damanik Asan	Istituto Nazionale di Fisica Nucleare	asandamanik11@gmail.com
Das Chitta	CERN	das@theor.jinr.ru
	Sanata Dharma University	
	JINR	
Dattoli Giuseppe	ENEA Frascati Fusion Department	giuseppe.dattoli@enea.it
Demidov Sergei	INR RAS	demidov@ms2.inr.ac.ru
Denisov Sergei	Logunov Institute for High Energy Physics	denisov@ihep.ru
Djama Fares	CPPM Marseille	djama@cppm.in2p3.fr
Dokuchaev Vyacheslav	INR RAS	dokuchaev@inr.ac.ru
Dolgov Alexander	Novosibirsk State University and ITEP	dolgov@fe.infn.it
Drachnev Ilia	St. Petersburg Nuclear Physics Institute NRC Kurchatov Institute	drachnev_is@pnpi.nrcki.ru
Dutta Bhaskar	Texas A&M University	dutta@physics.tamu.edu
Dvornikov Maxim	IZMIRAN	maxdvo@izmiran.ru
Dzyuba Alexey	Petersburg Nuclear Physics Institute	adzyuba@cern.ch
Ermakov Nikita	MEPhI	coffe92@gmail.com
Ereditato Antonio	Univ. of Bern	antonio.ereditato@cern.ch
Esmaili Arman	Pontificia Universidade Católica do Rio de Janeiro	arman@puc-rio.br
Evans Justin	University of Manchester	justin.evans@manchester.ac.uk
Fabbri Federica	INFN	federica.fabbri@cern.ch
Fantechi Riccardo	INFN - Sezione di Pisa and	fantechi@cern.ch
Fermani Paolo	CERN	paolo.fermani@lnf.infn.it
Ferreiro Iachellini Nahuel	INFN Max Planck Institute for Physics	ferreiro@mpp.mpg.de
Fil'chenkov Michael	Peoples' Friendship Univ. of Russia	fmichael@mail.ru
Garat Alcides	Universidad de la Republica	garat@fisica.edu.uy
Gariazzo Stefano	INFN & Univ. of Turin	gariazzo@to.infn.it
Garzelli Maria Vittoria	Univ. of Hamburg	garzelli@to.infn.it
Gavela Belen	Universidad Autonoma de Madrid	belen.gavela@uam.es

Ghosh Monojit	Tokyo Metropolitan University	mnjtghosh8@gmail.com
Giammarchi Marco	INFN	marco.giammarchi@mi.infn.it
Giovanni Guglielmi	INFN-Padova	alberto.guglielmi@pd.infn.it
Godizov Alexandre	Physico-Energy Institute University of Kentucky	gohn@pa.uky.edu
Gohn Wesley	NRNU MEPHI	oleg.golosov@gmail.com
Golosov Oleg	NRNU MEPhI	yulyagornaya@mail.ru
Gornaya Julia	Physical Research Laboratory	sruba@prl.res.in
Goswami Srubabati	MSU	grats@phys.msu.ru gratta@stanford.edu
Grats Yuri	Stanford University	andrei.gridin@cern.ch
Gratta Giorgio	JINR	dgrigoriou@chemeng.upatras.gr
Gridin Andrei	University of Patras	
Grigoriou Dimitrios		
Gromov Maxim	SINP-MSU	gromov@physics.msu.ru
Guglielmi Alberto	INFN-Padova	alberto.guglielmi@pd.infn.it
Guinn Ian	University of Washington	iguinn@uw.edu
Hagedorn Claudia	University of Southern Denmark	hagedorn@cp3.sdu.dk
Ieki Kei	University of Tokyo	iekikei@icepp.s.u-tokyo.ac.jp
Issadykov Aidos	JINR	issadykov.a@gmail.com
Jeans Daniel	High Energy Accelerator Research Organization (KEK)	richard.jacobsson@cern.ch daniel.jeans@kek.jp
Karkkainen Timo	University of Helsinki	timo.j.karkkainen@helsinki.fi
Kataev Andrei	INR RAS	kataev@ms2.inr.ac.ru
Kharlanov Oleg	MSU	okharl@mail.ru
Khyzhniak Yevheniia	MEPhI	eugenia.sh.el@gmail.com
Kirsanov Mikhail	INR RAS	Mikhail.Kirsanov@cern.ch
Kiselev Sergey	ITEP	sergey.Kiselev@cern.ch
Knect Marc	Centre National de la Recherche Scientifique	knecht@cpt.univ-mrs.fr
Knoepfle Karl	Max-Planck-Institutfuer Kernphysik	ktkno@mpi-hd.mpg.de
Kobayashi Takeshi	SISSA	takeshi.kobayashi@sissa.it
Kopeliovich Vladimir	INR RAS	kopelio@inr.ru
Koryukin Valery	Mari State Univ.	vmkoryukin@gmail.com
Kostromin Sergei	JINR	kostromin@jinr.ru
Kouchner Antoine	University Paris Diderot	kouchner@apc.univ-paris7.fr
Kouvaris Christoforos	University of Southern Denmark	kouvaris@gmail.com
Kravtsova Galina	MSU	gakr@chtc.ru
Krutenkova Anna	ITEP	krutenkova@itep.ru
Kudenko Yury	INR RAS	kudenko@inr.ru

Kulikova Anastasiya	Tomsk State Univ.	anasta-kulikova@yandex.ru
Lewis Peter	University of Hawaii	trogonpete@gmail.com
Li Yufeng	IHEP, China	liyufeng@ihep.ac.cn
Long Kenneth	Imperial College London	k.long@imperial.ac.uk
Luchuk Stanislav	INR RAS	sluchuk@yandex.ru
Liu Wei	IHEP, China	
Ludhova Livia	Forschungzentrum Juelich	ludhova@gmail.com
Lukin Peter	BINP&NSU,Novosibirsk,Russia	P.A.Lukin@inp.nsk.su
Malek Matthew	Univ. of Sheffield	m.malek@sheffield.ac.uk
Maret Violette Lucie	Université de Genève	lucie.violette.maret@cern.ch
Maris Ioana	IIHE-Universite Libre de Bruxelles	ioana.Maris@ulb.ac.be
Marsella Giovanni	University del Salento	giovanni.marsella@le.infn.it
Matsubara Tsunayuki	Tokyo Metropolitan University	matsubara@hepmail.phys.se.tmu.ac.jp
Matsuo Tomokazu	TOHO University	tomokazu.matsuo@sci.toho-u.ac.jp
Mayburov Sergei	Lebedev institute of Physics	mayburov@sci.lebedev.ru
Mazurov Alexander	Univ. of Birmingham	alexander.mazurov@cern.ch
Mibe Tsutomu	Institute of Particle and Nuclear Studies, High Energy Accelerator Research Organization	mibe@post.kek.jp
Mihara Satoshi	KEK	satoshi.mihara@kek.jp
Nakadaira Takeshi	High Energy Accelerator Research Organization (KEK)	nakadair@neutrino.kek.jp
Naranjo Roger	DESY	roger.naranjo@desy.de
Nasri Salah	United Arab Emirates Univ.	snasri@uaeu.ac.ae
Nesterenko Alexander	JINR	nesterav@gmail.com
Okorokov Vitalii	MEPhI	VAOkorokov@mephi.ru
Organokov Mukharbek	Institut Pluridisciplinaire Hubert CURIEN (IPHC) du Centre National de la Recherche Scientifique (CNRS)	mukharbek.organokov@iphc.cnrs.fr
Ott Christian	California Institute of Technology	cott@tapir.caltech.edu
Pallavicini Marco	Universita' di Genova and INFN Genova	marco.pallavicini@ge.infn.it
Panasyuk Mikhail	SINP MSU	panasyuk@sinp.msu.ru
Parfenov Peter	MEPhI	terrylapard@gmail.com
Picozza Piergiorgio	INFN	piergiorgio.picozza@gmail.com
Pinfold James	University of Alberta	jpinfold@ualberta.ca
Pokorski Stefan	University of Warsaw	pokorski@fuw.edu.pl

Polyarush Alexander	INR RAS	polyarush@yandex.ru
Popov Sergei	Sternberg Astronomical Institute, MSU	sergepolar@gmail.com
Poslavsky Stanislav	Institute for High Energy Physics NRC "Kurchatov Institute"	stvlpos@mail.ru
Pozzo Walter	University of Pisa	walter.delpozzo@unipi.it
Prokhorov Dmitry	Wits University	dmitry.prokhorov@lnu.se
Prokoshin Fedor	Universidad Tecnica Federico Santa Maria	fedor.prokoshin@usm.cl
Qian Sen	Institute of High Energy Physics, Chinese Academy of Sciences	qians@ihep.ac.cn
Rangarajan Raghavan	Physical Research Laboratory, New York	raghavan@prl.res.in
Rebelo Margarida N.	Instituto Superior Tecnico, U. Lisboa	rebelo@ist.utl.pt
Rodionov Vasily	Plekhanov Russian University of Economics	rodyvn@mail.ru
Romero Adam Elena	IFIC - CSIC	elena.romero.adam@cern.ch
Ronzani Manfredi	Albert-Ludwigs-Universität Freiburg	nicola.rossi@lngs.infn.it
Salamanna Giuseppe	Roma Tre University	giuseppe.salamanna@cern.ch
Saito Naohito	KEK / J-PARC	naohito.saito@kek.jp
Salucci Paolo	SISSA	salucci@sissa.it
Sasao Noboru	Okayama Univ.	sasao@okayama-u.ac.jp
Schneekloth Uwe	DESY	uwe.schneekloth@desy.de
Scholberg Kate	Duke University	schol@phy.duke.edu
Segarra Tamarit Alejandro	Instituto de Fisica Corpuscular	Alejandro.Segarra@uv.es
Selyuzhenkov Ilya	EMMI/GSI	ilya.selyuzhenkov@gmail.com
Semikoz Victor	IZMIRAN	semikoz@yandex.ru
Seo Seon-Hee	Seoul National University	shseo@phya.snu.ac.kr
Serednyakov Sergei	INR RAS	seredn@inp.nsk.su
Serpico Pasquale	LAPTh	serpico@lapth.cnrs.fr
Shakhmanov Vikentii	MSU	shakhmanov@physics.msu.ru
Shi Xin	Institute of High Energy Physics, Chinese Academy of Sciences	shixin@ihep.ac.cn
Shitova Anastasia		pick@mail.ru
Shpetim Nazarko	P.G. Demidov Yaroslavl State University Albania, Konica	shpetim.nazarko@gmail.com
Sizin Pavel	NUST MISiS	mstranger@list.ru
Smy Michael	University of California	smy@michael.ps.uci.edu
Sola Joan	Universitat de Barcelona	sola@fqa.ub.edu
Soldatov Evgeny	MEPhI	Evgeny.Soldatov@cern.ch

Smy Michael	Univ. of California, Irvine	msmy@uci.edu
Soualah Rachik	University of Sharjah	rsoualah@sharjah.ac.ae
Sparvoli Roberta	Univ. of Rome Tor Vergata	roberta.sparvoli@roma2.infn.it
Srivastava Yogendra	Northeastern University	yogendra.srivastava@gmail.com
Stanco Luca	INFN - Padova	stanco@pd.infn.it
Starobinsky Alexei	Landau Inst.	alstar@landau.ac.ru
Stoikou Aikaterini	Thessaloniki Greece	Kaitistoikou@yahoo.gr
Sultanov Renat	St. Cloud State University	rasultanov@stcloudstate.edu
Suzuki Kazuhito	Nagoya University	kazuhito@hepl.phys.nagoya-u.ac.jp
Swain Sanjay	National Institute of Science Education and Research	sanjay@niser.ac.in
Taketa Akimichi	Univ. of Tokyo	akimichi.taketa@gmail.com
Taranenko Arkadiy	MEPhI	AVTaranenko@mephi.ru
Tarazona Carlos	Universidad Manuela Beltran	caragomezt@unal.edu.co
Tokarev Mikhail	JINR	tokarev@jinr.ru
Tomassetti Nicola	Università degli Studi di Perugia:	nicola.tomassetti@gmail.com
Tunnell Christopher	University of Chicago	tunnell@uchicago.edu
van den Brand Johannes	Nikhef and Vrije Universiteit Amsterdam	cgvayenas@upatras.gr
Vayenas Constantinos	University of Patras	wilson@colostate.edu
Vernov Sergey	MSU	svernov@theory.sinp.msu.ru
Vladimir Solovov	LIP	solovov@coimbra.lip.pt
Wang Min-Zu	National Taiwan University	mwang@phys.ntu.edu.tw
Xu Jilei	Institute of High Energy Physics, Chinese Academy of Sciences	xujl@ihep.ac.cn
Xu Zhangbu	BNL/SDU	xzb@bnl.gov
Yang Litao	Tsinghua University	Pauli127@163.com
Yokoyama Masashi	The University of Tokyo	masashi@phys.s.u-tokyo.ac.jp
Yoshimura Motohiko	Okayama University	yoshim@okayama-u.ac.jp
Zaborov Dmitry	Centre de physique des particules de Marseille	zaborov@cppm.in2p3.fr
Zalesak Jaroslav	Institute of Physics, Czech Academy of Sciences	zalesak@fzu.cz
Zenaiev Oleksandr	Deutsches Elektronen-Synchrotron	oleksandr.zenaiev@desy.de
Zhao Zhen-hua	Liaoning Normal University	zhzhao@itp.ac.cn
Zherebtsova Elisaveta	MEPhI	ESZherebtsova@mephi.ru
Zhezher Yana	INR RAS	zhezher.yana@physics.msu.ru
Zhou Shun	Institute of High Energy Physics	zhoush@ihep.ac.cn
Studenikin Alexander	MSU & JINR	studenik@srd.sinp.msu.ru

CPSIA information can be obtained
at www.ICGtesting.com
Printed in the USA
BVHW040537020619
549643BV00004BA/14/P